Texts and Monographs in Computer Science

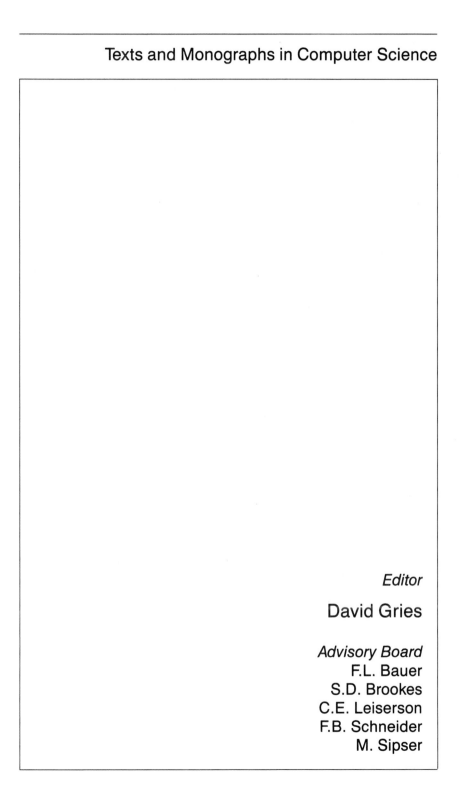

Editor

David Gries

Advisory Board
F.L. Bauer
S.D. Brookes
C.E. Leiserson
F.B. Schneider
M. Sipser

Texts and Monographs in Computer Science

Suad Alagic
Object-Oriented Database Programming
1989. XV, 320 pages, 84 illus.

Suad Alagic
Relational Database Technology
1986. XI, 259 pages, 114 illus.

Suad Alagic and Michael A. Arbib
The Design of Well-Structured and Correct Programs
1978. X, 292 pages, 68 illus.

S. Thomas Alexander
Adaptive Signal Processing: Theory and Applications
1986. IX, 179 pages, 42 illus.

Michael A. Arbib, A.J. Kfoury, and Robert N. Moll
A Basis for Theoretical Computer Science
1981. VIII, 220 pages, 49 illus.

Friedrich L. Bauer and Hans Wössner
Algorithmic Language and Program Development
1982. XVI, 497 pages, 109 illus.

Kaare Christian
A Guide to Modula-2
1986. XIX, 436 pages, 46 illus.

Edsger W. Dijkstra
Selected Writings on Computing: A Personal Perspective
1982. XVII, 362 pages, 13 illus.

Edsger W. Dijkstra and Carel S. Scholten
Predicate Calculus and Program Semantics
1990. XII, 220 pages

W.H.J. Feijen, A.J.M. van Gasteren, D. Gries, and J. Misra, Eds.
Beauty Is Our Business: A Birthday Salute to Edsger W. Dijkstra
1990. XX, 453 pages, 21 illus.

P.A. Fejer and D.A. Simovici
Mathematical Foundations of Computer Science, Volume I:
Sets, Relations, and Induction
1990. X, 425 pages, 36 illus.

Melvin Fitting
First-Order Logic and Automated Theorem Proving
1990. XIV, 242 pages, 26 illus.

Nissim Francez
Fairness
1986. XIII, 295 pages, 147 illus.

continued after index

Mathematical Foundations of Computer Science

Volume I:

Sets, Relations, and Induction

**Peter A. Fejer
Dan A. Simovici**

With 36 Illustrations

Springer-Verlag
New York Berlin Heidelberg London
Paris Tokyo Hong Kong Barcelona

Peter A. Fejer
Department of Math
  and Computer Science
University of Massachussetts
  at Boston
Boston, MA 02125
USA

Dan A. Simovici
Department of Math
  and Computer Science
University of Massachusetts
  at Boston
Boston, MA 02125
USA

Series Editor
David Gries
Department of Computer Science
Cornell University
Ithaca, NY 14853
USA

QA
76.9
M35
F45
1991
V.1

Library of Congress Cataloging-in Publication Data
Fejer, Peter A.
    Mathematical foundations of computer science / Peter A. Fejer, Dan
A. Simovici.
        p.   cm. -- (Texts and monographs in computer science)
    Includes bibliographical references and index.
    Contents: v. 1. Sets, relations, and induction.
    ISBN 0-387-97450-4. -- ISBN 3-540-97450-4 (Berlin)
    1. Computer science -- Mathematics.   I.   Simovici, Dan A.
II. Title.   III. Series
QA76.9.M35F45   1990
004'.01'51--dc20   90-49595

Printed on acid-free paper.

Photocomposed copy prepared from the authors' LaTeX file.
Printed and bound by R.R. Donnelley and Sons, Harrisonburg, Virginia.
Printed in the United States of America.

9 8 7 6 5 4 3 2 1

ISBN 0-387-97450-4 Springer-Verlag New York Berlin Heidelberg
ISBN 3-540-97450-4 Springer-Verlag Berlin Heidelberg New York

3 3001 00782 7644

Preface

As the title suggests, *Mathematical Foundations of Computer Science* deals with those topics from mathematics that have proven to be particularly relevant in computer science. The present volume treats basic topics, mostly of a set-theoretical nature: sets, relations and functions, partially ordered sets, induction, enumerability, and diagonalization. The next volume will discuss topics having a logical nature. Further volumes dealing with algebraic foundations of computer science are also contemplated.

We present the material in a way that is systematic, rigorous, and complete. Our approach is straightforward and, we hope, clear, but we do not avoid more difficult topics or sweep subtle points under the rug. Our goal is to make the subject, as Einstein said, "as simple as possible, but not simpler."

In Chapter 1, we discuss set theory from an intuitive point of view, but we indicate how difficulties arise and how an axiomatic approach might solve these problems.

Chapter 2 presents relations and functions, starting from the notion of the ordered pair. We emphasize the use of relations and functions as structuring devices for data, particularly for relational databases.

In Chapter 3, we provide an introduction to partially ordered sets. We define complete partial orders and prove results about fixed points of continuous functions, which are important for the semantics of programming languages. In the final section of the chapter, we analyze Zorn's Lemma. This proposition may appear to be of remote interest for computer science; nevertheless, results of real interest in computer science, such as connections between various types of partially ordered sets and fixed point results, are based on the use of this lemma.

Chapter 4 is dedicated to the study of mathematical induction. We present several versions of induction: induction on the natural numbers, inductively defined sets, well-founded induction, and fixed-point induction. Mathematical objects, such as formulas of propositional logic, grammars, and recursive functions, important for computer science, receive special attention in view of the role played by induction in their study.

In Chapter 5, we examine mathematical tools for investigating the limits of the notion of computability. We concentrate on diagonalization, a proof method originating in set theory that is an essential tool for obtaining limitative results in the theory of computation.

This volume is organized by mathematical area, which means that material on the same Computer Science topic appears in more than one place. Readers will find useful applications in algorithms, databases, semantics of programming languages, formal languages, theory of computation, and program verification.

There are few specific mathematical prerequisites for understanding the material in this volume, but it is written at a level that assumes the mathematical maturity gained from a good mathematics or computer science undergraduate major. Many of the applications require some exposure to introductory computer science.

Each chapter contains a large number of exercises, many with solutions (which we regard as supplements).

We would like to thank Lynn Montz, Suzanne Anthony and Natalie Johnson of Springer-Verlag for their attention to our manuscript and Karl Berry, Rick Martin, and James Campbell of the Computer Science Laboratory at UMass-Boston for maintaining the systems which allowed us to produce this book. Finally, the authors would like to acknowledge the many judicios remarks and suggestions made by Professor David Gries.

Contents

Preface   **vii**

1 Elementary Set Theory   **1**
    1.1 Introduction 1
    1.2 Sets, Members, Subsets 1
    1.3 Building New Sets 8
    1.4 Exercises and Supplements 13
    1.5 Bibliographical Comments 22

2 Relations and Functions   **23**
    2.1 Introduction 23
    2.2 Relations 24
    2.3 Functions 31
    2.4 Sequences, Words, and Matrices 40
    2.5 Images of Sets Under Relations 49
    2.6 Relations and Directed Graphs 51
    2.7 Special Classes of Relations 62
    2.8 Equivalences and Partitions 64
    2.9 General Cartesian Products 67
    2.10 Operations 74
    2.11 Representations of Relations and Graphs 78
    2.12 Relations and Databases 82
    2.13 Exercises and Supplements 104
    2.14 Bibliographical Comments 125

3 Partially Ordered Sets   **127**
    3.1 Introduction 127
    3.2 Partial Orders and Hasse Diagrams 127
    3.3 Special Elements of Partially Ordered Sets 131
    3.4 Chains 135
    3.5 Duality 138
    3.6 Constructing New Posets 139
    3.7 Functions and Posets 144
    3.8 Complete Partial Orders 148
    3.9 The Axiom of Choice and Zorn's Lemma 157

3.10 Exercises and Supplements 166
3.11 Bibliographical Comments 175

4 Induction   **177**
4.1 Introduction 177
4.2 Induction on the Natural Numbers 179
4.3 Inductively Defined Sets 199
4.4 Proof by Structural Induction 206
4.5 Recursive Definitions of Functions 214
4.6 Constructors 230
4.7 Simultaneous Inductive Definitions 246
4.8 Propositional Logic 257
4.9 Primitive Recursive and Partial Recursive Functions 264
4.10 Grammars 269
4.11 Peano's Axioms 285
4.12 Well-Founded Sets and Induction 290
4.13 Fixed Points and Fixed Point Induction 293
4.14 Exercises and Supplements 304
4.15 Bibliographical Comments 337

5 Enumerability and Diagonalization   **339**
5.1 Introduction 339
5.2 Equinumerous Sets 339
5.3 Countable and Uncountable Sets 344
5.4 Enumerating Programs 357
5.5 Abstract Families of Functions 374
5.6 Exercises and Supplements 383
5.7 Bibliographical Comments 416

References   **417**

Index   **421**

1

Elementary Set Theory

1.1 Introduction
1.2 Sets, Members, Subsets
1.3 Building New Sets
1.4 Exercises and Supplements
1.5 Bibliographical Comments

1.1 Introduction

The concept of set and the abstract study of sets (known as set theory) are cornerstones of contemporary mathematics and, therefore, are essential components of the mathematical foundations of computer science. For the computer scientist, set theory is not an exotic, remote area of mathematics but an essential ingredient in a variety of disciplines ranging from databases and programming languages to artificial intelligence.

Set theory as it is used by working mathematicians and computer scientists was formulated by Georg Cantor[1] in the last quarter of the 19th century. Cantor's approach led to difficulties that we will mention briefly in this chapter. The apparent solution to these difficulties requires an axiomatic approach, which we will allude to but not cover in detail.

In this chapter, we discuss sets and membership and examine ways of defining sets. Then, we introduce methods of building new sets starting from old ones and study properties of these methods.

1.2 Sets, Members, Subsets

Cantor attempted to define the notion of *set* as a collection into a whole[2] of definite, distinct objects of our intuition or our thought. The objects are called *elements* or *members* of the set.

---

[1] The German mathematician Georg F. L. P. Cantor was born on March 3, 1840, in St. Petersburg, Russia, and died on January 6, 1918, in Halle, Germany. He was affiliated with the University of Halle beginning in 1869. Cantor's main contribution was the initial development of modern mathematical set theory.

[2] Zusammenfassung zu einem Ganzen.

1

This definition does not satisfy the normal requirements of logic, which insist that a newly defined concept be a particularization of an already defined more general concept. Indeed, the term "collection" used in the Cantorian definition is hardly different from the defined term "set." Therefore, we shall regard the notion of set as being a primitive concept, i.e., a general notion that is understood intuitively but not defined precisely and can be used in defining other more particular notions. Hence, motivated by Cantor's "definition," we adopt the rather vague idea that a set is a collection of "things" that are called the elements of the set.

The primary concepts on which set theory is based are set and membership. Since we are viewing a set as a collection of objects, for any set S and object a, either a is one of the objects in S or it is not. In the former case, we use any of the following phrases: "a is a member of S," "a is an element of S," "a is contained in S," or "S contains a," and we write $a \in S$. This use of the symbol $\in$ was introduced by the Italian mathematician Giuseppe Peano[3] because the symbol $\in$ is similar to the first letter of the Greek word ἐστί which means "is." We write $a \notin S$ to denote that a is not a member of the set S.

Note that we did not consider the notion of "object" among the primary notions of set theory. Mathematicians have found that mathematics can be developed based on set theory, assuming that every element of every set is itself a set. For example, we shall see later (in Chapter 4) that every natural number can be considered to be a set. A similar point of view can be adopted for all the other common mathematical objects. Therefore, no other objects than sets need be considered, and when we use the term "object," this term can be interpreted to mean "set."

Sometimes, in order to emphasize that the elements of a set $\mathcal{C}$ are themselves sets, we refer to $\mathcal{C}$ as a *collection* of sets.

Two sets are the same if they have the same elements. Although this fact seems intuitively clear, it is important enough to be singled out as a principle of set theory.

Principle of Extensionality. Let S and T be two sets. If for every object a we have $a \in S$ if and only if $a \in T$, then $S = T$.

Sets can be specified in several ways. One method is to list explicitly the members of the set. If $x_1, \ldots, x_n$ are the elements of S, we denote S by

$$\{x_1, \ldots, x_n\}.$$

---

[3] Giuseppe Peano was born on August 27, 1858, in Cuneo, Italy, and died on April 20, 1932, in Turin. He taught mathematics at the University of Turin starting in 1884. Peano is one of the founders of symbolic logic and made important contributions to the general theory of functions. His main work, *Formulario Mathematico*, published between 1894 and 1908, was an inspiration for further work in the foundations of mathematics done by Russell and Whitehead and the Bourbaki group.

This notation is justified by the principle of extensionality because any other set that has the same elements is the same set. For instance, consider the set whose members are $1, 4, 9$, and 16. We denote this set by

$$\{1, 4, 9, 16\}.$$

If we cannot explicitly list all of the elements of a set we can use various suggestive extensions of the notation just given. For example, the set **N** of natural numbers can be denoted by

$$\{0, 1, 2, \ldots\}.$$

On the other hand, a set can also be specified by stating a *characteristic condition*, that is, a condition satisfied by all members of S and not satisfied by any other object. Consider, for instance, the condition

n is a natural number less than 20 that is a perfect square.

It is easy to see that this is a characteristic condition for the members of the set S defined above: all members of S satisfy the condition, and every object that satisfies the condition is a member of S. Also, note that we may have several characteristic conditions for a set. For instance, the set S can alternatively be specified as consisting of those sums less than 20 of consecutive odd natural numbers starting with 1.

By the principle of extensionality

1. the set $\{1, 4, 9, 16\}$,

2. the set that consists of all natural numbers that are perfect squares and are less than 20, and

3. the set of natural numbers that are sums of consecutive odd natural numbers starting with 1 and are less than 20,

are the same.

In (2) and (3) we have already used implicitly another principle, namely, the

Principle of Abstraction. Given a condition that objects satisfy or do not satisfy, there is a set that consists of the objects that satisfy the condition.

If $\mathcal{K}$ is a characteristic condition that allows us to define the set S, we could denote S by

$$\{x \text{ such that } x \text{ satisfies } \mathcal{K}\}.$$

In practice, we replace the phrase "such that x satisfies" by "|" and thus we write

$$\{x \mid \mathcal{K}\}.$$

The principle of abstraction is a working tool for everyday mathematics. However, its unrestrained application generates contradictions. Suppose, for instance, that, using this principle, we attempt to define the "set" R of all sets that are not elements of themselves,

$$R = \{x \mid x \notin x\}.$$

We can ask whether R belongs to itself. There are two possible answers:

1. $R \in R$ or

2. $R \notin R$,

and we can prove that both yield contradictions. Indeed, in the first case, the definition of the set R implies that $R \notin R$, which conflicts with the premise of this case. In the second case, the same definition implies $R \in R$; hence, we again obtain a contradiction.

The fact that the principle of abstraction allows this definition, which leads to an immediate contradiction, is known as Russell's paradox after the logician and philosopher Bertrand Russell [4] who discovered it.

Logicians have formulated a more restrictive version of this principle, which appears to eliminate these difficulties, namely, the

> **Principle of Comprehension.** Given a condition that objects satisfy or do not satisfy, and a set U, there is a set S that consists of the elements of U that satisfy the condition.

The difference between the principles of abstraction and comprehension is that in the latter we build a new set starting from an existing set U by collecting those members of U that satisfy the characteristic condition, while in the former we collect all objects satisfying a characteristic condition without restricting the search to the members of a set.

When using the principle of comprehension, we denote the set of those elements of U that satisfy the condition $\mathcal{K}$ by

$$\{x \in U \mid \mathcal{K}\}.$$

---

[4] Bertrand A. W. Russell, 3rd Earl Russell, was born in 1872 in Trelleck, Monmouthshire ,and died in 1970 near Penrhyndendraeth, Marioneth, in Wales. Russell was one of the major figures of 20th-century philosophy. His work is especially important for philosophical logic and for the theory of knowledge. His most important mathematical work, *Principia Mathematica*, written with A. N. Whitehead, was published between 1910 and 1913.

For instance, we could denote the set S considered above either as

$$\{n \in \mathbf{N} \mid n \text{ is a perfect square and } n < 20\}$$

or

$$\{n \in \mathbf{N} \mid n \text{ is a sum of consecutive odd natural}$$
$$\text{numbers starting with 1 and } n < 20\}.$$

Note that using the principle of comprehension, one cannot duplicate Russell's paradox because of the need to circumscribe the definition of R to the elements of some set U. In fact, for each set U, one can define the set

$$R_U = \{x \in U \mid x \notin x\};$$

however, the existence of this set does not lead to immediate contradiction. If $R_U \in R_U$, then $R_U \in U$ and $R_U \notin R_U$, which is impossible, but if $R_U \notin R_U$, then *either* $R_U \notin U$ *or* $R_U \in R_U$, and this is not contradictory; we merely conclude that $R_U \notin U$ (and $R_U \notin R_U$). Note that we have just shown that for any set U there is another set (namely, R_U) that is not a member of U. Consequently, there is no universal set V such that every set is a member of V.

In formal set theory, the principle of abstraction is rejected, because it leads to Russell's paradox, and the principle of comprehension is used instead. The paradoxical nature of the "set" R is taken to show that there is no such set. Of course, using the principle of comprehension requires, in some cases, additional principles to assert the existence of the set U. This leads to very tedious arguments. To avoid such arguments for now, we continue to use the principle of abstraction; however, in all of our arguments, the use of the principle of abstraction can be replaced by the principle of comprehension plus additional set existence principles.

Definition 1.2.1 *If S and T are two sets such that every element of S is an element of T, then we say that S is* included *in T or that S is a* subset *of T, and we write $S \subseteq T$.*

If S is not a subset of T, we write $S \not\subseteq T$.

Theorem 1.2.2 *For any sets S, T, and U,*

1. $S \subseteq S$;

2. if $S \subseteq T$ and $T \subseteq S$, then $S = T$;

3. if $S \subseteq T$ and $T \subseteq U$, then $S \subseteq U$.

Proof: Suppose that $S \subseteq T$ and $T \subseteq S$. Then, every element of S is an element of T, and every element of T is an element of S. This means that S and T have the same elements, and hence, by the principle of extensionality,

$S = T$. This shows the second part of the theorem. The other two parts are even easier and are left to the reader. ∎

Part (2) of Theorem 1.2.2 provides the standard way of showing that two sets are equal; namely, show that each set is a subset of the other.

Definition 1.2.3 *If S and T are sets such that $S \subseteq T$ and $S \neq T$, then we say that S is* strictly included *in T. We shall denote this by $S \subset T$. If S is not strictly included in T, we write $S \not\subset T$.*

Theorem 1.2.4 *For any sets S, T, and U,*

  1. $S \not\subset S$;

  2. if $S \subset T$ and $T \subset U$, then $S \subset U$;

  3. if $S \subset T$, then $T \not\subset S$.

Proof: Since $S = S$, we cannot have $S \subset S$.

Suppose that $S \subset T$ and $T \subset U$. Then, $S \subseteq T$ and $T \subseteq U$, so, by the third part of the previous theorem, we have $S \subseteq U$. If $S = U$, then we have $S \subseteq T$ and $T \subseteq S$, and hence by the second part of the previous theorem, $S = T$, which contradicts $S \subset T$. Therefore, $S \subset U$.

Finally, suppose that $S \subset T$. If $T \subset S$, then we have $S \subseteq T$ and $T \subseteq S$, which implies $S = T$, contradicting $S \subset T$. ∎

There exists a set with no members. This set is called the *empty set* and is denoted by $\emptyset$. The principle of extensionality implies that there is only one empty set. Furthermore, the empty set is a subset of every set.

If S is a set, we can build a new set by considering the set whose unique member is S. We denote this set by $\{S\}$, and we refer to it as a *singleton*. In particular, we can form the set $\{\emptyset\}$, and this set is not empty since it contains $\emptyset$ as a member.

Definition 1.2.5 *If S is a set, then the* power set *of S is the set which consists of all the subsets of S. We denote the power set of S by $\mathcal{P}(S)$.*

Example 1.2.6 Let $S = \{a, b, c\}$. Then,

$$\mathcal{P}(S) = \{\emptyset, \{a\}, \{b\}, \{c\}, \{a, b\}, \{a, c\}, \{b, c\}, \{a, b, c\}\}.$$

We also have

$$\mathcal{P}(\emptyset) = \{\emptyset\}.$$

Note that for every set S, $\emptyset \in \mathcal{P}(S)$, and so $\mathcal{P}(S)$ is never empty.

We wish to define what is meant by a property of the elements of a set. There are two points of view: intensional and extensional. The intensional viewpoint considers a property of the elements of a set to be a characteristic condition that the elements of the set may or may not satisfy. We have seen already that different characteristic conditions may define the same set, i.e., have the same extension. The extensional viewpoint regards a property as being given by its extension, and this is the point of view we adopt.

Definition 1.2.7 *Let S be a set. By a* property *of the elements of S, we mean a subset of S.*

From this perspective, the power set of a set S consists of all properties of the elements of S. Note that we regard *any* subset of a set as giving a property of the elements of the set even if we have no way of expressing a characteristic condition for the property.

If P is a property of the elements of a set S, we frequently use the phrases "$P(x)$ is true" and "$P(x)$ holds" to mean $x \in P$. The phrase "we will show $P(x)$" means "we will show that $x \in P$".

Example 1.2.8 The property of being an odd natural number is given by the set

$$D = \{1, 3, 5, \ldots\}.$$

The property of being an even natural number is given by the set

$$E = \{0, 2, 4, 6, \ldots\}.$$

The property of being either equal to 0 or the sum of two odd natural numbers is also given by E, and therefore from our extensional point of view, these properties are the same, although from the intensional point of view they are different.

Example 1.2.9 Let M be the collection of all people. Having age 40 is a property of the elements of the set M while having average age 40 is a property of the elements of $\mathcal{P}(M)$, that is of the subsets of M. This is an important distinction in databases where one should differentiate between properties of individual objects and properties of aggregates of objects.

In addition to the notation **N** introduced for the set of natural numbers, we introduce here notations for several other important sets that we will use throughout this book.

P is the set $\{1, 2, 3, \ldots\}$ of positive natural numbers.

Z is the set $\{\ldots, -1, 0, 1, \ldots\}$ of integers.

Q is the set of rational numbers.

R is the set of real numbers.

Note that

$$\mathbf{P} \subset \mathbf{N} \subset \mathbf{Z} \subset \mathbf{Q} \subset \mathbf{R}.$$

1.3   Building New Sets

Given two objects a and b, we can form the set $\{a, b\}$, which we will call an *unordered pair*. By the principle of extensionality, $\{a, b\}$ and $\{b, a\}$ are the same unordered pair.

Sometimes when we pair up two objects, we want to consider them in a definite order, with one first and the other second. Such a pair is called an ordered pair. The following definition shows one way of capturing this idea using the tools of set theory.

Definition 1.3.1 *Let a and b be two objects. The* ordered pair, *or simply, the* pair, *(a, b) is the collection of sets $\{\{a\}, \{a, b\}\}$.*

Notice that the pair (a, a) is the singleton $\{\{a\}\}$. Conversely, if (a, b) is a singleton then $\{a, b\} = \{a\}$, so $b \in \{a\}$, that is, $b = a$.

The following theorem gives the essential property of ordered pairs; any other definition of the notion of ordered pair that satisfies this property is equally acceptable.

Theorem 1.3.2 *If $(a, b) = (c, d)$, then $a = c$ and $b = d$.*

Proof. Suppose that $(a, b) = (c, d)$. If $a = b$, then (a, b) is a singleton, $(a, b) = \{\{a\}\}$, hence, (c, d) is a singleton, which means that $c = d$ and $(c, d) = \{\{c\}\}$, and this gives $a = c$. Clearly, this also means $b = d$.

Assume now that $a \neq b$. The pairs $(a, b), (c, d)$ are not singletons. However, they both contain exactly one singleton, namely, $\{a\}$ and $\{c\}$, respectively, so $a = c$. They also contain the unordered pairs $\{a, b\}$ and $\{c, d\}$, which must be equal. Since $d \in \{a, b\}$ and $d \neq c = a$, we have $d = b$. ∎

We call a the *first component* of the pair (a, b), and b its *second component*. This terminology is justified by the previous theorem.

Definition 1.3.3 *Let A and B be two sets. The* Cartesian product *of A and B is the set $A \times B$, which consists of all pairs (a, b) such that $a \in A$ and $b \in B$.*

If either of the two sets A or B is empty, then so is their Cartesian product.

Example 1.3.4 Consider the sets $A = \{a_1, a_2, a_3\}$ and $B = \{b_1, b_2\}$. Their Cartesian product is the set

$$A \times B = \{(a_1, b_1), (a_1, b_2), (a_2, b_1), (a_2, b_2), (a_3, b_1), (a_3, b_2)\}.$$

Example 1.3.5 Suppose that $A = B = \mathbf{R}$, where $\mathbf{R}$ is the set of real numbers, and consider the Cartesian product $\mathbf{R} \times \mathbf{R}$. The elements of this new set can be represented geometrically by the points of a plane endowed with a system of coordinates. Namely, the pair $(a, b) \in \mathbf{R} \times \mathbf{R}$ will be represented by the point P whose x-coordinate is a and y-coordinate is b.

It is not difficult to see that if $A \subseteq B$ and $C \subseteq D$ then $A \times C \subseteq B \times D$. We refer to this property as the *monotonicity of the Cartesian product with respect to set inclusion*.

Definition 1.3.6 *Let C be a collection of sets. The* union *of C, denoted by $\bigcup C$, is the set defined by*

$$\bigcup C = \{x \mid x \in A \text{ for some } A \in C\}.$$

If $C = \{A, B\}$, we have $x \in \bigcup C$ if and only if $x \in A$ or $x \in B$. In this case, $\bigcup C$ will be denoted by $A \cup B$. We refer to the set $A \cup B$ as the *union* of A and B.

Several elementary properties of union are given in the following theorem.

Theorem 1.3.7 *For any sets A, B, C, we have*

1. $A \cup (B \cup C) = (A \cup B) \cup C = \bigcup\{A, B, C\}$ *(associativity of union)*,

2. $A \cup B = B \cup A$ *(commutativity of union)*,

3. $A \cup A = A$ *(idempotency of union), and*

4. $A \cup \emptyset = A$.

Proof. Let $a \in A \cup (B \cup C)$. In view of the definition of the union, we have $a \in A$ or $a \in B \cup C$. In the first case, $a \in \bigcup\{A, B, C\}$ because of Definition 1.3.6. In the second case, we have either $a \in B$ or $a \in C$, and both situations imply $a \in \bigcup\{A, B, C\}$. Conversely, let $a \in \bigcup\{A, B, C\}$. Definition 1.3.6 implies that $a \in A$, $a \in B$ or $a \in C$. In the first case, we clearly have $a \in A \cup (B \cup C)$; in the second or the third case, we have $a \in B \cup C$, which also implies $a \in A \cup (B \cup C)$. This proves that $A \cup (B \cup C) = \bigcup\{A, B, C\}$. A similar argument shows that $(A \cup B) \cup C = \bigcup\{A, B, C\}$.

The second equality follows immediately from the fact that $\{A, B\} = \{B, A\}$. The third equality is a consequence of the fact that $\{A, A\} = \{A\}$ and of Exercise 13. Finally, let $a \in A \cup \emptyset$. Since $\emptyset$ contains no element, we must have $a \in A$, and therefore, $A \cup \emptyset \subseteq A$. The reverse inclusion is obvious and this gives the last equality. ∎

Definition 1.3.8 *Let C be a nonempty collection of sets. The* intersection *of C, denoted by $\bigcap C$, is the set defined by*

$$\bigcap C = \{x \mid x \in A \text{ for every } A \in C\}.$$

If $C = \{A, B\}$, we denote $\bigcap C$ by $A \cap B$, and we refer to $A \cap B$ as the intersection of A and B. Then, $x \in A \cap B$ if and only if $x \in A$ and $x \in B$.

Theorem 1.3.9 *For any sets A, B, C, we have*

1. $A \cap (B \cap C) = (A \cap B) \cap C = \bigcap\{A, B, C\}$ *(associativity of intersection)*,

2. $A \cap B = B \cap A$ *(commutativity of intersection)*,

3. $A \cap A = A$ *(idempotency of intersection)*, and

4. $A \cap \emptyset = \emptyset$.

Proof. Let $a \in A \cap (B \cap C)$. We have $a \in A$ and $a \in B \cap C$ and this, in turn, implies $a \in B$ and $a \in C$. This shows that a is a member of every set of the collection $\{A, B, C\}$, so $a \in \bigcap\{A, B, C\}$. Conversely, if $a \in \bigcap\{A, B, C\}$, we have $a \in A, a \in B$, and $a \in C$, which implies $a \in B \cap C$. Therefore, we have $a \in A \cap (B \cap C)$. We proved that $A \cap (B \cap C) = \bigcap\{A, B, C\}$. The equality $(A \cap B) \cap C = \bigcap\{A, B, C\}$ can be proven using a similar argument.

The second equality is a consequence of the equality $\{A, B\} = \{B, A\}$, while the third follows from $\{A, A\} = \{A\}$ and Exercise 13.

For the last equality, note that no element a may belong to $A \cap \emptyset$ since this would mean $a \in \emptyset$. Therefore, $A \cap \emptyset = \emptyset$. ∎

Definition 1.3.10 *Two sets A, B are* disjoint *if $A \cap B = \emptyset$. A collection of sets C is called a* collection of pairwise disjoint sets *if for every A and B in C, if $A \neq B$, A and B are disjoint.*

Definition 1.3.11 *Let A, B be two sets. The* difference *of A and B is the set $A - B$ defined by*

$$A - B = \{x \in A \mid x \notin B\}.$$

Sometimes, when the set A is understood from the context, we write $\overline{B}$ for $A - B$, and we refer to the set $\overline{B}$ as the *complement* of B with respect to A or simply the complement of B.

The relationship between set difference and set union and intersection is given in the following theorem.

Theorem 1.3.12 *For every set A and nonempty collection C of sets, we have*

$$A - \bigcup C = \bigcap\{A - C \mid C \in C\},$$
$$A - \bigcap C = \bigcup\{A - C \mid C \in C\}.$$

Proof: Let $a \in A - \bigcup C$. We have $a \in A$ and $a \notin C$ for every $C \in C$. This implies that $a \in A - C$ for every $C \in C$. Hence, $a \in \bigcap\{A - C \mid C \in C\}$. Conversely, if $a \in \bigcap\{A - C \mid C \in C\}$, then $a \in A - C$ for every $C \in C$; hence, $a \in A$ (since C is nonempty), and a is not a member of any set that belongs to C. This implies that a is not a member of $\bigcup C$; hence, $a \in A - \bigcup C$, which proves the first equality.

We leave the second equality to the reader. ∎

Corollary 1.3.13 *For any sets A, B, C, we have:*

$$A - (B \cup C) = (A - B) \cap (A - C),$$
$$A - (B \cap C) = (A - B) \cup (A - C).$$

Proof. The corollary follows immediately from the previous theorem by choosing $\mathcal{C} = \{B, C\}$. ∎

If the set A is understood in Corollary 1.3.13, then we get the following.

Corollary 1.3.14 (De Morgan's Laws) [5]

$$\overline{B \cup C} = \overline{B} \cap \overline{C},$$
$$\overline{B \cap C} = \overline{B} \cup \overline{C}.$$

The linkage between union and intersection is given by the distributivity properties contained in the following theorem.

Theorem 1.3.15 *For any collection of sets $\mathcal{C}$ and set A, we have*

$$\left(\bigcup \mathcal{C}\right) \cap A = \bigcup \{C \cap A \mid C \in \mathcal{C}\}.$$

If $\mathcal{C}$ is nonempty, we also have

$$\left(\bigcap \mathcal{C}\right) \cup A = \bigcap \{C \cup A \mid C \in \mathcal{C}\}.$$

Proof. We shall prove only the first equality; the proof of the second one is left as an exercise to the reader.

Let $a \in (\bigcup \mathcal{C}) \cap A$. This means that $a \in \bigcup \mathcal{C}$ and $a \in A$. There is a set $C \in \mathcal{C}$ such that $a \in C$; hence, $a \in C \cap A$, which implies $a \in \bigcup \{C \cap A \mid C \in \mathcal{C}\}$.

Conversely, if $a \in \bigcup \{C \cap A \mid C \in \mathcal{C}\}$, there exists a member $C \cap A$ of this collection such that $a \in C \cap A$, so $a \in C$ and $a \in A$. It follows that $a \in \bigcup \mathcal{C}$, and this, in turn, gives $a \in (\bigcup \mathcal{C}) \cap A$. ∎

Corollary 1.3.16 *For any sets A, B, and C, we have*

$$(B \cup C) \cap A = (B \cap A) \cup (C \cap A),$$
$$(B \cap C) \cup A = (B \cup A) \cap (C \cup A).$$

Proof. The corollary follows immediately by choosing $\mathcal{C} = \{B, C\}$. ∎

We also have distributivity of the Cartesian product over union, intersection, and difference.

Theorem 1.3.17 *If Ω is one of $\cup, \cap,$ or $-$, then for any sets A, B, and C, we have*

$$(A\Omega B) \times C = (A \times C)\Omega(B \times C),$$
$$C \times (A\Omega B) = (C \times A)\Omega(C \times B).$$

---

[5] Augustus De Morgan was born in 1806 in Madurai, India, and died in 1871 in London. He taught at the University College in London beginning in 1828 and was president of the London Mathematical Society. De Morgan's main contributions were in the area of the foundations of modern logic.

Proof. We prove only that

$$(A - B) \times C = (A \times C) - (B \times C).$$

Let $(x, y) \in (A - B) \times C$. We have $x \in A - B$ and $y \in C$. Therefore, $(x, y) \in A \times C$ and $(x, y) \notin B \times C$, which show that

$$(x, y) \in (A \times C) - (B \times C).$$

Conversely, if $(x, y) \in (A \times C) - (B \times C)$, we obtain $x \in A$ and $y \in C$ and also $(x, y) \notin B \times C$. This implies $x \notin B$, so $(x, y) \in (A - B) \times C$. ∎

According to the principle of extensionality, two sets are equal if they have the same elements. For instance, $\{x, x\} = \{x\}$. Sometimes it is useful to be able to differentiate between sets based on not just what elements are in the set but also how often the elements occur. This can be done using the notion of multiset, which we now define.

Definition 1.3.18 *A multiset or bag* [6] *on a set A is a subset M of $A \times \mathbf{N}$ such that for every $a \in A$ there is an $n \in \mathbf{N}$ such that $(a, m) \in M$ if and only if $m < n$. The set of all multisets on A is denoted by $\mathcal{M}(A)$.*

If A is a set, M is a multiset on A, and $a \in A$, then the number n such that $(a, m) \in M$ if and only if $m < n$ is unique since it is the least number r such that $(a, r) \notin M$. We call this number n the *multiplicity* of a in M.

Example 1.3.19 The two numbers

$$300 = 2^2 3^1 5^2 \text{ and } 360 = 2^3 3^2 5^1$$

have the same set of prime divisors $\{2, 3, 5\}$. However, the multisets

$$\{(2, 0), (2, 1), (3, 0), (5, 0), (5, 1)\}$$

and
$$\{(2, 0), (2, 1), (2, 2), (3, 0), (3, 1), (5, 0)\}$$

of prime divisors of these two numbers are distinct, and in fact, each positive number is determined by its multiset of prime divisors.

A multiset is determined by the multiplicities of its elements as the following theorem shows.

Theorem 1.3.20 *Let A be a set and let M and N be multisets on A. If for every $a \in A$, the multiplicity of a in M is the same as the multiplicity of a in N, then $M = N$.*

---

[6] Recently, the term "suite" has been suggested as an alternative to "multiset." Whether or not this term will catch on remains to be seen.

Proof: Both M and N are subsets of $A \times \mathbf{N}$. For each $(a, n) \in A \times \mathbf{N}$, $(a, n) \in M$ if and only if n is less than the multiplicity of a in M and $(a, n) \in N$ if and only if n is less than the multiplicity of a in N. Since these two multiplicities are the same, $(a, n) \in M$ if and only if $(a, n) \in N$. Thus, M and N have the same elements and hence by extensionality, $M = N$. ∎

Let A be a set and suppose that for each $a \in A$ we have specified a natural number m_a. Then, there is a unique multiset M on A such that for every $a \in A$, the multiplicity of a in M is m_a. Indeed, $M = \{(a, m) \in A \times \mathbf{N} \mid m < m_a\}$ is such a multiset, and the uniqueness follows from the previous theorem.

We denote a multiset by using square brackets instead of braces. For example, the two multisets of Example 1.3.19 can be written as $[2, 2, 3, 5, 5]$ and $[2, 2, 2, 3, 3, 5]$. Note that while multiplicity counts in a multiset, order does not matter; therefore, the multiset $[2, 2, 3, 5, 5]$ could also be denoted by $[5, 2, 3, 2, 5]$ or $[3, 2, 5, 5, 2]$. We use the abbreviation $n \cdot a$ in a multiset to mean n occurrences of a. For example, the multiset of letters appearing in the word Mississippi is $[1 \cdot \mathrm{M}, 4 \cdot \mathrm{i}, 4 \cdot \mathrm{s}, 2 \cdot \mathrm{p}]$.

In the exercises, we will introduce methods of building new multisets from old ones analogous to the methods we have defined for sets.

1.4 Exercises and Supplements

Sets, Members, Subsets

1. Show that if $a \in \{\{b\}\}$, then $b \in a$.

2. What is $\{x \mid x \in A\}$?

3. Suppose that in a country there are regional libraries and a central library. Each library compiles an inventory in book form of its holdings. This book is called the catalog of the library. Some of the regional libraries choose to have their catalog contain a reference to itself; others do not. Each regional library sends a copy of its catalog to the central library. The central library decides to produce a catalog of all of its catalogs that do not list themselves.

   (a) Prove that the central library's project is an unattainable task.

   (b) Describe the relationship of this project to Russell's paradox.

4. Give another argument for the nonexistence of a universal set V by showing that if V exists, then the set R of Russell's paradox exists.

   Hint: Use the principle of comprehension.

5. Let S, T, U be three sets. Prove that

(a) if $S \subseteq T$ and $T \subset U$, then $S \subset U$;

(b) if $S \subset T$ and $T \subseteq U$, then $S \subset U$; and

(c) $S \subset T$ if and only if $S \subseteq T$ and there is an $a \in T$ such that $a \notin S$.

6. Let S and T be two sets. Prove that if for every set U, $U \subseteq S$ if and only if $U \subseteq T$, then $S = T$.

7. Let S and T be two sets.

(a) Prove that $S \subseteq T$ if and only if $\mathcal{P}(S) \subseteq \mathcal{P}(T)$.

(b) Conclude that $S = T$ if and only if $\mathcal{P}(S) = \mathcal{P}(T)$.

Building New Sets

8. Consider the following possible definitions of the ordered pair (a, b):

(a) $(a, b) = \{\{a\}, \{b, \emptyset\}\}$,

(b) $(a, b) = \{a, \{b, \emptyset\}\}$,

(c) $(a, b) = \{\{a, 0\}, \{b, 1\}\}$, and

(d) $(a, b) = \{\{b\}, \{a, b\}\}$.

Determine which of these proposed definitions satisfies Theorem 1.3.2 and hence is an acceptable definition.

9. For any x, y, z, define (x, y, z) to be the pair $((x, y), z)$. Prove that if $(x, y, z) = (x', y', z')$ then $x = x', y = y'$, and $z = z'$.

10. Is there a set A for which $A \subseteq A \times A$?

11. Consider the sets $A = \{1, 2, 3, 4\}$ and $B = \{2, 3\}$. List the pairs of the following sets: $A \times B, B \times A, (A \times B) \cap (B \times A), (A \times B) - (B \times A), (B \times A) - (A \times B)$.

12. Compute

(a) $\bigcup \emptyset$,

(b) $\bigcup \{\emptyset\}, \bigcap \{\emptyset\}$,

(c) $\bigcup \{\{\emptyset\}\}, \bigcap \{\{\emptyset\}\}$, and

(d) $\bigcup \{\emptyset, \{\emptyset\}\}, \bigcap \{\emptyset, \{\emptyset\}\}$.

13. (a) Show that for any set A, $\bigcup \{A\} = A$.

(b) Show that for any set A, $\bigcap \{A\} = A$.

14. (a) Show that for any set A, $\bigcup (\bigcup \{\{A\}\}) = A$.

(b) Show that for any set A, $\bigcap (\bigcap \{\{A\}\}) = A$.

15. If $\mathcal{C} = \{\{x\} \mid x \in B\}$, show that $\bigcup \mathcal{C} = B$.

16. Let $\mathcal{C}$ be a collection of sets.

    (a) Show that for all $A \in \mathcal{C}$, $A \subseteq \bigcup \mathcal{C}$.

    (b) Show that for all sets B, if $A \subseteq B$ for every $A \in \mathcal{C}$, then $\bigcup \mathcal{C} \subseteq B$.

17. Let $\mathcal{C}$ be a nonempty collection of sets.

    (a) Show that for all $A \in \mathcal{C}$, $\bigcap \mathcal{C} \subseteq A$.

    (b) Show that for all sets B, if $B \subseteq A$ for every $A \in \mathcal{C}$, then $B \subseteq \bigcap \mathcal{C}$.

18. Let S and T be two sets. Show that $S - T$ is the largest subset of S which is disjoint from T; that is, $S - T \subseteq S$, $(S - T) \cap T = \emptyset$, and for every U such that $U \subseteq S$ and $U \cap T = \emptyset$, we have $U \subseteq S - T$.

19. Let $\mathcal{C}$ and $\mathcal{D}$ be collections of sets.

    (a) Show that if $\mathcal{C} \subseteq \mathcal{D}$, then $\bigcup \mathcal{C} \subseteq \bigcup \mathcal{D}$.

    (b) Show that if for every $C \in \mathcal{C}$ there is a $D \in \mathcal{D}$ such that if $C \subseteq D$, then $\bigcup \mathcal{C} \subseteq \bigcup \mathcal{D}$.

    (c) Show that if $\emptyset \subset \mathcal{C} \subseteq \mathcal{D}$, then $\bigcap \mathcal{D} \subseteq \bigcap \mathcal{C}$.

    (d) Show that if $\mathcal{C}$ and $\mathcal{D}$ are nonempty and for every $C \in \mathcal{C}$ there is a $D \in \mathcal{D}$ such that $D \subseteq C$, then $\bigcap \mathcal{D} \subseteq \bigcap \mathcal{C}$.

20. Let S be a set. Prove that the collection that consists of all x such that $x \notin S$ is not a set.

    Solution: If the set $T = \{x \mid x \notin S\}$ exists, then every set is a member of $S \cup T$, so $S \cup T$ is a universal set, contradicting Exercise 4.

21. In the definition of the intersection of a collection of sets, we required the collection $\mathcal{C}$ to be nonempty. Why?

    Solution: If $\bigcap \emptyset$ existed, it would consist of those objects a such that for every $S \in \emptyset$, $a \in S$. Since there are no such sets S, this condition is vacuously satisfied by every object a and thus $\bigcap \emptyset$ would be a universal set, which we have shown cannot exist.

22. (a) Let $\mathcal{C}$ and $\mathcal{D}$ be two collections of sets. Prove the following identities:

$$\bigcup \mathcal{C} \cup \bigcup \mathcal{D} = \bigcup (\mathcal{C} \cup \mathcal{D}),$$
$$\bigcup \mathcal{C} \cap \bigcup \mathcal{D} = \bigcup \{C \cap D \mid C \in \mathcal{C} \text{ and } D \in \mathcal{D}\}.$$

In addition, if $\mathcal{D}$ is nonempty, then for each $D \in \mathcal{D}$, define $F_D = \bigcup \{C - D \mid C \in \mathcal{C}\}$. Prove that

$$\bigcup \mathcal{C} - \bigcup \mathcal{D} = \bigcap \{F_D \mid D \in \mathcal{D}\}.$$

Finally, prove that if $\mathcal{C}$ and $\mathcal{D}$ are both nonempty, then

$$\bigcup \mathcal{C} \cup \bigcup \mathcal{D} = \bigcup \{C \cup D \mid C \in \mathcal{C} \text{ and } D \in \mathcal{D}\}.$$

(b) Prove that the first three of these identities generalize associativity of union (Theorem 1.3.7, Part 1), the first distributivity property of Theorem 1.3.15, and the first equality of Theorem 1.3.12, respectively.

(c) Find examples of collections $\mathcal{C}$ and $\mathcal{D}$ such that

$$\bigcup \mathcal{C} \cap \bigcup \mathcal{D} \neq \bigcup (\mathcal{C} \cap \mathcal{D}),$$
$$\bigcup \mathcal{C} - \bigcup \mathcal{D} \neq \bigcup (\mathcal{C} - \mathcal{D}).$$

23. (a) Let $\mathcal{C}$ and $\mathcal{D}$ be two nonempty collections of sets. Prove the following identities:

$$\bigcap \mathcal{C} \cup \bigcap \mathcal{D} = \bigcap \{C \cup D \mid C \in \mathcal{C} \text{ and } D \in \mathcal{D}\},$$
$$\bigcap \mathcal{C} \cap \bigcap \mathcal{D} = \bigcap (\mathcal{C} \cup \mathcal{D})$$
$$= \bigcap \{C \cap D \mid C \in \mathcal{C} \text{ and } D \in \mathcal{D}\}.$$

For each $D \in \mathcal{D}$, define $G_D = \bigcap \{C - D \mid C \in \mathcal{C}\}$. Prove that

$$\bigcap \mathcal{C} - \bigcap \mathcal{D} = \bigcup \{G_D \mid D \in \mathcal{D}\}.$$

(b) Prove that these identities generalize the second distributivity property of Theorem 1.3.15, associativity of intersection (Theorem 1.3.9, Part 1), and the second equality of Theorem 1.3.12, respectively.

(c) Find examples of collections $\mathcal{C}$ and $\mathcal{D}$ such that

$$\bigcap \mathcal{C} \cup \bigcap \mathcal{D} \neq \bigcap (\mathcal{C} \cup \mathcal{D}),$$
$$\bigcap \mathcal{C} \cap \bigcap \mathcal{D} \neq \bigcap (\mathcal{C} \cap \mathcal{D}),$$
$$\bigcap \mathcal{C} - \bigcap \mathcal{D} \neq \bigcap (\mathcal{C} - \mathcal{D}).$$

24. Let A, B be two sets. Prove that if any two of the statements

(a) A and B are disjoint,

(b) $A \subseteq B$, and

(c) $A = \emptyset$

are true then the third one is also true.

25. Prove that for all sets A, B the following statements are equivalent:

(a) $A \subseteq B$,

(b) $A - B = \emptyset$,

(c) $A \cup B = B$, and

(d) $A \cap B = A$.

26. Prove that for all sets A, B, C we have

$$(A \cup C) - (B \cup C) \subseteq A - B.$$

Give conditions under which the above inclusion becomes an equality.

27. Let A, B, C, D be four sets.

(a) Prove that if $C \neq \emptyset$ and $A \times C \subseteq B \times C$ then $A \subseteq B$; similarly, $C \neq \emptyset$ and $C \times A \subseteq C \times B$ implies $A \subseteq B$.

(b) Prove that

$$(A \times B) - (C \times D) = ((A - C) \times B) \cup (A \times (B - D)).$$

28. Let S be a set. Prove that for every a the following three statements are equivalent:

(a) $a \in S$,

(b) $\{a\} \subseteq S$, and

(c) $\{a\} \cap S \neq \emptyset$.

Conclude that for every set S, $S \in S$ if and only if $S \cap \{S\} \neq \emptyset$.

29. Consider the following statement:

> **Principle of Foundation.** For every nonempty set S, there is an $x \in S$ such that $x \cap S = \emptyset$.[7]

Use this principle to prove that for every set S, $S \notin S$. (Hence, Russell's set R, if it existed, would be a universal set.)

Hint: Apply the principle of foundation to $\{S\}$ and use Exercise 28.

30. Use the principle of foundation to show that there is no nonempty set A such that $A \subseteq A \times A$.

Solution: Let A be nonempty and define $A' = A \cup \{\{a\} \mid a \in A\}$. Then A' is nonempty, so by the principle of foundation there is $x \in A'$ with $x \cap A' = \emptyset$. We cannot have $x = \{a\}$ for some $a \in A$ since this would give $a \in x \cap A'$, so $x \in A$. If $x \in A \times A$, say $x = \{\{a\}, \{a, a'\}\}$ with both a and a' in A, then $\{a\} \in x \cap A'$, which is impossible. Thus, x is an element of A that is not in $A \times A$ and $A \not\subseteq A \times A$.

---

[7] The principle of foundation (also known as the principle of regularity) is a generally accepted principle of set theory, but it is less used than the other principles mentioned in this chapter, and in fact, most of mathematics can be done without it.

31. Let A, B be two sets. Prove that

$$A \times B \subseteq \mathcal{P}(\mathcal{P}(A \cup B)).$$

32. (a) Let U be a set and let A and B be subsets of U. Prove that the equation $A \cap X = B$ has a solution $X \in \mathcal{P}(U)$ if and only if $B \subseteq A$. Show that, in this case, X is a solution if and only if there is a $P \subseteq U - A$ with $X = B \cup P$.

    (b) Prove that the equation $A \cup X = B$ has a solution in X if and only if $A \subseteq B$. In this case, show that X is a solution if and only if $B - A \subseteq X \subseteq B$.

    (c) Let U be a set and let A, B, and C be subsets of U. Find necessary and sufficient conditions for the equation $(A \cap X) \cup B = C$ to have a solution $X \in \mathcal{P}(U)$. Find the general form of the solution if solutions exist.

33. Prove that sets A, B, and C satisfy the equality

$$\bigcap \{A \cup B, B \cup C, C \cup A\} = \bigcup \{A, B, C\}$$

if and only if there exist sets K, L, and M such that $A = K \cup L, B = L \cup M$, and $C = M \cup K$.

34. Let A, B, W be three sets. If $A \cap W = B \cap W$ and $A \cup W = B \cup W$, prove that $A = B$.

35. For each inequality, give an example of sets A and B that satisfy the inequality (and thereby show that $-$ and $\times$ are not commutative).

$$A - B \neq B - A,$$
$$A \times B \neq B \times A.$$

36. For each inequality, give an example of sets A, B and C that satisfy the inequality (and thereby show that $-$ and $\times$ are not associative).

$$A - (B - C) \neq (A - B) - C,$$
$$A \times (B \times C) \neq (A \times B) \times C.$$

37. Show that for any sets A, B, and C, the following equalities are true (and hence various distributive laws hold).

$$A \cup (B \cup C) = (A \cup B) \cup (A \cup C),$$
$$A \cap (B \cap C) = (A \cap B) \cap (A \cap C),$$
$$A \cap (B - C) = (A \cap B) - (A \cap C),$$
$$(B \cup C) - A = (B - A) \cup (C - A),$$
$$(B \cap C) - A = (B - A) \cap (C - A),$$
$$(B - C) - A = (B - A) - (C - A).$$

38. For each inequality, give examples of sets A, B, and C that satisfy the inequality (and thereby show that various possible distributive laws do not hold).

$$A \cup (B - C) \neq (A \cup B) - (A \cup C),$$
$$A \cup (B \times C) \neq (A \cup B) \times (A \cup C),$$
$$A \cap (B \times C) \neq (A \cap B) \times (A \cap C),$$
$$A - (B \cup C) \neq (A - B) \cup (A - C),$$
$$A - (B \cap C) \neq (A - B) \cap (A - C),$$
$$A - (B - C) \neq (A - B) - (A - C),$$
$$A - (B \times C) \neq (A - B) \times (A - C),$$
$$(B \times C) - A \neq (B - A) \times (C - A),$$
$$A \times (B \times C) \neq (A \times B) \times (A \times C),$$
$$(B \times C) \times A \neq (B \times A) \times (C \times A).$$

39. For any two sets A and B, we define the *symmetric difference* $A \oplus B$ of A and B to be $(A - B) \cup (B - A)$. (Another common notation for the symmetric difference of A and B is $A \bigtriangleup B$.) Prove the following for all sets A, B, C.

   (a) $A \oplus B = (A \cup B) - (A \cap B)$.

   (b) If $B \subseteq A$, then $A \oplus B = A - B$.

   (c) If A and B are disjoint, then $A \oplus B = A \cup B$.

   (d) $A \oplus \emptyset = \emptyset \oplus A = A$.

   (e) $A \oplus B = \emptyset$ if and only if $A = B$.

   (f) $A \oplus B = B \oplus A$.

   (g) $A \oplus (B \oplus C) = (A \oplus B) \oplus C$.

   (h) $A \cap (B \oplus C) = (A \cap B) \oplus (A \cap C)$.

   (i) $(B \oplus C) - A = (B - A) \oplus (C - A)$.

   (j) $A \times (B \oplus C) = (A \times B) \oplus (A \times C)$.

   (k) $(B \oplus C) \times A = (B \times A) \oplus (C \times A)$.

   (l) $A \cup B = A \oplus (B \oplus (A \cap B))$.

   (m) If $A - C = B - C$, then $A \oplus B \subseteq C$.

   (n) If $A \oplus B = A \oplus C$, then $B = C$.

40. For each inequality, give examples of sets A, B, and C that satisfy the inequality (and thereby show that various possible distributive

laws do not hold).

$$
\begin{aligned}
A \cup (B \oplus C) &\neq (A \cup B) \oplus (A \cup C), \\
A - (B \oplus C) &\neq (A - B) \oplus (A - C), \\
A \oplus (B \cup C) &\neq (A \oplus B) \cup (A \oplus C), \\
A \oplus (B \cap C) &\neq (A \oplus B) \cap (A \oplus C), \\
A \oplus (B - C) &\neq (A \oplus B) - (A \oplus C), \\
A \oplus (B \times C) &\neq (A \oplus B) \times (A \oplus C), \\
A \oplus (B \oplus C) &\neq (A \oplus B) \oplus (A \oplus C).
\end{aligned}
$$

41. Prove that for any sets A, B, and C we have

    (a) $A - (B \cup C) = (A - B) - C$,
    (b) $(A \cap B) - C = A \cap (B - C) = (A - C) \cap B$.

42. Show that for any set A, $\emptyset \in \mathcal{M}(A)$ (i.e., $\emptyset$ is a multiset on A).

43. Let A and B be two sets.

    (a) Prove that $A \subseteq B$ if and only if $\mathcal{M}(A) \subseteq \mathcal{M}(B)$.
    (b) Conclude that $A = B$ if and only if $\mathcal{M}(A) = \mathcal{M}(B)$.

44. Let A be a set and let M and N be multisets on A.

    (a) Show that $M \cup N$ is a multiset on A and that if $a \in A$ has multiplicity m in M and n in N, then a has multiplicity $\max(m, n)$ in $M \cup N$.
    (b) Show that $M \cap N$ is a multiset on A and that if $a \in A$ has multiplicity m in M and n in N, then a has multiplicity $\min(m, n)$ in $M \cap N$.

According to the previous exercise, union and intersection provide ways of combining multisets to get new multisets. Another way to do this is given in the next definition.

Definition 1.4.1 *Let A be a set and let M and N be multisets on A. The sum of M and N (denoted $M + N$) is the multiset on A such that for each $a \in A$, the multiplicity of a in $M + N$ is the sum of a's multiplicities in M and N.*

45. Let $M = [2, 3, 3, 5, 5, 5]$ and $N = [2, 2, 2, 5, 5, 7]$. What are $M \cup N$, $M \cap N$, and $M + N$?

46. Let M and N be the multisets of prime factors for the positive integers m and n. Which integers have as their multisets of prime divisors $M \cup N$, $M \cap N$, and $M + N$?

47. Let A be a set. For each subset S of A, we can define a multiset M_S on A by defining

$$M_S = \{(a, 0) \mid a \in S\}$$

(so if $a \in S$, then a has multiplicity 1 in M_S, and if $a \notin S$, then a has multiplicity 0 in M_S). Show that for any two subsets S and T of A we have

$$M_{S \cup T} = M_S \cup M_T,$$
$$M_{S \cap T} = M_S \cap M_T.$$

48. Give an example of a set A and multisets M and N on A such that neither $M - N$ nor $M \oplus N$ are multisets on A.

We have already seen many identities involving union and intersection in the chapter and exercises, and of course, each of these identities still holds when attention is restricted to multisets on some set A. We now give some further identities involving the sum.

49. Let A be a set. Show that for any three multisets M, N, and P on A we have

$$\emptyset + M = M,$$
$$M + N = N + M,$$
$$M + (N + P) = (M + N) + P,$$
$$M + (N \cup P) = (M + N) \cup (M + P),$$
$$M + (N \cap P) = (M + N) \cap (M + P),$$
$$M + N = (M \cup N) + (M \cap N).$$

50. For each of the following inequalities, give an example of a set A and multisets M, N, and P on A that satisfies the inequality (and thereby show that various possible distributive laws do not hold).

$$M \cup (N + P) \neq (M \cup N) + (M \cup P),$$
$$M \cap (N + P) \neq (M \cap N) + (M \cap P),$$
$$M + (N + P) \neq (M + N) + (M + P).$$

51. (a) Let M be a multiset on A and let N be a multiset on B. Show that $M \cup N$ is a multiset on $A \cup B$ and $M \cap N$ is a multiset on $A \cap B$.

    (b) Let M be a multiset on A and let N be a multiset on B. Show that M and N are both multisets on $A \cup B$.

    (c) Let M and N be multisets on A and also on B. Show that $M + N$ considered to be a multiset on A is the same (as a set of ordered pairs) as $M + N$ considered as a multiset on B.

It is natural to define the phrase "M is a multiset" (independent of any set A that M is a multiset on) by specifying that M is a multiset if there is a set A such that M is a multiset on A.

52. Show that M is a multiset if and only if M is a set of ordered pairs such that the second component of every pair in M is a natural number and in addition $(a, n) \in M$ and $m < n$ for some natural number m imply $(a, m) \in M$.

Exercise 51 allows us to use $\cup$, $\cap$, and $+$ to combine any two multisets. Specifically, according to part (a), the union and intersection of two multisets are again multisets, and we can define the sum of two multisets M and N as follows: by part (b) of the exercise, we can consider M and N to be multisets over a common set, and by part (c), we can form $M + N$ and get a result that is independent of the set chosen. The identities and nonidentities of Exercises 49 and 50 still hold in this point of view, and Exercise 47 can be phrased more elegantly without the extraneous set A.

1.5   Bibliographical Comments

There are a large number of introductory works in set theory. We recommend [Hal87], [End77], [HJ84], and [Roi90]. (The last two of these reach some of the more advanced topics as well.) Another interesting reference, which does not cover the most elementary material but assumes no more set theory than is covered in this chapter, is [Ham82].

Cantor's original papers in set theory can be read in [Can55]. Russell discovered his paradox in 1901 and described it in a 1902 letter to the German mathematician Gottlob Frege who was just finishing the second volume of a two volume work on set theory. These volumes were the culmination of Frege's life work, and Russell showed that work to be based on inconsistent assumptions, yet Frege responded with intellectual curiosity rather than bitterness. The correspondence between Russell and Frege can be found in [vH67].

2

Relations and Functions

2.1   Introduction
2.2   Relations
2.3   Functions
2.4   Sequences, Words, and Matrices
2.5   Images of Sets Under Relations
2.6   Relations and Directed Graphs
2.7   Special Classes of Relations
2.8   Equivalences and Partitions
2.9   General Cartesian Products
2.10 Operations
2.11 Representations of Relations and Graphs
2.12 Relations and Databases
2.13 Exercises and Supplements
2.14 Bibliographical Comments

2.1   Introduction

In mathematics, as in everyday life, one often speaks about relationships between objects and, in particular, of the idea of two objects being related or associated with each other in some way. In this chapter, we will study relations, a way of making precise the idea of an asssociation between objects. A relation will be defined to be a set of ordered pairs, the idea being that a relation will consist of exactly those ordered pairs (a, b) such that a is associated with b under whatever relationship we wish to study.

Another fundamental concept is that of an assignment, such as the assignment of a grade to each member of a class or of a salary to each employee of a company. We can regard such an assignment as an association where each object is associated with the object assigned to it and no others. The mathematical notion corresponding to the idea of assignment is function. Motivated by the considerations just given, we will define a function to be a relation under which each object is related to at most one other object.

After defining the notions of relation and function, we will study ways of combining relations and functions to get new ones and consider special classes of relations and functions of particular importance to computer science.

Relations and functions are important for computer science from at least two points of view: they serve as programming paradigms and they help provide precise mathematical models for the ways in which data are structured. We have focused this chapter on the latter role; the use of functions to define words, sequences, and matrices is presented in Section 2.4, while Section 2.12 contains an introduction to the relational database model.

2.2 Relations

Definition 2.2.1 *A relation is a set of ordered pairs. If A and B are sets and ρ is a relation, then we call ρ a relation from A to B if $\rho \subseteq A \times B$.*

Note that $\mathcal{P}(A \times B)$ is the set of all relations from A to B. Among the relations from A to B, we distinguish the *empty relation*, $\emptyset$, and the *full relation*, $A \times B$.

If $(a, b) \in \rho$, we sometimes denote this fact by $a \rho b$, and we write $a \not\rho b$ to denote $(a, b) \notin \rho$. A relation from A to A is called a *relation on A*.

Example 2.2.2 Consider the set C of all cities and the set S of all states in the U.S. The relation $\rho_{in} \subseteq C \times S$, where $(c, s) \in \rho_{in}$ if city c is in state s, is a relation from C to S.

Example 2.2.3 Consider the set P of patients who are hospitalized in a certain hospital and the set S of surgeons of that hospital. For each date d, define the relation σ_d by $(p, s) \in \sigma_d$ if patient p was operated on by surgeon s on date d.

Example 2.2.4 For any set A, we can consider the *identity relation* $\iota_A \subseteq A \times A$ defined by

$$\iota_A = \{(x, x) \mid x \in A\}.$$

Note that if $A \subseteq B$, then $\iota_A \subseteq \iota_B$.

Definition 2.2.5 *The* domain *of a relation ρ is the set*

$$Dom(\rho) = \{a \mid (a, b) \in \rho \text{ for some } b\}.$$

The range *of ρ is the set*

$$Ran(\rho) = \{b \mid (a, b) \in \rho \text{ for some } a\}.$$

It follows easily that if ρ is a relation and A and B are sets, then ρ is a relation from A to B if and only if $Dom(\rho) \subseteq A$ and $Ran(\rho) \subseteq B$. Naturally, ρ is always a relation from $Dom(\rho)$ to $Ran(\rho)$.

Note that if ρ and σ are relations and $\rho \subseteq \sigma$, then we have $Dom(\rho) \subseteq Dom(\sigma)$ and $Ran(\rho) \subseteq Ran(\sigma)$. We also remark that $Dom(A \times B) = A$ unless $B = \emptyset$. In the latter case, we have $Dom(A \times \emptyset) = \emptyset$. Similarly, $Ran(A \times B) = B$ if $A \neq \emptyset$. For $A = \emptyset$, we have $Ran(\emptyset \times B) = \emptyset$.

Example 2.2.6 The domain of the relation ρ defined in Example 2.2.2 is the set of all cities that are located in some state. For instance, Boston is in the domain of ρ because (Boston, Massachusetts)$\in \rho$. On the other hand, Washington D.C. does not belong to $\text{Dom}(\rho)$.

For the relation σ_d considered in Example 2.2.3, a patient is in $\text{Dom}(\sigma_d)$ if and only if that patient had surgery in the hospital on date d. A surgeon s belongs to the range of σ_d if and only if s operated on a patient in the hospital on date d.

Let us now consider examples that have a mathematical nature.

Example 2.2.7 Let A be a subset of $\mathbf{R}$, the set of real numbers. The relation "less than" on A is given by

$$\{(x,y) \mid x, y \in A \text{ and } y = x + z \text{ for some positive real } z\}.$$

Example 2.2.8 Consider the relation $\nu \subseteq \mathbf{Z} \times \mathbf{Q}$ given by

$$\nu = \{(n,q) \mid n \in \mathbf{Z}, \ q \in \mathbf{Q}, \text{ and } n \leq q < n + 1\}.$$

We have $(-3, -2.3) \in \nu$ and $(2, 2.3) \in \nu$. Clearly, $(n, q) \in \nu$ if and only if n is the integral part of the rational number q.

Example 2.2.9 We can define a relation $\delta_{\mathbf{Z}} \subseteq \mathbf{Z} \times \mathbf{Z}$, where $(m, n) \in \delta_{\mathbf{Z}}$ if there is $k \in \mathbf{Z}$ such that $n = mk$. In other words, $(m, n) \in \delta_{\mathbf{Z}}$ if m divides n evenly.

Example 2.2.10 Let S be a set. Then, a relation from S to $\mathcal{P}(S)$ is given by

$$\{(x, X) \mid X \in \mathcal{P}(S) \text{ and } x \in X\}.$$

We next want to consider ways of getting new relations from old ones. Since relations are sets of ordered pairs, we can use the ways of getting new sets from old sets introduced in Chapter 1. In particular, if ρ and σ are relations, then so are $\rho \cup \sigma$, $\rho \cap \sigma$, and $\rho - \sigma$, and in fact, if ρ and σ are both relations from A to B, then so will be these new relations.

We now consider some ways of getting new relations from old ones that are specific to relations.

Definition 2.2.11 *Let ρ be a relation. The* inverse *of ρ is the relation ρ^{-1} given by*

$$(y, x) \in \rho^{-1} \text{ if and only if } (x, y) \in \rho,$$

that is,

$$\rho^{-1} = \{(y, x) \mid (x, y) \in \rho\}.$$

The basic properties of the inverse relation are given in the following theorem.

Theorem 2.2.12 *Let ρ and σ be relations.*

1. $Dom(\rho^{-1}) = Ran(\rho)$.

2. $Ran(\rho^{-1}) = Dom(\rho)$.

3. *If ρ is a relation from A to B, then ρ^{-1} is a relation from B to A.*

4. $(\rho^{-1})^{-1} = \rho$.

5. *If $\rho \subseteq \sigma$, then $\rho^{-1} \subseteq \sigma^{-1}$ (montonicity of the inverse).*

Proof: We leave these arguments to the reader. ∎

Definition 2.2.13 *Let ρ and σ be relations. The product of ρ and σ is the relation $\rho \bullet \sigma$, where*

$$\rho \bullet \sigma = \{(x, z) \mid \text{ for some } y, \ (x, y) \in \rho, \text{ and } (y, z) \in \sigma\}.$$

The product of two relations ρ and σ is also called the *composition* of ρ and σ, and we also use the alternative notation $\sigma \circ \rho$ for the relation product $\rho \bullet \sigma$.

Example 2.2.14 Let P, L, and C be the set of points, the set of lines, and the set of circles of a given plane, respectively. Let ρ be the relation that consists of those pairs (p, l), where $p \in P$, $l \in L$, and p is on l. Also, consider the relation $\sigma = \{(l, c) \mid l \in L, c \in C, \text{ and } l \text{ is tangent to } c\}$.

We have $(p, c) \in \rho \bullet \sigma$ if and only if there is a line l such that $(p, l) \in \rho$ and $(l, c) \in \sigma$, that is, if and only if there exists a line l passing through p and tangent to the circle c. Clearly, this happens if and only if p is not inside c.

The important properties of the relation product are given in the following theorem.

Theorem 2.2.15 *Let ρ, σ, and θ be relations.*

1. $Dom(\rho \bullet \sigma) \subseteq Dom(\rho)$.

2. $Ran(\rho \bullet \sigma) \subseteq Ran(\sigma)$.

3. *If ρ is a relation from A to B and σ is a relation from C to D, then $\rho \bullet \sigma$ is a relation from A to D.*

4. $\rho \bullet (\sigma \bullet \theta) = (\rho \bullet \sigma) \bullet \theta$ (associativity of relation product).

5.

$$\begin{aligned}
\rho \bullet (\sigma \cup \theta) &= (\rho \bullet \sigma) \cup (\rho \bullet \theta), \\
(\rho \cup \sigma) \bullet \theta &= (\rho \bullet \theta) \cup (\sigma \bullet \theta)
\end{aligned}$$

(distributivity of relation product over union).

6. $(\rho \bullet \sigma)^{-1} = \sigma^{-1} \bullet \rho^{-1}$.

7. If $\sigma \subseteq \theta$, then

$$\rho \bullet \sigma \subseteq \rho \bullet \theta,$$
$$\sigma \bullet \rho \subseteq \theta \bullet \rho$$

   (monotonicity of relation product).

8. If A and B are any sets, then $\iota_A \bullet \rho \subseteq \rho$ and $\rho \bullet \iota_B \subseteq \rho$. Furthermore, $\iota_A \bullet \rho = \rho$ if and only if $Dom(\rho) \subseteq A$, and $\rho \bullet \iota_B = \rho$ if and only if $Ran(\rho) \subseteq B$. (Thus, ρ is a relation from A to B if and only if $\iota_A \bullet \rho = \rho = \rho \bullet \iota_B$).

Proof: We prove (4), (5), and (7) and leave the other parts as exercises.
(4) Let $(a, d) \in \rho \bullet (\sigma \bullet \theta)$. There is a b such that $(a, b) \in \rho$ and $(b, d) \in \sigma \bullet \theta$. This means that there exists c such that $(b, c) \in \sigma$ and $(c, d) \in \theta$. Therefore, we have $(a, c) \in \rho \bullet \sigma$, which implies $(a, d) \in (\rho \bullet \sigma) \bullet \theta$. This shows that

$$\rho \bullet (\sigma \bullet \theta) \subseteq (\rho \bullet \sigma) \bullet \theta.$$

Conversely, let $(a, d) \in (\rho \bullet \sigma) \bullet \theta$. There is a c such that $(a, c) \in \rho \bullet \sigma$ and $(c, d) \in \theta$. This implies the existence of a b for which $(a, b) \in \rho$ and $(b, c) \in \theta$. For this b, we have $(b, d) \in \sigma \bullet \theta$, which gives $(a, d) \in \rho \bullet (\sigma \bullet \theta)$. We have proven the reverse inclusion

$$(\rho \bullet \sigma) \bullet \theta \subseteq \rho \bullet (\sigma \bullet \theta),$$

which gives the associativity of relation product.
(5) Let $(a, c) \in \rho \bullet (\sigma \cup \theta)$. Then, there is a b such that $(a, b) \in \rho$ and $(b, c) \in \sigma$ or $(b, c) \in \theta$. In the first case, we have $(a, c) \in \rho \bullet \sigma$; in the second, $(a, c) \in \rho \bullet \theta$. Therefore, we have $(a, c) \in (\rho \bullet \sigma) \cup (\rho \bullet \theta)$ in either case and this proves

$$\rho \bullet (\sigma \cup \theta) \subseteq (\rho \bullet \sigma) \cup (\rho \bullet \theta).$$

Let $(a, c) \in (\rho \bullet \sigma) \cup (\rho \bullet \theta)$. We have either $(a, c) \in \rho \bullet \sigma$ or $(a, c) \in \rho \bullet \theta$. In the first case, there is a b such that $(a, b) \in \rho$ and $(b, c) \in \sigma \subseteq \sigma \cup \theta$. Therefore, $(a, c) \in \rho \bullet (\sigma \cup \theta)$. The second case is handled similarly. This establishes

$$(\rho \bullet \sigma) \cup (\rho \bullet \theta) \subseteq \rho \bullet (\sigma \cup \theta).$$

The other distributivity property has a similar argument.
(7) Let σ, θ be such that $\sigma \subseteq \theta$. Since $\sigma \cup \theta = \theta$, we obtain from (5) that

$$\rho \bullet \theta = (\rho \bullet \sigma) \cup (\rho \bullet \theta),$$

which shows that $\rho \bullet \sigma \subseteq \rho \bullet \theta$. The second inclusion is proven similarly. ∎

Definition 2.2.16 A relation ρ is called a function if for all $a, b_1,$ and b_2, $(a, b_1) \in \rho$ and $(a, b_2) \in \rho$ imply $b_1 = b_2$; ρ is a one-to-one relation if for all $a_1, a_2,$ and b, $(a_1, b) \in \rho$ and $(a_2, b) \in \rho$ imply $a_1 = a_2$.

Note that $\emptyset$ is a function (referred to in this context as the *empty function*). This remark can be justified by examining the condition under which a relation ρ may fail to be a function. In order for this to happen, we should have $a \in \text{Dom}(\rho)$ and $b_1 \neq b_2$ such that both $(a, b_1) \in \rho$ and $(a, b_2) \in \rho$. Clearly, if ρ is empty, no such elements can be found; hence, $\rho = \emptyset$ is indeed a function because $\emptyset$ satisfies vacuously the defining condition for being a function.

Example 2.2.17 For every set A, the relation ι_A is a function and also a one-to-one relation. The relation ρ_{in} from Example 2.2.2 is also a function, but it is not a one-to-one relation. The relation ν from Example 2.2.8 is a one-to-one relation, but it is not a function.

Theorem 2.2.18 *For any relation ρ,*

1. *ρ is a function if and only if ρ^{-1} is a one-to-one relation.*

2. *ρ is a one-to-one relation if and only if ρ^{-1} is a function.*

Proof: Suppose that ρ is a function and let $(b_1, a), (b_2, a) \in \rho^{-1}$. Definition 2.2.11 implies that $(a, b_1), (a, b_2) \in \rho$; hence, $b_1 = b_2$ because ρ is a function. This proves that ρ^{-1} is one-to-one.

Conversely, assume that ρ^{-1} is one-to-one and let $(a, b_1), (a, b_2) \in \rho$. Applying Definition 2.2.11, we obtain $(b_1, a), (b_2, a) \in \rho^{-1}$ and, since ρ^{-1} is one-to-one, we have $b_1 = b_2$. This shows that ρ is a function.

The second part of the theorem is an immediate consequence of the first part and of Theorem 2.2.12, Part (4). ■

Example 2.2.19 We observed that the relation ν introduced in Example 2.2.8 is one-to-one. Therefore, its inverse $\nu^{-1} \subseteq \mathbf{Q} \times \mathbf{Z}$ is a function. In fact, ν^{-1} associates to each rational number q its integer part $\lfloor q \rfloor$.

Observe that $\emptyset^{-1} = \emptyset$. Hence, since $\emptyset$ is a function, by Theorem 2.2.18, $\emptyset$ is also a one-to-one relation.

Definition 2.2.20 *A relation ρ from A to B is called* total *if $\text{Dom}(\rho) = A$ and is called* onto *if $\text{Ran}(\rho) = B$.*

Any relation ρ is a total and onto relation from $\text{Dom}(\rho)$ to $\text{Ran}(\rho)$. If A and B are nonempty, than $A \times B$ is a total, onto relation from A to B.

Theorem 2.2.21 *Let ρ be a relation from A to B.*

1. *ρ is a total relation from A to B if and only if ρ^{-1} is an onto relation from B to A.*

2. *ρ is an onto relation from A to B if ρ^{-1} is a total relation from B to A.*

Proof: The theorem follows immediately from Theorem 2.2.12. ∎

We remark that if ρ is a relation then one can determine whether or not ρ is a function or is one-to-one just by looking at the ordered pairs of ρ. Whether ρ is a total or onto relation from A to B depends on what A and B are.

In the next theorem, using the relation product and inverse, we give an equivalent condition for each of the four properties of relations we have just discussed.

Theorem 2.2.22 *Let ρ be a relation.*

1. ρ is a function if and only if

$$\rho^{-1} \bullet \rho \subseteq \iota_{Ran(\rho)}.$$

2. ρ is a one-to-one relation if and only if

$$\rho \bullet \rho^{-1} \subseteq \iota_{Dom(\rho)}.$$

Let ρ be a relation from A to B.

3. ρ is a total relation from A to B if and only if

$$\iota_A \subseteq \rho \bullet \rho^{-1}.$$

4. ρ is an onto relation from A to B if and only if

$$\iota_B \subseteq \rho^{-1} \bullet \rho.$$

Proof: We prove only the first statement of the theorem.

Let ρ be a function. If $(b_1, b_2) \in \rho^{-1} \bullet \rho$, there exists a such that $(b_1, a) \in \rho^{-1}$ and $(a, b_2) \in \rho$. This implies $(a, b_1) \in \rho$, and since ρ is a function, we obtain $b_1 = b_2$, that is, $(b_1, b_2) \in \iota_{Ran(\rho)}$.

Conversely, assume that

$$\rho^{-1} \bullet \rho \subseteq \iota_{Ran(\rho)}.$$

If $(a, b_1), (a, b_2) \in \rho$, we have $(b_1, b_2) \in \rho^{-1} \bullet \rho$, and this implies $(b_1, b_2) \in \iota_{Ran(\rho)}$; hence, $b_1 = b_2$, which shows that ρ is a function. ∎

Theorem 2.2.23 *Let ρ and σ be relations.*

1. If ρ and σ are functions, then $\rho \bullet \sigma$ is also a function.

2. If ρ and σ are one-to-one relations, then $\rho \bullet \sigma$ is also a one-to-one relation.

Let ρ be a relation from A to B and σ be a relation from B to C.

3. *If ρ is a total relation from A to B and σ is a total relation from B to C, then $\rho \bullet \sigma$ is a total relation from A to C.*

4. *If ρ is an onto relation from A to B and σ is an onto relation from B to C, then $\rho \bullet \sigma$ is an onto relation from A to C.*

Proof: All of the parts have simple direct proofs and can also be given interesting algebraic proofs. To show part (1) directly, suppose that ρ and σ are both functions and that (a, c_1) and (a, c_2) both belong to $\rho \bullet \sigma$. Then, there exists a b_1 such that $(a, b_1) \in \rho$ and $(b_1, c_1) \in \sigma$, and there exists a b_2 such that $(a, b_2) \in \rho$ and $(b_2, c_2) \in \sigma$. Since ρ is a function, $b_1 = b_2$, and hence, since σ is a function, $c_1 = c_2$, as desired.

Part (2) can be shown directly or algebraically, but can also be derived easily from Part (1). Suppose that relations ρ and σ are one-to-one (and hence, that ρ^{-1} and σ^{-1} are both functions). To show that $\rho \bullet \sigma$ is one-to-one, it suffices to show that $(\rho \bullet \sigma)^{-1} = \sigma^{-1} \bullet \rho^{-1}$ is a function. This follows immediately from Part (1).

Rather than give the direct proof for (3), we give an algebraic one. Suppose that ρ is a total relation from A to B and σ is a total relation from B to C. We have

$$
\begin{aligned}
(\rho \bullet \sigma) \bullet (\rho \bullet \sigma)^{-1} &= (\rho \bullet \sigma) \bullet (\sigma^{-1} \bullet \rho^{-1}) \\
&= (\rho \bullet (\sigma \bullet \sigma^{-1})) \bullet \rho^{-1} \\
&\supseteq (\rho \bullet \iota_B) \bullet \rho^{-1} && \text{by Theorem 2.2.22,} \\
& && \text{Part (3)} \\
& && \text{and monotonicity} \\
&= \rho \bullet \rho^{-1} && \text{by Theorem 2.2.15,} \\
& && \text{Part (8)} \\
&\supseteq \iota_A,
\end{aligned}
$$

which shows that $\rho \bullet \sigma$ is a total relation from A to C.

Part (4) follows from Part (3) in the same way that Part (2) follows from Part (1). ∎

Definition 2.2.24 *Let ρ be a relation and let S be a set. The restriction of ρ to S, $\rho|_S$, is the set of all pairs in ρ whose first and second components are in S, that is,*

$$\rho|_S = \rho \cap (S \times S).$$

Note that $\rho|_S$ is always a relation on S.

Example 2.2.25 The less than relation on $\mathbf{N}$ is the restriction of the less than relation on $\mathbf{R}$ to $\mathbf{N}$. The restriction of the relation $\delta_{\mathbf{Z}}$ of Example 2.2.9 to the set P of prime natural numbers is ι_P, the identity relation on P.

2.3   Functions

In the previous section, we defined what it means for a relation to be a function. The notion of function is very important, and in this section, we study it in more detail.

We have seen that a function is a relation ρ such that for every a in Dom(ρ), there is only one b such that $(a, b) \in \rho$. In other words, a function assigns a unique value to each member of its domain. For example, if we are given the set of 50 ordered pairs (Alabama, Montgomery),....,(Wyoming, Cheyenne) that for each state contains the ordered pair whose first component is the state and whose second component is the capital city of that state, then this set of ordered pairs is a function. Given a state, we look in the set of ordered pairs for one whose first component is that state. The second component of that pair gives the value that the function assigns to the state.

If ρ is a function, then, for each a in Dom(ρ), we let $\rho(a)$ denote the unique b with $(a, b) \in \rho$, and we sometimes call $\rho(a)$ the *image of a under* ρ.

Definition 2.3.1 *Let A and B be sets. A* partial function *from A to B is a relation from A to B that is a function. A* total function *from A to B (also called a* function *from A to B or a* mapping *from A to B) is a partial function from A to B that is a total relation from A to B.*

Note that when we say that ρ is a relation from A to B this does not necessarily mean that ρ is a total relation from A to B, but when we say that ρ is a function from A to B, this automatically means that ρ is total in addition to being a relation from A to B, which is a function. This usage is somewhat inconsistent, but is fairly standard. (Some authors, however, use the phrase "ρ is a function from A to B" in the same sense as our use of the phrase "ρ is a partial function from A to B.")

Example 2.3.2 Programs may generate partial functions. Consider, for instance, the following PASCAL function:

```
function f(x: integer):integer;
        var y:integer;
        begin
        y := 1;
                while (x <> 0) do
                        begin
                        x := x − 2;
                        y := y ∗ 2
                        end;
            f := y
        end;
```

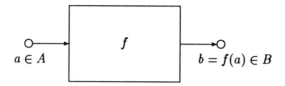

FIGURE 2.1. The black box image of f.

It is fairly easy to see that a call $f(2k)$, with $k \geq 0$, returns the number 2^k. On any other input, the computation will not stop. Therefore, this program fragment computes the function $\rho = \{(2k, 2^k) \mid k \in \mathbf{N}\}$, which is a partial function from $\mathbf{Z}$ to $\mathbf{Z}$.

Example 2.3.3 The relation $\rho \subseteq \mathbf{R} \times \mathbf{R}$, given by $\rho = \{(x, y) \mid x, y \in \mathbf{R} \text{ and } y = \sqrt{x + 1}\}$, is a partial function from $\mathbf{R}$ to $\mathbf{R}$. Its domain is the set $\{x \mid x \geq -1\}$, while its range is the set of all non-negative reals.

From now on, we will use the letters f, g, h, and k to denote functions, and we will denote the identity relation ι_A, which we have already remarked is a function, by 1_A.

We denote the set of all partial functions from A to B by $A \rightsquigarrow B$ and the set of all total functions from A to B by $A \longrightarrow B$. Note that for any sets A and B, we have $A \longrightarrow B \subseteq A \rightsquigarrow B$.

The fact that f is a partial function from A to B is indicated by writing $f : A \rightsquigarrow B$, rather than $f \in A \rightsquigarrow B$. Similarly, instead of writing $f \in A \longrightarrow B$, we use the notation $f : A \longrightarrow B$.

A function $f : A \longrightarrow B$ can be thought of as a black box equipped with an input wire (which can receive an element of A) and an output wire (which is capable of generating elements of B). Namely, when a is applied to the input wire, the output wire generates the element $b = f(a)$. (See Figure 2.1.)

We usually define a function by giving its domain and specifying for each domain element x the value that the function takes on x. (A formal justification for this is given in Exercise 9.) A common way to give the value of a function f on a domain element x is to write $f(x) = \mathcal{E}$, where $\mathcal{E}$ is some expression that can be evaluated for each domain element x. (For example, $f(x) = x^2 + x + 1$.) Another notation, called the *lambda notation*, is to write $(\lambda x.\mathcal{E})$ for the function whose value on x is given by $\mathcal{E}$. [Thus, the example just given can be rewritten as $f = (\lambda x.x^2 + x + 1)$.] Note that an expression $\mathcal{E}$ does not by itself fully define a function; the domain must also be specified.

For a function $f : D_1 \times D_2 \longrightarrow D$, we use the notation $(\lambda(m, n).\mathcal{E})$. For instance, the function $f : \mathbf{R} \times \mathbf{N} \longrightarrow \mathbf{R}$, given by $f(x, n) = x^{n+1}$ could also be denoted by $(\lambda(x, n).x^{n+1})$.

Example 2.3.4 Functions defined on finite sets are represented by arrays in programming languages. For instance, let W be the set of names for the days of the week, $W = \{$monday, tuesday, wednesday, thursday, friday, saturday, sunday$\}$. In representing the number of hours worked every day, we can use a function $nh : W \longrightarrow \mathbf{R}$. If we put in 5 hours of work on Friday, we define $f(\text{friday})=5$.

Consider now the following type definitions:

type
    weekday = (monday, tuesday, wednesday, thursday, friday,
                saturday, sunday);
    wh = **array** [weekday] **of real**;

We can represent the number of hours worked by an employee by using the declaration

$$nh : \text{wh};$$

Following this declaration, we can assign specific values for the function nh as in
    nh[monday]=5; nh[tuesday]=7.5; nh[wednesday]=7.5;

Note that for any sets A and B, we have $\emptyset \in A \rightsquigarrow B$. If either A or B is empty, then $\emptyset$ is the only partial function from A to B. If $A = \emptyset$, then the empty function is a total function from A to any B. Thus, for any sets A and B, we have

$$A \rightsquigarrow \emptyset = \{\emptyset\},$$
$$\emptyset \rightsquigarrow B = \{\emptyset\},$$
$$\emptyset \longrightarrow B = \{\emptyset\}.$$

Furthermore, if A is nonempty, then there can be no (total) function from A to the empty set, so we have

$$A \longrightarrow \emptyset = \emptyset \ (\text{if } A \neq \emptyset).$$

Definition 2.3.5 *A* one-to-one *function is called an* injection. *A function* $f : A \rightsquigarrow B$ *is called a* surjection *(from A to B) if f is an onto relation from A to B, and it is called a* bijection *(from A to B) or a one-to-one correspondance between A and B if it is total, an injection, and a surjection.*

Using our notation for functions, we can restate the definition of injection as follows: f is an injection if for all $a_1, a_2 \in \text{Dom}(f)$, $f(a_1) = f(a_2)$ implies $a_1 = a_2$. Likewise, $f : A \rightsquigarrow B$ is a surjection if for every $b \in B$ there is an $a \in A$ with $f(a) = b$.

Example 2.3.6 Let A, B be two sets and assume that $A \subseteq B$. The *containment mapping* $c : A \longrightarrow B$ defined by $c(a) = a$ for $a \in A$ is an injection. We denote such a containment by $c : A \hookrightarrow B$.

Example 2.3.7 Let S be the set of sectors of a magnetic disk pack and let C be the set of cylinders of that pack. We can define a function $a : S \longrightarrow C$, where $a(s)$ is the cylinder that contains sector s. Clearly, each cylinder is the image under a of the sectors it contains. Therefore, a is a surjection.

Example 2.3.8 Let $m \in \mathbf{N}$ be a natural number, $m \geq 2$. Consider the function $r_m : \mathbf{N} \longrightarrow \{0, \ldots, m-1\}$, where $r_m(n)$ is the remainder when n is divided by m. Obviously, r_m is well defined since the remainder p when a natural number is divided by m satisfies $0 \leq p \leq m - 1$. The function r_m is onto because of the fact that for any $p \in \{0, \ldots, m-1\}$ we have $r_m(km + p) = p$ for any $k \in \mathbf{N}$.

For instance, if $m = 4$, we have $r_4(0) = r_4(4) = r_4(8) = \cdots = 0$, $r_4(1) = r_4(5) = r_4(9) = \cdots = 1$, $r_4(2) = r_4(6) = r_4(10) = \cdots = 2$ and $r_4(3) = r_4(7) = r_4(11) = \cdots = 3$.

Since a function is a relation, the ideas introduced in the previous section for relations in general can be equally well applied to functions. In particular, we can consider the inverse of a function and the product of two functions.

If f is a function, then, by Theorem 2.2.18, f^{-1} is a one-to-one relation; however, f^{-1} is not necessarily a function. In fact, by the same theorem, if f is a function, then f^{-1} is a function if and only if f is an injection.

Suppose now that $f : A \rightsquigarrow B$ is an injection. Then, $f^{-1} : B \rightsquigarrow A$ is also an injection. Further, by Theorem 2.2.21, $f^{-1} : B \rightsquigarrow A$ is total if and only if $f : A \rightsquigarrow B$ is a surjection, and $f^{-1} : B \rightsquigarrow A$ is a surjection if and only if $f : A \rightsquigarrow B$ is total. It follows that $f : A \rightsquigarrow B$ is a bijection if and only if $f^{-1} : B \rightsquigarrow A$ is a bijection.

If f and g are functions, then we will always use the alternative notation $g \circ f$ for the product $f \bullet g$, and we will call $g \circ f$ the composition of f and g rather than the product. We will sometimes write just gf for $g \circ f$.

By Theorem 2.2.23, the composition of two functions is a function. In fact, it follows from the definition of composition that

$$\mathrm{Dom}(g \circ f) = \{a \in \mathrm{Dom}(f) \mid f(a) \in \mathrm{Dom}(g)\},$$

and for all $a \in \mathrm{Dom}(g \circ f)$,

$$g \circ f(a) = g(f(a)).$$

This explains why we use $g \circ f$ rather than $f \bullet g$. If we used the other notation, the previous equation would become $f \bullet g(a) = g(f(a))$, which is rather confusing. If we used the notation $(a)f$ in place of $f(a)$, then we could write $(a)f \bullet g = ((a)f)g$ and would not need to use different notations

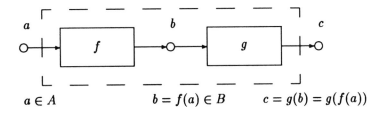

$$a \in A \qquad\qquad b = f(a) \in B \qquad c = g(b) = g(f(a))$$

FIGURE 2.2. The black box image of $g \circ f$.

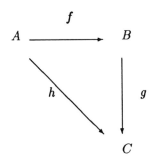

FIGURE 2.3. Commutative diagram showing $g \circ f = h$.

for function composition and relation product. Several authors have used the $(a)f$ notation, but it has not caught on in the United States.

Using Theorem 2.2.23, we can derive several facts about function composition. First of all, the composition of two injections is an injection. Also, let $f : A \rightsquigarrow B$ and $g : B \rightsquigarrow C$. Then, $g \circ f : A \rightsquigarrow C$. If f and g are both total, then $g \circ f$ is total (i.e., if $f : A \longrightarrow B$ and $g : B \longrightarrow C$, then $g \circ f : A \longrightarrow C$). If f and g are both surjections, then $g \circ f$ is also a surjection. Finally, if f and g are both bijections, then $g \circ f$ is also a bijection.

It follows from Theorem 2.2.15, Part (8), that if $f : A \longrightarrow B$, then $f \circ 1_A = f = 1_B \circ f$.

Using the black box picture of a function, one can easily illustrate the meaning of the composition of a function $f : A \longrightarrow B$ and a function $g : B \longrightarrow C$. This is done in Figure 2.2.

Certain facts about compositions of particular functions can be expressed graphically using diagrams. For instance, if $f : A \longrightarrow B$, $g : B \longrightarrow C$, and $h : A \longrightarrow C$, then when we say that the diagram in Figure 2.3 is commutative; we mean simply that $g \circ f = h$.

Also, if $f : A \longrightarrow B$, $g : B \longrightarrow C$, $l : A \longrightarrow D$, and $k : D \longrightarrow C$ are four functions such that $g \circ f = k \circ l$, then we say that the diagram in Figure 2.4 is commutative.

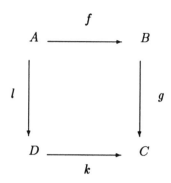

FIGURE 2.4. Commutative diagram showing $g \circ f = k \circ l$.

Let f be an injection. Then, by Theorem 2.2.22, Parts (1) and (2), $f \circ f^{-1} \subseteq 1_{\text{Ran}(f)}$ and $f^{-1} \circ f \subseteq 1_{\text{Dom}(f)}$. Since f is total and onto as a function from $\text{Dom}(f)$ to $\text{Ran}(f)$, we get from Parts (3) and (4) of the same theorem the reverse inclusions. Thus, when f is an injection, we have $f \circ f^{-1} = 1_{\text{Ran}(f)}$ and $f^{-1} \circ f = 1_{\text{Dom}(f)}$. This means that for all $a \in \text{Dom}(f)$

$$f^{-1}(f(a)) = a,$$

and for all $b \in \text{Ran}(f)$, we have

$$f(f^{-1}(b)) = b.$$

If $f : A \longrightarrow B$ is a bijection, we thus have $f^{-1} \circ f = 1_A$ and $f \circ f^{-1} = 1_B$.

Definition 2.3.9 *Let $f : A \longrightarrow B$. A* left inverse *(relative to A and B) for f is a function $g : B \longrightarrow A$ such that $g \circ f = 1_A$. A* right inverse *(relative to A and B) for f is a function $g : B \longrightarrow A$ such that $f \circ g = 1_B$.*

Example 2.3.10 In an encryption scheme, a sender composes a message and wishes to send it securely to a receiver. The sender takes the message (called clear text) and applies an encryption method to it to produce a coded message (called cypher text). The coded message is transmitted to the receiver who applies a decryption method to it to convert it back to the clear text. Mathematically, if A is the set of all clear text messages the sender might want to send and B is the set of all cypher text messages that can be transmitted, then we have an encryption function $\text{enc} : A \longrightarrow B$ and a decryption function $\text{dec} : B \longrightarrow A$. For the scheme to work, we must have $\text{dec}(\text{enc}(a)) = a$ for messages $a \in A$, i.e., dec must be a left inverse for enc and enc must be a right inverse for dec.

Theorem 2.3.11 *Let $f : A \longrightarrow B$.*

1. *f is a surjection if and only if f has a right inverse (relative to A and B).*

2. *If A is nonempty, then f is an injection if and only if f has a left inverse (relative to A and B).*

Proof: To see (1), suppose first that $f : A \longrightarrow B$ is a surjection. Define a function $g : B \longrightarrow A$ as follows: For each $b \in B$, let $g(b)$ be some arbitrarily chosen element $a \in A$ such that $f(a) = b$.[1] (Such elements a exist because f is surjective.) Then, by definition, $f(g(b)) = b$ for all $b \in B$, so g is a right inverse for f. Conversely, suppose that f has a right inverse g. Let $b \in B$ and let $a = g(b)$. Then, we have $f(a) = f(g(b)) = 1_B(b) = b$. Thus, f is surjective.

To see (2), first suppose that $f : A \longrightarrow B$ is an injection and A is nonempty. Let a_0 be some fixed element of A. Define a function $g : B \longrightarrow A$ as follows: If $b \in \mathrm{Ran}(f)$, then, since f is an injection, there is a unique element $a \in A$ such that $f(a) = b$. Define $g(b)$ to be this a. If $b \in B - \mathrm{Ran}(f)$, define $g(b) = a_0$. Then, it is immediate from the definition of g that, for all $a \in A$, $g(f(a)) = a$, so g is a left inverse for f. Conversely, suppose that f has a left inverse g. For all $a_1, a_2 \in A$, if $f(a_1) = f(a_2)$, we have $a_1 = 1_A(a_1) = g(f(a_1)) = g(f(a_2)) = 1_A(a_2) = a_2$. Hence, f is an injection. ∎

Theorem 2.3.12 *Let $f : A \longrightarrow B$. Then, the following statements are equivalent.*

1. *f is a bijection.*

2. *There is a function $g : B \longrightarrow A$ that is both a left and a right inverse for f.*

3. *f has a left inverse and f has a right inverse.*

Furthermore, if f is a bijection, then f^{-1} is the only left inverse that f has, and it is the only right inverse that f has.

Proof. Suppose first that $f : A \longrightarrow B$ is a bijection. Then, we have already remarked that $f^{-1} : B \longrightarrow A$ is both a left and a right inverse for f. It is obvious that (2) implies (3). If f has both a left inverse and a right inverse and A is nonempty, then it follows immediately from Theorem 2.3.11 that f is both injective and surjective, so f is a bijection. If A is empty, then the existence of a left inverse function from B to A implies that B is also empty; this means that f is the empty function, which is a bijection from the empty set to itself.

---

[1] See Section 3.9 for more information on this step of the proof.

Finally, suppose that $f : A \longrightarrow B$ is a bijection and that $g : B \longrightarrow A$ is a left inverse for f. Then, we have

$$f^{-1} = 1_A \circ f^{-1} = (g \circ f) \circ f^{-1} = g \circ (f \circ f^{-1}) = g \circ 1_B = g.$$

Thus, f^{-1} is the unique left inverse for f. A similar proof shows that f^{-1} is the unique right inverse for f. ∎

If we want to show that $f : A \longrightarrow B$ is a bijection, one way to do this is to show directly that f is both one-to-one and onto. Theroem 2.3.12 provides another way to show that f is a bijection. If we define a function $g : B \longrightarrow A$ and show that g is both a left and a right inverse for f, then f is a bijection and $g = f^{-1}$.

Since a function is a special type of relation, Definition 2.2.24 gives a meaning to the phrase "the restriction of f to K" whenever f is a function and K is a set. However, in ordinary mathematical practice, this phrase has another meaning, which we give in the following definition. This double definition of "the restriction of f to K" is unfortunate but unavoidable if we are to follow standard terminology. The problem arises only because we identify a function as being a particular type of relation. In practice, it will always be clear from the context which definition of restriction is meant.

Definition 2.3.13 *Let f be a function and let K be a set. The* restriction *of f to K, $f{\restriction}K$, is the set of all pairs in f whose first component is in K, that is,*

$$f{\restriction}K = f \cap (K \times Ran(f)).$$

It is easy to see that for any function f and set K, $f{\restriction}K = f \circ 1_K$. Hence, any restriction of a function is a function. In fact, the domain of $f{\restriction}K$ is $Dom(f) \cap K$, and for each $a \in Dom(f{\restriction}K)$, $f{\restriction}K(a) = f(a)$. If $f : A \rightsquigarrow B$, then $f{\restriction}K : A \cap K \rightsquigarrow B$. If $f : A \longrightarrow B$ and $K \subseteq A$, then $f{\restriction}K : K \longrightarrow B$. In addition, any restriction of an injection is an injection.

Example 2.3.14 Consider the function $f : \mathbf{N} \longrightarrow \mathbf{R}$ given by $f(n) = n/2$ for every $n \in \mathbf{N}$. Its restriction $f{\restriction}E : E \longrightarrow \mathbf{R}$ to the set E of even natural numbers maps every even natural number $2m$ to m. Note that we also have $f{\restriction}E : E \longrightarrow \mathbf{N}$.

If f and g are functions such that $f \subseteq g$, then we call g an *extension* of f. Thus, g is an extension of f if and only if $Dom(f) \subseteq Dom(g)$ and $f(a) = g(a)$ for every $a \in Dom(f)$. Observe that g is an extension of f if and only if f is the restriction of g to some set.

Let f be a function and a and b be two objects. By $[a \rightarrow b]f$, we mean the function obtained from f by removing the pair whose first component is a (if there is such a pair) and adding (a, b). In other words, $[a \rightarrow b]f$ is the function f' with domain $Dom(f) \cup \{a\}$ such that $f'(a) = b$ and $f'(x) = f(x)$ for all x in $Dom(f) - \{a\}$. If $a \notin Dom(f)$, then $[a \rightarrow b]f$ is an extension of f; otherwise, $[a \rightarrow b]f$ is the function obtained from f by changing its value on a to b. Exercise 23 gives elementary properties of this idea.

The next definition provides another way of viewing a subset of a fixed set M.

Definition 2.3.15 *Let M be a set. A* characteristic function over M *is a function $\chi : M \longrightarrow \{0,1\}$. If P is a subset of M, then the* characteristic function *of P (as a subset of M) is the function $\chi_P : M \longrightarrow \{0,1\}$ given by*

$$\chi_P(x) = \begin{cases} 1 & \text{if } x \in P \\ 0 & \text{otherwise,} \end{cases}$$

for every $x \in M$.

The relationship between subsets and characteristic functions is discussed in the following.

Theorem 2.3.16 *There is a bijection $\Psi : \mathcal{P}(M) \longrightarrow (M \longrightarrow \{0,1\})$ between the set of subsets of M and the set of characteristic functions defined on M.*

Proof. For $P \in \mathcal{P}(M)$ define $\Psi(P) = \chi_P$. The mapping Ψ is one-to-one. Indeed, assume that $\chi_P = \chi_Q$, where $P, Q \in \mathcal{P}(M)$. We have $x \in P$ if and only if $\chi_P(x) = 1$, which is equivalent to $\chi_Q(x) = 1$. This happens if and only if $x \in Q$; hence, $P = Q$ so Ψ is one-to-one.

Let $f : M \longrightarrow \{0,1\}$ be an arbitrary function. Define the set $T_f = \{x \in M \mid f(x) = 1\}$. It is easy to see that f is the characteristic function of the set T_f; hence, $\Psi(T_f) = f$ which shows that the mapping Ψ is also onto; hence, it is a bijection.

As an alternative to this proof that Ψ is a bijection, we can use Theorem 2.3.12. Define a function $\Phi : (M \longrightarrow \{0,1\}) \longrightarrow \mathcal{P}(M)$ by setting $\Phi(f) = \{x \in M \mid f(x) = 1\}$ for all $f \in M \longrightarrow \{0,1\}$. Then, for every $f \in M \longrightarrow \{0,1\}$, we have $\Psi(\Phi(f)) = \chi_{\{x \in M \mid f(x) = 1\}}$, and this last function is easily seen to be equal to f. Also, for each $P \in \mathcal{P}(M)$, we have $\Phi(\Psi(P)) = \{x \in M \mid \chi_P(x) = 1\} = P$. Thus, by Theorem 2.3.12, $\Psi : \mathcal{P}(M) \longrightarrow (M \longrightarrow \{0,1\})$ is a bijection, and Φ is its inverse. ∎

In some circumstances, alternative terminology is used for characteristic functions. We take $\mathbf{T}$ to be another notation for the integer 1 and $\mathbf{F}$ to be another notation for the integer 0. Then, $\mathbf{T}$ and $\mathbf{F}$ are called *truth values*, for reasons which will be discussed in Section 4.8. We let $\mathbf{Bool} = \{\mathbf{T}, \mathbf{F}\}$.[2]

---

[2] The name **Bool**, used for the set of truth values, honors George Boole. George Boole, a British mathematician and logician, was born on November 2, 1815 in Lincoln, Lincolnshire, England, and died on December 8, 1864 in Ballintemple, County Cork, Ireland. Boole taught at Queen's College, County Cork, and was elected Fellow of the Royal Society. His main contributions are in the algebraization of logic.

Definition 2.3.17 *A predicate on a set M is a mapping $p : M \longrightarrow$ Bool.*

Under our identification of the truth values **T** and **F** with 1 and 0, predicate is just another name for characteristic function. The *characteristic predicate* of a subset P of M is the predicate $p_P : M \longrightarrow$ **Bool** defined by

$$p_P(x) = \begin{cases} \mathbf{T} & \text{if } x \in P \\ \mathbf{F} & \text{otherwise,} \end{cases}$$

for every $x \in M$. Hence, the characteristic predicate of P is the same as the characteristic function of P.

If p is a predicate on M, let

$$T_p = \{x \in M \mid p(x) = \mathbf{T}\}.$$

Then, T_p is a subset of M (and hence, is a property of the elements of M), and it follows from the proof of Theorem 2.3.16 that p is the characteristic predicate of T_p. Furthermore, for $P, Q \subseteq M$, we have $P = Q$ if and only if $p_P = p_Q$.

Several elementary properties of characteristic functions of subsets of a set are given in Exercise 29.

2.4   Sequences, Words, and Matrices

Functions can confer a structure to amorphous sets by casting their elements in a mold that allows new possibilities for processing these elements. In this section, we present three such structuring concepts: sequences, words, and matrices.

Let I be a set. A *family of elements indexed by I* is a function s whose domain is I. If $s : I \longrightarrow D$, then s is called *a family of elements of D indexed by I*.

Let s be a family of elements indexed by I. If $i \in I$, then the element $s(i)$ is denoted by s_i, and i is called an *index* for s_i under s. Note that an element can have several indices under s since s need not be an injection. The range of s is denoted by $\{s_i \mid i \in I\}$. If the value of s_i is given by some expression $\mathcal{E}$ involving i, then the range of s is commonly denoted by $\{\mathcal{E} \mid i \in I\}$, thereby avoiding the need to name s. For example, $\{2n + 1 \mid n \in \mathbf{N}\}$ is, in fact, $\{s_n \mid n \in \mathbf{N}\}$, where $s : \mathbf{N} \longrightarrow \mathbf{P}$ is defined by $s(n) = 2n + 1$.

A function s with domain I is called an *indexing* for $\{s_i \mid i \in I\}$. Every set D can be indexed; for instance, we can use the identity mapping $1_D : D \longrightarrow D$ to index D. Therefore, in dealing with a set, we can assume without restricting the generality that the set is indexed.

A special case arises when the index set is $\{k \in \mathbf{N} \mid k < n\}$ for some $n \in \mathbf{N}$. We will adopt the more suggestive notation $\{0, \ldots, n-1\}$ for this set. Note that when $n = 0$, $\{0, \ldots, n-1\}$ denotes the empty set, and if n and

m are two different natural numbers, then $\{0,\ldots,n-1\}$ and $\{0,\ldots,m-1\}$ are different sets (since if $n > m$, then m belongs to the first set but not the second and similarly if $n < m$).

Definition 2.4.1 *A* finite sequence *is a function whose domain is* $\{0,\ldots,$ $n-1\}$ *for some* $n \in \mathbf{N}$. *If* s *is a finite sequence with domain* $\{0,\ldots,n-1\}$, *then* n *is called the* length *of* s *and is denoted by* $|s|$, *and* s *is called a* sequence of length n; *the elements of the range of* s *are called the* entries *of* s.

If D is a set, then a finite sequence of elements of D *or a* finite sequence over D *is a finite sequence whose range is a subset of* D.

If s is a sequence of length n, then s is denoted by $(s_0,\ldots,s_{n-1})$, where $s_i = s(i)$, for $0 \le i \le n-1$.

Example 2.4.2 Consider the sequence $s : \{0,1,2,3\} \longrightarrow \mathbf{R}$ given by $s(0) = -2.12$, $s(1) = 0.33$, $s(2) = -10$, and $s(3) = 4.67$. Using the convention just introduced, we denote s by $(-2.12, 0.33, -10, 4.67)$.

Note that we now use the notation (d, d') to represent a sequence of length 2 even though we previously defined (d, d') to be the ordered pair $\{\{d\},\{d,d'\}\}$, which is a different mathematical object. Nevertheless, it is still true, in this new context, that $(d, d') = (d_1, d'_1)$ implies $d = d_1$ and $d' = d'_1$, so this ambiguity of notation is harmless. (It is tempting to suggest that the definition of ordered pair be postponed to this point in the development and that an ordered pair should simply be defined to be a sequence of length 2. This idea is incorrect. We have defined a sequence to be a particular kind of function, a function to be a particular kind of relation, and a relation to be a set of ordered pairs. Thus, we need a definition of ordered pair before we can define the notion of sequence.)

A sequence of length 0 is a function whose domain is the empty set. As we have seen, there is exactly one such function, namely, the empty function, so there is only one sequence of length 0. We call this sequence *the null sequence* or *the empty sequence*. According to the notation for sequences introduced previously, () denotes the empty sequence. We will more often denote the empty sequence by λ. Other common notations for the empty sequence are e and ϵ.

A sequence t of length 1 is completely determined by its value $t(0)$. We will sometimes find it convenient to regard an object d as representing the sequence (d) of length 1 whose only entry is d, even though d and (d) are not in fact the same.[3]

Functions also allow us to formalize the notion of infinite sequence.

---

[3] Note that in a list-based programming language, such as LISP, such an identification is not made, as the atom a and the list (cons 'a nil) are not the same nor are the list l and the list (list 'l).

Definition 2.4.3 *An* infinite sequence *is a function with domain* **N**. *The* elements *of the range of an infinite sequence* s *are called the* entries *of* s. *If* D *is a set, then an* infinite sequence of elements of D *or an* infinite sequence over D *is an infinite sequence whose range is contained in* D.

An infinite sequence s will be denoted by $(s_0, s_1, \ldots, s_n, \ldots)$, where $s_n = s(n)$ for $n \in \mathbf{N}$.

Example 2.4.4 Consider the infinite sequence $s : \mathbf{N} \longrightarrow \mathbf{P}$, where $s(n) = 2n + 1$, for every natural number n. Then, s is denoted by $(1, 3, 5, \ldots)$. The entries of this sequence are all the odd natural numbers.

If D is a set, then we use both $\mathrm{Seq}_n(D)$ and D^n to denote the set of all sequences of length n over D. We also use $\mathrm{Seq}(D)$ to denote the set of all finite sequences over D, $\mathrm{Seq}^+(D)$ to denote the set of all finite non-null sequences over D, and $\mathrm{ISeq}(D)$ to denote the set of all infinite sequences over D.

The following equalities are an immediate consequence of the previous notations.

$$
\begin{aligned}
\mathrm{Seq}_n(D) &= \{0, \ldots, n-1\} \longrightarrow D, \\
\mathrm{Seq}(D) &= \bigcup_{n \geq 0} \mathrm{Seq}_n(D), \\
\mathrm{Seq}^+(D) &= \bigcup_{n \geq 1} \mathrm{Seq}_n(D), \\
\mathrm{ISeq}(D) &= \mathbf{N} \longrightarrow D.
\end{aligned}
$$

Observe that $D^0 = \{\lambda\}$, and, as discussed previously, D^1 can be identified with D and D^2 can be identified with $D \times D$. (More precisely, there are natural bijections between D^1 and D and between D^2 and $D \times D$. For instance, we have the bijection $\Psi : D^1 \longrightarrow D$ given by $\Psi((d)) = d$ for every $(d) \in D^1$.)

If $D = \emptyset$, then $D^0 = \{\lambda\}$ and, for all $n > 0$, $D^n = \emptyset$; hence, $\mathrm{Seq}(D) = \{\lambda\}$.

Definition 2.4.5 *Let* s_1 *be a sequence of length* n *and let* s_2 *be a sequence of length* m. *The* product *or* concatenation *of the sequences* s_1, s_2 *is the sequence* z *of length* $n + m$ *defined by*

$$
z(i) = \begin{cases} s_1(i) & \text{if } 0 \leq i \leq n-1 \\ s_2(i-n) & \text{if } n \leq i \leq n+m-1. \end{cases}
$$

The concatenation of the sequences s_1 and s_2 is denoted by $s_1 \cdot s_2$. Of course, if $s_1, s_2 \in \mathrm{Seq}(D)$, then so is $s_1 \cdot s_2$.

Example 2.4.6 Let $s_1 \in \mathrm{Seq}_3(\mathbf{R})$ and $s_2 \in \mathrm{Seq}_2(\mathbf{R})$ be the sequences $s_1 = (1.1, -3, -600)$ and $s_2 = (2, 8)$. The concatenation of these sequences is $z = (1.1, -3, -600, 2, 8) \in \mathrm{Seq}_5(\mathbf{R})$.

The basic properties of concatenation are given in the next theorem.

Theorem 2.4.7 *Let u, v, w be finite sequences. We have*

1. $u \cdot (v \cdot w) = (u \cdot v) \cdot w$ *(associativity of concatenation);*

2. *if $u \cdot v = u \cdot w$, then $v = w$ (left cancellation);*

3. *if $v \cdot u = w \cdot u$, then $v = w$ (right cancellation);*

4. $\lambda \cdot u = u = u \cdot \lambda$.

Proof. Let $|u| = n$, $|v| = m$, $|w| = p$. To see associativity, let $y = u \cdot (v \cdot w)$ and $z = (u \cdot v) \cdot w$. Clearly, we have $|v \cdot w| = m + p$ and $|u \cdot v| = n + m$. Applying Definition 2.4.5, we have

$$y(i) = \begin{cases} u(i) & \text{if } 0 \le i \le n - 1 \\ v \cdot w(i - n) & \text{if } n \le i \le n + m + p - 1, \end{cases}$$

and

$$v \cdot w(j) = \begin{cases} v(j) & \text{if } 0 \le j \le m - 1 \\ w(j - m) & \text{if } m \le j \le m + p - 1. \end{cases}$$

This, in turn, results in

$$y(i) = \begin{cases} u(i) & \text{if } 0 \le i \le n - 1 \\ v(i - n) & \text{if } n \le i \le n + m - 1 \\ w(i - n - m) & \text{if } n + m \le i \le n + m + p - 1. \end{cases}$$

For the sequence $u \cdot v$, we have

$$u \cdot v(i) = \begin{cases} u(i) & \text{if } 0 \le i \le n - 1 \\ v(i - n) & \text{if } n \le i \le n + m - 1, \end{cases}$$

and therefore, we obtain $z(i) = y(i)$ for $0 \le i \le n + m + p - 1$.

For left cancellation, note first that the equality $u \cdot v = u \cdot w$ implies $m = p$. Applying the definition of word product, we have

$$u \cdot v(i) = \begin{cases} u(i) & \text{if } 0 \le i \le n - 1 \\ v(i - n) & \text{if } n \le i \le n + m - 1, \end{cases}$$

and

$$u \cdot w(i) = \begin{cases} u(i) & \text{if } 0 \le i \le n - 1 \\ w(i - n) & \text{if } n \le i \le n + m - 1. \end{cases}$$

Therefore, $u \cdot v(i) = u \cdot w(i)$ for $0 \le i \le n + m - 1$ means that $v(i - n) = w(i - n)$ for $n \le i \le m + n - 1$, which means that $v = w$.

Right cancellation is proven similarly to left cancellation, and the fourth part of the theorem is trivial. ∎

Theorem 2.4.8 *Sequence concatenation is not commutative; in fact,* $u \cdot v = v \cdot u$ *for all* $u, v \in Seq(D)$ *if and only if* D *is empty or a singleton.*

Proof. If $D = \emptyset$, then $\mathrm{Seq}(D) = \{\lambda\}$, and the commutativity of sequence concatenation for sequences in $\mathrm{Seq}(D)$ is trivial.

If $D = \{a\}$, $|u| = n$, and $|v| = m$, then we have $u \cdot v(i) = v \cdot u(i) = a$ for all i, $0 \le i \le n + m - 1$, which implies $u \cdot v = v \cdot u$, showing that sequence concatenation is commutative for all sequences in $\mathrm{Seq}(D)$.

Conversely, assume that D is nonempty and is not a singleton. Then, D contains two distinct elements, say, a, b. Choosing $u = (a)$ and $v = (b)$, we obtain $u \cdot v = (a, b) \neq (b, a) = v \cdot u$, so we do not have commutativity of concatenation for all sequences in $\mathrm{Seq}(D)$. ∎

The notion of sequence allows us to define the notion of finite set.

Definition 2.4.9 *A set* A *is* finite *if there exists a natural number* n *such that there is a bijection* $f : \{0, \ldots, n - 1\} \longrightarrow A$. *If a set is not finite, then we call it* infinite.

In other words, a set A is finite if we can list all its elements without repetitions as a sequence $(f(0), \ldots, f(n - 1))$. Different bijections between the set $\{0, \ldots, n-1\}$ and A will provide distinct ways of listing the elements of A. When we use notations such as $A = \{a_0, \ldots, a_{n-1}\}$, we are indicating that the function $f(i) = a_i$ is a bijection from $\{0, \ldots, n - 1\}$ to A and, hence, that A is finite.

If A is finite, then there is, in fact, a unique number n such that there is a bijection from $\{0, \ldots, n - 1\}$ to A. This statement is intuitively clear, but we postpone the proof until Chapter 4, where we study the method necessary to give the proof. (See Corollary 4.2.8.) For now, we assume the result and define this unique number to be the *cardinality* of A and denote it by $|A|$. Two elementary but important facts about the cardinalities of finite sets are given in the following theorems.

Theorem 2.4.10 *If* A *is a finite set and* $f : A \longrightarrow B$ *is a bijection, then* B *is finite and* $|A| = |B|$.

Proof: Let $|A| = n$ and let $g : \{0, \ldots, n-1\} \longrightarrow A$ be a bijection. Then, $f \circ g : \{0, \ldots, n - 1\} \longrightarrow B$ is a bijection, so B is finite and $|B| = n = |A|$. ∎

Theorem 2.4.11 *Let* A, B *be two finite sets that are disjoint. Then,* $A \cup B$ *is finite and*

$$|A \cup B| = |A| + |B|.$$

Proof. Let $f : \{0, \ldots, n - 1\} \longrightarrow A$ and $g : \{0, \ldots, m - 1\} \longrightarrow B$ be two bijections. We can regard f and g as being sequences.

Let $h : \{0, \ldots, n + m - 1\} \longrightarrow A \cup B$ be the concatenation of f and g. Since both f and g are surjective, so is h. The injectivity of f and g and the fact that $A \cap B = \emptyset$ imply that h is injective. Since the length of h is $n + m = |A| + |B|$, we have $|A \cup B| = |A| + |B|$. ∎

In practice, we often use the notation $\{a_1, \ldots, a_n\}$ for a finite set having n elements rather than $\{a_0, \ldots, a_{n-1}\}$. Of course, the former notation simply indicates the existence of a bijection $f : \{0, \ldots, n-1\} \longrightarrow A$ given by $f(i) = a_{i+1}$.

In computer science, we have a special interest in the study of strings of symbols and of sets of such strings. Programs written in any programming language are strings of symbols that are put together according to some precise rules and to which we ascribe certain meanings. Data that are processed by such programs are also represented by strings of characters, and ultimately, the content of the memory of any computing system consists of strings of 0s and 1s. Therefore, we need a mathematical concept that will formalize the notion of a string of symbols. We, in fact, have such a concept already, since a string of symbols is just a sequence of symbols. However, there are some changes in terminology and notation that take place in the standard treatment of strings of symbols, and we take these up now.

First of all, we define an *alphabet* to be any finite nonempty set, and we refer to the elements of the set as *symbols*. (Note that we do not regard symbols as being nonmathematical objects, nor do we make any attempt to single out a particular class of objects as being symbols. In our model, any mathematical object, no matter how complicated, can be considered to be part of an alphabet and hence to be a symbol.)

Definition 2.4.12 *A* word *of length n on an alphabet V is a sequence of length n of symbols of this alphabet (i.e., an element of $Seq_n(V)$).*

We will write $w(0) \ldots w(n-1)$ for the word $(w(0), \ldots, w(n-1))$. Under this notation, a word w of length 1 is denoted by $w(0)$. (Hence, we use the same notation for a symbol from the alphabet and the word of length one whose entry is the symbol. This is in line with the identification of V and V^1 mentioned earlier.) Our notation for strings puts certain restrictions on our choices for names for individual symbols. In practice, one chooses single letters or digits, possibly with subscripts, as names for symbols in an alphabet in order to avoid ambiguity.

Since words are sequences, we have the *null word* λ, which is a word over any alphabet, each word has a length, and we can take the concatenation or product of two words over an alphabet to get another word over the alphabet. In line with our notation for words as a juxtaposition of the names of the individual symbols, we write uv for the concatenation of the two words u and v rather than $u \cdot v$.

If V is an alphabet, then we write V^* for $Seq(V)$, the set of all words over V, and we write V^+ for $Seq^+(V)$, the set of all non-null words over V. Note that we have $V^* = V^+ \cup \{\lambda\}$.

Definition 2.4.13 *A* prefix *of* $u \in V^*$ *is a word t such that* $u = tw$ *for some* $w \in V^*$.

A suffix *of u is a word v such that* $u = sv$ *for some* $s \in V^*$.

Example 2.4.14 Let $V = \{a, b, c\}$ and consider the word $u = accabac$. The word acc is a prefix of u, while the word $abac$ is a suffix of the same word.

Since $\lambda x = x \lambda = x$, every word x is a prefix and suffix of itself. Similarly, the null word is a prefix and a suffix of every word from V^*.

Definition 2.4.15 *A* proper prefix *(proper suffix) of a word x is a prefix (suffix) of x that is different from both x and* λ.

Another related notion is introduced in the following definition.

Definition 2.4.16 *An* infix *of a word x over V (also called a* substring *or* subword *of x) is a word t such that* $x = utv$ *for some* $u, v \in V^*$. *The infix is* proper *if both* $u, v \neq \lambda$.

Definition 2.4.17 *If V is an alphabet, then a* language *over V is a subset of* V^*.

In other words, a language over V is any set of words over this alphabet. For instance, $\{a, ab, abba\}$ is a finite language over the alphabet $\{a, b\}$. Similarly, $L = \{a, aa, aaa, \ldots\}$ is an infinite language over the same alphabet.

Since we identify words of length 1 with the symbols of V, the set V itself is a language over V. Also among the languages over V are the *empty language* $\emptyset$ and the *full language* V^*.

Of course, if L is a language over an alphabet V and $V \subseteq V'$, then L is also a language over the alphabet V'.

Since languages are sets of words we can apply union, intersection, difference, etc. If $L \subseteq V^*$, the complement of L with respect to the alphabet V is $\overline{L}_V = V^* - L$. If V is understood from the context, we denote the complement $\overline{L}_V$ simply by $\overline{L}$.

Definition 2.4.18 *The* product *of two languages L and K over an alphabet V is the language*

$$LK = \{xy | x \in L \text{ and } y \in K\}.$$

The product $\{u\}L$ will be written as uL; likewise, we write Lu instead of $L\{u\}$.

The reader will easily verify the following elementary properties of language product:

$$K(LH) = (KL)H,$$
$$\emptyset L = \emptyset = L\emptyset,$$
$$\{\lambda\}L = L = L\{\lambda\},$$

for all languages L, K, H over V.

Theorem 2.4.19 *The product is distributive with respect to union, that is,*

$$K(L \cup H) = KL \cup KH \text{ and } (L \cup H)K = LK \cup HK,$$

for all $K, L, H \subseteq V^*$.

Proof. We prove only the first equality; the proof of the second is similar, and it is left to the reader.

Let $x \in K(L \cup H)$. We have $x = uv$, where $u \in K$ and $v \in L \cup H$. If $v \in L$, then $x \in KL$; if $v \in H$, then $x \in KH$; hence, in either case, $x \in KL \cup KH$.

If $x \in KL \cup KH$, then either $x \in KL$ or $x \in KH$. In the first case, $x = uv$, where $u \in K$ and $v \in L$. Therefore, $v \in L \cup H$, and this imples $x \in K(L \cup H)$. The case when $x \in KH$ implies the same conclusion; hence, we obtain the first equality of the proposition. ∎

Corollary 2.4.20 *Let* K, L, H *be three languages over an alphabet* V. *If* $L \subseteq H$, *then*

$$KL \subseteq KH \text{ and } LK \subseteq HK.$$

This property is called the monotonicity of language product.

Proof. If $L \subseteq H$, then $L \cup H = H$; therefore, using the previous proposition, we have $KH = KL \cup KH$, that is, $KH \subseteq KL$. The second inclusion of the corollary can be proved similarly. ∎

Definition 2.4.21 *Let* L *be a language. The language* L^*, *the* star closure *or* Kleene closure *of* L, *is the set of all words that can be written as a product of zero or more words of* L.

The language L^+, *the* positive closure *of* L, *is the set of words that can be written as a product of one or more words from* L.

(This definition will be reformulated more precisely in Example 4.5.32.)

The fact that we consider in L^* the product of zero words from L means that the null word λ is a member of L^* for any language L. On the other hand, the above definition also implies that $L \subseteq L^+ \subseteq L^*$ and $LL^* = L^*L = L^+$. Furthermore, if $u, v \in L^*$, then $uv \in L^*$.

Example 2.4.22 Let $L = \{a, bab\}$ be a language over the alphabet $V = \{a, b\}$. We have in L^* the words λ, a, bab, $abab$, $babbab$, aa, etc. and L^+ consists of the same words except for λ.

In view of the definition of L^*, it is easy to see that we have the following properties for any languages L and H:

$$(L^*)^* = L^*,$$
$$L \subseteq H \text{ implies } L^* \subseteq H^*.$$

Note that when V is considered to be an alphabet over itself (by identifying V^1 and V), then V^*, in the sense of Definition 2.4.21, is Seq(V). Thus, our notation V^* for Seq(V), when V is an alphabet, anticipated Definition 2.4.21. Observe, however, that, in general, for a language L over V, L^* is not the same thing as Seq(L).

One final note on words and languages. We have defined languages only over alphabets (i.e., *finite* sets) because this is the standard terminology. Mathematically, there is no reason why we could not define a language to be any set all of whose elements are finite sequences and to call a language L a language over a (possibly infinite) set V if $L \subseteq$ Seq(V). We could still define the product of two languages and the star closure and positive closure of a language, and all of the properties given above and in the exercises would still be true.

We assume that the reader is familiar with the notion of matrix over the set of real numbers. Using functions, we provide a formal definition for matrices over arbitrary sets.

Definition 2.4.23 *A $p \times q$ matrix on a set D is a mapping $M : \{1, \ldots, p\} \times \{1, \ldots, q\} \longrightarrow D$. We denote $M(i, j)$ by m_{ij}, for $1 \leq i \leq p$ and $1 \leq j \leq q$.*

A $p \times q$ matrix over a set D can be represented graphically as a rectangular array:

$$\begin{pmatrix} m_{11} & m_{12} & \cdots & m_{1q} \\ m_{21} & m_{22} & \cdots & m_{2q} \\ \vdots & \vdots & \vdots & \vdots \\ m_{p1} & m_{p2} & \cdots & m_{pq} \end{pmatrix}.$$

The first subscript of m_{ij} gives the row where m_{ij} occurs, while the second shows the column of the same element. A common notation for the set of $p \times q$ matrices over a set D is $M_{p \times q}(D)$.

Matrices can be used to represent relations over finite sets. If $\rho \subseteq A \times B$, where $A = \{a_1, \ldots, a_n\}$ and $B = \{b_1, \ldots, b_m\}$, then ρ can be represented by the matrix $m(\rho) \in M_{n \times m}(\{0, 1\})$ defined by

$$m(\rho)(i, j) = \begin{cases} 1 & \text{if } (a_i, b_j) \in \rho \\ 0 & \text{otherwise.} \end{cases}$$

Example 2.4.24 Consider the sets $A = \{a_1, a_2, a_3\}$ and $B = \{b_1, b_2\}$. The relation $\rho = \{(a_1, b_2), (a_2, b_1), (a_2, b_2), (a_3, b_1)\}$ is represented by the matrix

$$m(\rho) = \begin{pmatrix} 0 & 1 \\ 1 & 1 \\ 1 & 0 \end{pmatrix}.$$

Conversely, given two sets, $A = \{a_1, \ldots, a_n\}$, $B = \{b_1, \ldots, b_m\}$, and a matrix, $M \in M_{n \times m}(\{0, 1\})$, there is a relation $r(M) \subseteq A \times B$ such that $M = m(r(M))$, namely, $r(M) = \{(a_i, b_j) \mid M(i,j) = 1\}$. It is also easy to see that $r(m(\rho)) = \rho$ for every $\rho \subseteq A \times B$. Consequently, the mapping $m : \mathcal{P}(A \times B) \longrightarrow M_{n \times m}(\{0,1\})$ is a bijection with inverse $r : M_{n \times m}(\{0,1\}) \longrightarrow \mathcal{P}(A \times B)$.

2.5   Images of Sets Under Relations

In this section, we study how relations and functions behave when applied to sets of elements rather than to single elements.

Definition 2.5.1 *Let ρ be a relation and K be a set. The* image *of K under ρ is the set*

$$\rho(K) = \{b \mid (a,b) \in \rho \text{ for some } a \in K\}.$$

When f is a function, the image of a set K under the relation f^{-1} is called the *preimage* of K under f.

The notation $f(K)$ for a function f is quasi-standard but rather unfortunate since K may be an element of the domain of f and a subset of this domain, which makes $f(K)$ ambiguous. For example, if f is the function $\{(0,1),(\{0\},2)\}$, then $f(\{0\})$ is 2 if $\{0\}$ is interpreted as an element of the domain of f and is $\{1\}$ if $\{0\}$ is regarded as a subset of the domain. In practice, this ambiguity does not cause problems since the context will make clear how K is to be interpreted.[4]

Theorem 2.5.2 *For any relation ρ and collection of sets $\mathcal{C}$, we have*

$$\rho(\bigcup \mathcal{C}) = \bigcup \{\rho(C) \mid C \in \mathcal{C}\}.$$

If $\mathcal{C}$ is nonempty, we have

$$\rho(\bigcap \mathcal{C}) \subseteq \bigcap \{\rho(C) \mid C \in \mathcal{C}\}.$$

In addition, if $\mathcal{C}$ is nonempty and ρ is one-to-one, we have

$$\rho(\bigcap \mathcal{C}) = \bigcap \{\rho(C) \mid C \in \mathcal{C}\}.$$

Proof: Let $b \in \bigcup \{\rho(C) \mid C \in \mathcal{C}\}$. There is some $C \in \mathcal{C}$ and $a \in C$ with $(a,b) \in \rho$. Since $a \in C \subseteq \bigcup \mathcal{C}$, we can say that $b \in \rho(\bigcup \mathcal{C})$. We conclude that

$$\rho(\bigcup \mathcal{C}) \supseteq \bigcup \{\rho(C) \mid C \in \mathcal{C}\}.$$

---

[4]An alternative notation for the image of a set K under a relation ρ is $\rho[K]$. With this notation, there is no ambiguity if ρ is a function.

Conversely, if $b \in \rho(\bigcup \mathcal{C})$, there exists $a \in \bigcup \mathcal{C}$ such that $(a, b) \in \rho$. The element a belongs to at least one of the sets $C \in \mathcal{C}$, which means that $b \in \bigcup \{\rho(C) \mid C \in \mathcal{C}\}$. This implies the reverse inclusion

$$\rho(\bigcup \mathcal{C}) \subseteq \bigcup \{\rho(C) \mid C \in \mathcal{C}\},$$

which shows the first equality of the theorem.

Let $b \in \rho(\bigcap \mathcal{C})$. There is an a that is an element of every $C \in \mathcal{C}$ such that $(a, b) \in \rho$. We obtain $b \in \rho(C)$ for every $C \in \mathcal{C}$, which proves that $\rho(\bigcap \mathcal{C}) \subseteq \bigcap \{\rho(C) \mid C \in \mathcal{C}\}$.

Suppose now that ρ is one-to-one and let $b \in \bigcap \{\rho(C) \mid C \in \mathcal{C}\}$. Then, in every set $C \in \mathcal{C}$, there is an element a_C such that $(a_C, b) \in \rho$. Since ρ is one-to-one, there is an a such that $a_C = a$ for every $C \in \mathcal{C}$, which shows that b is in $\rho(\bigcap \mathcal{C})$, which, in combination with the inclusion of the previous paragraph, gives the last part of the theorem. ∎

Corollary 2.5.3 *For any relation ρ and any sets K_1 and K_2, we have*

$$\rho(K_1 \cup K_2) = \rho(K_1) \cup \rho(K_2),$$
$$\rho(K_1 \cap K_2) \subseteq \rho(K_1) \cap \rho(K_2).$$

In addition, if ρ is one-to-one, then we have

$$\rho(K_1 \cap K_2) = \rho(K_1) \cap \rho(K_2).$$

Proof. It suffices to take $\mathcal{C} = \{K_1, K_2\}$ in Theorem 2.5.2. ∎

Corollary 2.5.4 *For $f : A \longrightarrow B$ and any $K_1, K_2 \subseteq A$, $H_1, H_2 \subseteq B$, we have*

$$f(K_1 \cup K_2) = f(K_1) \cup f(K_2),$$
$$f(K_1 \cap K_2) \subseteq f(K_1) \cap f(K_2),$$
$$f^{-1}(H_1 \cup H_2) = f^{-1}(H_1) \cup f^{-1}(H_2),$$
$$f^{-1}(H_1 \cap H_2) = f^{-1}(H_1) \cap f^{-1}(H_2).$$

Proof. This follows immediately from Corollary 2.5.3, since the inverse of a function is a one-to-one relation. ∎

Note that the second part of the corollary cannot be strengthened to an equality as the following example shows.

Example 2.5.5 Let $A = \{a_1, a_2\}$ and $B = \{b\}$. Consider the mapping $f : A \longrightarrow B$ given by $f(a_1) = f(a_2) = b$. For the subsets $K_1 = \{a_1\}$ and $K_2 = \{a_2\}$, we have $K_1 \cap K_2 = \emptyset$. However, $f(K_1) \cap f(K_2) = \{b\}$, which proves that the inclusion may be strict.

Theorem 2.5.6 *If $K_1 \subseteq K_2$ and ρ is a relation, then $\rho(K_1) \subseteq \rho(K_2)$.*

Proof. If $K_1 \subseteq K_2$, we have $K_1 \cup K_2 = K_2$, and this gives $\rho(K_2) = \rho(K_1) \cup \rho(K_2)$, which implies $\rho(K_1) \subseteq \rho(K_2)$. ∎

Let ρ be a relation from A to B, where A, B are two finite sets. In data modeling (and therefore, in databases and artificial intelligence), it is important to classify such relations based on the number of elements in sets of the form $\rho(\{a\})$ and $\rho^{-1}(\{b\})$, where $a \in A$ and $b \in B$.

Let $\mathbf{N}_\infty = \mathbf{N} \cup \{\infty\}$, where $\infty \notin \mathbf{N}$. We extend arithmetic on $\mathbf{N}$ by taking $n \leq \infty$, for every $n \in \mathbf{N}$, $n\infty = \infty n = \infty$ for $n \neq 0$, $0\infty = \infty 0 = 0$, and finally $\infty\infty = \infty$.

Definition 2.5.7 *Let A and B be two finite nonempty sets and let $p, m \in \mathbf{N}$ and $q, n \in \mathbf{N}_\infty$.*

A relation $\rho \subseteq A \times B$ is of type $\langle p : q, m : n \rangle$ *from A to B if for every $a \in A$ and $b \in B$ we have*

$$m \leq |\rho(\{a\})| \leq n,$$
$$p \leq |\rho^{-1}(\{b\})| \leq q.$$

Many properties of relations on finite sets can be expressed in terms of their types. For a relation ρ from A to B we have

1. ρ is total if and only if it is of type $\langle 0 : \infty, 1 : \infty \rangle$;

2. ρ is onto if and only if it is of type $\langle 1 : \infty, 0 : \infty \rangle$;

3. ρ is a partial function if and only if it is of type $\langle 0 : \infty, 0 : 1 \rangle$;

4. ρ is one-to-one if and only if it is of type $\langle 0 : 1, 0 : \infty \rangle$.

Example 2.5.8 For a college, let S and C be the set of students and the set of courses, respectively. Suppose that each student must register for at least one and for not more than four courses and that a course can be offered if at least seven student register for that course. The maximum capacity of a course is thirty students.

Let ρ be the relation from S to C that consists of all pairs (s, c) such that student s is registered for course c. The above restrictions on students and courses can be expressed by requiring that ρ be of type $\langle 7 : 30, 1 : 4 \rangle$.

2.6   Relations and Directed Graphs

This section contains introductory concepts of graph theory to the extent to which directed graphs constitute a tool for representing relations.

From an intuitive point of view, a directed graph is a set of *vertices*, represented graphically by points, and a set of *arcs* represented by arrows joining one vertex to another.

In Figure 2.5 we represent a directed graph whose set of vertices is $V = \{a, b, c, d, e\}$ and whose set of arcs is $U = \{u_1, u_2, u_3, u_4, u_5, u_6, u_7, u_8, u_9\}$.

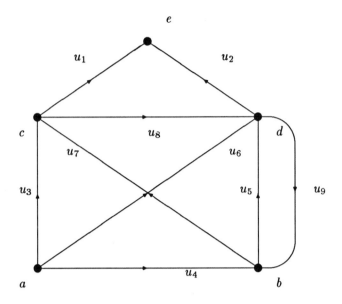

FIGURE 2.5. The Graph $G = (V, U)$.

The position of the vertices in the drawing of the directed graph is immaterial; only the way in which vertices are joined by the arcs is significant.

We can now give a formal definition of the notion of directed graph.

Definition 2.6.1 *A directed graph, or digraph, is a pair $G = (V, U)$, where $U \subseteq V \times V$. We refer to the elements of V as the vertices of G and to the elements of U as the arcs of G.*

Directed graphs can be used for representing relations on a set since every arc is an ordered pair. For instance, the arc u_8 is the pair (c, d) and the directed graph from Figure 2.5 represents the relation

$$\rho = \{(c, e), (d, e), (a, c), (a, b), (b, d), (a, d), (b, c), (c, d), (d, b)\}$$

on the set $\{a, b, c, d, e\}$. We have listed the pairs in the order of the subscripts of the arcs: $u_1, \ldots, u_9$. If ρ is a relation on a set V, we denote by $G(\rho)$ the digraph $G = (V, \rho)$ when V is understood from the context.

Let G be a digraph whose set of vertices is $V = \{v_1, \ldots, v_n\}$. The set of arcs U of G is a relation on V, and the matrix that represents this relation (see Section 2.4) also represents the digraph G. In this capacity, the matrix is called the *incidence matrix of the digraph*.

For an arc $u = (a, b)$, the vertices a, b are its *endpoints*; vertex a is the *initial endpoint*, while b is the *final endpoint* of u. They are denoted by $\alpha(u)$ and $\omega(u)$, respectively.

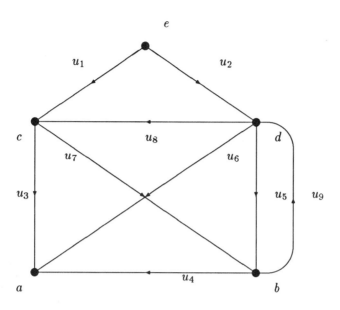

FIGURE 2.6. The converse of the Graph $G = (V, U)$.

If $a = \alpha(u)$ and $b = \omega(u)$, we say that u is *incident out of the vertex a* and *incident in the vertex b*. The number of arcs incident out of a vertex a is called the *out-degree* of the vertex a; the number of arcs incident in the vertex a is the *in-degree* of a. These two numbers are denoted by $d_G^+(a)$ and $d_G^-(a)$, respectively.

An arc of the form (a, a) is called a *loop*.

Definition 2.6.2 *Vertex b is an* immediate successor *of vertex a in the digraph $G = (V, U)$ if there is an arc $u = (a, b)$ in the set U. In this case, a is an* immediate predecessor *of b. The set of all immediate successors of a in the digraph G is denoted by $\Gamma_G(a)$. The set of immediate predecessors of b is denoted by $\Gamma_G^{-1}(b)$.*

The digraph $G(\rho^{-1})$ is referred to as *the converse digraph* of $G(\rho)$. The converse digraph is obtained from $G(\rho)$ by reversing the direction of the arcs.

Example 2.6.3 The converse digraph of the digraph represented in Figure 2.5 is given in Figure 2.6.

Definition 2.6.4 *Let $G = (V, U)$ be a digraph and let $A \subseteq V$ be a set of vertices of G.*

The subgraph *of G generated by set A is the digraph $G_A = (A, U_A)$, where $U_A = U \mid_A$.*

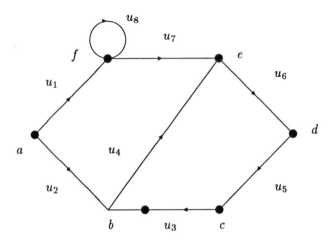

FIGURE 2.7. The digraph $G' = (\{a, b, c, d, e, f\}, U)$.

Example 2.6.5 Consider the digraph G' given in Figure 2.7. The subgraph of the digraph G' generated by the set $\{b, c, d, e\}$ is represented in Figure 2.8.

Definition 2.6.6 *Let W be a set of arcs from the digraph $G = (V, U)$, $W \subseteq U$.*

The partial digraph of G generated by W is the digraph $G_{[W]} = (V, W)$, i.e., the digraph G without the arcs from $U - W$.

Let $n \in \mathbf{N}$ be a natural number and let $G = (V, U)$ be a graph.

Definition 2.6.7 *A path of length n in G is a sequence of vertices $v_0, \ldots, v_n$ such that for each i, $1 \leq i \leq n$, $(v_{i-1}, v_i) \in U$.*

The arcs used by the path are $(v_0, v_1), \ldots, (v_{n-1}, v_n)$. The source of the path is v_0, and the destination of the path is v_n. We say that this path joins v_0 to v_n.

A path is elementary *if no vertex appears twice. If no arc is used twice in the path, then the path is called* simple. *Every elementary path is simple; however, a simple path need not be elementary.*

A circuit is a simple path whose source and destination are the same.

For a finite digraph $G = (V, U)$, the length of an elementary path in the digraph is at most $|V| - 1$, since no vertex can appear twice on the path.

If there is a path joining a vertex v to a vertex v', we say that v' is *a successor* of v and v is *a a predecessor* of v'. The *set of successors* of v is denoted by $\Gamma_G^+(v)$; the set of predecessors of v' is denoted by $\Gamma_G^-(v')$.

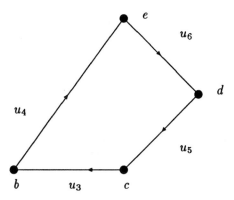

FIGURE 2.8. The subdigraph of G' generated by $\{b, c, d, e\}$.

In a path of length 0, the initial and the final vertex coincide.

Example 2.6.8 In the digraph from Figure 2.7, we have the path (u_1, u_7, u_6) of length 3 joining the vertices a and d and the circuit (u_4, u_6, u_5, u_3).

Let $\rho \subseteq V \times V$ and let $G(\rho) = (V, U)$ be the digraph representing the relation ρ. For $n \geq 1$, we consider the following

Definition 2.6.9 *The relation $\rho^n \subseteq V \times V$ is given by $(x, y) \in \rho^n$ if there exists a path of length n that joins x to y.*

In other words, $(x, y) \in \rho^n$ if we have $a_0, \ldots, a_n \in A$ such that $x = a_0, a_n = y$, and $(a_0, a_1), (a_1, a_2), \ldots, (a_{n-1}, a_n) \in \rho$.

From the previous definition of ρ^n, it follows that $\rho^1 = \rho$ and $\rho^0 = \iota_A$. Moreover, we have the following proposition.

Proposition 2.6.10 *For $m, n \in N$, we have*

$$\rho^m \bullet \rho^n = \rho^{m+n}. \tag{2.1}$$

Proof. Suppose that $(x, y) \in \rho^m \bullet \rho^n$. There is $z \in A$ such that $(x, z) \in \rho^m$ and $(z, y) \in \rho^n$. Consequently, we have the paths

$$(P_1) : x = a_0, (a_0, a_1) \in \rho, \ldots, (a_{m-1}, a_m) \in \rho, a_m = z,$$
$$(P_2) : z = b_0, (b_0, b_1) \in \rho, \ldots, (b_{n-1}, b_n) \in \rho, b_n = y.$$

Putting together the paths P_1 and P_2 we obtain a path of length $m + n$ joining x and y; hence, $(x, y) \in \rho^{m+n}$. We have obtained $\rho^m \bullet \rho^n \subseteq \rho^{m+n}$.

Conversely, let $(x,y) \in \rho^{m+n}$. From the existence of the path $x = c_0, (c_0, c_1) \in \rho, \ldots, (c_{m-1}, c_m) \in \rho, \ldots, (c_{m+n-1}, c_{m+n}) \in \rho, c_{m+n} = y$, it follows that $(x, c_m) \in \rho^m$ and $(c_m, y) \in \rho^n$, which show that $(x, y) \in \rho^m \bullet \rho^n$. Therefore, $\rho^{m+n} \subseteq \rho^m \bullet \rho^n$. ∎

Let $\rho \subseteq V \times V$ be a relation on the set V and let $G(\rho) = (V, U)$ be the digraph of this relation.

Definition 2.6.11 *The relation ρ^+ is defined by $(x, y) \in \rho^+$ if there is a path of positive length in the digraph $G(\rho)$ joining x to y.*

The relation ρ^* is defined by $(x, y) \in \rho^+$ if there is a path in the digraph $G(\rho)$ joining x to y.

From the definition of the relations ρ^n we obtain

$$\rho^+ = \bigcup_{n \geq 1} \rho^n \text{ and } \rho^* = \bigcup_{n \geq 0} \rho^n.$$

Therefore, $\rho^* = \rho^+ \cup \iota_A$.

Example 2.6.12 For relation ρ introduced by the digraph from Figure 2.5 we have $(a, e) \in \rho^2$ since (u_6, u_2) is a path of length 2 joining a to e. Therefore, $(a, e) \in \rho^+$.

Definition 2.6.13 *A directed acyclic digraph is a directed digraph without circuits. An acyclic relation is a binary relation $\rho \subseteq V \times V$ such that $\rho^n \cap \iota_V = \emptyset$ for every $n \in P$.*

It is easy to see that a relation ρ is acyclic if and only if its digraph $G(\rho)$ is acyclic.

Proposition 2.6.14 *In a finite acyclic digraph G, there is a vertex v such that $d_G^-(v) = 0$.*

Proof. Let $G = (V, U)$ be a finite acyclic digraph and let P be an elementary path of maximum length in G terminating at the vertex y. Suppose that the initial vertex of P is x. If there is a vertex z on the path P such that $(z, x) \in U$, then the digraph G has a circuit, contradicting the fact that G is acyclic. If x is the final endpoint of any arc u whose initial endpoint is not on P, then there exists a longer elementary path in G ending in y, contradicting the maximality of P. Thus, there is no vertex w such that $(w, x) \in U$. ∎

Let ρ be an acyclic relation on a *finite* set V, where $|V| = n$.

Definition 2.6.15 *A topological sorting of the set V with respect to ρ is a bijection $t : \{0, \ldots, n-1\} \longrightarrow V$ such that if $(t(i), t(j)) \in \rho$ then $i < j$.*

An algorithm for obtaining a topological sort on the set A can be formulated using Proposition 2.6.14.

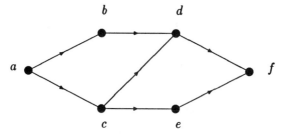

FIGURE 2.9. Graph to be topologically sorted.

Algorithm 2.6.16 Input: *A finite set* $V = \{v_0, \ldots, v_{n-1}\}$ *and an acyclic relation* ρ *on* V, *given by its digraph* $G(\rho)$.

   Output: *A topological sort* $t : \{0, \ldots, n-1\} \longrightarrow V$ *of set* V *with respect to the relation* ρ.

   Method: *By Proposition 2.6.14, the digraph* $G(\rho)$ *has at least one vertex* v *with* $d_G^-(v) = 0$. *Choose one such vertex* v *and define* $W = V - \{v\}$. *The subgraph* G_W *of* $G(\rho)$ *determined by* W *is also acyclic and, therefore, this process can be repeated.*

$i := 0;$
$W := V;$
while $W \neq \emptyset$ **do**
begin
        choose w *a vertex of* W *such that* $d_{G_W}^-(w) = 0;$
        $t(i) = w;$
        $W := W - \{w\};$
        $i := i + 1;$
end

Example 2.6.17 Consider the digraph given in Figure 2.9.
   The application of the topological sort is summarized by the following table

i	$t(i)$	W
0	a	$\{a, b, c, d, e, f\}$
1	b	$\{b, c, d, e, f\}$
2	c	$\{c, d, e, f\}$
3	d	$\{d, e, f\}$
4	e	$\{e, f\}$
5	f	$\{f\}$
6	$-$	$\emptyset$

Graph concepts can be organized in pairs of dual concepts. Specifically, the dual of a digraph concept is obtained when that concept is applied to the converse digraph. Examples of such dual concepts are given in the table below.

Dual Graph Concepts.

Concept	Dual Concept
out-degree	in-degree
in-degree	out-degree
predecessor	successor
successor	predecessor
initial endpoint	final endpoint
final endpoint	initial endpoint

The following remark enables us to avoid trivial repetitions of arguments made for the duals of certain digraph concepts.

Principle of Duality for Digraphs. For each statement valid for *all* digraphs, there is a corresponding valid statement obtained by replacing each concept by its dual concept. As an example of the application of the duality principle, we have the following

Proposition 2.6.18 *In a finite acyclic digraph G, there is a vertex v such that $d_G^+(v) = 0$.*

Proof. The argument is obtained by applying the duality principle to Proposition 2.6.14. ∎

In the rest of this section we present a class of digraphs that is used, in general, in handling hierarchical aspects of data organization.

Definition 2.6.19 *A rooted tree is a pair $T = (G, v_0)$, where $G = (V, U)$ is an acyclic digraph and $v_0 \in V$ is a special vertex of the digraph called the root of T such that there is a unique path joining v_0 to any other vertex of the digraph.*

Example 2.6.20 Consider the pair (G, v_0), where

$$G = (\{v_0, v_1, v_2, v_3, v_4, v_5, v_6\}, U),$$

given in Figure 2.10. A brief inspection of the digraph will convince the reader that G has no circuits and that every vertex of the tree is accessible from v_0 through a unique path. For instance, vertex v_6 can be reached from the root by following the path v_0, v_2, v_4, v_6.

Let (G, v_0) be a rooted tree.

Definition 2.6.21 *The height of a vertex $v \neq v_0$ is defined as the length of the path the from root to that vertex. The height of the vertex v is denoted by $h(v)$; clearly, $h(v_0) = 0$.*

Example 2.6.22 The heights of the vertices of the tree from Figure 2.10 are given by $h(v_0) = 0$, $h(v_1) = h(v_2) = 1$, $h(v_3) = h(v_4) = h(v_5) = 2$, and $h(v_6) = 4$.

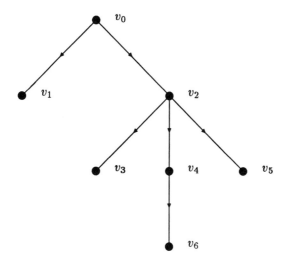

FIGURE 2.10. Rooted tree.

Let $T = (G, v_0)$ be a rooted tree, where $G = (V, U)$ and let $h : V \longrightarrow \mathbf{N}$ be its height function.

Definition 2.6.23 *The* level n *of the tree is the set*

$$V_n = \{v \mid v \in V, h(v) = n\}.$$

Example 2.6.24 Level 0 of the tree from Figure 2.10 is $V_0 = \{v_0\}$; level 2 of that tree is $V_2 = \{v_2, v_3, v_4\}$.

Example 2.6.25 The file systems of many operating systems (e.g., UNIX or MS-DOS) are organized as rooted trees. To keep track of files, the operating system uses *directories*, which are also files. A directory contains one entry for each subordinate file.

The root of the tree representing the file system shown in Figure 2.11 is a directory denoted by "/. " The rest of the vertices are denoted by words over appropriate alphabets. These vertices can be *subdirectories* containing information about other subdirectories or files or *files* containing actual data and programs.

Paths are usually denoted in a clever way that justifies the notation chosen for the root of the tree. Namely, if we have a sequence of vertices $v_0, v_1, \ldots, v_k$ that is a path in the rooted tree, we denote this path by $v_0/v_1/\ldots/v_k$. For instance, we have the path $ra/ds/b/2.tex$. If the path

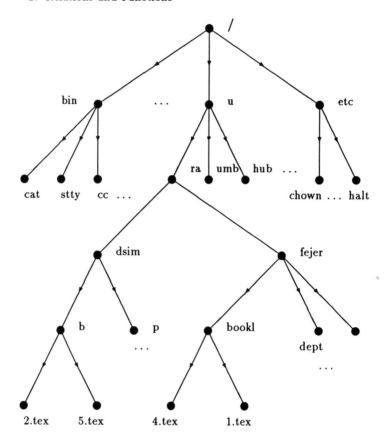

FIGURE 2.11. Rooted tree representing a file system.

originates in the root (denoted by "/"), we write a single "/" rather than "//"; an example is the notation of the path joining the root to 1.tex: $/u/ra/fejer/bookl/1.tex$.

Example 2.6.26 Let V be a finite alphabet. We define the digraph $G_V = (V^*, U)$, where $(u, v) \in U$ if there exists a symbol $a \in V$ such that $v = ua$. We leave to the reader the verification of the acyclicity of G_V.

The digraph G_V is indeed a rooted tree having λ as its root since for any word $v = a_{i_1} \ldots a_{i_n}$ we have the unique path $(\lambda, a_{i_1}), (a_{i_1}, a_{i_1} a_{i_2}), \ldots,$ $(a_{i_1} \ldots a_{i_{n-1}}, a_{i_1} \ldots a_{i_{n-1}} a_{i_n})$, joining λ to the word v.

The rooted tree representing the words over the alphabet $V = \{a, b\}$ is given in Figure 2.12.

Let $T = (G, v_0)$ be a rooted tree, where $G = (V, U)$.

Definition 2.6.27 *A vertex v is an* immediate descendant *of a vertex u*

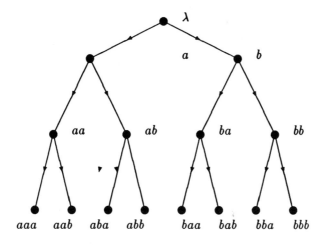

FIGURE 2.12. Rooted tree representing V^*.

(and u is an immediate ancestor of the vertex v) if there exists an arc
(u, v) in G. The vertex v is a descendant *of a vertex u if there is a path*
originating in u and ending in v. In this case, u is an ancestor *of v.*

Example 2.6.28 In the rooted tree from Figure 2.10, the vertex v_6 is an
immediate descendant of the vertex v_4 and a descendant of the vertices v_2
and v_0.

Definition 2.6.29 *A* leaf *of the rooted tree is a vertex of the tree without*
descendants.

Example 2.6.30 The leaves of the tree from Figure 2.10 are v_1, v_3, v_6, v_5.

Definition 2.6.31 *The u-subtree of the rooted tree $T = (G, v_0)$ is the*
rooted tree $T_u = (G_u, u)$, where G_u is the subgraph generated by the set of
descendants of the vertex u.

Example 2.6.32 The v_2-subtree of the tree from Figure 2.10 is represented in Figure 2.13.

A property of rooted trees that is important for its many applications in
logic and computer science is given below.

Theorem 2.6.33 (König Lemma) *Let $T = (G, v_0)$ be a rooted tree. If*
every vertex has a finite number of descendants and all the paths in T are
finite, then G is a finite graph.

Proof. Suppose that the digraph G is infinite. Consider the set $\Gamma(v_0)$ of
descendants of the root v_0. There is at least one vertex v_1 in this set for
whom the v_1-subtree is infinite since, otherwise, the digraph G would be

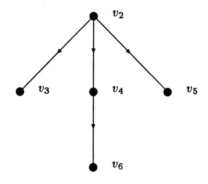

FIGURE 2.13. The v_2-subtree of the tree T.

finite. Let $T_1 = (G_1, v_1)$ be this subtree. Applying the same argument to T_1 , there is a v_2, descendant of v_1, for which the v_2-subtree is infinite, etc. In this manner, it is possible to obtain an infinite path $(v_0, v_1, v_2, \ldots)$ in G which contradicts the hypothesis of the theorem. We conclude that the digraph G is finite. ∎

2.7   Special Classes of Relations

The properties of relations that we have studied so far allow us to introduce and study several classes of relations that play a very important role in computer science.

Let A be a set and consider a relation $\rho \subseteq A \times A$.

Definition 2.7.1 *The relation ρ is*

1. reflexive, *if $(a, a) \in \rho$ for every $a \in A$;*

2. irreflexive, *if $(a, a) \notin \rho$ for every $a \in A$;*

3. symmetric, *if $(a_1, a_2) \in \rho$ implies $(a_2, a_1) \in \rho$ for $a_1, a_2 \in A$;*

4. antisymmetric, *if $(a_1, a_2), (a_2, a_1) \in \rho$ implies $a_1 = a_2$ for $a_1, a_2 \in A$;*

5. asymmetric, *if $(a_1, a_2) \in \rho$ implies $(a_2, a_1) \notin \rho$;*

6. transitive, *if $(a_1, a_2), (a_2, a_3) \in \rho$ implies $(a_1, a_3) \in \rho$.*

It is easy to see that we have the following equivalent formulation for the properties introduced in Definition 2.7.1. The relation $\rho \subseteq A \times A$ is

1. *reflexive*, if $\iota_A \subseteq \rho$;

2. *irreflexive*, if $\iota_A \cap \rho = \emptyset$;

3. *symmetric*, if $\rho^{-1} = \rho$;

4. *antisymmetric*, if $\rho \cap \rho^{-1} \subseteq \iota_A$;

5. *asymmetric*, if $\rho^{-1} \cap \rho = \emptyset$.

6. *transitive*, if $\rho^2 \subseteq \rho$.

Example 2.7.2 The relation ι_A is reflexive, symmetric, antisymmetric, and transitive for any set A.

Example 2.7.3 Consider the divisibility relation $\delta_{\mathbf{N}} \subseteq \mathbf{N} \times \mathbf{N}$, where $(m, n) \in \delta_{\mathbf{N}}$ if there is $p \in \mathbf{N}$ such that $mp = n$. $\delta_{\mathbf{N}}$ is reflexive since $n \cdot 1 = n$ for any $n \in \mathbf{N}$.

Suppose that $(m, n), (n, m) \in \delta_{\mathbf{N}}$. There are $p, q \in \mathbf{N}$ such that $mp = n$ and $nq = m$. If $n = 0$, then this also implies $m = 0$; hence, $m = n$. Let us assume that $n \neq 0$. The previous equalities imply $nqp = n$, and since $n \neq 0$, we have $qp = 1$. In view of the fact that both p and q belong to $\mathbf{N}$, we have $p = q = 1$; hence, $m = n$, which proves the antisymmetry of ρ.

Let $(m, n), (n, r) \in \delta_{\mathbf{N}}$. We can write $n = mp$ and $r = nq$ for some $p, q \in \mathbf{N}$, which gives $r = mpq$. This means that $(m, r) \in \delta_{\mathbf{N}}$, which shows that $\delta_{\mathbf{N}}$ is also transitive.

The properties of the relations mentioned in the previous definition have obvious correspondents for graphs. For instance, if ρ is reflexive, all vertices of the graph $\Gamma(\rho)$ have loops. If, on the other hand, ρ is irreflexive, then the graph $\Gamma(\rho)$ has no loops.

Assuming that ρ is transitive, if there is a path of length greater than 1 in $\Gamma(\rho)$ from x to y, then there is an arc from x to y. Indeed, let $\rho \subseteq A \times A$ be a transitive relation. Since $\rho^2 \subseteq \rho$, by applying the monotonicity of the relation product, we have $\rho \bullet \rho^2 \subseteq \rho \bullet \rho$; hence, $\rho^3 \subseteq \rho^2$. Continuing in the same manner, we obtain the descending sequence of relations

$$\rho \supseteq \rho^2 \supseteq \rho^3 \supseteq \rho^n \supseteq \cdots$$

for any transitive relation ρ.

If ρ is an arbitrary relation on the set A, the relation ρ^+, introduced before, is *the least transitive relation* containing the relation ρ. This is formally stated in the following.

Proposition 2.7.4 *For any relation $\rho \subseteq A \times A$, the relation ρ^+ is transitive. Furthermore, if θ is a transitive relation such that $\rho \subseteq \theta$, then $\rho^+ \subseteq \theta$.*

Proof. Suppose that $(x, y) \in \rho^+$ and $(y, z) \in \rho^+$. There exists a path P_1 in the graph $G(\rho)$ joining x to y and a path P_2 joining y to z. Putting these two paths together, we obtain a path joining x to z, which shows that $(x, z) \in \rho^+$. This proves that ρ^+ is indeed, transitive.

Consider a transitive relation θ such that $\rho \subseteq \theta$. If $(x, y) \in \rho^+$, then there is a path $x = a_1, (a_1, a_2), \ldots, (a_n, a_{n+1}), a_{n+1} = y$ in the graph $G(\rho)$ that joins x to y. Consequently, we have $(a_1, a_2), \ldots, (a_n, a_{n+1}) \in \theta$, which means that $(x, y) \in \theta$ because of the transitivity of this relation. Therefore, we have $\rho^+ \subseteq \theta$, which concludes our argument. ∎

The relation ρ^+ is called the *transitive closure of the relation* ρ.

Proposition 2.7.5 *The relation ρ^* is both transitive and reflexive and $\rho \subseteq \rho^*$. Furthermore, for any transitive and reflexive relation $\theta \subseteq A \times A$, we have $\rho^* \subseteq \theta$.*

Proof. The proof is similar to the argument of the previous proposition and it is left to the reader as an exercise. ∎

The previous proposition justifies referring to ρ^* as the reflexive and transitive closure of ρ.

2.8   Equivalences and Partitions

Definition 2.8.1 *An* equivalence relation, *or simply an* equivalence, *on set A is a relation that is reflexive, symmetric, and transitive.*

The set of equivalences on A is denoted by $\mathrm{Eq}(A)$.

A very important example of equivalence relation is presented in the following definition. Let A, B be two sets and consider a function $f : A \longrightarrow B$.

Definition 2.8.2 *The relation $\ker(f) \subseteq A \times A$, called the kernel of f, is given by*

$$\ker(f) = \{(a_1, a_2) \mid a_1, a_2 \in A \text{ and } f(a_1) = f(a_2)\}.$$

In other words, $(a_1, a_2) \in \ker(f)$ if f maps both a_1 and a_2 into *the same* element of B. The definition also shows that $\ker(f) = f \bullet f^{-1}$.

It is easy to verify that the relation introduced above is an equivalence. Indeed, since $f(a) = f(a)$ for $a \in A$, we have $(a, a) \in \ker(f)$ for any $a \in A$, which shows that $\iota_A \subseteq \ker(f)$.

The relation $\ker(f)$ is symmetric, since $(a_1, a_2) \in \ker(f)$ means that $f(a_1) = f(a_2)$; hence, $f(a_2) = f(a_1)$, which implies $(a_2, a_1) \in \ker(f)$.

Suppose that $(a_1, a_2), (a_2, a_3) \in \ker(f)$. Then, we have $f(a_1) = f(a_2)$ and $f(a_2) = f(a_3)$, which gives $f(a_1) = f(a_3)$. This shows that $(a_1, a_3) \in \ker(f)$; hence, $\ker(f)$ is transitive.

Example 2.8.3 Let $m \in \mathbf{N}$ be a positive natural number. Define the function $f_m : \mathbf{Z} \longrightarrow \mathbf{N}$ by $f_m(n) = r$ if r is the remainder of the division of n by m. The range of the function f_m is the set $\{0, \ldots, m-1\}$.

The relation $\mathbf{ker}(f_m)$ is usually denoted by $\equiv_m$. We have $(p, q) \in \equiv_m$ if and only if $p-q$ is divisible by m; if $(p, q) \in \equiv_m$, we also write $p \equiv q(\mathrm{mod}\ m)$.

Let ρ be an equivalence on a set A and let $x \in A$.

Definition 2.8.4 *The* equivalence class *of x is the set $[x]_\rho$ given by*

$$[x]_\rho = \{y \mid y \in A, (x, y) \in \rho\}.$$

When there is no risk of confusion, we write simply $[x]$ instead of $[x]_\rho$.

Notice that an equivalence class $[x]$ of an element x is never empty, since $x \in [x]$ because of the reflexivity of ρ.

Proposition 2.8.5 *Let ρ be an equivalence on a set A and let $x, y \in A$. The following three statements are equivalent:*

1. $(x, y) \in \rho$;

2. $[x] = [y]$;

3. $[x] \cap [y] \neq \emptyset$.

Proof. *(1) implies (2)* Suppose that $(x, y) \in \rho$ and let $z \in [x]$. We have $(x, z) \in \rho$, and using the symmetry of ρ, we obtain $(y, x) \in \rho$. Following from the transitivity of ρ, we have $(y, z) \in \rho$, which means that $z \in [y]$; hence, $[x] \subseteq [y]$. Conversely, if $z \in [y]$, we have $(y, z) \in \rho$, which implies $(x, z) \in \rho$ by the transitivity of ρ; hence, $[y] \subseteq [x]$. Therefore, $[x] = [y]$.

(2) implies (3) This is immediate.

(3) implies (1) Assume that $[x] \cap [y] \neq \emptyset$. There is $t \in [x] \cap [y]$, which means that $(x, t) \in \rho$ and $(y, t) \in \rho$. Consequently, we have $(t, y) \in \rho$; hence, $(x, y) \in \rho$ by the symmetry and transitivity of ρ. ∎

A notion closely related to the notion of equivalence is considered in in the following definition.

Definition 2.8.6 *A* partition *of set A is a collection of subsets of A, $\pi = \{B_i \mid i \in I\}$ such that*

1. $B_i \neq \emptyset$ for $i \in I$;

2. $B_i \cap B_j = \emptyset$ for $i, j \in I$, $i \neq j$;

3. $\bigcup_{i \in I} B_i = B$.

We refer to the subsets B_i as the blocks of the partition π.

The set of partitions of a set A will be denoted by Part(A).

Proposition 2.8.5 shows that two equivalence classes either coincide or are disjoint. Therefore, starting from an equivalence ρ on a set A, we can build a partition of the set A, namely, $\pi_\rho = \{[x]_\rho \mid x \in A\}$. We create in this manner *a new set*: the set of partition blocks of the equivalence ρ over A.

Definition 2.8.7 *The* quotient set *of the set A with respect to the equivalence ρ is the set A/ρ, where*

$$A/\rho = \{[x]_\rho \mid x \in A\}.$$

Moreover, we can prove that any partition defines an equivalence.

Proposition 2.8.8 *Let $\pi = \{B_i \mid i \in I\}$ be a partition of the set A. Define the relation ρ_π by $(x,y) \in \rho_\pi$ if there is a set $B_i \in \pi$ such that $\{x,y\} \subseteq B_i$. The relation ρ_π is an equivalence.*

Proof. Let B_i be the block of the partition that contains x. Since $\{x\} \subseteq B_i$, we have $(x,x) \in \rho_\pi$ for any $x \in A$, which shows that ρ_π is reflexive.

The relation ρ_π is clearly symmetric. To prove the transitivity of ρ_π, consider $(x,y),(y,z) \in \rho_\pi$. We have the blocks B_i, B_j such that $\{x,y\} \subseteq B_i$ and $\{y,z\} \subseteq B_j$. Since $y \in B_i \cap B_j$, we obtain $B_i = B_j$ by the definition of partitions; hence, $(x,z) \in \rho_\pi$. ∎

The previous propositions allow us to prove the following.

Corollary 2.8.9 *For any equivalence ρ on the set A, we have $\rho = \rho_{\pi_\rho}$. For any partition π of the set A, we have $\pi = \pi_{\rho_\pi}$.*

Proof. The justification of the above corollary is left to the reader as an exercise. ∎

The previous corollary amounts to the fact that there is a bijection ϕ : Eq(A) $\longrightarrow$ Part(A), where $\phi(\rho) = \pi_\rho$. The inverse of this mapping, Ψ : Part(A) $\longrightarrow$ Eq(A) is given by $\psi(\pi) = \rho_\pi$.

Proposition 2.8.10 *For any mapping $f : A \longrightarrow B$, there exists a bijection $h : A/\ker(f) \longrightarrow f(A)$.*

Proof. Consider the $\ker(f)$ class of an element $x \in A$, $[x]$ and define $h([x]) = f(x)$. The mapping h is well defined for, if $x' \in [x]$, then $(x,x') \in \ker(f)$, which gives $f(x) = f(x')$.

Furthermore, h is onto since, if $y \in f(A)$, then there is $x \in A$ such that $f(x) = y$, and this gives $y = h([x])$.

To prove the injectivity of h assume that $h([x]) = h([y])$. This means that $f(x) = f(y)$; hence, $(x,y) \in \ker(f)$, which means, of course, that $[x] = [y]$. ∎

An important consequence of the previous proposition is the following decomposition theorem for mappings.

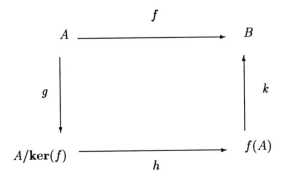

FIGURE 2.14. Decomposition of $f : A \longrightarrow B$.

Theorem 2.8.11 *Every mapping $f : A \longrightarrow B$ can be decomposed as a composition of three mappings: a surjection $g : A \longrightarrow A/\mathbf{ker}(f)$, a bijection $h : A/\mathbf{ker}(f) \longrightarrow f(A)$, and a one-to-one mapping $k : f(A) \hookrightarrow B$ (which is actually an inclusion mapping).*

Proof. The mapping $g : A \longrightarrow A/\mathbf{ker}(f)$ is defined by $g(a) = [a]$, for $a \in A$, while $k : f(A) \hookrightarrow B$ is the inclusion mapping given by $k(b) = b$ for all $b \in f(A)$. Therefore, $k(h(g(a))) = k(h([a])) = k(f(a)) = f(a)$ for all $a \in A$, which shows that the diagram from Figure 2.14 is commutative. ∎

2.9   General Cartesian Products

We extend the Cartesian product of two sets to an operation involving arbitrary families of sets.

Let $\mathcal{F} = \{D_i \mid i \in I\}$ be a family of sets.

Definition 2.9.1 *The* Cartesian product *of the family $\mathcal{F}$ is the set*

$$\prod_{i \in I} D_i = \{t \mid t : I \longrightarrow \bigcup_{i \in I} D_i, t(i) \in D_i \text{ for all } i \in I\}.$$

For a finite family of sets $\mathcal{F} = \{D_0, \dots, D_{n-1}\}$, the Cartesian product is written as $D_0 \times \cdots \times D_{n-1}$.

This represents a redefinition of the notion introduced previously in Definition 1.3.3. The reader is entitled to a discussion of the consistency of these two definitions for the case of the Cartesian product of two sets.

Let D_0, D_1 be two sets. According to Definition 2.9.1, the Cartesian product $D_0 \times D_1$ consists of those mappings $t : \{0, 1\} \longrightarrow D_0 \cup D_1$ for which $t(0) \in D_0$ and $t(1) \in D_1$. Therefore, it is transparent that t is merely a representation of the ordered pair $(t(0), t(1))$. Conversely, every ordered pair (d_0, d_1), where $d_i \in D_i$ for $i = 0, 1$, can be represented as a mapping $t : \{0, 1\} \longrightarrow D_0 \cup D_1$, where $t(i) = d_i \in D_i$ for $i = 0, 1$. Whichever definition we choose for ordered pairs, we still have the essential property (that is, the equality of two ordered pairs implies the equality of their respective components). An obvious question is why we did not just define the ordered pair in the first place as a sequence of length two and thus avoid this redefinition. The answer is that we defined a function to be a particular type of relation, that is, a particular type of set of ordered pairs. If we now define an ordered pair to be a sequence, then we have circular definitions. Thus, it was important originally to define the notion of ordered pair in a way that does not use the idea of sequence. But, now that we have the idea of a sequence, we do not need to differentiate between the two possible definitions of an ordered pair.

Note that for us the Cartesian products $D_0 \times D_1$ and $D_1 \times D_0$ are the same since they both correspond to the family of sets $\{D_0, D_1\}$.

In the case of a family that contains n sets $D_0, \ldots, D_{n-1}$, the Cartesian product $D_0 \times \cdots \times D_{n-1}$, consists of n-tuples. If $t : \{0, \ldots, n-1\} \longrightarrow \bigcup_{0 \leq i \leq n-1} D_i$ is a member of $D_0 \times \cdots \times D_{n-1}$ we denote t by the sequence $(d_0, \ldots, d_{n-1})$, where $d_i = t(i)$ for $0 \leq i \leq n-1$. The advantage of this notation is that it constitutes a natural generalization for the notation used for ordered pairs introduced before.

Clearly, an n-tuple t is nothing but a sequence of elements of the set $\bigcup_{1 \leq i \leq n} D_i$, which satisfies the additional condition $t(i) \in D_i$ for $1 \leq i \leq n$. For certain values of n, we refer to n-tuples using specific terms; for instance, instead of 2-tuples, 3-tuples, and 4-tuples we use the terms *pairs*, *triples*, *quadruples*, etc.

Let $I = \{0, \ldots, n-1\}$ be a finite set, $n \geq 1$, and assume that $D_0 = \cdots = D_{n-1} = D$. The Cartesian product $\prod_{0 \leq i \leq n-1} D_i$ is denoted by D^n. The set D^n is referred to as the n-th Cartesian power of D.

The set D^0 is defined as $D^0 = \{\emptyset\}$ for any set D. In other words, D^0 contains exactly one element: the empty set.

For $n = 1$, an element d of D is different from the 1-tuple (d). However, we sometimes ignore this difference and think of d and (d) as being the same, even though, literally speaking, they are not. If we do this, then we have identified the sets D and D^1. More formally, if D is a set, then the function $\phi : D \longrightarrow D^1$ given by $\phi(d) = (d)$ is a bijection with the inverse $\psi : D^1 \longrightarrow D$ given by $\psi((d)) = d$ for every $(d) \in D^1$. Since this bijection is so natural, we often ignore the difference between D and D^1.

Example 2.9.2 Let $D = \{0, 1, 2, 3\}$. Then, $D^1 = \{(0), (1), (2), (3)\}$. This is not the same as D, but if we identify (d) with d, then these two sets are identified with each other.

Example 2.9.3 The Cartesian product is used in some programming languages for defining record types. For instance, consider the following type definitions in Pascal:

type

weekday =
 (monday, tuesday, wednesday, thursday,
 friday, saturday, sunday);
day   = 1..31;
month =(jan, feb, mar, apr, may, jun, jul, aug,
 sep, oct, nov, dec);
year   =1900..1999;
datetype =
 record
  wd: weekday;
  d: day;
  m: month;
  y: year;
 end;

The newly defined type *datetype* is the Cartesian product of four sets:
weekday ={ monday, tuesday, wednesday, thursday, friday, saturday, sunday },
day= $\{n \mid n \in \mathbf{N}, 1 \leq n \leq 31\}$,
month={ jan, feb, mar, apr, may, jun, jul, aug, sep, oct, nov, dec},
year=$\{n \mid n \in \mathbf{N}, 1900 \leq n \leq 1999\}$.
The declaration

$$today: datetype;$$

defines the variable *today* as ranging over this Cartesian product. In view of this fact, a 4-tuple, such as (friday, 1, jul, 1988), may be a value of the variable *today*.

Definition 2.9.4 *An n-ary relation on the sets $D_0, \ldots, D_{n-1}$ is a subset ρ of the Cartesian product $D_0 \times \cdots \times D_{n-1}$.*

If $D_0 = \cdots = D_{n-1} = D$, we say that ρ is an *n*-ary relation on the set D.

Example 2.9.5 Consider a plane and let C be the set of circles of this plane, T the set of triangles, and P the set of points of this plane. We can define the 3-ary (or the ternary) relation $\rho \subseteq C \times T \times P$, where $(c, t, p) \in \rho$ if p is the center of mass of the triangle t inscribed in the circle c.

On another hand, we can consider the 4-ary relation $\theta \subseteq P^4$, where $(p_1, p_2, p_3, p_4) \in \theta$ if there exists a circle containing all points p_1, p_2, p_3, p_4.

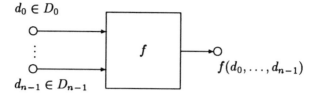

FIGURE 2.15. The black box image of $f : D_0 \times \cdots \times D_{n-1} \longrightarrow D$.

A special case is the one involving the 1-ary (or unary) relations. An unary relation is defined starting from one single set D and consists of a set of mappings $t : \{1\} \longrightarrow D$. Since we identified the elements of D with the mappings of D^1, we identify the unary relations on D with the subsets of D.

For a family of n sets $\{D_0, \ldots, D_{n-1}\}$ and a set D, we can consider the set of functions $D_0 \times \cdots \times D_{n-1} \longrightarrow D$. If $f : D_0 \times \cdots \times D_{n-1} \longrightarrow D$, we write $f(d_0, \ldots, d_{n-1})$ rather than $f(t)$, where $t \in D_0 \times \cdots \times D_{n-1}$ and $t(i) = d_i$, $0 \le i \le n - 1$. This notation justifies referring to f as *a function of n arguments*. Using the black box image introduced before, a function of n arguments can be viewed as a box with n input wires capable of receiving elements from the sets $D_0, \ldots, D_{n-1}$; when the n-tuple $(d_0, \ldots, d_{n-1})$ is applied to the input wires, the black box generates the element $f(d_0, \ldots, d_{n-1})$ at its output wire (see Figure 2.15).

Let $f : D_0 \times \cdots \times D_{n-1} \longrightarrow D$ be a function of n arguments and consider the functions $g_i : E_0 \times \cdots \times E_{m-1} \longrightarrow D_i$, for $0 \le i \le n - 1$.

Definition 2.9.6 *The* composition *of the functions* $f, g_0, \ldots, g_{n-1}$ *is the function*

$$f \circ (g_0, \ldots, g_{n-1}) : E_0 \times \cdots \times E_{m-1} \longrightarrow D,$$

where

$$f \circ (g_0, \ldots, g_{n-1})(x_0, \ldots, x_{m-1})$$
$$= f(g_0(x_0, \ldots, x_{m-1}), \ldots, g_{n-1}(x_0, \ldots, x_{m-1})),$$

for every $(x_0, \ldots, x_{m-1}) \in E_0 \times \cdots \times E_{m-1}$.

When $n = 1$, the composition reduces to the functional product $f \circ g_0$. In keeping with our previous convention, we shall sometimes write $f(g_0, \ldots, g_{n-1})$ instead of $f \circ (g_0, \ldots, g_{n-1})$.

If $f : E_0 \times \cdots \times E_{n-1} \longrightarrow D$ and $E_0 = \cdots = E_{n-1} = E$, then f is a function of n arguments defined on the set E.

If $n = 0$, we obtain the case of the *zero-ary functions*, $f : E^0 \longrightarrow D$. Since $E^0 = \{\emptyset\}$ for any set E, it follows that every zero-ary function f is defined by its value $f() \in D$.

Consider the family $\mathcal{F} = \{D_i \mid i \in I\}$ and let j be a member of I.

Definition 2.9.7 *The j-th* projection mapping *is the mapping*

$$p_j : \prod_{i \in I} D_i \longrightarrow D_j$$

given by $p_j(t) = t(j)$ *for all* $t \in \prod_{i \in I} D_i$.

If $D_i \neq \emptyset$ for every $i \in I$, then it is clear that every projection mapping $p_j : \prod_{i \in I} D_i \longrightarrow D_j$ is onto. This ceases to be true if at least one of the sets D_i is empty. For instance, the projection $p_1 : D_1 \times \emptyset \longrightarrow D_1$ is the empty mapping (since $D_1 \times \emptyset = \emptyset$).

Example 2.9.8 We presented (in Example 2.9.3) the method of implementing Cartesian products in programming languages by using the **record** construction. Projections can be realized by using *selectors* or *field names*. For instance, for the variable TODAY ranging over the Cartesian product

$$weekday \times day \times month \times year$$

the projection

$$p_{month}: weekday \times day \times month \times year \longrightarrow month$$

can be specified by the notation *today.month*.
A series of assignments such as
today.weekday =friday;
today.day=1;
today.month=july;
today.year=1988;
will result in the creation of the quadruple (friday, 1,jul,1988).

Projections play an important role in characterizing the Cartesian product of a family of sets.

Theorem 2.9.9 *Let* $\{D_i \mid i \in I\}$ *be a family of sets, and let* M *be a set such that for each* $i \in I$ *we have a mapping* $q_i : M \longrightarrow D_i$. *Then, there is a unique mapping* $\phi : M \longrightarrow \prod_{i \in I} D_i$ *such that* $q_i = p_i \phi$ *for all* $i \in I$.

Proof. Let $m \in M$. Consider the collection $\{q_i(m) \mid i \in I\}$ and consider the mapping $t : I \longrightarrow \bigcup_{i \in I} D_i$ defined by $t(i) = q_i(m)$ for $i \in I$.
Since $t \in \prod_{i \in I} D_i$, we can define ϕ by $\phi(m) = t$ (see Figure 2.16). This gives $p_i(\phi(m)) = p_i(t) = t(i) = q_i(m)$, for any $m \in M$ and $i \in I$.
To prove the uniqueness of the mapping ϕ, suppose that there is another mapping $\psi : M \longrightarrow \prod_{i \in I} D_i$ such that $q_i = p_i \psi$ for all $i \in I$. This gives $p_i(\psi(m)) = p_i(\phi(m))$ for all $m \in M$ and $i \in I$. If $\psi(m) = t$ and $\phi(m) = s$, this implies $t(i) = s(i)$ for all $i \in I$, and therefore, $t = s$. This shows that $\psi(m) = \phi(m)$ for any $m \in M$, hence $\psi = \phi$. ∎
The previous theorem is known as the *universal property of the Cartesian product*.

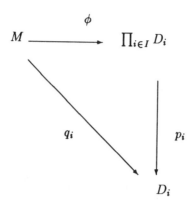

FIGURE 2.16. Characterization of Cartesian product.

Cartesian products of posets can be equipped with interesting partial orders in several ways.

Let $\mathcal{F} = \{D_i \mid i \in I\}$ be a family of sets indexed by the set I.

Definition 2.9.10 *The* disjoint sum *of $\mathcal{F}$ is the set*

$$\sum \mathcal{F} = \bigcup \{D_i \times \{i\} \mid i \in I\}.$$

Example 2.9.11 Consider the family $\{D_1, D_2\}$, where $D_1 = \{a, b\}$ and $D_2 = \{b, c, d\}$. The disjoint sum of these sets is

$$\sum \{D_1, D_2\} = \{(a, 1), (b, 1), (b, 2), (c, 2), (d, 2)\}.$$

Notice that we have two "copies" of b in the disjoint sum: $(b, 1)$ and $(b, 2)$, each corresponding to an occurrence of b in one of the sets D_1 and D_2, respectively.

For a family $\mathcal{F} = \{D_i \mid i \in I\}$ and its disjoint sum $\sum \{D_i \mid i \in I\}$ we consider a family of standard mappings introduced in the next definition.

Definition 2.9.12 *The* canonical injection $j_i : D_i \longrightarrow \sum \{D_i \mid i \in I\}$ *is given by $j_i(x) = (x, i)$ for $x \in D_i$.*

A characterization of disjoint sums is given below.

Theorem 2.9.13 *Let $\mathcal{F} = \{D_i \mid i \in I\}$ be a family of sets, let S be a set and assume that for each $i \in I$ there is a mapping $g_i : D_i \longrightarrow S$.*

There is a unique mapping $\phi : \sum_{i \in I} D_i \longrightarrow S$ such that $\phi j_i = g_i$ for all $i \in I$.

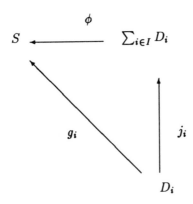

FIGURE 2.17. Universality of disjoint sums.

Proof. The statement of the theorem amounts to the fact that there exists a unique mapping $\phi : \sum_{i \in I} D_i$ that makes commutative any of the diagrams given in Figure 2.17 for $i \in I$.

For $(x, i) \in \sum_{i \in I} D_i$, we have $x \in D_i$; we define $\phi(x, i)$ as $\phi(x, i) = g_i(x)$. Since $j_i(x) = (x, i)$, we obtain $\phi(j_i(x)) = g_i(x)$ for $x \in D_i$ which proves the commutativity of any of the diagrams given in Figure 2.17.

Furthermore, assume that $\psi : \sum_{i \in I} D_i \longrightarrow S$ is another mapping such that $\psi j_i = g_i$ for all $i \in I$. Consider an arbitrary element $(x, i) \in \sum_{i \in I} D_i$. We have $\psi(x, i) = \psi(j_i(x)) = g_i(x) = \phi(j_i(x)) = \phi(x, i)$, which gives $\phi = \psi$. ∎

The above theorem is known as the *universal property of the disjoint sum*.

There are constructions in programming languages that implement disjoint sums. For instance, PASCAL has the construction of record with variants. Consider the following type definitions:

type
coordtype=(rectangular, polar);

points =
    record
    tag: coordtype;
            case coordtype **of**
            rectangular: (x,y:**real**);
            polar: (r,t:**real**);
    end;

Let **R** be the set of numbers that can be represented as reals. Type *coordtype* is represented by the set $\mathbf{R} \times \mathbf{R}$; furthermore, type *points* represents the disjoint sum of $\{(\mathbf{R} \times \mathbf{R})_{rectangular}, (\mathbf{R} \times \mathbf{R})_{polar}\}$.

The above definition of type *points* introduces the subscript *tag*. With the declaration v: points, we can write

with v **do**
    if tag = rectangular **then**
        writeln ('rectang. coordinates x=',x,'y=',y)
    else
        writeln ('polar coordinates r=',r,'t=',t);

When $v.tag$ = "rectangular" we operate with the components $v.x$, $v.y$, which represent an element of the set $(\mathbf{R} \times \mathbf{R})_{rectangular}$, i.e., a pair of numbers interpreted as the rectangular coordinates of a point; similarly, if $v.tag$ = "polar", we deal with the pairs from the set $(\mathbf{R} \times \mathbf{R})_{polar}$.

2.10   Operations

A notion that we already used informally (and we use frequently throughout this book) is introduced in the next definition.

Definition 2.10.1 *An n-ary operation on a set A is a mapping* $f : A^n \longrightarrow A$.

Example 2.10.2 Consider the ternary operation $f : \mathbf{R}^3 \longrightarrow \mathbf{R}$ defined by

$$f(x_1, x_2, x_3) = \frac{x_1 + x_2 + x_3}{3}.$$

Most of the operations considered so far are binary.

Example 2.10.3 Let M be a set. We considered in the first chapter several binary operations on $\mathcal{P}(M)$. Union, for instance, can be regarded as a binary operation $f_\cup : \mathcal{P}(M)^2 \longrightarrow \mathcal{P}(M)$, where $f_\cup(K, L) = K \cup L$, for $K, L \in \mathcal{P}(M)$. Similarly, intersection and the difference of sets can be regarded as binary operations.

Following standard mathematical practice, we use the notation xfy instead of $f(x, y)$ when dealing with binary operations. Also, these operations are usually designated by special symbols, such as

$$\vee, \wedge, +, \cdot, -, \star, \bullet, \circ.$$

In keeping with the previous convention, we write $x \star y$ rather than $\star(x, y)$.

We reconsider here several properties of binary operations that we already used informally.

Definition 2.10.4 *A binary operation $\star$ on a set M is*

1. commutative *if $x \star y = y \star x$ for all $x, y \in M$*;

2. associative *if $x \star (y \star z) = (x \star y) \star z$ for all $x, y, z \in M$*;

3. idempotent *if $x \star x = x$ for every $x \in M$*.

Example 2.10.5 Both set union and set intersection are commutative, associative, and idempotent. The relation product is associative; however, it is neither commutative nor idempotent.

Example 2.10.6 Define the operation $*$ on $\mathbf{R}$ by

$$x * y = \frac{x + y}{2}.$$

This operation is commutative and idempotent but not associative since $x * (y * z) = 0.5x + 0.25y + 0.25z$ and $(x * y) * z = 0.25x + 0.25y + 0.5z$.

Example 2.10.7 Consider the two-element set $B = \{0, 1\}$. Define the operations $\vee$ and $\wedge$ on B by

$$0 \vee 0 = 0, \ 0 \vee 1 = 1,$$
$$1 \vee 0 = 1, \ 1 \vee 1 = 1,$$

and

$$0 \wedge 0 = 0, \ 0 \wedge 1 = 1,$$
$$1 \wedge 0 = 1, \ 1 \wedge 1 = 1.$$

Both these operations are associative, commutative, and idempotent (exercise).

Similar operations are defined on the set $\mathbf{Bool} = \{\mathbf{F}, \mathbf{T}\}$. Define the operations $\vee, \wedge$, and $\neg$ on the set $\mathbf{Bool}$ as follows:

$$
\begin{array}{llll}
\mathbf{F} \vee \mathbf{F} & = & \mathbf{F}, & \mathbf{F} \vee \mathbf{T} & = & \mathbf{T}, \\
\mathbf{T} \vee \mathbf{F} & = & \mathbf{T}, & \mathbf{T} \vee \mathbf{T} & = & \mathbf{T}, \\
\mathbf{F} \wedge \mathbf{F} & = & \mathbf{F}, & \mathbf{F} \wedge \mathbf{T} & = & \mathbf{F}, \\
\mathbf{T} \wedge \mathbf{F} & = & \mathbf{F}, & \mathbf{T} \wedge \mathbf{T} & = & \mathbf{T}, \\
\neg \mathbf{T} & = & \mathbf{F}, & \neg \mathbf{F} & = & \mathbf{T}.
\end{array}
$$

The reader can prove, as an exercise, that $\vee$ and $\wedge$ are associative, commutative, and idempotent.

Furthermore, the operations $\vee$ and $\wedge$ satisfy the *absorption laws*:

$$x \vee (x \wedge y) = x,$$
$$x \wedge (x \vee y) = x,$$

for any $x, y \in \mathbf{Bool}$, and *De Morgan's laws*:

$$\neg(x \vee y) = (\neg x) \wedge (\neg y),$$
$$\neg(x \wedge y) = (\neg x) \vee (\neg y),$$

for every $x, y \in$ **Bool**.

A binary operation defines certain special elements of a set on which it acts. Let M be a set and assume that $\star$ is a binary operation on M.

Definition 2.10.8 *An element u of the set M is*

1. a left identity *of the set M if* $u \star x = x$ *for all* $x \in M$;

2. a right identity *of the set M if* $x \star u = x$ *for all* $x \in M$;

3. an identity *if it is both a left and a right identity.*

Example 2.10.9 Addition on the set of reals **R** has 0 as an identity; the identity of the multiplication is 1. The operation $\ast$ introduced in Example 2.10.6 has no left or right identity.

If an operation has an identity then that identity is unique. Suppose, for instance, that both u and v are both identities for the relation $\star$. Then,

$$u \star x = x \star u = x,$$
$$v \star x = x \star v = x,$$

for all $x \in M$.

Taking $x = v$ in the first equality, we obtain $u \star v = v$; taking $x = u$ in the second, we have $u \star v = u$, which shows that $u = v$.

By repeatedly using binary operations, we can build expressions that involve more than two elements. If a binary operation is not associative, the result of the computation is dependent on the order in which the elements get involved with the operation, as can be seen from Example 2.10.6. For an associative operation "$\star$" on a set M, the terms $x \star (y \star z)$ and $(x \star y) \star z$ refer to the same element. Therefore, it is natural to denote their common value by $x \star y \star z$, that is, to eliminate parentheses. An expression such as $(x \star y) \star (x \star (z \star v))$ yields the same element as $((x \star y) \star x) \star (z \star v)$. Applying again the associativity, we obtain the equivalent expression $(((x \star y) \star x) \star z) \star v$. Clearly, the value of all these expressions depends on the nature of the elements and on the order they appear in the expression. The order in which the operation itself is applied is immaterial, and we can use the notation $x \star t \star x \star z \star v$ to designate all these expressions. In the chapter on induction, we discuss a formal treatment of the notations for associative operations.

A binary operation $\ast$ defined on a finite set $M = \{x_1, \ldots, x_n\}$ can be defined by using an $n \times n$ matrix (also called *the Cayley table of the operation* $\ast$).

Definition 2.10.10 *The* Cayley table of the operation $\ast$ *is the matrix* C : $\{1, \ldots, n\} \times \{1, \ldots, n\} \longrightarrow M$ *defined by* $C(i, j) = x_i \ast x_j$ *for* $1 \leq i, j \leq n$.

Example 2.10.11 Consider the operation $*$ on the set $\{x_1, x_2, x_3\}$ defined by the matrix

$$\begin{pmatrix} x_1 & x_2 & x_3 \\ x_2 & x_2 & x_3 \\ x_3 & x_3 & x_3 \end{pmatrix}.$$

It is customary to write element x_i in front of the i-th line and x_j at the top of the j-th column. With this convention, the operation $*$ is represented by the following table

$x_i * x_j$	x_1	x_2	x_3
x_1	x_1	x_2	x_3
x_2	x_2	x_2	x_3
x_3	x_3	x_3	x_3

The associativity of a binary operation on a finite set containing n elements can be verified by constructing $2n$ tables and comparing these tables.

For a binary operation $*$ on the set $\{x_1, \ldots, x_n\}$, define the $2n$ operations $*_j$ and $*^j$ by

$$x_i *_j x_k = (x_i * x_j) * x_k,$$

and

$$x_i *^j x_k = x_i * (x_j * x_k).$$

Clearly, $*$ is associative if and only if for each j, $1 \le j \le n$, the table of $*_j$ coincides with the table for $*^j$.

Example 2.10.12 For the operation $*$ introduced in Example 2.10.11, we need to consider three pairs of tables

$*_1$	x_1	x_2	x_3		$*^1$	x_1	x_2	x_3
x_1	x_1	x_2	x_3		x_1	x_1	x_2	x_3
x_2	x_2	x_2	x_3		x_2	x_2	x_2	x_3
x_3	x_3	x_3	x_3		x_3	x_3	x_3	x_3

$*_2$	x_1	x_2	x_3		$*^2$	x_1	x_2	x_3
x_1	x_2	x_2	x_3		x_1	x_2	x_2	x_3
x_2	x_2	x_2	x_3		x_2	x_2	x_2	x_3
x_3	x_3	x_3	x_3		x_3	x_3	x_3	x_3

$*_3$	x_1	x_2	x_3		$*^3$	x_1	x_2	x_3
x_1	x_3	x_3	x_3		x_1	x_3	x_3	x_3
x_2	x_3	x_3	x_3		x_2	x_3	x_3	x_3
x_3	x_3	x_3	x_3		x_3	x_3	x_3	x_3

Comparing the three pairs of arrays corresponding to the operations $*_j$ and $*^j$, we conclude that $*$ is associative.

Let $*$ be a binary operation on the set M having the identity u.

Definition 2.10.13 *A left (right) inverse of the element $x \in M$ is an element $x' \in M$ such that $x' * x = u$ $(x * x' = u)$.*
The element x' is inverse *for x if it is both a right and a left inverse.*

If $*$ is an associative operation then, if x has a left inverse x_1 and a right inverse x_2, $x_1 = x_2$ and x has the inverse $x' = x_1 = x_2$. Indeed, we have $x_1 = x_1 * u = x_1 * (x * x_2) = (x_1 * x) * x_2 = u * x_2 = x_2$.

Furthermore, if $*$ is an *associative operation* having the identity u and x has an inverse y, then this inverse is unique. To justify this remark, observe that if both y and z are inverses of x then y is a left inverse and z is a right inverse; hence, $y = z$.

Taking $n = 0$ in Definition 2.10.1, we obtain the case of the *zero-ary* operation. Such an operation $f : M^0 \longrightarrow M$ will single out an element $f(\emptyset)$ of the set M.

For $n = 1$, we obtain *the unary operation*. Such an operation $f : M \longrightarrow M$ is often called a *transformation* of the set M.

A more general type of operation is introduced below.

Definition 2.10.14 *A n-ary partial operation on a set D is a partial mapping $f : D^n \rightsquigarrow D$.*

We use the same notations for partial binary operations as for usual binary operations.

Example 2.10.15 Consider the real number division $/ : \mathbf{R}^2 \rightsquigarrow \mathbf{R}$. This is a partial operation because $(x, 0)$ does not belong to $Dom(/)$ for any $x \in \mathbf{R}$.

2.11   Representations of Relations and Graphs

At this point in the chapter, the reader has been exposed to a variety of applications of relations and functions to computer science. In this section, we discuss methods for representing relations and graphs.

Let $\rho \subseteq A \times B$ and $\sigma \subseteq B \times C$ be two relations, where $A = \{x_1, \ldots, x_m\}$, $B = \{y_1, \ldots, y_n\}$, and $C = \{z_1, \ldots, z_p\}$. These relations are represented by the $m \times n$ matrix $M(\rho)$ and by the $n \times p$ matrix $M(\sigma)$, respectively.

Proposition 2.11.1 *The relation $\rho \bullet \sigma$ is represented by the $m \times p$ matrix M, where*

$$M_{ij} = (M(\rho)_{i1} \wedge M(\sigma)_{1j}) \vee \cdots \vee (M(\rho)_{in} \wedge M(\sigma)_{nj}), \qquad (2.2)$$

$1 \le i \le m$ *and* $1 \le j \le p$.

Proof. We need to prove that $(x_i, z_j) \in \rho \bullet \sigma$ if and only if $M_{ij} = 1$. Suppose that $M_{ij} = 1$. Since

$$M_{ij} = (M(\rho)_{i1} \wedge M(\sigma)_{1j}) \vee \cdots \vee (M(\rho)_{in} \wedge M(\sigma)_{nj})$$

for $1 \leq i \leq m$ and $1 \leq j \leq p$, at least one of the expressions $M(\rho)_{ik} \wedge M(\sigma)_{kj}$ must equal 1. If this happens for $k = k_0$, then we have $M(\rho)_{ik_0} = 1$ and $M(\sigma)_{k_0 j} = 1$, which means that $(x_i, y_{k_0}) \in \rho$ and $(y_{k_0}, z_j) \in \sigma$. This implies $(x_i, z_j) \in \rho \bullet \sigma$.

Conversely, if $(x_i, z_j) \in \rho \bullet \sigma$, then there is $y_k \in B$ such that $(x_i, y_k) \in \rho$ and $(y_k, z_j) \in \sigma$, which means that $M(\rho)_{ik} = 1$ and $M(\sigma)_{kj} = 1$; hence, $M(\rho)_{ik} \wedge M(\sigma)_{kj} = 1$. This implies $M_{ij} = 1$. ∎

Formula 2.2 can also be written as

$$M_{ij} = \bigvee_{1 \leq k \leq n} (M(\rho)_{ik} \wedge M(\sigma)_{kj}).$$

The last proposition introduces a partial operation on the set of matrices ranging over the two-element set $B = \{0, 1\}$. Let $M \in M_{m \times n}(\{0, 1\})$ and $P \in M_{n \times p}(\{0, 1\})$.

Definition 2.11.2 *The product of matrices M and P is the matrix $Q \in M_{m \times p}(\{0, 1\})$ given by*

$$Q_{ij} = \bigvee_{1 \leq k \leq n} (M_{ik} \wedge P_{kj}). \tag{2.3}$$

The product of the matrices M and P, introduced above, is denoted by MP. Under this notation, Proposition 2.11.1 implies

$$M(\rho \bullet \sigma) = M(\rho)M(\sigma). \tag{2.4}$$

Formula 2.3, for computing the matrix product for matrices over the set $M_{m \times p}(\{0, 1\})$, is similar to the formula giving the elements of the product of two matrices ranging over the set $\mathbf{R}$ of real numbers; here, the sum has been replaced by $\vee$, while the product has been replaced by $\wedge$.

The matrix product is a partial operation, since for two matrices involved in a product, the number of columns of the first matrix has to be equal to the number of rows of the second matrix. If we limit the product to the set of square matrices $M_{m \times m}(\{0, 1\})$, the matrix product is defined for any pair of matrices from this set.

We noticed that for a matrix $M \in M_{n \times n}(\{0, 1\})$ there exists a set $A = \{x_1, \ldots, x_n\}$ and a relation ρ on this set such that $M = M(\rho)$. This makes possible the transfer of certain properties of the relation product to the matrix product. For instance, since the relation product is associative (see Theorem 2.2.15), the product of $\{0, 1\}$-matrices is also associative, that is, we have $M(PQ) = (MP)Q$ for any matrices $M \in M_{m \times n}(\{0, 1\})$, $P \in M_{n \times p}(\{0, 1\})$, and $Q \in M_{p \times q}(\{0, 1\})$.

The n-th power of a matrix $M \in M_{n \times n}(\{0,1\})$ is defined by $M^n = M(\rho^n)$, where ρ is the relation corresponding to M. In view of equality 2.4, we have

$$M(\rho)^n = \underbrace{M(\rho) \cdots M(\rho)}_{n}.$$

Furthermore, we have $M(\rho)^n(i,j) = 1$ if and only if in the graph $G(\rho)$ there is a path of length n originating in x_i and ending in x_j.

The zero-th power of M is the *diagonal matrix:*

$$M^0 = M(\iota_A) = \begin{pmatrix} 1 & 0 & \cdots & 0 \\ 0 & 1 & \cdots & 0 \\ \vdots & \vdots & \cdots & \vdots \\ 0 & 0 & \cdots & 1 \end{pmatrix}.$$

For two matrices $M, P \in M_{p \times q}(\{0,1\})$, we consider the matrices $M \vee P$ and $M \wedge P$ defined by

$$\begin{aligned} (M \vee P)_{ij} &= M_{ij} \vee P_{ij}, \\ (M \wedge P)_{ij} &= M_{ij} \wedge P_{ij}. \end{aligned}$$

It is not difficult to see that

$$\begin{aligned} M(\rho \cup \sigma) &= M(\rho) \vee M(\sigma), \\ M(\rho \cap \sigma) &= M(\rho) \wedge M(\sigma). \end{aligned}$$

The reader can easily verify that the operations $\vee$ and $\wedge$ defined on the set $M_{m \times n}(\{0,1\})$ are associative, commutative, and idempotent. Also, we have

$$\begin{aligned} M(P \vee Q) &= MP \vee MQ, \\ (P \vee Q)L &= PL \vee QL. \end{aligned}$$

We can now formulate an algorithm for computing the transitive closure of a relation over a finite set. This algorithm is rather inefficient, and we shall discuss later more efficient algorithms for performing this computation.

Let $G = (V, U)$ be a finite graph representing the relation $\rho \subseteq V \times V$ and assume that $|V| = n$. As was noticed before, we have

$$\rho^+ = \rho \cup \rho^2 \cup \cdots \rho^n \cdots.$$

Note that if there is a path of length $m > n$ in G joining vertex x to vertex x' then the same vertices can be joined by a path of length at most n. This path can be obtained by removing the repeated vertices from the longer path. In terms of relations, this means that if $(x, x') \in \rho^m$ with $m > n$ then $(x, y) \in \rho^n$. Therefore, we can write

$$\rho^+ = \rho \cup \rho^2 \cup \cdots \cup \rho^n. \tag{2.5}$$

This allows us to formulate the following.

Algorithm 2.11.3 Input: *The incidence matrix of the finite digraph* $G = (V, U)$ *representing the relation* ρ.

Output: *The matrix* $M(\rho^+)$ *representing the transitive closure of the relation* ρ.

Method: *Compute the matrix* $M(\rho^+)$ *by*

$$M(\rho^+) = M(\rho) \vee \cdots \vee M(\rho)^n, \tag{2.6}$$

where $n = |V|$.

The justification of the algorithm is immediate because of equality (2.5).

Note that $2n-1$ operations are required in order to compute an element of the product MP of two matrices $M \in M_{m \times n}(\{0,1\}$ and $P \in M_{n \times p}(\{0,1\})$ (more exactly, n operations $\wedge$ and $n-1$ operations $\vee$). Furthermore, $(2n-1)mp$ operations are needed in order to compute all the elements of the product. Consequently, $(2n-1)n^2$ operations are required to compute the product of two matrices from $M_{n \times n}(\{0,1\})$. This number of operations is required for computing each of the matrices $M(\rho)^k$ that appear in equation (2.6), and this gives $(2n-1)n^2(n-1)$ operations. In addition, $n^2(n-1)$ $\vee$-operations are needed to compute $M(\rho^+)$, which shows that the number of required operations is $O(n^4)$. Of course, this implies that if the size of the set doubles, the time required by the algorithm grows 16 times, which makes this algorithm quite inefficient.

For a relation $\rho \subseteq V \times V$, where V is a finite set, the matrix $M(\rho)$ contains $|V|^2$ entries even if the relation ρ may contain a small number of pairs. To eliminate this disadvantage, we can represent the relation as a word $\alpha(\rho)$ over the alphabet $V \cup \{:, ;\}$. If $V = \{v_1, \ldots, v_n\}$, then

$$\alpha(\rho) = v_1 : v_{11} \cdots v_{1m_1}; v_2 : v_{21} \cdots v_{2m_2}; \cdots; v_n : v_{n1} \cdots v_{nm_n}$$

if $v_{i1}, \ldots, v_{im_i}$ is the list of all vertices that are final endpoints of arcs originating in v_i.

Example 2.11.4 Consider the graph represented in Figure 2.5. We can represent the relation defined by the graph by the word

$$\alpha(\rho) = a : bcd; b : cd; c : de; d : e; e :; .$$

The word $\alpha(\rho)$ can be implemented by a collection of linked lists. Namely, for the set $V = \{x_1, \ldots, x_n\}$, we shall consider an array A containing n entries, one for each vertex. The element $A[j]$ contains a pointer toward the linked list containing the immediate descendents of the vertex x_j.

Example 2.11.5 Figure 2.18 contains the linked list implementation of the directed graph from Figure 2.5.

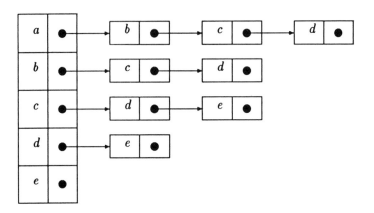

FIGURE 2.18. Linked list representation of a graph.

2.12   Relations and Databases

A database is a repository for data. The need for efficiently managing large amounts of data generated an interest in the development of a special kind of programming systems known as *database management systems*. In order to achieve this efficacy and to allow the user to conceptualize the organization of the database, the data must be structured along a certain database model. Currently, the most important database model is the relational model, which emerged from Codd's work done in the early 1970s. In a relational database system, data are structured in tables, which are essentially a convenient representation of finite relations. Unless otherwise specified, we assume in this section that all relations are finite.

Let $\{D_{i_1}, \ldots, D_{i_n}\}$ be a collection of n finite sets. Any finite n-ary relation $\rho \subseteq \prod\{D_i \mid i \in \{i_1, \ldots, i_n\}\}$,

$$\rho = \{(x_1, \ldots, x_n), \ldots, (z_1, \ldots, z_n)\},$$

can be graphically represented by the *table* τ given below; the *name* of this table is a word m from a certain alphabet V. We assume that V contains at least the small and the capital letters of the Latin alphabet as well as some special symbols such as

$$\cup \quad \cap \quad - \quad \times \quad [\quad] \quad \bowtie \quad \div \quad .$$

In addition, we consider in V any subscripted symbols we might need in a certain context. V will be referred to as *the basic alphabet*.

$$
\begin{array}{|c|c|c|}
\hline
i_1 & \cdots & i_n \\
\hline
x_1 & \cdots & x_n \\
\vdots & \cdots & \vdots \\
z_1 & \cdots & z_n \\
\hline
\end{array}
$$

$$m$$

The subscripts $i_1, \ldots, i_n$ of the n sets involved in the Cartesian product are the *attributes* of the table τ. We assume that these subscripts are *also words on* V. Throughout this section, we require *the attributes of any table to be pairwise distinct*.

The set denoted by D_{i_m} is *the domain of the attribute* i_m, $1 \leq m \leq n$. We denote D_{i_m} by $\mathrm{Dom}(i_m)$.

Definition 2.12.1 *A* database schema *is a pair* $S = (I, \mathcal{D})$, *where* $I \subseteq V^*$ *is the* set of attributes *and* $\mathcal{D} = \{D_i \mid i \in I\}$ *is a collection of sets indexed by the set* I.

The set D_i *is the* domain *of the attribute* i.

During the lifetime of a database, attributes might be added or removed from the schema.

The notion of table will also be formalized.

Definition 2.12.2 *A* table *of the database schema* $S = (I, \mathcal{D})$ *is a triple* $\tau = (m, H, \rho)$, *where* $m \in V^*$ *is called* the name of the table, $H = \{i_1, \ldots, i_n\} \subseteq I$ *is the* the schema of the table *or* (the heading of the table), *and* ρ *is a relation,* $\rho \subseteq \prod_{i \in H} D_i$.

In keeping with the practice of database theory, we denote the schema of the table τ by $\mathrm{sch}(\tau) = i_1 \cdots i_n$. Also, if $\tau = (m, H, \rho)$, we refer to $\mathrm{sch}(\tau)$ as the *schema* of the relation ρ, and we use the notation $\mathrm{sch}(\rho)$ instead of $\mathrm{sch}(\tau)$.

A table τ can be regarded as a set of rows having its own name, whose structure is described by the same heading. If $\tau = (m, H, \rho)$, the relation ρ will be denoted by $\rho(\tau)$.

Let $S = (I, \mathcal{D})$ be a database schema.

Definition 2.12.3 *A* relational database *is a finite collection* $\mathbf{D}$ *of tables,* $\mathbf{D} = \{\tau_1, \ldots, \tau_p\}$ *over the schema* S. *The schema* S *of the database* $\mathbf{D}$ *is denoted by* $\mathrm{sch}(\mathbf{D})$.

Example 2.12.4 Consider the database of a college department $\mathbf{D}_{dept}$ over the schema

$$
\begin{aligned}
S = (&\{\mathrm{C\#}, \mathrm{CNAME}, \mathrm{CREDITS}, \mathrm{ROOM}, \mathrm{PNAME}, \mathrm{OFFICE}, \\
&\quad \mathrm{STUDENTN}, \mathrm{STUDENT\#}, \mathrm{GRADE}\}, \\
&\{D_{C\#}, D_{CNAME}, D_{CREDITS}, D_{ROOM}, D_{PNAME}, D_{OFFICE}, \\
&\quad D_{STUDENTN}, D_{STUDENT\#}, D_{GRADE}\}).
\end{aligned}
$$

Our database consists of the tables shown below. We represented a *snapshot* of the actual data contained by the database; needless to say, the actual content of the database varies in time, due to deletions, insertions, or updates performed on the records.

COURSES

C#	CNAME	CREDITS	ROOM
110	Introduction to Programming	4	W1100
210	Intermediate Programming	4	M3210
240	Computer Architecture	3	S1120
310	Alg. and Data Structures	3	S2063
320	Discrete Mathematics	3	W2615
440	Operating Systems	3	S2063

PROFS

PNAME	C#	OFFICE
Smith	110	S556
Wesson	240	S173
Brown	210	S220
Smith	310	S556
McNeil	440	S230
Jones	310	S041

GRADES

C#	STUDENTN	STUDENT#	GRADE
110	John Albert	2341	A
110	Mary Simpson	5641	C
110	Kathy Driscoll	9008	A
110	John Albert	2341	F
210	Donald Trump	4430	B
210	Ron Wallace	2345	C
210	Paul Gibson	7655	D
310	Jane Brokaw	0988	A
310	Tom Rather	5448	B

Let τ_1, τ_2, τ_3 be the tables named COURSES, PROFS, GRADES, respectively. We have

$$\text{sch}(\tau_1) = \text{C\# CNAME CREDITS ROOM},$$
$$\text{sch}(\tau_2) = \text{PNAME C\# OFFICE},$$
$$\text{sch}(\tau_3) = \text{C\# STUDENTN STUDENT\# GRADE}.$$

Of course, $D_{C\#} = \text{Dom}(C\#)$ consists of the course numbers of all courses taught at that college. Likewise, $D_{PNAME} = \text{Dom}(PNAME)$ contains all names of professors affiliated with the college, etc.

Each table has a *unique name*. We introduce an operation allowing the creation of an *alias* of a table with the possible renaming of its attributes; in keeping with the uniqueness of the names, we assume that each time a table is created or an alias of a table is defined we have a new name, *different from all names we have used so far*.

Let $\tau = (m, H, \rho)$ be a table, where $\text{sch}(\tau) = H$, $H = i_1 \cdots i_n$. Assume that H' is another set of attributes and $H' = j_1 \cdots j_n$ such that $\text{Dom}(i_p) = \text{Dom}(j_p)$ for $1 \leq p \leq n$.

Definition 2.12.5 *The table $\tau' = (m', H', \rho)$ is obtained by* renaming *from the table $\tau = (m, H, \rho)$ if they represent the same relation ρ, while the attribute j_p replaces i_p in τ' for $1 \leq p \leq n$.*
We denote the definition of the table τ' by

$$m'(j_1 \cdots j_n) := m.$$

If $H = H'$, the table τ' is obtained by changing the name of the table τ to τ'. The definition of τ' is denoted in this case by $m' := m$.

Example 2.12.6 Consider the table named TEACHERS, where

$$\text{TEACHERS(TNAME CNO ROOM)} := \text{PROFS.}$$

The table TEACHERS will be obtained from the table named PROFS from Example 2.12.4.

TEACHERS

TNAME	CNO	ROOM
Smith	110	S556
Wesson	240	S173
Brown	210	S220
Smith	310	S556
McNeil	440	S230
Jones	310	S041

In order to define the necessary data-retrieval capabilities of relational database systems, Codd has introduced a collection of operations on relations called *relational algebra*. In the sequel, we define these operations and explore various relationships existing among them. Specifically, we consider a collection of eight operations: four "set-theoretical" operations, that is, operations that use the fact that relations are *sets* of tuples (union, intersection, difference, product) and four special operations (selection, projection, join, and division). More operations will be discussed under Exercises and Supplements. Not all operations we are about to introduce are independent. As we shall see, it is possible to express all operations of relational algebra through a set of five operations (union, difference, product, selection, projection).

The operations of the relational algebra generate *new* relations. In keeping with the uniqueness of names for tables, we assume that each time we introduce a relations we give a *new* name to the table representing this relation; the table whose name is m will be denoted by $\tau(m)$.

Unary operations (selection and projection) have priority over the rest of the operations. No other priority rule is specified for the operations of relational algebra.

Definition 2.12.7 *The relations ρ and θ are* compatible *if they are subsets of the same Cartesian product $\prod_{i \in I} D_i$.*

This concept allows us to introduce several set-theoretical operations. Let ρ, θ be two compatible relations.

Definition 2.12.8 *The* union (intersection, difference) *of the relations ρ, θ is the relation $\rho \cup \theta$ ($\rho \cap \theta$, $\rho - \theta$, respectively).*

If $\tau = (m, H, \rho)$, $\tau' = (m', H, \theta)$ are the tables representing the relations ρ and θ, then the relation $\rho \omega \theta$ will be represented by the table $(m \omega m', H, \rho \omega \theta)$, where $\omega \in \{\cup, \cap, -\}$.

Observe that the union (intersection, difference) of two compatible relations is also compatible with ρ and θ.

For a table $\tau = (m, I, \rho)$, where $I = i_1 \cdots i_n$, we consider the attributes $m.i_1, \ldots, m.i_n$; we refer to these attributes as *the m-qualified attributes* of I.

Let ρ, θ be two relations represented by two distinct tables (m, I, ρ) and (m', J, θ), respectively, where $I = i_1 \cdots i_p$ and $J = j_1 \cdots j_q$.

Consider the schemas $I^m = m.i_1 \cdots m.i_p$, $J^{m'} = m'.j_1 \cdots m'.j_q$ and observe that $I^m \cap J^{m'} = \emptyset$ since $m \neq m'$ even if I and J are not necessarily disjoint.

We also assume that $\text{Dom}(m.i_k) = \text{Dom}(i_k)$ and $\text{Dom}(m'.j_l) = \text{Dom}(j_l)$ for $1 \leq k \leq p$ and $1 \leq l \leq q$, respectively.

Definition 2.12.9 *The* product *of the relations ρ, θ is the relation $\rho \times \theta$, where $sch(\rho \times \theta) = I^m \cup J^{m'}$ and $w \in \rho \times \theta$ if there are $r \in \rho$ and $t \in \theta$ such that $w(m.i_k) = r(i_k)$ and $w(m'.j_h) = t(j_h)$ for $i_k \in I$, $j_h \in J$.*

The table representing the relation $\rho \times \theta$ is $\tau = (m \times m', I^m \cup J^{m'}, \rho \times \theta)$.

Let $\rho \subseteq \prod_{i \in I} D_i$ be a relation on the schema I and assume that J is a subset of its attributes, $J \subseteq \{i_1, \ldots, i_n\}$. Consider the *inclusion mapping* $c_J : J \hookrightarrow I$ given by $c_J(i) = i$ for all $i \in J$.

Definition 2.12.10 *The* projection *of the tuple* $t \in \rho$ *is the tuple* $z : J \longrightarrow \bigcup_{i \in J} D_i$ *defined by* $z = t_{c_J}$. *The new tuple will be denoted by* $t[J]$.
  The projection *of the relation* ρ *on the set of attributes* J *is the relation* $\rho[J] = \{t[J] \mid t \in \rho\}$.
  If $\tau = (m, I, \rho)$ *is the table representing* ρ, *then the relation* $\rho[J]$ *will be represented by the table* $\tau' = (m[J], J, \rho[J])$.

Example 2.12.11 Consider the relation GRADES. If grades have to be posted, we want to eliminate the student names and give only the course numbers, the student numbers, and their respective grades. This can be accomplished through the projection

GRADES [C# STUDENT# GRADE].

The table describing this projection is given here.

GRADES [C# STUDENT# GRADE]

C#	STUDENT#	GRADE
110	2341	A
110	5641	C
110	9008	A
110	2341	F
210	4430	B
210	2345	C
210	7655	D
310	0988	A
310	5448	B

Let ρ be a relation, represented by $\tau = (m, I, \rho)$ and assume that ζ is a relation such that $\text{sch}(\zeta) = H$, where $H = h_1 \cdots h_p \subseteq I$.

Definition 2.12.12 *The* ζ-selection *of the relation* ρ *is the relation*

$$sel_\zeta(\rho) = \{t \mid t \in \rho, t[H] \in \zeta\}.$$

This relation will be represented by $(m_\zeta, I, sel_\zeta(\rho))$.

Consider an attribute $i \in I$ and let $a \in \text{Dom}(i)$. Let ζ be a relation having the schema $\{i\}$ and assume that $\zeta = \{a\}$. The result of the selection $sel_\zeta(\rho)$ is the set of all tuples of ρ whose i-th component equals a, $\{t \mid t \in \rho \text{ and } t(i) = a\}$. This special selection will be denoted by $sel_{i=a}(\rho)$; the name of the table representing this is $sel_{i=a}(m)$.

Example 2.12.13 Suppose that we need to determine the courses that offer four credits. This can be done by applying a selection. Namely, the result of the selection $sel_{CREDITS=4}(\text{COURSES})$ is the relation represented by the following table.

$$\text{sel}_{CREDITS=4}(COURSES)$$

C#	CNAME	CREDITS	ROOM
110	Introduction to Programming	4	W1100
210	Intermediate Programming	4	M3210

Let ρ be a relation such that i, j are two attributes for which $\text{Dom}(i) = \text{Dom}(j) = D$. If α is a relation on D ($\text{sch}(\alpha) = ij$), then we denote the selection $\text{sel}_\alpha(\rho)$ by $\text{sel}_{i\alpha j}(\rho)$.

Example 2.12.14 Suppose that we have a table recording the size of the pictures of a collection. The pictures are referred by their inventory number $INV\#$, and the table reflects their horizontal and vertical dimensions in centimeters (HD and VD, respectively).

COLLECTION

INV#	HD	VD
1221	30	50
1278	40	70
3411	50	40
4670	120	80
5182	80	60

The selection $\text{sel}_{HD>VD}(COLLECTION)$ will result in a table describing the works whose horizontal dimension is larger than the vertical dimension:

$$\text{sel}_{HD>VD}(COLLECTION)$$

INV#	HD	VD
3411	50	40
4670	120	80
5182	80	60

Theorem 2.12.15 *Selection operators commute under composition, that is, for $H, K \subseteq I = \text{sch}(\rho)$ we have:*

$$\text{sel}_\zeta(\text{sel}_\xi(\rho)) = \text{sel}_\xi(\text{sel}_\zeta(\rho)) \tag{2.7}$$

for any relations ζ, ξ with $\text{sch}(\zeta) = H$ and $\text{sch}(\xi) = K$.

Furthermore, selection is distributive over union, intersection, and difference, that is,

$$\text{sel}_\zeta(\rho \, \omega \, \theta) = \text{sel}_\zeta(\rho) \, \omega \, \text{sel}_\zeta(\theta),$$

where $\omega \in \{\cup, \cap, -\}$.

Proof. The commutativity of the selections follows from the fact that both sides of equation (2.7) are equal to $\{t \mid t \in \rho, t[H] \in \zeta,$ and $t[K] \in \xi\}$.

We prove the distributivity of selection over difference; the other cases are left to the reader as exercises.

Let $t \in \text{sel}_\zeta(\rho - \theta)$. We have $t \in \rho - \theta$ and $t[H] \in \zeta$. Therefore, $t \in \text{sel}_\zeta(\rho)$ and $t \notin \text{sel}_\zeta(\theta)$; hence, $t \in \text{sel}_\zeta(\rho) - \text{sel}_\zeta(\theta)$. This shows that

$$\text{sel}_\zeta(\rho - \theta) \subseteq \text{sel}_\zeta(\rho) - \text{sel}_\zeta(\theta).$$

Conversely, let $t \in \text{sel}_\zeta(\rho) - \text{sel}_\zeta(\theta)$. This implies $t \in \rho$ and $t[H] \in \zeta$. Since $t \notin \text{sel}_\zeta(\theta)$, this also means that $t \notin \theta$; hence, $t \in \rho - \theta$. Therefore, $t \in \text{sel}_\zeta(\rho - \theta)$, and we have the reverse inclusion

$$\text{sel}_\zeta(\rho - \theta) \supseteq \text{sel}_\zeta(\rho) - \text{sel}_\zeta(\theta),$$

hence the needed equality. ∎

Since $\text{sel}_\zeta(\text{sel}_\xi(\rho)) = \text{sel}_\xi(\text{sel}_\zeta(\rho))$, we denote the common value of these relations by $\text{sel}_{\zeta \text{ and } \xi}(\rho)$. We combine this notation with the previous notation introduced for special selections.

Example 2.12.16 The relation *COLLECTION* was introduced in Example 2.12.14. The expression

$$\text{sel}_{INV\#>2000 \text{ and } HD>VD}(COLLECTION)$$

designates the pictures whose inventory number exceeds 2000 and whose horizontal dimension is larger than the vertical dimension.

Let ρ, θ be two relations:

$$\rho \subseteq \prod_{i \in I} D_i, \text{ and } \theta \subseteq \prod_{i \in J} D_i,$$

represented by the tables $\tau = (m, I, \rho)$ and $\tau' = (m', J, \theta)$, respectively. Assume that $I \cap J = K$ and $I \cup J = L$.

Definition 2.12.17 *The tuples $t \in \rho, s \in \theta$ are* joinable *if $t(i) = s(i)$ for all $i \in K$. Their* join *is the tuple $z \in \prod_{i \in L} D_i$, where*

$$z(i) = \begin{cases} t(i) & \text{if } i \in I, \\ s(i) & \text{if } i \in J. \end{cases}$$

The join *of the relations ρ and θ is the relation $\rho \bowtie \theta$ on the schema $I \cup J$ that consists of all the joins of the joinable tuples of the two relations. The table representing $\rho \bowtie \theta$ is $(m \bowtie m', I \cup J, \rho \bowtie \theta)$.*

Notice that the definition of the joinability makes the definition of the join a correct one.

In terms of projections, t, s are joinable if $t[K] = s[K]$. Their join z will satisfy both $z[I] = t[I]$ and $z[J] = s[J]$.

Proposition 2.12.18 *The join of two compatible relations coincides with their intersection.*

Proof. Let $\rho, \sigma \subseteq \prod_{i \in I} D_i$ be two compatible relations on the schema I and let $z \in \rho \bowtie \sigma$. There exist two joinable tuples $t \in \rho$ and $s \in \sigma$ such that $z(i) = t(i)$ and also $z(i) = s(i)$ for all $i \in I$, because of the definition of joinability. This immediately gives $z = t = s \in \rho \cap \sigma$.

Conversely, if $u \in \rho \cap \sigma$, then u is joinable with itself, and this gives $u \in \rho \bowtie \sigma$. ∎

The join has a number of useful properties; some of those are mentioned below.

Theorem 2.12.19 *The join operation is associative, commutative, and idempotent, that is,*

$$\rho \bowtie (\mu \bowtie \theta) = (\rho \bowtie \mu) \bowtie \theta,$$
$$\rho \bowtie \mu = \mu \bowtie \rho,$$
$$\rho \bowtie \rho = \rho,$$

for any relations ρ, μ, θ.

Join is distributive with respect to union and to intersection, that is,

$$\rho \bowtie (\mu \cup \theta) = (\rho \bowtie \mu) \cup (\rho \bowtie \theta),$$
$$\rho \bowtie (\mu \cap \theta) = (\rho \bowtie \mu) \cap (\rho \bowtie \theta).$$

Proof. We prove only the associativity of join; the other properties are left to the reader as exercises.

Before presenting the argument for the associativity, notice that if ρ_1, ρ_2 are two relations on the schemas I_1, I_2, respectively, $t_1 \in \rho_1$ and $t_2 \in \rho_2$, respectively, and $t_1[P] = t_2[P]$ for $P \subseteq I_1 \cap I_2$ then we have $t_1[Q] = t_2[Q]$, for any $Q \subseteq P$. Also, $t_1[R_1 \cup R_2] = t_2[R_1 \cup R_2]$ if and only if $t_1[R_1] = t_2[R_1]$ and $t_2[R_1] = t_2[R_2]$.

Let ρ, μ, θ be three relations on the schemas I, J, K, respectively. Let $u \in \rho \bowtie (\mu \bowtie \theta)$. Then, there are two joinable tuples t, s such that $t \in \rho$ and $s \in \mu \bowtie \theta$ and $u[I] = t[I]$, $u[J \cup K] = s[J \cup K]$. Because of the previous remark, we also have $u[J] = s[J]$ and $u[K] = s[K]$. The joinability of t and s means that $t[I \cap (J \cup K)] = s[I \cap (J \cup K)]$, and this, in turn, gives $t[I \cap J] = s[I \cap J]$ and $t[I \cap K] = s[I \cap K]$.

Since $s \in \mu \bowtie \theta$, there are $z \in \mu$ and $w \in \theta$ such that $s[J] = z[J]$, $s[K] = w[K]$, and z, w are joinable, that is, $z[J \cap K] = w[J \cap K]$.

The tuples t, z are joinable since $t[I \cap J] = u[I \cap J] = s[I \cap J] = z[I \cap J]$. Their join is a tuple $y \in \rho \bowtie \mu$, and we have $y[I] = t$, $y[J] = z$, and, of course, $y \in \rho \bowtie \mu$.

Note that the tuples y and w are joinable since $y[(I \cup J) \cap K] = w[(I \cup J) \cap K]$. Indeed, $y[I \cap K] = t[I \cap K] = u[I \cap K]$ and $w[I \cap K] = s[I \cap K] = u[I \cap K]$; likewise, $y[J \cap K] = z[J \cap K] = s[J \cap K] = u[J \cap K]$, and this gives

$$y[(I \cup J) \cap K] = y[(I \cap K) \cup (J \cap K)]$$
$$= w[(I \cap K) \cup (J \cap K)] = w[(I \cup J) \cap K].$$

Let v be the join of y and w, $v \in (\rho \bowtie \mu) \bowtie \theta$. We have $v[I] = y[I] = t[I] = u[I]$, $v[J] = y[J] = z[J] = s[J] = u[J]$, and $v[K] = w[K] = s[K] = u[K]$. Therefore, $v[I \cup J \cup K] = u[I \cup J \cup K]$; hence, $v = u$, which means that $u \in (\rho \bowtie \mu) \bowtie \theta$. We proved the inclusion $\rho \bowtie (\mu \bowtie \theta) \subseteq (\rho \bowtie \mu) \bowtie \theta$. A similar argument can be used to prove the converse inclusion. ∎

Let ρ, θ be two relations such that $sch(\theta) \subseteq sch(\rho)$. Assume that the two relations are represented by the tables $\tau = (m, I \cup J, \rho)$ and $\tau' = (m', J, \theta)$, respectively , where $I \cap J = \emptyset$, $I = i_1 \cdots i_n$ and $J = j_1 \cdots j_m$.

Definition 2.12.20 *The quotient of the relations ρ and θ is the relation $\mu = \rho \div \theta$, where $sch(\mu) = I$ and $t \in \mu$ if $\{t\} \bowtie \theta \subseteq \rho$.*
The relation μ is represented by the table $(m \div m', I, \mu)$.

The operation $\div$ will be called *division*. In order for a tuple t to qualify for membership in $\rho \div \theta$, t must be joinable with *all* tuples of θ, and the result of these joins must belong to the relation ρ.

Let $\mathbf{D}$ be a relational database.

Definition 2.12.21 *A computation of the relation ρ in relational algebra is a sequence of names of tables $m_1, \ldots, m_n$ such that there exists a sequence of tables $\{\tau_i \mid \tau_i = (m_i, I_i, \rho_i), 1 \leq i \leq n\}$, and for every i, $1 \leq i \leq n$, we have $\rho_i \in \mathbf{D}$, or ρ_i is obtained from some of its predecessors in the sequence by renaming or by applying one of the operations of relational algebra.*

Proposition 2.12.22 *The join of two relations can be computed in relational algebra by using the operations of product, selection, and projection.*

Proof. Let $\mathbf{D}$ be a relational database and let ρ, σ be two relations represented by the tables (r, I, ρ) and (s, J, σ), respectively, where $I \cap J = K$, $I = i_1 \cdots i_p k_1 \cdots k_m$, $J = k_1 \cdots k_m j_1 \cdots j_q$ and $K = k_1 \cdots k_m$. Consider the sets $K^r = r.k_1 \cdots r.k_m$ and $K^s = s.k_1 \cdots s.k_m$.

Let ζ be the relation on $\prod_{k \in K^r \cup K^s} \mathrm{Dom}(k)$ defined by

$$\zeta = \{t \mid t \in \prod_{k \in K^r \cup K^s} \mathrm{Dom}(k) \text{ and } t(r.k_n) = t(s.k_n) \text{ for } k_n \in K\}.$$

Consider the following computation in relational algebra:

$$\begin{aligned}
&r, \\
&s, \\
&m := r \times s, \\
&u := \mathrm{sel}_\zeta(m)[r.i_1, \ldots, r.i_p, r.k_1, \\
&\qquad\qquad \ldots, r.k_m, s.j_1, \ldots, s.j_q], \\
v(i_1, \ldots, i_p, k_1, \ldots, k_m, j_1, \ldots, j_q) &:= u.
\end{aligned}$$

For

$$t_r = (t_r(i_1), \ldots, t_r(i_p), t_r(k_1), \ldots, t_r(k_m)) \in r$$

and

$$t_s = (t_s(k_1), \ldots, t_s(k_m), t_s(j_1), \ldots, t_s(j_q)) \in s,$$

the table m will contain the tuple

$$(t_r(i_1), \ldots, t_r(i_p), t_r(k_1), \ldots, t_r(k_m), t_s(k_1), \ldots, t_s(k_m), t_s(j_1), \ldots, t_s(j_q)).$$

Such a tuple will be retained by the selection sel_ζ if and only if $t_r(k_1) = t_s(k_1), \ldots, t_r(k_m) = t_s(k_m)$, and this is exactly the joinability condition for the tuples t_r and t_s. The last step of the computation will drop the unnecessary attributes, and we shall be left with the join of the relations ρ and σ. ∎

The standard query language SQL[5] computes the join of two relations in the manner described above. A discussion of the semantics of retrieval operations in SQL using relational algebra can be found at the end of this section.

Theorem 2.12.23 *All operations of relational algebra can be expressed through five basic operations: union, difference, product, projection, and selection.*

Proof. Proposition 2.12.22 shows that the join can be expressed through product, selection, and projection. Also, the identity contained by Exercise 101 shows that the divison can be expressed through product, projection and difference. To conclude the argument, observe that intersection can be expressed through join as shown in Proposition 2.12.18. ∎

Relational algebra is used as a standard against which we measure any relational database system. A database system is called *relationally complete* if it is capable of performing all the operations of relational algebra. The last theorem shows that in order to prove relational completeness it will suffice to show that the five fundamental operations can be implemented.

We now consider several queries based on the database $\mathbf{D}_{dept}$ and the computations that solve these queries.

Example 2.12.24 Query: Find the professors who are teaching student Albert in a three credit course.

The information required by this query can be found in three tables: COURSES (for C# and the number of credits), PROFS (for the association between courses and professors) and GRADES (for the association between

---

[5]SQL (Structured Query Language) is a data manipulation language developed initially by IBM for a database management system prototype called System R. IBM and others have produced a number of systems that support SQL, and it is expected that SQL will become the standard relational language.

courses and students). We use the following computation:

$$
\begin{aligned}
&COURSES; \\
&PROFS; \\
&GRADES; \\
T1 :=\ &COURSES \bowtie PROFS \bowtie GRADES; \\
T2 :=\ &\mathrm{sel}_{STUDENTN='Albert',CREDITS=4}(T1); \\
T3 :=\ &T2[PNAME];.
\end{aligned}
$$

The purpose of computing $T1$ is to bring together (by the joining operation) all the information required by the query; then, in $T2$, only the tuples involving student Albert are retained. Eventually, in $T3$, we project these tuples on the professors' names, and we obtain the desired answer as a table.

The same query can be solved through the application of the product; this alternative solution presents a special interest in view of the standard relational query language SQL. This solution is achieved by the following computation:

$$
\begin{aligned}
&COURSES; \\
&PROFS; \\
&GRADES; \\
T1 :=\ &COURSES \times PROFS \times GRADES; \\
T2 :=\ &\mathrm{sel}_{COURSES.C\#=PROFS.C\#\ \mathrm{and}\ PROFS.C\#=GRADES.C\#}(T1); \\
T3 :=\ &\mathrm{sel}_{GRADES.STUDENTN='Albert'\ \mathrm{and}\ COURSES.CREDITS=4}(T2); \\
T4 :=\ &T3[PROFS.PNAME];.
\end{aligned}
$$

The formation of the product of a relation with itself is needed in many occasions when we need to compare the members of the same relation. This construction requires some special precautions; indeed, we have assumed that the attributes of any table are pairwise distinct. If $\tau = (m, H, \rho)$ is a table representing the relation ρ with $\mathrm{sch}(\rho) = i_1 \cdots i_n$ then, if we try to compute $\nu = \rho \times \rho$, we obtain a table whose schema is going to be $m.i_1 \cdots m.i_n m.i_1 \cdots m.i_n$, which is unacceptable, in view of the previous assumption.

The remedy for this problem is to define an alias $\tau' = (m', I', \rho)$ of the table $\tau = (m, I, \rho)$, where $I \cap I' = \emptyset$, and then to form the product $m \times m'$. We consider this technique in the next two examples.

Example 2.12.25 Query: Find all pairs of course numbers taught by the same professors.

Consider the following computation:

$$PROFS;$$
$$TA := PROFS[PNAME, C\#];$$
$$TB := TA;$$
$$T1 := TA \times TB;$$
$$T2 := \text{sel}_{TA.C\#=TB.C\#}(T2);$$
$$T3 := T2[TA.PNAME, TB.PNAME];$$
$$T4 := \text{sel}_{TA.PNAME<TB.PNAME}(T3);.$$

The definition of the table TA retains that part of the table $PROFS$ that is useful for solving the query; next, TB, an alias of TA, is created, and this allows the computation of the product contained by the third step.

In $T2$, we keep only those tuples that refer to the same course; the names of the professors are selected in $T3$. The last step is used in order to discard redundant pairs. For instance, if professors Smith and Jones are teaching the same course, we obtain in the answer four pairs: (Jones, Jones), (Jones, Smith), (Smith, Jones), and (Smith, Smith). The last step allows us to retain only the pair (Jones, Smith).

Example 2.12.26 Query: Find the name of the students who obtained the highest grade in course 110.

In order to solve this query with the tools of relational algebra, we initially identify the students who *did not* obtain the highest grade in course 110, that is, the students for whom there are other students in course 110 who have a higher grade. This can be done through the following computation:

$$GRADES;$$
$$G := \text{sel}_{C\#=110}(GRADES);$$
$$GA := G[STUDENTN, GRADE];$$
$$GB := GA;$$
$$T1 := GA \times GB;$$
$$T2 := \text{sel}_{GA.GRADE>GB.GRADE}(T1);$$
$$T3 := T2[GB.STUDENTN];$$
$$T4(STUDENTN) := T3;.$$

A list of all students who receive grades is obtained through

$$T5 := GRADES[STUDENTN];,$$

and the computation can be concluded by eliminating from $T5$ those student names that appear in $T4$:

$$T6 := T5 - T4;.$$

Example 2.12.27 Query: Find the students who take all courses taught by Professor Smith.

We determine initially the courses taught by Professor Smith. Afterwards we extract, by projection, the names of the students and the courses they take from the table GRADES. Finally, the desired result will be obtained by using division. These operations are accomplished by the following program:

$$PROFS;$$
$$T1 := sel_{PNAME='Smith'}(PROFS);$$
$$T2 := T1[C\#];$$
$$GRADES;$$
$$T3 := GRADES[C\#, STUDENTN];$$
$$T4 := T3 \div T2;.$$

SQL is the standard query language for relational database systems. Queries are expressed through constructions called *SELECT phrases*. The answer returned by a query is a table, and we present the effect of such queries in terms provided by relational algebra.

SQL contains "data definition " facilities allowing the creation or elimination of tables and other related database objects. The user may indicate certain domains for the attributes of the tables chosen from a list of domains that is specific to each implementation of SQL. We mention below several of the domains available in SQL for the INGRES system.[6]

vchar(n) The set of strings of up to 2000 characters.

c(n) The set of strings of up to 255 printing ASCII characters.

float4 The set of floating-point numbers in simple precision.

float8 The set of floating-point numbers in double precision.

integer1 The set of one-byte integers.

integer2 The set of two-byte integers.
or smallint

integer4 The set of four-byte integers.
or integer

A statement such as

CREATE TABLE COURSES (C# **INTEGER2**,
                       CNAME **VCHAR(40)**,
                       CREDITS **INTEGER1**, ROOM **VCHAR(5)**)

---

[6] INGRES is a trademark of Relational Technologies, Inc.

creates the table (COURSES, C# CNAME CREDITS ROOM $,\emptyset$), that is, it creates an empty shell where the user can start inserting the tuples of the relation. We have $D_{C\#} = \textbf{INTEGER2}$, $D_{CNAME} = \textbf{VCHAR}(40)$, $D_{CREDITS} = \textbf{INTEGER1}$, and $D_{ROOM} = \textbf{VCHAR}(5)$.

In order to create the table $\tau = (m, i_1 \cdots i_n, \emptyset)$ we write the statement

$$\textbf{CREATE TABLE } m(i_1 \ d_1, \ldots, i_n \ d_n),$$

where d_k is the domain for i_k for $1 \leq k \leq n$.

The general form of a SELECT phrase is

$$\textbf{SELECT } \mathcal{A} \textbf{ FROM } \mathcal{T} \textbf{ WHERE } \mathcal{C}.$$

Here, $\mathcal{T}$ is a list $l_1, \ldots, l_k$, where l_j can be either a table name m_j or it may define an alias $\hat{m}_j$ of the table m_j. In the last case, l_j is $l_j = m_j \ \hat{m}_j$, where the two names are separated by a space, *not* by a comma.

$\mathcal{A}$ is a list of qualified attributes called *the target list* and $\mathcal{C}$ is a list of conditions as they appear as the subscript of a selection operation. When there is no ambiguity involving the tables where the attributes occur, the qualifications may be omitted.

The presence of the list of conditions $\mathcal{C}$ is optional. This means that

$$\textbf{SELECT } \mathcal{A} \textbf{ FROM } \mathcal{T}$$

is also a legal SELECT phrase.

If only one table is involved in a **SELECT** phrase and we need to consider all its attributes in the target list, then we can replace $\mathcal{A}$ by $*$.

The execution of an SQL phrase consists, at least from a conceptual point of view,[7] of the execution of three steps:

1. Compute the product $m = l_1 \times \cdots \times l_k$; if $l_j = m_j \ \hat{m}_j$, consider in the product the alias $\hat{m}_j$ rather than the table m_j.

2. Apply the selection specified by $\mathcal{C}$.

3. Project on the fields contained by the list $\mathcal{A}$.

Example 2.12.28 For the query discussed in Example 2.12.24, we use the following SELECT phrase in order to implement the second solution discussed in that example:

SELECT PNAME **FROM** COURSES, PROFS, GRADES
               WHERE COURSES.C#=PROFS.C# **AND**
               PROFS.C#=GRADES.C# **AND**
               GRADES.STUDENTN='Albert' **AND**
               COURSES.CREDITS = 4.

---

[7]The real execution plan of the query is set by a component of the database system called the *query optimizer*. This is, however, not of immediate interest for the database user.

Example 2.12.29 The query from Example 2.12.25 is solved by the following SELECT phrase:

SELECT PA.C#, PB.C# **FROM** PROFS PA, PROFS PB
     WHERE PA.C# = PB.C# **AND**
     PA.PNAME < PB.PNAME.

Note that two aliases of the table PROFS, PA and PB, are created in this phrase. All rows of the new table are compared, and only those rows of the product $PA \times PB$ containing the same course numbers and distinct professors' names (in alphabetical order) are retained.

SELECT phrases can appear as *subselects* in the condition C of another SELECT phrase. We assume that a SELECT phrase enclosed between parentheses returns a *set of values* rather than a table.

Example 2.12.30 Consider the following subselect:

     ... (**SELECT** STUDENTN **FROM** GRADES **WHERE** C#=110).

This would return to us the set of all student names enrolled in course 110.

Sets generated by subqueries can be involved in selections. A condition such as

$$v \text{ IN (SELECT } ...)$$

will be satisfied by those tuples whose component corresponding to the attribute v will be a member of the set generated by the subselect.

Example 2.12.31 This technique is illustrated by the following phrase:

SELECT STUDENTN **FROM** GRADES
     WHERE C# **IN**
     (**SELECT** C# **FROM** PROFS
     WHERE PNAME ='Jones')

The subselect generates the set of courses taught by Professor Jones; a row of the table GRADES satisfies the condition of the selection if its $C\#$-component is a member of the set of course numbers taught by Professor Jones, which means that the main SELECT phrase will return the names of students who get a grade in a course taught by Professor Jones.

Since a course may be taught by several professors, note that a student need not get his or her grade from Professor Jones in order to have his or her name on the list.

Subselects can be made dependent on a piece of information transmitted from the calling SELECT phrase.

Example 2.12.32 This feature of SQL is illustrated by another solution to the query considered in the previous example.

SELECT STUDENTN **FROM** GRADES
               WHERE 'Jones' IN
               (**SELECT** PNAME **FROM** PROFS
               WHERE C# = GRADES.C#).

For each student whose name appears in the GRADES table, we compute in the subselect the set of professors' names who are teaching the course that the student takes. The course numbers are transmitted to the subselect by the main query through the parameter GRADES.C# of the subselect.

Example 2.12.33 Assume that we choose a student name from the table GRADES, and we denote this student name by GRADES.STUDENTN. To determine the courses in which this student is enrolled, we operate on another copy GRADESA of the table GRADES. The following subselect

(**SELECT** C# **FROM** GRADES GRADESA
               WHERE GRADESA.STUDENTN =
               GRADES.STUDENTN)

depends on GRADES.STUDENTN. The solution of this query is given by

SELECT STUDENTN **FROM** GRADES
               WHERE C# IN
               (**SELECT** C# **FROM** GRADES GRADESA
               WHERE GRADESA.STUDENTN =
               GRADES.STUDENTN).

Subselects can be tested in SQL for emptiness by

$$\textbf{NOT EXISTS (SELECT } \dots)$$

or for nonemptiness by

$$\textbf{EXISTS (SELECT } \dots).$$

Such tests help in implementing in SQL the difference operation from relational algebra.

Example 2.12.34 Determine the names of students who do not take any course with Professor Jones. In relational algebra, we can solve this problem by the following computation:

$$GRADES;$$
$$T1 := GRADES[STUDENTN];$$
$$T2 := sel_{PNAME='Jones'}(PROFS);$$
$$T3 := GRADES \bowtie T2;$$
$$T4 := T3[STUDENTN];$$
$$T5 := T1 - T4; .$$

In SQL, this can be done by requiring that the set of courses taken by a student with Professor Jones be empty:

SELECT STUDENTN **FROM** GRADES
               WHERE NOT EXISTS
               (SELECT STUDENTN **FROM**
               GRADES GRADESA, PROFS **WHERE**
               GRADESA.STUDENTN = GRADES.STUDENTN
               AND GRADESA.C# = PROFS.C# **AND**
               PROFS.PNAME='Jones').

Two SQL phrases that return the same target lists can be involved in a UNION operation.

Example 2.12.35 Suppose we need to determine the courses that give four credits or are taught by Professor Smith. This can be solved in SQL by the following construction:

SELECT C# **FROM** COURSES
               WHERE CREDITS=4
UNION
SELECT C# **FROM** PROFS
               WHERE PNAME='Smith'.

SQL is capable of performing insertion, deletions, or updates on tables. In order to perform an insertion into a table $\tau = (m, i_1 \cdots i_n, \rho)$ the following INSERT command must be used:

      INSERT INTO $m(i_1, \ldots, i_n)$ **VALUES** $(v_1, \ldots, v_n)$.

This inserts in the table m whose schema is $i_1 \cdots i_n$ the n-tuple $(v_1, \ldots, v_n)$.

Example 2.12.36 In order to insert the 4-tuple

$$(330, \text{Databases}, 3, \text{W2310})$$

into the table COURSES, we write

INSERT INTOCOURSES (C#,CNAME,CREDITS,ROOM)
VALUES   (330,'Databases',3,'W2310').

Tuples may be added to a table starting from existing data. This can be accomplished by an INSERT having the following form:

INSERT INTO$m(i_1, \ldots, i_n)$
SELECT   ...,

where the SELECT is extracting data from an existing table.

Example 2.12.37 We can create a table for the three-credit courses by

CREATE TABLE C3CR (C# **INTEGER2**, CNAME **VCHAR(40)**
               ROOM **VCHAR(5)**).

Then, the triples corresponding to all three-credit courses existent in the table COURSES can be added to the table C3CR by

INSERT INTOC3CR(C#, CNAME, ROOM)
SELECT C# ,CNAME, ROOM **FROM** COURSES
WHERE CREDITS=3.

Data can be removed from a table m using a DELETE statement. The general form of such a statement is

DELETE FROM m WHERE $\mathcal{C}$,

where $\mathcal{C}$ is a condition that qualifies the tuples to be deleted. The presence of the condition is optional; in other words, we can have an unqualified deletion:

DELETE FROM m.

The effect of this statement is to delete *all* rows of the table m.

Example 2.12.38 The following DELETE can be used to remove the tuples that refer to the student whose number is 9008 from the table GRADES:

DELETE FROM GRADES **WHERE** STUDENTN=9008.

Example 2.12.39 To remove from the table COURSES all courses taught by professors Smith or Wesson, we can execute the following DELETE:

DELETE FROM COURSES **WHERE**
C# **IN (SELECT** C# **FROM** PROFS
WHERE PNAME='Smith' **OR**
PNAME='Wesson').

The SQL UPDATE command changes values in existing tables. The general form of an update performed on a table $\tau = (m, H, \rho)$ is

UPDATE m
SET $i_1 = e_1, \ldots, i_n = e_n$
WHERE $\mathcal{C}$,

where $\mathcal{C}$ is an optional condition, $H = i_1 \cdots i_n$ and $e_1, \ldots, e_n$ are values or expressions to be assigned to the corresponding components of the tuples of m that satisfy the condition $\mathcal{C}$.

Example 2.12.40 In order to increase by 1 the number of credits of courses above the 300 level, we may execute the following UPDATE:

UPDATE COURSES
SET CREDITS = CREDITS + 1
WHERE C# > 300.

Example 2.12.41 To change the grades of all students taking a course with Professor Smith to an A, we can execute the following update:

UPDATE GRADES
>   **SET** GRADE='A'
>   **WHERE** C# **IN**
>   (SELECT C# **FROM** PROFS
>   **WHERE**
>   PNAME='Smith').

The relationship between SQL and relational algebra is stated by the following theorem.

Theorem 2.12.42 *Let* $\tau' = (m', G, \rho)$ *and* $\tau'' = (m'', H, \sigma)$ *be two tables and assume that the table* $\tau = (m, K, \theta)$ *is constructed starting from the set of tables* $\{\tau', \tau''\}$ *by applying one of the operations of relational algebra or by renaming. In each case, the same table can be constructed in SQL.*

Proof. We need to consider six cases corresponding to renaming and to each of the five basic operations of relational algebra.

1. Suppose that τ is obtained from τ' by renaming, $K = i_1 \cdots i_n$ and $G = j_1 \cdots j_n$. In this case, we create τ and then insert the rows of τ' into τ:

CREATE TABLE $m(i_1\ d_1, \ldots, i_n\ d_n)$;
INSERT INTO $m(i_1, \ldots, i_n)$
>   **SELECT** $j_1, \ldots, j_n$ **FROM** m'.

2. If τ is created from τ' and τ'' through union, then the tables τ and τ'' must be compatible, that is, $G = H = K = i_1 \cdots i_n$. After creating the table τ, we can fill in this table with the tuples that belong to either τ' or to τ'' by using the following INSERT statements:

INSERT INTO $m(i_1, \ldots, i_n)$
>   **SELECT** $i_1, \ldots, i_n$ FROM m';
INSERT INTO $m(i_1, \ldots, i_n)$
>   **SELECT** $i_1, \ldots, i_n$ FROM m''.

3. For the case when τ is built from τ' and τ'' by difference, we must have again the compatibility between the tables τ' and τ''. After creating the table τ, we use two SQL statements: an insertion followed by a deletion:

INSERT INTO $m(i_1, \ldots, i_n)$
>   **SELECT** $i_1, \ldots, i_n$ **FROM** m';
DELETE FROM m **WHERE**
>   **EXISTS(SELECT * FROM** m''
>   **WHERE** $i_1 = m''.i_1$ **AND**
>   ...**AND** $i_n = m''.i_n$).

4. Let us assume now that τ is created from τ' and τ'' by product. If $G = g_1 \cdots g_m$ and $H = h_1 \cdots h_n$, then $K = m'.g_1 \cdots m'.g_m m''.h_1 \cdots m''.h_n$. In this case, after creating the table τ, we use the following INSERT statement:

INSERT INTO $m(m'.g_1 \cdots m'.g_m m''.h_1 \cdots m''.h_n)$
SELECT $m'.g_1 \cdots m'.g_m m''.h_1 \cdots m''.h_n$
          FROM m', m''.

5. When τ is obtained from τ' by projection, we have $K \subseteq G$. If $G = i_1 \cdots i_n$ and $K = i_{p_1} \cdots i_{p_r}$, then we create the table τ and then use the following INSERT:

INSERT INTO $m(i_{p_1}, \cdots, i_{p_r})$
          SELECT $i_{p_1} \cdots i_{p_r}$ **FROM**
          m'.

6. Finally, if τ is obtained from τ' by a selection specified by the condition $\mathcal{C}$, then we can use the insert

INSERT INTO $m(i_1, \cdots, i_n)$
          SELECT $i_1, \ldots, i_n$ **FROM** m'
          WHERE $\mathcal{C}$. ∎

SQL, like many other relational query languages, offers the possibility of using a collection of *built-in functions*. Among the most important ones are sum, avg, max, min, count, which compute the sum, the average, the maximum, and the minimum of a set of values or count the number of tuples of a certain set, respectively.

These functions can be used in SELECT phrases.

Example 2.12.43 Let us determine how many courses the student whose number is 2341 is taking; we can use the following phrase:

SELECT COUNT(C#) **FROM** GRADES
          WHERE STUDENT# $=$'2341'.

Consider the relation ρ on the scheme R and let $K \subseteq R$. Define the equivalence θ_K on ρ (that is, the set of n-tuples that constitute the relation ρ) as $(u, v) \in \theta_K$ if $u[K] = v[K]$. The equivalence classes of the set ρ/θ_K are *the groups* determined by the attributes of K. Namely, the group of a tuple t consists of all tuples that have the same components on the attributes of K as t does. SQL is capable of partitioning a relation based on any of its attributes and then computing a single-valued characteristic for any of these classes using the option **GROUP BY** in the **SELECT** phrases. If the option **WHERE** is used, then the option **GROUP BY** may be used after the option **WHERE**. In this situation, the quotient set will be computed after applying the selection specified by **WHERE**.

Example 2.12.44 Suppose that we need to determine the number of courses passed by the students in the database. Of course, we group the tuples of the table in the equivalence classes of the relation $\theta_{STUDENTN}$. This is accomplished using the option GROUP BY of the select phrase:

SELECT STUDENTN, COUNT(C#) **FROM** GRADES
                              WHERE GRADE !='F'
                              GROUP BY STUDENTN.

If the table GRADES has the content as shown,

GRADES

C#	STUDENTN	STUDENT#	GRADE
110	John Albert	2341	A
110	Mary Gallagher	5641	D
110	Kathy Driscoll	9008	B
110	John Albert	2341	F
210	Mary Gallagher	4430	F
210	James Dudley	2345	A
210	James Dudley	7655	A
310	Kathy Driscoll	0988	A
310	James Dudley	5448	A

then, from a conceptual point of view, the table is divided into the following groups after eliminating those tuples whose *GRADE* component is *F*.

GRADES

C#	STUDENTN	STUDENT#	GRADE
110	John Albert	2341	A
110	Mary Gallagher	5641	D
110	Kathy Driscoll	9008	B
310	Kathy Driscoll	0988	A
210	James Dudley	2345	A
210	James Dudley	7655	A
310	James Dudley	5448	A

As a result, we will obtain the following table.

STUDENTN

John Albert	1
Mary Gallagher	1
Kathy Driscoll	2
James Dudley	3

SQL requires that the the queries return *simple values* for each group. For instance, the SQL phrase

SELECT STUDENTN, C# **FROM** GRADES
                  WHERE GRADE !='F'
                  GROUP BY STUDENTN

is unacceptable because the answer would return *a set of course numbers* instead of the cardinality of that set (which is a "simple" piece of data).

Once a grouping is accomplished through the use of GROUP BY, we can choose certain groups using the HAVING option. If we need to have only the students who passed at least two courses, we can use the following SELECT phrase

SELECT STUDENTN, COUNT(C#) **FROM** GRADES
                  WHERE GRADE !='F'
                  GROUP BY STUDENTN
                  HAVING COUNT(C#) > =2.

As a result, we have the following table

STUDENTN

Kathy Driscoll	2
James Dudley	3

2.13   Exercises and Supplements

Relations

1. What are $\text{Dom}(\emptyset)$ and $\text{Ran}(\emptyset)$?

2. Let $M = \{a, b, c\}$. Consider the relations

$$\rho = \{(a,b), (b,c), (c,a)\} \text{ and } \sigma = \{(a,c), (b,a)\}$$

on M. Compute $\rho \bullet \sigma$, $\sigma \bullet \rho$, $\rho \bullet \rho$, and $\sigma \bullet \sigma$. Verify that $(\rho \bullet \sigma)^{-1} = \sigma^{-1} \bullet \rho^{-1}$ and that $\rho \bullet (\sigma \bullet \rho) = (\rho \bullet \sigma) \bullet \rho$.

3. Let ρ be a relation.

   (a) Show that ρ is a function if and only if there exists a set A such that
$$\rho^{-1} \bullet \rho \subseteq \iota_A.$$

   (b) Show that ρ is one-to-one if and only if there exists a set A such that
$$\rho \bullet \rho^{-1} \subseteq \iota_A.$$

4. Let ρ be a relation.

(a) Show that for any set A, $A \subseteq \text{Dom}(\rho)$ if and only if $\iota_A \subseteq \rho \bullet \rho^{-1}$. (This result generalizes Theorem 2.2.22, part 3.) Conclude that for any relation ρ, $\iota_{\text{Dom}(\rho)} \subseteq \rho \bullet \rho^{-1}$.

(b) Use the previous part and part 2 of Theorem 2.2.22 to conclude that ρ is one-to-one if and only if $\rho \bullet \rho^{-1} = \iota_{\text{Dom}(\rho)}$.

(c) Show that for any set B, $B \subseteq \text{Ran}(\rho)$ if and only if $\iota_B \subseteq \rho^{-1} \bullet \rho$. (This result generalizes Theorem 2.2.22, part 4.) Conclude that for any relation ρ, $\iota_{\text{Ran}(\rho)} \subseteq \rho \bullet \rho^{-1}$.

(d) Use the previous part and part 1 of Theorem 2.2.22 to conclude that ρ is a function if and only if $\rho^{-1} \bullet \rho = \iota_{\text{Ran}(\rho)}$.

5. Let ρ and σ be relations and B and C sets. Prove the following results, which generalize parts 3 and 4 of Theorem 2.2.23.

(a) If $A \subseteq \text{Dom}(\rho)$ and $\text{Ran}(\rho) \subseteq \text{Dom}(\sigma)$, then $A \subseteq \text{Dom}(\rho \bullet \sigma)$.

(b) If $C \subseteq \text{Ran}(\sigma)$ and $\text{Dom}(\sigma) \subseteq \text{Ran}(\rho)$, then $C \subseteq \text{Ran}(\rho \bullet \sigma)$.

6. Using Exercise 3, prove part 1 of Theorem 2.2.23.

Solution: Suppose that ρ and σ are functions. Then, we have

$$
\begin{aligned}
(\rho \bullet \sigma)^{-1} \bullet (\rho \bullet \sigma) &= (\sigma^{-1} \bullet \rho^{-1}) \bullet (\rho \bullet \sigma) \\
&= \sigma^{-1} \bullet ((\rho^{-1} \bullet \rho) \bullet \sigma) \\
&\subseteq \sigma^{-1} \bullet (\iota_{\text{Ran}(\rho)} \bullet \sigma) && \text{by Theorem 2.2.22,} \\
& && \text{part 1 and} \\
& && \text{monotonicity} \\
&\subseteq \sigma^{-1} \bullet \sigma && \text{by Theorem 2.2.15,} \\
& && \text{part (8) and} \\
& && \text{monotoncity} \\
&\subseteq \iota_{\text{Ran}(\sigma)}.
\end{aligned}
$$

Hence, by Exercise 3, $\rho \bullet \sigma$ is a function.

7. Let ρ and σ be two relations. Prove that

(a) $\rho \bullet \iota_{\text{Dom}(\sigma)} \subseteq \iota_{\text{Dom}(\rho \bullet \sigma)} \bullet \rho$,

(b) $\iota_{\text{Ran}(\rho)} \bullet \sigma \subseteq \sigma \bullet \iota_{\text{Ran}(\rho \bullet \sigma)}$.

8. Let ρ be a relation and let S be a set. Show that $\rho|_S = (\iota_S \bullet \rho) \bullet \iota_S$. Using Theorem 2.2.23, conclude that if ρ is a function, then so is $\rho|_S$, and if ρ is a one-to-one relation, then so is $\rho|_S$.

Functions

9. (a) Show that for any function f we have

$$f = \{(a, f(a)) \mid a \in \text{Dom}(f)\}.$$

(b) Prove that for any functions f and g we have $f = g$ if and only if $\text{Dom}(f) = \text{Dom}(g)$, and for every $a \in \text{Dom}(f)$ we have $f(a) = g(a)$.

10. Under which conditions on A and B do we have $A \longrightarrow B = A \rightsquigarrow B$?

11. Prove that the following mappings are bijections:

(a) $f : \mathbf{Z} \longrightarrow \mathbf{Z}$, where $f(n) = (-1)^{|n|}n$ for every $n \in \mathbf{Z}$.

(b) $g : \mathbf{N} \longrightarrow \mathbf{N}$, where

$$g(n) = \begin{cases} n+1, & \text{if } n \text{ is even,} \\ n-1, & \text{if } n \text{ is odd.} \end{cases}$$

(c) $h : \mathbf{N} \times \mathbf{N} \longrightarrow \mathbf{N}$, where $f(m,n) = 2^m(2n+1)-1$, for $m, n \in \mathbf{N}$.

12. A function $f : \mathbf{R} \longrightarrow \mathbf{R}$ is called strictly increasing (decreasing) if $a < b$ implies $f(a) < f(b)$ $(f(a) > f(b)$, respectively). Prove that if f is strictly increasing or strictly decreasing then f is one-to-one.

13. Let a, b, c be three real numbers. Consider the function $g : \mathbf{R} \longrightarrow \mathbf{R}$ defined by $g(x) = ax^2 + bx + c$ for every $x \in \mathbf{R}$.

(a) Prove that if $a \neq 0$ then g is not one-to-one and if $a = 0$ and $b \neq 0$ then g is one-to-one.

(b) Prove that the range of g is the set $\{x \in \mathbf{R} \mid \frac{4ac-b^2}{4a} \leq x\}$ if $a > 0$. Formulate a similar result for $a < 0$.

14. Let $f : \mathbf{R} \longrightarrow \mathbf{R}$ be the function defined by

$$f(x) = \frac{b}{1 + \sqrt{a}} + x\sqrt{a},$$

where $a, b \in \mathbf{R}$ and $a > 0$. Prove that $(f \circ f)(x) = ax + b$, for $x \in \mathbf{R}$.

15. Prove that the following assertions are equivalent for any mapping $f : A \longrightarrow B$:

(i) f is one-to-one;

(ii) for every set C and mappings $g_1, g_2 : C \longrightarrow A$, the equality $fg_1 = fg_2$ implies $g_1 = g_2$.

16. Prove that the following assertions are equivalent for any mapping $f : A \longrightarrow B$:

(i) f is onto;

(ii) for every set D and mappings $k_1, k_2 : B \longrightarrow D$, the equality $k_1f = k_2f$ implies $k_1 = k_2$.

17. Prove that if a mapping $f : \mathbf{N} \longrightarrow \mathbf{N}$ has more than one left inverse then it has an infinite number of left inverses.

18. Construct a mapping $f : \mathbf{N} \longrightarrow \mathbf{N}$ that has exactly two right inverses.

    Hint: Consider the function $f : \mathbf{N} \longrightarrow \mathbf{N}$ defined by $f(n) = 0$ if $n \leq 1$ and $f(n) = n - 1$ if $n \geq 2$.

19. Let $f : A \longrightarrow B$. Prove that f is a bijection if and only if f has a unique right inverse. In addition, show that if A is not a one-element set then f is a bijection if and only if f has a unique left inverse.

20. Show that if $S \subseteq T$ and f is a function then $f{\upharpoonright}S = (f{\upharpoonright}T){\upharpoonright}S$.

21. Consider the function $f : \mathbf{R} \longrightarrow \mathbf{R}$ such that $f(x) = x/2$. Show that there is a set S such that $f|_S \neq f{\upharpoonright}S$.

    Solution: Take $S = \mathbf{N}$. Then, $f{\upharpoonright}\mathbf{N}$, is a total function on $\mathbf{N}$, while $f|_\mathbf{N}$ is a partial function on $\mathbf{N}$ whose domain is the set of even natural numbers.

22. Two functions f and g are *compatible* if $f(a) = g(a)$ for all $a \in \mathrm{Dom}(f) \cap \mathrm{Dom}(g)$.

    (a) Prove that two functions whose domains are disjoint are compatible.

    (b) A collection of functions $\mathcal{F}$ is called a *family of compatible functions* if any two functions $f, g \in \mathcal{F}$ are compatible. Prove that a family of functions $\mathcal{F}$ is a family of compatible functions if and only if $\bigcup \mathcal{F}$ is a function that is an extension of every member of $\mathcal{F}$.

23. Let f be a function.

    (a) Show that for all $a \in \mathrm{Dom}(f)$, $[a \rightarrow f(a)]f = f$;

    (b) What is $[a \rightarrow b]([a \rightarrow c]f)(a)$?

    (c) Prove that if $a_1 \neq a_2$ then
    $$[a_1 \rightarrow b_1]([a_2 \rightarrow b_2]f) = [a_2 \rightarrow b_2]([a_1 \rightarrow b_1]f).$$

24. Simplify the following expressions:

    (a) $(\lambda n.n + 2)(5)$,

    (b) $(\lambda(m, n).m^2 + n)(2, 3)$,

    (c) $(\lambda m.(\lambda n.m^2 + n))(2)(3)$,

    (d) $(\lambda n.(\lambda m.m + 2)(3))(12)$,

    (e) $(\lambda x.f(x))(y^2)$,

(f) $(\lambda x.f(x^2))(y)$,

(g) $(\lambda x.f(y))(z)$, and

(h) $(\lambda x.f(x))$.

25. Let A, B, C be three sets. Define a function

$$\Psi : (A \times B \longrightarrow C) \longrightarrow (A \longrightarrow (B \longrightarrow C))$$

by letting, for each $f : A \times B \longrightarrow C$, $\Psi(f)$ be the function $g : A \longrightarrow (B \longrightarrow C)$, where, for all $a \in A$, $g(a)$ is the function $h : B \longrightarrow C$, where $h(b) = f(a, b)$ for all $b \in B$. More concisely, $\Psi(f)(a)(b) = f(a, b)$. Prove that Ψ is a bijection. (On page 155, we return to this idea.)

Hint: Define

$$\Phi : (A \longrightarrow (B \longrightarrow C)) \longrightarrow (A \times B \longrightarrow C)$$

by letting $\Phi(f)(a, b) = f(a)(b)$ for all $f : A \longrightarrow (B \longrightarrow C)$, $a \in A$ and $b \in B$, and show that Φ is both a right and left inverse for Ψ.

26. The function Ψ of Exercise 25 can be expressed using lambda notation as

$$\Psi = (\lambda f.(\lambda a.(\lambda b.f(a, b)))).$$

Express Φ, the inverse of Ψ, in lambda notation and then use these expressions to demonstrate again that $\Psi(\Phi(f)) = f$ and $\Phi(\Psi(g)) = g$ for all $f \in A \longrightarrow (B \longrightarrow C)$ and $g \in A \times B \longrightarrow C$.

27. (a) Prove that for any sets A and B, there is a bijection between the sets $\mathcal{P}(A \times B)$ and $A \longrightarrow \mathcal{P}(B)$.

    Hint: For every $R \in \mathcal{P}(A \times B)$, define $\Psi(R)(a) = \{b \in B \mid (a, b) \in R\}$ and show that Ψ is a bijection.

    (b) Prove that the existence of the bijection mentioned above can be obtained from Exercise 25 and Theorem 2.3.16.

28. Let A and B be two sets and let $\perp$ be an object that does not belong to B. Define $\Psi : (A \rightsquigarrow B) \longrightarrow (A \longrightarrow (B \cup \{\perp\}))$ by letting

$$\Psi(f)(a) = \begin{cases} f(a) & \text{if } a \in \text{Dom}(f), \\ \perp & \text{otherwise,} \end{cases}$$

for each $f \in A \rightsquigarrow B$ and $a \in A$. Show that Ψ is a bijection. What is Ψ^{-1}?

29. Let M be a set. Prove that for $P_1, P_2 \in \mathcal{P}(M)$ and $x \in M$

we have the following identities (where characteristic functions are taken with respect to M):

$$\chi_{P_1 \cap P_2}(x) = \chi_{P_1}(x)\chi_{P_2}(x),$$
$$\chi_{P_1 \cup P_2}(x) = \chi_{P_1}(x) + \chi_{P_2}(x) - \chi_{P_1}(x)\chi_{P_2}(x),$$
$$\chi_{M-P_1}(x) = 1 - \chi_{P_1}(x).$$

30. Programs written in high-level languages involve variable names, usually called *identifiers*. Identifiers can be regarded as mere notations for memory locations. In turn, locations may contain values.

    Let I be the set of identifiers of a program P. Assume that the set of memory locations of the computing system is L and the set of values that can be contained by the locations is V. An *environment* of P is a mapping $e : I \longrightarrow L$; a *store* is a mapping $s : L \longrightarrow V$.

    For an identifier i, $e(i)$ is the location denoted by i; for a location l, $s(l)$ is the content of l. Therefore, $s(e(i))$ is the value corresponding to the identifier i. We assume that the computing system has infinitely many locations (which, of course, is nothing but a theoretical hypothesis) and that there exists an *addressing mapping* defined as a bijection $a : L \longrightarrow \mathbf{N}$.

    Let T be a nonempty set of data items and assume that

    $$(\mathbf{N} \cup \{-1\}) \cup (T \times (\mathbf{N} \cup \{-1\})) \subseteq V.$$

    (a) Let $x, y \in I$. Explain why the execution of the assignment statement $x \leftarrow y$ in environment e and store s results in the new store $[e(x) \rightarrow s(e(y))]s$.

    (b) An identifier p can be considered to be a pointer if $s(e(p)) \in \mathbf{N}$. What is the meaning of $s(a^{-1}(s(e(p))))$ if p is a pointer?

    (c) An identifier i represents a linked list of elements $t_0, \ldots, t_{\ell-1}$ of T if either $\ell = 0$ and $s(e(i)) = -1$ or $\ell > 0$, and there is a sequence of locations $(l_0, \ldots, l_{\ell-1})$ such that $s(e(i)) = a(l_0)$, $s(l_k) = (t_k, a(l_{k+1}))$ for every k, $0 \le k \le \ell - 2$, and $s(l_{\ell-1}) = (t_{\ell-1}, -1)$.
    Let p be a pointer such that $s(e(p)) = a(l_0)$. Consider the sequence of stores $(s_0, \ldots, s_m)$, where $m < \ell - 1$, $s_0 = s$, and

    $$s_{k+1} = [e(p) \rightarrow \pi_2(s(a^{-1}(s(e(p)))))]s_k,$$

    where $\pi_2 : T \times \mathbf{N} \longrightarrow \mathbf{N}$ is given by $\pi_2(t, n) = n$ for every $(t, n) \in T \times \mathbf{N}$. What does p point to in the store s_k?
    Modify the definition of the linked list given above to represent a circular linked list.

Sequences, Words, and Matrices

31. Show that a sequence of length n is a finite set with cardinality n. (Hence, the notation $|x|$ for the length of a sequence x does not conflict with the notation used to denote cardinalities.)

32. Consider an infinite sequence of subsets of a set A:

$$(A_0, A_1, \ldots, A_n, \ldots),$$

and the sequences $(B_0, B_1, \ldots, B_n, \ldots)$ and $(C_0, C_1, \ldots, C_n, \ldots)$ defined by

$$B_n = \bigcup\{A_p \mid p \geq n\},$$
$$C_n = \bigcap\{A_p \mid p \geq n\},$$

for every $n \in \mathbf{N}$.

(a) Prove that if $m \leq n$ then $B_m \supseteq B_n$ and $C_m \subseteq C_n$.

(b) Prove that

$$\bigcup\{C_n \mid n \in \mathbf{N}\} \subseteq \bigcap\{B_n \mid n \in \mathbf{N}\}$$

and give an example of a sequence of sets $(A_0, A_1, \ldots, A_n, \ldots)$ for which this is a strict inclusion.

The sets $\bigcup\{C_n \mid n \in \mathbf{N}\}$ and $\bigcap\{B_n \mid n \in \mathbf{N}\}$ are known as the *lower limit of the sequence* $(A_0, A_1, \ldots, A_n, \ldots)$ and the *upper limit of the sequence* $(A_0, A_1, \ldots, A_n, \ldots)$, and they are denoted by $\lim\inf_{n\in\mathbf{N}} A_n$ and by $\lim\sup_{n\in\mathbf{N}} A_n$, respectively.

(c) For $a \in A$, define the set $I_a = \{n \mid a \in A_n\}$. Prove that $a \in \lim\sup_{n\in\mathbf{N}} A_n$ if and only if the set I_a is infinite. Prove that $a \in \lim\inf_{n\in\mathbf{N}} A_n$ if and only if the set $\mathbf{N} - I_a$ is finite.

(d) A sequence $(A_0, A_1, \ldots, A_n, \ldots)$ is convergent if

$$\lim\inf_{n\in\mathbf{N}} A_n = \lim\sup_{n\in\mathbf{N}} A_n.$$

Prove that if $A_n \subseteq A_{n+1}$ for every $n \in \mathbf{N}$ then the sequence $(A_0, A_1, \ldots, A_n, \ldots)$ is convergent. The same holds if $A_n \supseteq A_{n+1}$ for every $n \in \mathbf{N}$.

33. If x is a finite sequence, (for example, $|x| = n$), then we define the *reversal* of x, x^R to be the sequence of length n such that

$$x^R(i) = x(n - i - 1)$$

for $0 \leq i \leq n - 1$.

Prove that

(a) $\lambda^R = \lambda$,

(b) $(x^R)^R = x$ for all finite sequences x, and

(c) $(x \cdot y)^R = y^R \cdot x^R$ for all finite sequences x and y.

34. A word on an alphabet V is *square-free* if it contains no infix of the form xx, where $x \in V^+$.

   (a) List all square-free words of length three over the alphabet $\{a, b\}$.

   (b) Show that for the alphabet $\{a, b\}$ there are no square-free words of length at least equal to 4.

   (c) Let $v \in V^*$. A word $w \in W^*$ is *an interpretation* of v in W if $|w| = |v|$ and $v(i) \neq v(j)$ implies $w(i) \neq w(j)$ for every i and j, $0 \leq i, j \leq |v| - 1$. For instance, if $V = \{a_1, a_2\}$ and $W = \{b_1, b_2, b_3\}$, the word $w = b_1 b_3 b_2 b_2 b_3$ is an interpretation of $v = a_1 a_2 a_1 a_1 a_2$, while $w' = b_1 b_3 b_2 b_3 b_2$ is not because $v(1) \neq v(3)$ while $w'(1) = w'(3)$.

   Prove that if x is square-free then so is every interpretation of x.

35. Prove that for every language L, $L^* = L^+ \cup \{\lambda\}$. Conclude that if $\lambda \in L$, then $L^* = L^+$. Also, if $\lambda \notin L$, then $L^+ = L^* - \{\lambda\}$.

36. (a) Under what conditions on a language L is $L^+ = L$?

    (b) Under what conditions on a language L is $L^* = L$?

37. Prove each of the following identities:

$$\begin{aligned}
\emptyset^* &= \{\lambda\}, \\
\{\lambda\}^* &= \{\lambda\}, \\
\emptyset^+ &= \emptyset, \\
\{\lambda\}^+ &= \{\lambda\}, \\
\overline{\emptyset}_V &= V^*, \\
\overline{\{\lambda\}}_V &= V^+.
\end{aligned}$$

38. Let V be an alphabet and let L and K be languages over V. Prove that the following identities hold.

$$\begin{aligned}
(L^*)^+ &= L^*, \\
(L^+)^* &= L^*, \\
(L^+)^+ &= L^+, \\
(L \cup K)^* &= (L^* K^*)^* = (K^* L^*)^*.
\end{aligned}$$

39. Prove that for any languages L, K, and H the following inclusions hold (assume that complementation is with respect to the alphabet V):

$$(L \cap K)H \subseteq LH \cap KH,$$
$$H(L \cap K) \subseteq HL \cap HK,$$
$$(L - K)H \supseteq LH - KH,$$
$$H(L - K) \supseteq HL - HK,$$
$$(L \cup K)^* \supseteq L^* \cup K^*,$$
$$(L \cap K)^* \subseteq L^* \cap K^*,$$
$$(L \cup K)^+ \supseteq L^+ \cup K^+,$$
$$(L \cap K)^+ \subseteq L^+ \cap K^+,$$
$$\overline{L^*} \subseteq (\overline{L})^*,$$
$$\overline{L^+} \subseteq (\overline{L})^+,$$
$$(LH)^* \subseteq (L \cup H)^*.$$

Give examples of languages for which these inclusions are strict.

40. If L is a language, then we define a language L^R by

$$L^R = \{x^R \mid x \in L\}.$$

Prove the following identities involving reversal of languages, where L and K can be any languages and complementation is with respect to some alphabet V.

$$\emptyset^R = \emptyset,$$
$$\{\lambda\}^R = \{\lambda\},$$
$$(L^R)^R = L,$$
$$(LK)^R = K^R L^R,$$
$$(L \cup K)^R = L^R \cup K^R,$$
$$(L \cap K)^R = L^R \cap K^R,$$
$$(L - K)^R = L^R - K^R,$$
$$(L^*)^R = (L^R)^*,$$
$$(L^+)30g^R = (L^R)^+,$$
$$(\overline{L})^R = \overline{L^R}.$$

41. Let L and K be two languages over an alphabet V. Define $f : L \times K \longrightarrow LK$ by $f(u, v) = uv$ for all $(u, v) \in L \times K$. Show that f is surjective. Give examples to show that f can be, but is not necessarily, a bijection.

42. A language L is *catenatively independent* if no word in L can be expressed as a product of at least two words in L. Prove that the language $\{a, aba, baba, bb, bbba\}$ is catenatively independent. Also, the language $\{x \in V^* \mid |x| = n\}$ is catenatively independent for any n. Show that no catenatively independent language may contain λ.

43. A language L over an alphabet V is a *code* if for any words $x_1, \ldots, x_n$, $y_1, \ldots, y_m$ from L the equality $x_1 \cdots x_n = y_1 \cdots y_m$ implies $m = n$ and $x_i = y_i$ for $1 \leq l \leq n$.

If L is a code, then every word from L^* can be decoded in a unique way as a product of words of L.

Prove that a language L over the alphabet V is a code if and only if L is catenatively independent and $L^* w \cap L^* \neq \emptyset$, $w L^* \cap L^* \neq \emptyset$ for a word $w \in V^*$ imply $w \in L^*$.

Solution: Assume that L satisfies the conditions given. We need to show that L is a code. Note that $\lambda \notin L$ because of the catenative independence of L. Thus, if L would not be a code, we would have words $x_1, \ldots, x_n, y_1, \ldots, y_m$ from L with $x_1 \cdots x_n = y_1 \cdots y_m$ and $x_1 \neq y_1$. Indeed, if this is not the case with the initial words, we could use the cancellation property to arrive at the same conclusion on a shorter equality. Therefore, one of the words x_1, y_1 must be a proper prefix of the other. Without restricting the generality, we assume that $y_1 = x_1 z$, which gives, by the cancellation property $x_2 \cdots x_n = z y_2 \cdots y_m$. This means that $L^* z \cap L^* \neq \emptyset$ and $z L^* \cap L^* \neq \emptyset$; hence, $z \in L^*$ and $z \neq \lambda$. Since $y_1 = x_1 z$, this contradicts the catenative independence of L.

Conversely, assume that L is a code. The catenative independence of L is immediate. Suppose that $L^* w \cap L^* \neq \emptyset$ and $w L^* \cap L^* \neq \emptyset$ for a word $w \in V^*$. This means that we have words $u_1, \ldots, u_m, v_1, \ldots, v_n$ and $x_1, \ldots, x_p, y_1, \ldots, y_q$ in L such that

$$u_1 \cdots u_m w = v_1 \cdots v_n,$$
$$w x_1 \cdots x_p = y_1 \cdots y_q.$$

From the above equation, we obtain the equality

$$u_1 \cdots u_m y_1 \cdots y_q = v_1 \cdots v_n x_1 \cdots x_p.$$

The fact that L is a code implies $m + q = n + p$, and in addition, $u_1 = v_1, \ldots, y_q = x_p$.

We must have $m \leq n$, because if $m > n$, then $u_{n+1} \ldots u_m w = \lambda$, and this would imply $u_{n+1} = \cdots = u_m = w = \lambda$, which contradicts the catenative independence of the language L.

If $m = n$, then $w = \lambda \in L^*$; otherwise, $m < n$, and this implies $w = v_{m+1} \cdots v_n$, which gives $w \in L^*$.

44. Let $M \in M_{m \times n}(D)$ be a matrix. Find sets $S, T \subseteq \{1, \ldots, m\} \times \{1, \ldots, n\}$ such that the i-th row of M is the restriction of M to S and the j-th column of M is the restriction of M to T.

45. The *transpose* of an $m \times n$ matrix M on a set D, written M', is the $n \times m$ matrix whose i-th row is the i-th column of the matrix M. In other words, $m'_{i,j} = m_{j,i}$ for $1 \leq i \leq n$ and $1 \leq j \leq m$.

Prove that

(a) $(M + P)' = M' + P'$, for $M, P \in M_{m \times n}(\mathbf{R})$;

(b) $(MP)' = P'M'$, for $M \in M_{m \times n}(\mathbf{R})$ and $P \in M_{n \times p}(\mathbf{R})$.

Find similar properties for matrices in $M_{m \times n}(\{0, 1\})$.

Images of Sets Under Relations

46. Prove that for every relation ρ and sets A and B, we have

$$\rho^{-1}(\rho(A)) \supseteq A \cap \mathrm{Dom}(\rho),$$
$$\rho(\rho^{-1}(B)) \supseteq B \cap \mathrm{Ran}(\rho).$$

Conclude that if $A \subseteq \mathrm{Dom}(\rho)$ and $B \subseteq \mathrm{Ran}(\rho)$, then

$$\rho^{-1}(\rho(A)) \supseteq A,$$
$$\rho(\rho^{-1}(B)) \supseteq B.$$

47. Let ρ be a relation. Prove that

$$\rho(A \cap B) = \rho(A) \cap \rho(B),$$

for all sets A and B if and only if ρ is one-to-one.

48. Show that for any relation ρ and sets A and B, $\rho(A - B) \supseteq \rho(A) - \rho(B)$. Prove that ρ is one-to-one if and only if we have $\rho(A - B) = \rho(A) - \rho(B)$ for all sets A and B.

49. Show that for every relation ρ and sets A and B,

$$\rho(A \oplus B) \supseteq \rho(A) \oplus \rho(B).$$

Prove that ρ is one-to-one if and only if we have

$$\rho(A \oplus B) = \rho(A) \oplus \rho(B)$$

for all sets A and B.

50. Let ρ be a relation from A to B. Prove that ρ is onto if and only if $B - \rho(K) \subseteq \rho(A - K)$ for every $K \subseteq A$.

51. Every relation $\rho \subseteq A \times B$ generates a function $F_\rho : \mathcal{P}(A) \longrightarrow \mathcal{P}(B)$ defined by $F_\rho(K) = \rho(K)$ for every $K \in \mathcal{P}(A)$.

Prove that a function $F : \mathcal{P}(A) \longrightarrow \mathcal{P}(B)$ is generated by a relation $\rho \subseteq A \times B$ if and only if for any family $\mathcal{C}$ of subsets of A, we have

$$f\left(\bigcup \mathcal{C}\right) = \bigcup\{f(C) \mid C \in \mathcal{C}\}.$$

52. A mapping $f : A \longrightarrow A$ is called an *involution* if $f(f(x)) = x$ for all $x \in A$.

   (a) Prove that every involution on a set A is a bijection on A.

   (b) For an involution $f : A \longrightarrow A$, the mapping $Q_f : \mathcal{P}(A) \longrightarrow \mathcal{P}(A)$ is given by $Q_f(K) = A - f(K)$ for $K \in \mathcal{P}(A)$. Prove that

$$Q_f(Q_f(K)) = K,$$
$$Q_f(K \cup H) = Q_f(K) \cap Q_f(H),$$
$$Q_f(K \cap H) = Q_f(K) \cup Q_f(H).$$

   (c) Give a characterization of those subsets K of A for which

$$Q_f(K) \cap K = \emptyset.$$

53. Let A, B be two finite sets and let $\rho \subseteq A \times B$ be a relation of type $\langle p : q, m : n \rangle$.

   (a) Prove that ρ^{-1} is of type $\langle m : n, p : q \rangle$ from B to A.

   (b) If $p' \leq p, q \leq q', m' \leq m$, and $n \leq n'$, prove that ρ is also of type $\langle p' : q', m' : n' \rangle$.

   (c) For every $k_1, k_2 \in \mathbf{N}$, define

$$k_1 \star k_2 = \begin{cases} k_1 & \text{if } k_2 \neq 0, \\ 0 & \text{otherwise.} \end{cases}$$

   Let $\sigma \subseteq B \times C$ be a relation of type $\langle p' : q', m' : n' \rangle$. Prove that the relation $\rho \bullet \sigma$ is of type $\langle p \star p' : qq', m' \star m : nn' \rangle$ from A to C.

Relations and Directed Graphs

54. Let $G = (V, U)$ be a finite digraph. Prove that $\sum \{d_G^+(v) \mid v \in V\} = \sum \{d_G^- \mid v \in V\} = |U|$.

55. A digraph $G = (V, U)$ is *functional* if there exists a functional relation $\rho \subseteq V \times V$ such that $G = G(\rho)$. A directed graph is *injective* if $d_G^-(x) \leq 1$ for all vertices x. Prove that a directed graph is functional if and only if its converse is injective.

56. Let $N_8 = \{n \mid n \in \mathbf{N}, 1 \leq n \leq 8\}$. Consider the mapping $f : N_8 \longrightarrow N_8$ defined by the following table.

n	1	2	3	4	5	6	7	8
$f(n)$	2	3	4	1	1	7	6	6

Construct the digraph $G(f)$.

57. Prove that for every function $f : A \longrightarrow A$, where A is a finite set, the digraph $G(f)$ consists of circuits to which converse rooted trees may be attached through their roots. A converse rooted tree is the converse digraph of a rooted tree.

58. Prove that the finiteness of the directed graph is essential in Proposition 2.6.14. In other words, show that if the set V is infinite then there exists an acyclic digraph $G = (V, U)$ such that G has no vertex v with $d_G^-(v) = 0$.

59. Prove that for a finite directed graph with no circuits $G = (V, U)$ there exists a vertex r such that (G, r) is a rooted tree if and only if for every pair of vertices x, y there exists a vertex z such that both x and y are successors of z.

    Solution: If (G, r) is a finite rooted tree, the condition is obviously necessary. Let us prove that the condition is sufficient. Let $V = \{x_1, \ldots, x_n\}$ be the set of vertices of the graph G. There is z_2 such that both x_1 and x_2 are successors of z_2. There exists a vertex z_3 such that z_2 and x_3 are successors of z_3, which means, of course, that x_1, x_2, x_3 are all successors of z_3, etc. Eventually, we obtain the existence of the vertex z_n such that z_{n-1} and x_n are successors of z_n. The root of the tree is clearly z_n.

60. Define the n-th negative power of a relation $\rho \subseteq M \times M$ as $\rho^{-n} = (\rho^n)^{-1}$ for $n \in \mathbf{N}$. Prove that $\rho^m \bullet \rho^{-n} = \rho^{m-n}$ for $m, n \in \mathbf{N}$.

61. A rooted tree is m-*uniform* if each vertex that is not a leaf has exactly m descendants. Prove that there are m^k descendants of a vertex u situated at distance k from the vertex u. In other words, if u is a vertex situated on level n, then there are m^k descendants of u on level $n + k$ of the tree T.

62. Apply the topological sorting algorithm to the rooted tree defined in Example 2.6.20. What can be said, in general, about the application of topological sorting to rooted trees?

Special Classes of Relations

63. Let C be the set of all U.S. cities and let ρ be the relation $\rho \subseteq C \times C$ that consists of those pairs (c, c') such that the distance between c and c' is less than 300 miles. Determine whether or not ρ is reflexive, symmetric, transitive, antisymmetric, or asymmetric.

64. Let $f : A \longrightarrow A$ be a bijection on the set A, which is also a symmetric relation. Prove that $f(f(a)) = a$ for any $a \in A$.

65. Let $\rho \subseteq A \times A$ be a symmetric and transitive relation. If $\mathrm{Dom}(\rho) = A$, prove that ρ is reflexive.

66. Describe the relations $\rho \subseteq A \times A$ that are both symmetric and anti-symmetric.

67. Let ρ, σ be two binary relations on the set A.

 (a) Prove that

$$(\rho \cup \sigma)^2 = \rho^2 \cup \sigma^2 \cup (\rho \bullet \sigma) \cup (\sigma \bullet \rho).$$

 (b) If ρ, σ are equivalences on A, then $\rho \cup \sigma$ is an equivalence on A if and only if $\rho \bullet \sigma \subseteq \rho \cup \sigma$ and $\sigma \bullet \rho \subseteq \rho \cup \sigma$.

68. Let $\mathcal{R}$ be a class of relations. Determine if for $\rho, \sigma \in \mathcal{R}$ we have $\rho \cup \sigma \in \mathcal{R}$ when $\mathcal{R}$ is the class of symmetric, antisymmetric, asymmetric, reflexive, or transitive relations.

69. Prove that if $\sigma_1, \sigma_2 \subseteq B \times C$, $\rho \subseteq A \times B$, and $\theta \subseteq C \times D$ then $\sigma_1 \subseteq \sigma_2$ implies $\rho \bullet \sigma_1 \subseteq \rho \bullet \sigma_2$ and $\sigma_1 \bullet \theta \subseteq \sigma_2 \bullet \theta$.

70. Let ρ be a relation on the set A. Define $\hat{\rho}$ by $\hat{\rho} = \rho \cup \rho^{-1}$. Prove that $\hat{\rho}$ is reflexive. Furthermore, if θ is a reflexive relation on A such that $\rho \subseteq \theta$, then $\hat{\rho} \subseteq \theta$.

 The relation $\hat{\rho}$ is called *the reflexive closure* of ρ.

71. Let ρ, σ be two reflexive relations on a set M. If $\rho \bullet \sigma \subseteq \sigma \bullet \rho$, prove that $\rho \bullet \sigma = \sigma \bullet \rho$.

Equivalences and Partitions

72. Let $\{\rho_i \mid i \in I\}$ be a collection of relations on the set A. Define

$$\rho_I = \bigcap_{i \in I} \rho_i.$$

 Prove that if all relations from the collection $\{\rho_i \mid i \in I\}$ are reflexive (symmetric, antisymmetric, transitive) then so is ρ_I.

 This implies that the intersection of an arbitrary family of equivalences (partial orders) is an equivalence (a partial order, respectively).

73. Prove the equalities contained by Corollary 2.8.9.

74. Define the relation ρ on V^* by $(x, y) \in \rho$ if there is $u \in V^*$ such that $xu = uy$. In this case, the words x, y are said to be *conjugate*. Prove that ρ is an equivalence on V^*.

75. Let M be a set. For $H, K \in \mathcal{P}(M)$, define $K \equiv_P L$ if for their symmetric difference we have $H \oplus K \subseteq P$. Prove that for every set P the relation $\equiv_P$ is an equivalence on $\mathcal{P}(M)$.

76. Let $\rho, \theta \subseteq A \times A$ be two equivalences on A such that $\rho \subseteq \theta$. Define the relation ρ/θ on the set A/ρ by $([x], [y]) \in \rho/\theta$ if $(x, y) \in \theta$.

    Prove that ρ/θ is well defined and that it is an equivalence on the quotient set A/ρ.

77. Let ρ be a relation on A such that $\mathrm{Dom}(\rho) = A$. If $\rho \bullet \rho^{-1} \bullet \rho = \rho$ prove that both $\rho \bullet \rho^{-1}$ and $\rho^{-1} \bullet \rho$ are equivalence relations.

78. Let $\mathcal{F}$ be a family of subsets of a set I such that $K, H \in \mathcal{F}$ implies $K \cap H \in \mathcal{F}$ and $K \in \mathcal{F}$, $K \subseteq L$ imply $L \in \mathcal{F}$. Consider the relation $\equiv_{\mathcal{F}}$ on the set $I \longrightarrow M$, where $s \equiv_{\mathcal{F}} t$ if $\{i \mid i \in I, s(i) = t(i)\} \in \mathcal{F}$. Prove that $\equiv_{\mathcal{F}}$ is an equivalence on this set of mappings.

79. For a relation $\theta \subseteq A \times A$, define $\overline{\theta} = (A \times A) - \theta$. Prove that the relation $\rho \subseteq A \times A$ is an equivalence if and only if

$$\rho = \overline{\overline{\rho^{-1}} \bullet \rho} \cap \overline{\rho^{-1} \bullet \overline{\rho}}$$

80. Let $f : A \longrightarrow B$ and $g : A \longrightarrow C$ be two functions. Prove that there is a mapping $h : B \longrightarrow C$ such that $g = hf$ if and only if $\mathbf{ker}(f) \subseteq \mathbf{ker}(g)$.

81. The set $K \subseteq A$ is recognized by the mapping $f : A \longrightarrow B$ if $K = f^{-1}(f(K))$.

    (a) Prove that the following three statements are equivalent:

        i. K is recognized by $f : A \longrightarrow B$;
        ii. K is the union of some equivalence classes of the quotient set $A/\mathbf{ker}(f)$;
        iii. $\mathbf{ker}(f) \subseteq \mathbf{ker}(\chi_K)$.

    (b) If K is recognized by $f : A \longrightarrow B$ and $g : A \longrightarrow C$ is a mapping such that $f = hg$ for some mapping $h : C \longrightarrow B$, then K is also recognized by g.

    (c) Prove that K is recognized by f if and only if there is $L \subseteq B$ such that $K = f^{-1}(L)$.

    (d) If $K, H \subseteq A$, K is recognized by f, and H is recognized by g, then the sets $K \cup H$, $K \cap H$, and $K - H$ are all recognized by the mapping $h : A \longrightarrow B \times C$ given by $h(a) = (f(a), g(a))$ for $a \in A$.

    (e) If $K \subseteq A$ is recognized by the mapping $f : A \longrightarrow B$ and $g : C \longrightarrow A$ is a mapping, show that the set $L = g^{-1}(K)$ is recognized by the mapping fg.

82. Let $\rho \subseteq A \times A$ be an equivalence and assume that $\theta = (A \times A) - \rho$. Prove that $\rho \bullet \theta = \theta \bullet \rho = \theta$.

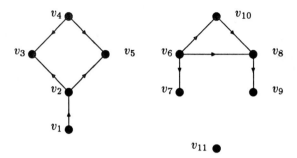

FIGURE 2.19. Connected components of a digraph.

83. A relation $\rho \subseteq A \times A$ is *circular* if $(a_1, a_2), (a_2, a_3) \in \rho$ implies $(a_3, a_1) \in \rho$. Prove that ρ is an equivalence if and only if it is both reflexive and circular.

84. A relation $\rho \subseteq A \times A$ has the *strong (weak) Church-Rosser property* if $x, y, z \in A$ and $(x, y) \in \rho$ and $(x, z) \in \rho$ implies the existence of $v \in A$ such that $(y, v), (z, v) \in \rho$ (or $(y, v), (z, v) \in \rho^*$, respectively).

   (a) Prove that ρ has the strong (weak) Church-Rosser property if and only if $\rho^{-1} \bullet \rho \subseteq \rho \bullet \rho^{-1}$ (or, $\rho^{-1} \bullet \rho \subseteq (\rho^*) \bullet (\rho^*)^{-1}$, respectively).

   (b) Prove that if ρ has the weak Church-Rosser property that its transitive closure has the strong Church-Rosser property.

   (c) An element $y \in A$ is a *normal form* of an element x of A with respect to the relation ρ if $(x, y) \in \rho^+$, and if $(y, z) \in \rho^*$, then $z = y$. Prove that if ρ has the weak Church-Rosser property then every element of A can have at most one normal form.

85. Let $G = (V, U)$ be a directed graph. A *walk* is a finite alternating sequence of vertices and edges $w = (v_0, u_1, v_1, \ldots, u_n, v_n)$, such that $u_i = (v_{i-1}, v_i)$ or $u_i = (v_i, v_{i-1})$. Such a walk is *joining* the vertices v_0 and v_n.

   (a) Define the relation $\equiv \subseteq V \times V$ by $v \equiv v'$ if $v = v'$ or there is a walk joining v to v'. Prove that $\equiv$ is an equivalence relation on the set of vertices of the graph. The equivalence classes of the quotient set $V/\equiv$ are *the connected components* of the graph $G = (V, U)$.

   (b) Verify that the connected components of the graph given in Figure 2.19 are $\{v_1, v_2, v_3, v_4, v_5\}$, $\{v_6, v_7, v_8, v_9, v_{10}\}$ and $\{v_{11}\}$.

(c) Two vertices v, v' are mutually connected in the graph G if $v = v'$ or there is a path in G joining v to v' and a path joining v' to v. Define the relation "$\sim$" by $v \sim v'$ if v and v' are mutually connected. Prove that "$\sim$" is an equivalence relation and $\sim \subseteq \equiv$. The equivalence classes of $V/\sim$ are the *strong components* of the graph G.

(d) Compute the strong components of the graph given in Figure 2.19.

86. Let S be the set of sequences of rational numbers, $S = \{f \mid f : P \longrightarrow Q\}$. Define the relation $\equiv_R$ by $(f, g) \in \equiv_R$ if for every positive rational number ϵ there exists a number $n_\epsilon \in R_+$ such that $\mid f(p) - g(p) \mid < \epsilon$ if $p \geq n_\epsilon$. Show that $\equiv_R$ is an equivalence on S.

Operations

87. Let "$\triangle$" be an operation defined on the set of positive real numbers R_+ by $x \triangle y = x^{\log y}$. Prove that $\triangle$ is both commutative and associative. Determine whether $\triangle$ has an identity element.

88. Prove that if the operation $\star$ has both a left identity u and a right identity v then $u = v$.

89. (a) Verify that the operations $\vee$ and $\wedge$ defined on the set **Bool** are associative, commutative, and idempotent.

(b) Show that these operations satisfy the absorbition laws and the De Morgan laws.

90. Consider the operation $\star$ on the set R of real numbers by $x \star y = axy + bx + cy + d$ for $x, y \in R$, where $a, b, c, d \in R$.

(a) Determine sufficient conditions on a, b, c, d for the associativity of $\star$.

(b) If $\star$ is associative, prove that it is also commutative.

(c) Prove that if $\star$ is idempotent then it is not associative.

91. Let $\diamond$ be an operation on a set A having the identity u, such that

$$x \diamond (y \diamond z) = (x \diamond z) \diamond y,$$

for all $x, y, z \in A$. Prove that $\diamond$ is commutative and associative.

92. An operation or a relation on a set B generates a corresponding operation or relation on the set of functions $A \longrightarrow B$. Indeed, if ω is an operation on the set B and $f, g : A \longrightarrow B$, then we can define $h = f\omega g$ by $h(a) = f(a)\omega g(a)$ for $a \in A$. Also, if $\rho \subseteq B \times B$ and $f_1, f_2 \in A \longrightarrow B$, then we have $f_1 \rho f_2$ if $(f_1(a), f_2(a)) \in \rho$ for all $a \in A$.

Prove that if $(b_1, b_2), (b_3, b_4) \in \rho$ imply $(b_1 \omega b_3, b_2 \omega b_4) \in \rho$ for $b_i \in B$ for $1 \leq i \leq 4$ then $(f, g) \in \rho$ implies $(k\omega f, k\omega g) \in \rho$ and $(f\omega l, g\omega l) \in \rho$ for $f, g, k, l \in A \longrightarrow B$.

93. Consider the relation $\equiv_{\mathbf{Z}} \subseteq \mathbf{N} \times \mathbf{N}$ defined by $((u, v), (x, y)) \in \equiv_{\mathbf{Z}}$ if $u + y = x + v$.

   (a) Prove that $\equiv_{\mathbf{Z}}$ is an equivalence on $\mathbf{N} \times \mathbf{N}$.

   (b) Prove that the mapping $f : (\mathbf{N} \times \mathbf{N})/\equiv_{\mathbf{Z}} \longrightarrow \mathbf{Z}$, where

   $$f([(x, y)]) = x - y$$

   is well defined and, furthermore, that it is a bijection.

   (c) Define the operation "+" on $(\mathbf{N} \times \mathbf{N})/\equiv_{\mathbf{Z}}$ by

   $$[(x_1, y_1)] + [(x_2, y_2)] = [(x_1 + x_2, y_1 + y_2)]$$

   for every $[(x_1, y_1)], [(x_2, y_2)] \in \mathbf{N} \times \mathbf{N}/\equiv_{\mathbf{Z}}$. Prove that the operation is well defined. For the bijection f previously defined, verify that

   $$f([(x_1, y_1)] + [(x_2, y_2)]) = f([(x_1, y_1)]) + f([(x_2, y_2)]).$$

   (d) Define the operation "·" on the set $(\mathbf{N} \times \mathbf{N})/\equiv_{\mathbf{Z}}$ by

   $$[(x_1, y_1)] \cdot [(x_2, y_2)] = [(x_1 x_2 + y_1 y_2, x_1 y_2 + x_2 y_1)].$$

   Prove that the operation · is well defined. For the bijection f, verify that

   $$f([(x_1, y_1)] \cdot [(x_2, y_2)]) = f([(x_1, y_1)])f([(x_2, y_2)]).$$

Solution: (a) The relation $\equiv_{\mathbf{Z}}$ is obviously reflexive and transitive. To prove that it is transitive consider the pairs $(u, v), (x, y), (z, w) \in \mathbf{N} \times \mathbf{N}$ such that $((u, v), (x, y)) \in \equiv_{\mathbf{Z}}$ and $((x, y), (z, w)) \in \equiv_{\mathbf{Z}}$, that is,

$$u + y = x + v \text{ and } x + w = z + y,$$

which, in turn, give $u + y + w = x + v + w = z + y + v$; hence, $u + w = z + v$. Therefore, $((u, v), (z, w)) \in \equiv_{\mathbf{Z}}$, which shows that $\equiv_{\mathbf{Z}}$ is transitive.

(b) Since the function f is defined by choosing an arbitrary element of an equivalence class, we need to prove first that it is well defined. Suppose that $(u, v) \in [(x, y)]$. We have $((u, v), (x, y)) \in \equiv_{\mathbf{Z}}$, which means that $u + y = x + v$. This implies $u - v = x - y$, which means that the value of the function f is the same for all members of the same equivalence class $[(x, y)]$. In other words, f is well defined.

The function f is onto; indeed, if k is an integer, then we have

$$f([k,0]) = k \text{ if } k \geq 0 \text{ and } f([0,-k]) = k \text{ if } k < 0.$$

To prove the injectivity of f, assume that $f([u,v]) = f([x,y]) = k$. This means that $u - v = x - y = k$, which gives $u + y = x + v$. This amounts to $((u,v),(x,y)) \in \equiv_Z$, or $[(u,v)] = [(x,y)]$, which proves that f is one-to-one.

The function f considered here shows that it is possible to construct the set of integers starting from the set of natural numbers, by regarding the integers as equivalence classes on the set of pairs of natural numbers.

(c) We need to show, again, that the operation "+" is well defined. Assume that $(u_1, v_1) \in [(x_1, y_1)]$ and $(u_2, v_2) \in [(x_2, y_2)]$. The correctness of the definition amounts to proving that $(u_1 + u_2, v_1 + v_2) \in [(x_1 + x_2, y_1 + y_2)]$.

Since $u_1 + y_1 = x_1 + v_1$ and $u_2 + y_2 = x_2 + v_2$, we have $u_1 + u_2 + y_1 + y_2 = x_1 + x_2 + v_1 + v_2$, which shows that "+" is, indeed, well defined.

The equality involving the function f amounts to

$$x_1 + x_2 - (y_1 + y_2) = x_1 - y_1 + x_2 - y_2,$$

which is clearly satisfied.

(d) We leave this part to the reader.

94. Let Z be the set of integers, $Z_1 = Z - \{0\}$ and Q be the set of rationals. Consider the relation $\equiv_Q$ on the set $Z \times Z_1$ defined by $((u,v),(x,y)) in \equiv_Q$ if $uy = xv$.

(a) Prove that $\equiv_Q$ is an equivalence on $Z \times Z_1$.

(b) Consider the function $g : (Z \times Z_1)/\equiv_Q \longrightarrow Q$ defined by $g([x,y]) = x/y$. Prove that g is well defined and is a bijection. The function g indicates the possibility of the reconstruction of the set of rational numbers by regarding the rational numbers as equivalence classes of pairs of integers.

(c) Define the operations "+" and "·" on $(Z \times Z_1)/\equiv_Q$ by

$$[(x_1, y_1)] \cdot [(x_2, y_2)] = [(x_1 x_2, y_1 y_2)]$$
$$\text{and}$$
$$[(x_1, y_1)] + [(x_2, y_2)] = [(x_1 y_2 + x_2 y_1, y_1 y_2)]$$

for every $[(x_1, y_1)], [(x_2, y_2)] \in (Z \times Z_1)/\equiv_Q$. Prove that both operations are well defined. Show that

$$g([(x_1, y_1)] \cdot [(x_2, y_2)]) = g([(x_1, y_1)])g([(x_2, y_2)]),$$
$$g([(x_1, y_1)] + [(x_2, y_2)]) = g([(x_1, y_1)]) + g([(x_2, y_2)]),$$

for every $[(x_1, y_1], [(x_2, y_2)] \in (\mathbf{Z} \times \mathbf{Z}_1)/ \equiv_Q$.

Representations of Relations and Graphs

95. Write the matrices representing the following relations on the set $M = \{1, \ldots, 5\}$:

   (a) $\rho = \{(k, h) \mid k, h \in M, k \leq h\}$;
   (b) $\sigma = \{(k, h) \mid k, h \in M, k \geq h\}$;
   (c) $\theta = \{(k, h) \mid k, h \in M, -1 \leq k - h \leq 1\}$.

96. For a matrix $M \in M_{p \times q}(B)$ (where $B = \{0, 1\}$), denote by $|M|$ the number of its entries equal to 1. Prove that $|M_1 M_2| \leq |M_1||M_2|$ for any matrices $M_1 \in M_{p \times q}$ and $M_2 \in M_{q \times r}(B)$.

   Hint: Consider three sets $P = \{x_1, \ldots, x_p\}$, $Q = \{y_1, \ldots, y_q\}$ and $R = \{z_1, \ldots, z_r\}$ and the relations $\rho_1 \subseteq P \times Q$ and $\rho_2 \subseteq Q \times R$, regarded as *sets of pairs*. Prove that there is an onto mapping ϕ : $\rho_1 \times \rho_2 \longrightarrow \rho_1 \bullet \rho_2$.

97. Consider the sets A, B, C and the relation $\rho \subseteq A \times B$ and $\sigma \subseteq B \times C$. If $\pi = \{B_i \mid 1 \leq i \leq k\}$ is a partition of the set B, define the relations $\rho_i = \rho \cap (A \times B_i)$ and $\sigma_i = \sigma \cap (B_i \times C)$ for $1 \leq i \leq k$.

   (a) Prove the equality

   $$\rho \bullet \sigma = (\rho_1 \bullet \sigma_1 \cup \cdots \cup \rho_k \bullet \sigma_k).$$

   (b) Write the matrix equality that corresponds to the previous decomposition of $\rho \bullet \sigma$ into a union of relation products.

98. The matrix representing a relation on two finite sets depends on the indexing of those sets.

   We shall define a reindexing of a finite set $A = \{a_1, \ldots, a_n\}$ as a bijection $h : \{1, \ldots, n\} \longrightarrow \{1, \ldots, n\}$. This, in turn, allows us to write A as $A = \{a_{h(1)}, \ldots, a_{h(n)}\}$. For instance, if we write the set $A = \{a, b, c, d\}$ as $A = \{a_1, a_2, a_3, a_4\}$ with $a_1 = a, a_2 = b, a_3 = c$, and $a_4 = d$, then the bijection $h : \{1, 2, 3, 4\} \longrightarrow \{1, 2, 3, 4\}$, given by $h(1) = 4, h(2) = 2, h(3) = 1, h(4) = 3$, corresponds to $A = \{a_{h(1)}, a_{h(2)}, a_{h(3)}, a_{h(4)}\} = \{d, b, a, c\}$.

   The mapping h, like any relation, can be represented by a matrix $M(h)$ from $M_{n \times n}(\{0, 1\})$. For instance, the mapping h defined above is represented by the matrix

   $$\begin{pmatrix} 0 & 0 & 0 & 1 \\ 0 & 1 & 0 & 0 \\ 1 & 0 & 0 & 0 \\ 0 & 0 & 1 & 0 \end{pmatrix}.$$

(a) Characterize the matrices from $M_{m \times n}(\{0,1\})$ representing partial functions, total functions, bijections.

(b) Let

$$h : \{1,\ldots,m\} \longrightarrow \{1,\ldots,m\} \text{ and } g : \{1,\ldots,n\} \longrightarrow \{1,\ldots,n\}$$

be two renumberings of the sets $A = \{a_1,\ldots,a_m\}$ and $B = \{b_1,\ldots,b_n\}$, respectively. Let $P, Q \in M_{m \times n}(\{0,1\})$ be the matrices representing the relation $\rho \subseteq M \times P$ before and after renumbering, respectively. Prove that $Q = (M(h))' P M(g)$.

99. Let ρ be a relation on $A \times B$. Prove that the relation ρ^{-1} is represented by the transposed matrix $(M(\rho)^{-1})$.

Relations and Databases

100. Let ρ, ζ be two relations on the schemas I, H, respectively, with $H \subseteq I$. Prove that $\text{sel}_\zeta(\rho) = \zeta \bowtie \rho$. This shows that selection can be realized through join.

101. Let ρ, θ be two relations such that $\text{sch}(\theta) \subseteq \text{sch}(\rho)$. Assume that $\text{sch}(\theta) = J$ and $\text{sch}(\rho) = I \cup J$, where $I \cap J = \emptyset$. Prove that

$$\rho \div \theta = \rho[I] - (\rho[I] \bowtie \theta - \rho)[I].$$

102. Let ρ, σ be two relations on the schemas R, S, respectively. Define the *semijoin operation* (denoted by $\vec{\bowtie}$) as $\rho \vec{\bowtie} \sigma = (\rho \bowtie \sigma)[R]$. Prove that $(\rho \vec{\bowtie} \sigma) \bowtie \sigma = \rho \bowtie \sigma$.

103. Consider the relational database $\mathbf{D}_{dept}$ introduced in Example 2.12.4. Solve the following queries in relational algebra.

(a) Find names of students taking courses with Professor Smith.

(b) Find all grades given by Professor Smith in a three-credit course.

(c) Find the students who are taking some four-credit courses.

(d) Find the students who are getting an A in a four-credit course.

(e) Find the office rooms of those professors who teach student Jones in a four-credit course.

(f) Find all pairs of student names obtaining the same grade in a three-credit course.

(g) Find the students who took all four-credit courses.

(h) Find those professors who give only A's.

(i) Find the students who got only A's.

(j) Find the students who take at least all courses taken by student Jones.

(k) Find the students who take some courses with a professor who teaches student Jones.

(l) Find all pairs of students who take no course together.

(m) Find the students who got the lowest grade in course 110.

(n) Find the professors who teach student Jones but do not teach course 110.

(o) Find the courses in which all students pass.

104. Solve in SQL the queries mentioned in the previous exercise.

105. Suppose that we add to the database $\mathbf{D}_{dept}$ the table NUMVAL, which contains the points corresponding to the letter grades:

NUMVAL

GRADE	VALUE
A	4
B	3
C	2
D	1
F	0

Solve the following queries in SQL.

(a) Determine the grade point average of the student whose number is 1234.

(b) Determine the grade average of all grades given by Professor Smith.

(c) Find the smallest and the largest grade obtained by the student whose number is 1234.

106. Using the GROUP BY option of SQL, solve the following queries.

(a) Determine the grade point average of all students.

(b) For every student, find the smallest and the largest grade.

2.14 Bibliographical Comments

A classical reference for the mathematical aspects of this chapter is [MB67]. More recently, Hamilton's book, [Ham82] touches many problems discussed in this chapter, and it is a useful reference for further chapters in this book.

There are many excellent books on graph theory, ranging from the classical reference [Ber73] to the more recent ones [Bol85, Cha77, BCLF81, Eve79].

Fundamental references on relational databases, with a thorough mathematical treatment of the subject, are [Mai83, Ull88].

3

Partially Ordered Sets

3.1   Introduction
3.2   Partial Orders and Hasse Diagrams
3.3   Special Elements of Partially Ordered Sets
3.4   Chains
3.5   Duality
3.6   Constructing New Posets
3.7   Functions and Posets
3.8   Complete Partial Orders
3.9   The Axiom of Choice and Zorn's Lemma
3.10 Exercises and Supplements
3.11 Bibliographical Comments

3.1   Introduction

In this section, we introduce and study partial orders, a class of relations of paramount importance for both mathematics and computer science, and sets equipped with such relations, later referred to as partially ordered sets or posets. Later in this book, we explore a variety of mathematical structures related to partial orders, and we discuss applications of partial orders in formal language theory, semantics of programming languages, concurrent access of data, etc.

After introducing and discussing chains, an important type of posets, we look into ways of building partially ordered sets starting from simpler ones. From the point of view of computer science, the chapter culminates with a section on complete partially ordered sets, an essential instrument for denotational semantics.

The last section is dedicated to Zorn's lemma and to a fundamental principle of set theory, the axiom of choice, which is equivalent to Zorn's lemma.

3.2   Partial Orders and Hasse Diagrams

The central notion of this chapter is introduced in the following definition.

Definition 3.2.1 *A partial order* on the set A is a relation $\rho \subseteq A \times A$ that *is reflexive, antisymmetric, and transitive. The pair* (A, ρ) *is referred to as a* partially ordered set *or, in short, as a* poset.

A strict partial order, *or simpler, a* strict order *on A, is a relation* $\rho \subseteq A \times A$ *that is irreflexive and transitive.*

Every strict partial order is also asymmetric. Indeed, let ρ be a strict partial order on M and assume that $(x, y) \in \rho$. If $(y, x) \in \rho$, then $(x, x) \in \rho$ due to the transitivity of ρ, which contradicts the irreflexivity of ρ. This shows that ρ is indeed asymmetric.

Note that a strict partial order is *not*, in general, a partial order since strict partial orders are irreflexive, while partial orders are reflexive.

Example 3.2.2 Consider the inclusion relation "$\subseteq$" on the set of subsets $\mathcal{P}(M)$ of the set M. We obtain the poset $(\mathcal{P}(M), \subseteq)$. The reader can easily verify that "$\subseteq$" is reflexive, symmetric, and transitive.

Example 3.2.3 The relation $\delta_{\mathbf{N}}$ introduced in Example 2.7.3 is a partial order on $\mathbf{N}$.

Example 3.2.4 A partial order relation can be introduced on the set of partial functions $A \rightsquigarrow B$. For $f, g : A \rightsquigarrow B$, we write $f \sqsubseteq g$ if $\text{Dom}(f) \subseteq \text{Dom}(g)$, and for $x \in \text{Dom}(f)$, we have $f(x) = g(x)$. It is easy to see that this happens if and only if $f \subseteq g$ (if we regard f and g as sets of ordered pairs). Therefore, "$\sqsubseteq$" is a partial order on $A \rightsquigarrow B$.

Example 3.2.5 Let V be an alphabet and let V^* be the set of all words over V. For $u, v \in V^*$, we write $u \trianglelefteq v$ if u is a prefix of v. Since every word is a prefix of itself, the relation $\trianglelefteq \subseteq V^* \times V^*$ is reflexive.

The relation "$\trianglelefteq$" is antisymmetric because, if $u \trianglelefteq v$ and $v \trianglelefteq u$, there are $s, t \in V^*$ such that $v = us$ and $u = vt$. This gives $u = ust$; hence, $st = \lambda$ (using Proposition 2.4.7). Therefore, $s = t = \lambda$, which gives $u = v$.

To prove the transitivity of "$\trianglelefteq$," consider the words $u, v, w \in V^*$ such that $u \trianglelefteq v$ and $v \trianglelefteq w$. There are $s, t \in V^*$ such that $v = us$ and $w = vt$. This, in turn, gives $w = ust$; hence, $u \trianglelefteq w$, and this shows that "$\trianglelefteq$" is a partial order on V^*, and we have the poset $(V^*, \trianglelefteq)$.

Proposition 3.2.6 *If ρ is a partial order on A and $\rho_1 = \rho - \iota_A$, then the relation ρ_1 is a strict partial order on A.*

Conversely, if ρ_1 is a strict partial order, then $\rho = \rho_1 \cup \iota_A$ is a partial order on A.

Proof. Since $\iota_A \cap \rho_1 = \emptyset$, the relation ρ_1 is irreflexive.

To prove the transitivity of ρ_1 consider $(a, b), (b, c) \in \rho_1$. Because of the transitivity of ρ, we have $(a, c) \in \rho$. On the other hand, we also have $a \neq c$. Indeed, if we assume that $a = c$, then we would have both $(c, b) \in \rho_1$ and $(b, c) \in \rho_1$, which is impossible by the asymmetry of ρ_1. Therefore, $(a, c) \in \rho - \iota_A = \rho_1$, which implies the transitivity of ρ_1.

Now, let ρ_1 be a strict partial order and let $\rho = \rho_1 \cup \iota_A$. The reflexivity of ρ is immediate.

To show that ρ is antisymmetric, assume that $(x, y), (y, x) \in \rho$. Because of the definition of ρ, we may have $(x, y) \in \rho_1$ or $(x, y) \in \iota_A$ (that is, $x = y$). In the first case, we have a contradiction. Indeed, if $(y, x) \in \rho_1$, this contradicts the asymmetry of ρ_1; if $(y, x) \in \iota_A$, we also have $(x, y) \in \iota_A$, and this contradicts the irreflexivity of ρ_1. Consequently, we must have $x = y$.

Let $(x, y), (y, x) \in \rho$. We have four cases:

1. If $(x, y), (y, z) \in \rho_1$, we have $(x, z) \in \rho_1$ because of the transitivity of ρ_1. This implies $(x, z) \in \rho$.

2. If $(x, y) \in \iota_A$ and $(y, z) \in \rho_1$, we have $x = y$; hence, $(x, z) \in \rho_1 \subseteq \rho$.

3. If $(x, y) \in \rho_1$ and $(y, z) \in \iota_A$, we follow an argument similar to the one used in the previous case.

4. If $(x, y), (y, z) \in \iota_A$, we have $(x, z) \in \iota_A$ because of the transitivity of ι_A; hence, $(x, z) \in \rho$.

We proved that ρ is also transitive, and this concludes our argument. ∎

Example 3.2.7 Consider the relation $\leq \subseteq R \times R$, which is a partial order. The strict partial order attached to it by the previous proposition is the relation "$<$."

We remind the reader that a relation $\rho \subseteq V \times V$ is acyclic if $\rho^n \cap \iota_V = \emptyset$ for every $n \in P$.

Acyclicity is *a hereditary property*; this means that if a relation $\sigma \subseteq V \times V$ is acyclic and $\theta \subseteq \sigma$ then θ is also acyclic.

Proposition 3.2.8 *Every strict partial order is acyclic. Furthermore, the transitive closure of an acyclic relation is a strict partial order.*

Proof. Let ρ be a strict partial order relation on V. Its transitivity implies the existence of the descending sequence:

$$\cdots \subseteq \rho^n \subseteq \cdots \subseteq \rho^2 \subseteq \rho.$$

Since ρ is irreflexive, we have $\rho \cap \iota_V = \emptyset$, and this implies $\rho^n \cap \iota_V = \emptyset$.

To prove the second part of the proposition, let θ be an acyclic relation. We need to show that the relation θ^+ is irreflexive.

Suppose that $(x, x) \in \theta^+$. Because of the definition of θ^+, there is $n \geq 1$ such that $(x, x) \in \theta^n$, which gives $\theta^n \cap \iota_V \neq \emptyset$; this contradicts the acyclicity of θ. ∎

Partial orders can be represented graphically by Hasse diagrams. Let ρ be a partial order on the set A.

Definition 3.2.9 *The* Hasse diagram *of ρ is the graph of the relation $\rho_1 - \rho_1^2$, where ρ_1 is the strict partial order corresponding to ρ.*

In view of the properties of acyclic relations discussed above, it is clear that the relation $\sigma = \rho_1 - \rho_1^2$ is acyclic; therefore, the Hasse diagram is always an acyclic directed graph.

Observe that $(x, y) \in \rho_1 - \rho_1^2$ if $x \neq y$, $(x, y) \in \rho$, and there is no $u \in A$ such that $(x, u) \in \rho$ and $(u, y) \in \rho$. In other words, if $(x, y) \in \sigma$, then y covers x directly, without any intermediate elements. The use of Hasse diagrams in representing posets is justified by the following.

Proposition 3.2.10 *If $\rho \subseteq A \times A$ is a partial order on the finite set A, ρ_1 is the strict partial order corresponding to ρ, and $\theta = \rho_1 - \rho_1^2$, then $\theta^* = \rho$.*

Proof. Let $(x, y) \in \rho$. If $x = y$, then we have $(x, y) \in \iota_A \subseteq \theta^*$.

Assume now that $(x, y) \in \rho$ and $x \neq y$, which means that $(x, y) \in \rho_1$. Consider the collection $\mathcal{C}_{xy}$ of all sequences of elements of A that can be "interpolated" between x and y:

$$\mathcal{C}_{xy} = \{(s(0), \ldots, s(n-1)) \mid x = s(0), s(n-1) = y, \text{ and} \\ (s_i, s_{i+1}) \in \rho_1 \text{ for } 0 \leq i \leq n-2, n \geq 2\}.$$

Note that $\mathcal{C}_{xy}$ is not empty since the sequence (x, y) belongs to $\mathcal{C}_{xy}$. Furthermore, no sequence from $\mathcal{C}_{xy}$ may contain a repetition because, if we would have $s(p) = s(q)$ for a sequence $(s(0), \ldots, s(n-1))$ with $1 \leq p < q \leq n$, by the transitivity of ρ_1, this would imply $(s(p), s(q)) \in \rho_1$. This is a contradiction in view of the fact that $\rho_1 \cap \iota_A = \emptyset$. Since A is a finite set, $\mathcal{C}_{xy}$ contains a finite number of sequences.

Let us consider a sequence of maximal length from $\mathcal{C}_{xy}$:

$$(s(0), s(2), \ldots, s(m-1)),$$

where $x = s(0)$ and $y = s(m-1)$. Observe that for no pair $(s(i), s(i+1))$ can we have $(s(i), s(i+1)) \in \rho_1^2$. Indeed, if $(s(j), s(j+1)) \in \rho_1^2$, then there is $s \in A$ such that $(s(i), s) \in \rho_1$, and $(s, s(i+1)) \in \rho_1$, and this contradicts the maximality of m. Therefore, $(s(i), s(i+1)) \in \rho_1 - \rho_1^2 = \theta$, and this shows that $(x, y) \in \theta^{m-1} \subseteq \theta^*$.

Conversely, if $(x, y) \in \theta^*$ there is $k \in \mathbf{N}$ such that $(x, y) \in \theta^k$, which means that there exists a sequence $(z(0), \ldots, z(k))$ such that

$$x = z(0), (z(i), z(i+1)) \in \theta \text{ for } 0 \leq i \leq k-1 \text{ and } y = z(k).$$

This implies $(z(i), z(i+1)) \in \rho$; hence, $(x, y) \in \rho^k \subseteq \rho$, because of the transitivity of ρ. ∎

The relation θ introduced in Proposition 3.2.10 is called the *transitive reduction* of the partial order ρ (see also Exercise 5 for a characterization of this relation).

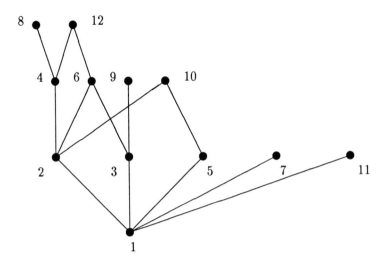

FIGURE 3.1. The Hasse diagram of the poset (M, δ).

Example 3.2.11 Consider the set $M = \{j \mid j \in \mathbf{N}, 1 \le j \le 12\}$ and the relation $\delta = (M \times M) \cap \delta_{\mathbf{N}}$. It is easy to verify that δ is indeed a partial order on M. Its Hasse diagram is given in Figure 3.1. In drawing this graph, we have used the following standard convention for representing posets. If (x, y) is an arc in the graph, we represent this arc by drawing y above x and joining x to y by a line rather than an arrow.

From the diagram, it is clear that $(2, 12) \in \delta$ since $(2, 4), (4, 12) \in \theta$.

Example 3.2.12 The Hasse diagram of the poset $(\mathcal{P}(M), \subseteq)$, where $M = \{a, b, c\}$, is given in Figure 3.2.

3.3   Special Elements of Partially Ordered Sets

Let (M, ρ) be a poset and let $K \subseteq M$.

Definition 3.3.1 *The* set of upper bounds *of the set K is the set K^s, which consists of those elements y of M such that $(x, y) \in \rho$ for all $x \in K$.*

The set of lower bounds *of the set K is the set K^i, which consists of those elements y of M such that $(y, x) \in \rho$ for all $x \in K$.*

Example 3.3.2 Consider the poset (M, δ) whose Hasse diagram is given in Figure 3.1. For $K = \{2, 3, 6\}$, the set of upper bounds is $K^s = \{6, 12\}$. The set of lower bounds of the set $H = \{4, 6, 10\}$ is $H^i = \{1, 2\}$. The set of upper bounds of $\{8, 12\}$ is empty, etc.

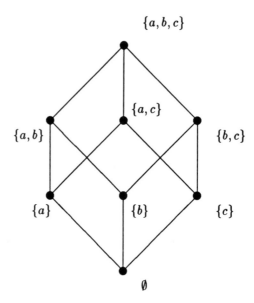

FIGURE 3.2. The Hasse diagram of the poset $(\mathcal{P}(M), \subseteq)$.

Proposition 3.3.3 *For any subset K of a poset (M, ρ), the sets $K \cap K^s$ and $K \cap K^i$ contain at most one element.*

Proof. Suppose that $y_1, y_2 \in K \cap K^s$. Since $y_1 \in K$ and $y_2 \in K^s$, we have $(y_1, y_2) \in \rho$. Reversing the roles of y_1 and y_2, that is, considering now that $y_2 \in K$ and $y_1 \in K^s$, we obtain $(y_2, y_1) \in \rho$. Therefore, we may conclude that $y_1 = y_2$ because of the antisymmetry of the relation ρ, which shows that $K \cap K^s$ contains at most one element.

A similar argument can be used for the second part of the proposition; we leave it to the reader. ∎

Definition 3.3.4 *The least (greatest) element of the subset K of the poset (M, ρ) is the unique element of the set $K \cap K^i$ ($K \cap K^s$, respectively) if such an element exists.*

Example 3.3.5 Not every subset of a poset has a least or a greatest element. Indeed, for the poset (M, δ) from Example 3.3.2, the subset $\{8, 12\}$ has neither the least nor the greatest element. The set $\{2, 3, 6\}$ has the greatest element 6 but no least element. The poset $\{2, 4, 6, 12\}$ has the least element 2 and the greatest element 12.

Applying Definition 3.3.4 to the set M, the *least (greatest) element* of the poset (M, ρ) is an element a of M such that $(a, x) \in \rho$ ($(x, a) \in \rho$, respectively) for all $x \in A$.

Example 3.3.6 For the poset introduced in Example 3.2.12, the greatest element is $\{a, b, c\}$, while the least element is $\emptyset$.

The poset introduced in Example 3.2.11 has no greatest element, while it has the least element 1. The poset $\{x \mid x \in R, 0 < x < 1\}$ has no least or greatest elements.

Definition 3.3.7 *The subset K of the poset (M, ρ) has a* least upper bound u *if* $K^s \cap (K^s)^i = \{u\}$.
K has the greatest lower bound v *if* $K^i \cap (K^i)^s = \{v\}$.

We note that a set can have *at most* one least upper bound and *at most* one greatest lower bound. Indeed, we have seen above that for any set U the set $U \cap U^i$ may contain an element or be empty. Applying this remark to the set K^s, it follows that the set $K^s \cap (K^s)^i$ may contain at most one element, which shows that K may have at most one least upper bound. A similar argument can be made for the greatest lower bound.

If the set K has a least upper bound, we denote it by $\sup K$. The greatest lower bound of a set will be denoted by $\inf K$. These notations come from the terms *supremum* and *infimum* used alternatively for the least upper bound and the greatest lower bound, respectively.

Example 3.3.8 Consider the poset $(\mathbf{N}, \delta_{\mathbf{N}})$ from Example 3.2.3 and let m, n be two *distinct* natural numbers, $m \neq n$.

We claim that any set $\{m, n\}$ has both an infimum and a supremum. Indeed, let p be the least common multiple of m and n. Since $(n, p), (m, p) \in \delta_{\mathbf{N}}$, it is clear that p is an upper bound of the set $\{m, n\}$. On the other hand, if k is an upper bound of $\{m, n\}$, then k is a multiple of both m and n. In this case, k must also be a multiple of p because, otherwise, we could write $k = pq + r$, with $0 < r < p$ by dividing k by p. This would imply $r = k - pq$; hence, r would be a multiple of both m and n, because both k and p have this property. However, this would contradict the fact that p is *the least* multiple that m and n share! This shows that the least common multiple of m and n coincides with the supremum of the set $\{m, n\}$.

The reader will prove that the infimum of $\{m, n\}$ coincides with the greatest common divisor of the numbers m and n.

Example 3.3.9 Consider a set M and the poset $(\mathcal{P}(M), \subseteq)$. Let K, H be two subsets of M. The set $\{K, H\}$ has an infimum and a supremum. Indeed, let $L = K \cap H$. Clearly, $L \subseteq K$ and $L \subseteq H$, so L is a lower bound of the set $\{K, H\}$. Furthermore, if $J \subseteq K$ and $J \subseteq H$, then $J \subseteq L$, by the definition of the intersection. This proves that the infimum of $\{K, H\}$ is the intersection $K \cap H$. A similar argument shows that $K \cup H$ is the supremum of $\{K, H\}$.

In the previous two examples, any two-element subset of the poset has both a supremum and an infimum.

For a one-element subset $\{x\}$ of a poset (M, ρ), we have $\sup\{x\} = \inf\{x\} = x$.

Definition 3.3.10 *A* minimal element *of a poset* (M, ρ) *is an element* $x \in M$ *such that* $\{x\}^i = \{x\}$. *A* maximal element *of* (M, ρ) *is an element* $y \in M$ *such that* $\{y\}^s = \{y\}$.

In other words, x is a minimal element of the poset (M, ρ) if there is no element less than or equal to x other than itself; similarly, x is maximal if there is no element greater than or equal to x other than itself.

Example 3.3.11 In the poset whose Hasse diagram is given in Figure 3.1 the elements $7, 8, 9, 10, 11$, and 12 are all maximal elements, while 1 is the single minimal element.

It is clear that if a poset has a least element u (as is the case with the poset from the previous example) then u is the unique minimal element of that poset. A similar statement holds for the greatest and the maximal elements.

We present now a construction that allows us to endow any set M with a very simple and natural partial order relation having a least element.

For a set M, define $M_\perp = M \cup \{\perp_M\}$, where $\perp_M \notin M$. If there is no risk of confusion, we write $\perp$ rather than $\perp_M$.

Definition 3.3.12 *The* flat partial order *on the set* $M_\perp$ *is the relation* $\sqsubseteq_M$ *defined by* $\perp_M \sqsubseteq x$ *for all* $x \in M_\perp$. *The poset* $(M_\perp, \sqsubseteq_M)$ *is the* M-flat poset.

When there is no risk of confusion, we omit the subscript M and, we denote the partial order simply by $\sqsubseteq$.

Example 3.3.13 For $M = \{a, b, c\}$, the Hasse diagram of the poset $M_\perp$ is given in Figure 3.3.

We use the notation " $\perp$ " for the least element of *any* poset (M, ρ) (if such an element exists), not only for flat posets.

Let I, J be two sets and let (M, ρ) a poset. The following technical result will be used later in this chapter.

Proposition 3.3.14 *Let* $\phi : I \times J \longrightarrow M$ *be an arbitrary function. Assume that for each* $i \in I$ *there exists* $b_i = \sup\{\phi(i, j) \mid j \in J\}$. *If* $u = \sup\{b_i \mid i \in I\}$ *exists, then* $v = \sup\{\phi(i, j) \mid i \in I, j \in J\}$ *exists, and furthermore,* $\sup\{b_i \mid i \in I\} = \sup\{\phi(i, j) \mid i \in I, j \in J\}$.

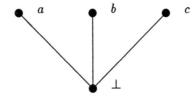

FIGURE 3.3. The Hasse diagram of the flat poset $\{a, b, c\}_\perp$.

Proof. We have $(b_i, u) \in \rho$ for every $i \in I$, and therefore, $(\phi(i,j), u) \in \rho$ for all $i \in I$ and $j \in J$. Consequently, u is an upper bound of the set $\{\phi(i,j) \mid i \in I, j \in J\}$.

Let t be an arbitrary upper bound of $\{\phi(i,j) \mid i \in I, j \in J\}$. We have $(b_i, u) \in \rho$ for $i \in I$ because of the definition of b_i, and this, in turn, implies $(u, t) \in \rho$ because of the definition of u. Consequently, u is $\sup\{\phi(i,j) \mid i \in I, j \in J\}$. ∎

A similar argument can be made for $c_j = \sup\{\phi(i,j) \mid i \in I\}$. If all these elements exist and, furthermore, there exists $w = \sup\{c_j \mid j \in J\}$, then $w = \sup\{\phi(i,j) \mid i \in I, j \in J\}$. The above proposition and this remark can be combined in the following corollary.

Corollary 3.3.15 *If $\phi : I \times J \longrightarrow M$ is a function whose range is in the poset (M, ρ) then, assuming that*

1. $\sup_{i \in I} \phi(i,j)$ *exists for every $j \in J$,*

2. $\sup_{j \in J} \phi(i,j)$ *exists for every $i \in I$,*

3. $\sup_{j \in J} \sup_{i \in I} \phi(i,j)$ *exists and*

4. $\sup_{i \in I} \sup_{j \in J} \phi(i,j)$ *exists,*

then we have the equality

$$\sup_{j \in J} \sup_{i \in I} \phi(i,j) = \sup_{i \in I} \sup_{j \in J} \phi(i,j).$$

3.4 Chains

Definition 3.4.1 *A partial order ρ on a set M is total or linear if $\rho \cup \rho^{-1} = M \times M$. In this case, (M, ρ) is called a chain.*

In other words, ρ is a total order if for every $x, y \in M$ we have either $(x, y) \in \rho$ or $(y, x) \in \rho$.

Example 3.4.2 The set of real numbers equipped with the usual partial order $(\mathbf{R}, \leq)$ is a chain, since for every $x, y \in \mathbf{R}$, we have either $x \leq y$ or $y \leq x$.

Let $V = \{a_1, \ldots, a_n\}$ be an alphabet containing n symbols. An important example of total order defined on the set V^* is given in the next example.

Example 3.4.3 For $x, y \in V^*$, we write $x \preceq y$ if $x \trianglelefteq y$ (that is, if x is a prefix of y) or if there exist $u, x', y' \in V^*$ and $a_i, a_j \in V$ such that $x = ua_i x'$, $y = ua_j y'$ with $i < j$. The reader will easily supply the proof of the fact that " $\preceq$ " is, indeed, a total order. This is the *lexicographic order* on the set V^*; if words of V^* were arranged in a dictionary, this would be the order they would appear.

For instance, for the alphabet $\{a_1, a_2\}$, we have

$$\lambda \preceq a_1 \preceq a_1 a_1 \preceq \cdots a_1 a_2 \cdots \preceq a_2 \preceq \cdots a_2 a_2 \preceq \cdots$$

Observe that $x \preceq y$ if and only if $ux \preceq uy$. However, we can have $x \preceq y$ without having $xv \preceq yv$, as shown by the following example.

Example 3.4.4 Consider the alphabet $V = \{a_1, a_2, a_3\}$. Clearly, we have $a_1 \preceq a_1 a_2$; however, $a_1 a_3 \succeq a_1 a_2 a_3$.

Not every poset is a chain. For instance, the poset $(\mathcal{P}(M), \subseteq)$ considered in Example 3.2.12 is not a chain; elements of $\mathcal{P}(M)$ such as $\{a, b\}$ and $\{b, c\}$ are *incomparable*, that is, we have neither $\{a, b\} \subseteq \{b, c\}$ nor $\{b, c\} \subseteq \{a, b\}$.

If (M, ρ) is a poset and $K \subseteq M$, then (K, ρ_K) is also a poset, where $\rho_K = \rho|_K$ is *the trace of ρ on K*. We leave to the reader the proof of this statement (see exercise 3).

A poset (M, ρ) that is not a chain may very well contain subsets that are chains with respect to the trace of the partial order of the set itself.

Example 3.4.5 The poset from Example 3.2.11 is not a chain since it contains incomparable elements (for instance, 4 and 6 are such elements). However, the subset $\{1, 2, 6, 12\}$ is a chain, as can be easily seen.

Definition 3.4.6 *A* well-ordered *poset is a poset for which every nonempty subset has a least element.*

Note that a well-ordered set is necessarily a chain. Indeed, consider the well-ordered set (M, ρ) and $x, y \in M$. Since the set $\{x, y\}$ must have a least element, we have either $(x, y) \in \rho$ or $(y, x) \in \rho$.

Example 3.4.7 The set of natural numbers is well-ordered. We shall encounter later this property of natural numbers as the *well-ordering principle*.

Another axiom of set theory, related to well-ordered sets is given in the following. As we shall see later in this chapter, this axiom is equivalent to yet another axiom of set theory known as the axiom of choice and to Zorn's lemma (see Section 3.9).

Axiom 3.4.8 (Well-Ordering Axiom) *Given any set M, there is a binary relation ρ that makes (M, ρ) a well-ordered set.*

The set $(\mathbf{R}, \leq)$ is not well ordered, despite the fact that it is a chain, since it contains subsets such as $\{x \mid x \in R, 0 < x < 1\}$ that do not have a least element.

Let ρ_1 be the strict partial order of the poset (M, ρ).

Definition 3.4.9 *An* infinite descending sequence *in a poset (M, ρ) is an infinite sequence $s : \mathbf{N} \longrightarrow M$ such that $(s(n+1), s(n)) \in \rho_1$ for all $n \in \mathbf{N}$.*

An infinite ascending sequence *in a poset (M, ρ) is an infinite sequence $s : \mathbf{N} \longrightarrow M$ such that $(s(n), s(n+1)) \in \rho_1$ for all $n \in \mathbf{N}$.*

A poset with no infinite descending sequences is called artinian.[1] *A poset with no infinite ascending sequences is called* noetherian.[2]

Clearly, the range of every infinite ascending or descending sequence is a chain.

Example 3.4.10 The poset $(\mathbf{N}, \delta_{\mathbf{N}})$ is artinian. Indeed, suppose that s would be an infinite descending sequence of natural numbers. If $s(0) \neq 0$, then the natural number $s(0)$ would have an infinite set of divisors $\{s(0), s(1), \ldots\}$. If $s(0) = 0$, in view of the fact that any natural number is a divisor of 0, we obtain the impossibility of an infinite descending sequence by applying the same argument to $s(1)$. However, this poset is not noetherian. For instance, the sequence $z : \mathbf{N} \longrightarrow \mathbf{N}$ defined by $z(n) = 2^n$ for $n \in \mathbf{N}$ is an infinite ascending sequence.

There are noetherian posets that are not artinian (see Section 3.10, Exercises and Supplements).

Example 3.4.11 Any flat poset is both artinian and noetherian.

A generalization of well-ordered posets is considered in the next definition.

Definition 3.4.12 *A* well-founded poset *is a partially ordered set where every nonempty subset has a minimal element.*

---

[1] An artinian poset is named for the german mathematician Emil Artin (1898-1962), who made contributions in algebra and algebraic geometry.

[2] A noetherian poset is named for Emmy Noether (1882-1935), one of the most important algebraists of modern times.

Since the least element of a subset is also a minimal element, it is clear that a well-ordered set is also well-founded. However, the inverse is not true; for instance, not every finite set is well ordered.

Theorem 3.4.13 *A poset* (M, ρ) *is well founded if and only if it is artinian.*

Proof. Let (M, ρ) be a well-founded poset and suppose that s is an infinite descending sequence in this poset. The set $S = \{s(n) \mid n \in \mathbf{N}\}$ has no minimal element since for every $s(k) \in S$ we have $(s(k+1), s(k)) \in \rho_1$, which contradicts the well-foundedness of (M, ρ).

Conversely, assume that (M, ρ) is artinian, that is there is no infinite descending sequence in (M, ρ). Suppose that K is a nonempty subset of M without minimal elements. Let x_0 be an arbitrary element of K. Such an element exists since K is not empty. Since x_0 is not minimal, there is $x_1 \in K$ such that $(x_1, x_0) \in \rho$. Since x_1 is not minimal, there is $x_2 \in K$ such that $(x_2, x_1) \in \rho$, etc., and this construction can continue indefinitely. In this way, we can build an infinite descending sequence $s : \mathbf{N} \longrightarrow M$, where $s(n) = x_n$ for $n \in \mathbf{N}$. ∎

Theorem 3.4.13 implies immediately that any finite poset is well-founded.

3.5   Duality

In this section, we discuss a general property of partial ordered sets that will allow us to obtain half of some of the arguments related to the properties of partial orders for free.

Proposition 3.5.1 *Let* (M, ρ) *be a poset. The inverse* ρ^{-1} *is also a partial order.*

Proof. Since $\iota_M \subseteq \rho$, we have $\iota_M = \iota_M^{-1} \subseteq \rho^{-1}$, because of the reflexivity of ι_M and Proposition 2.2.12. This shows that ρ^{-1} is reflexive.

The symmetry of ρ^{-1} follows from $(\rho^{-1})^{-1} = \rho = \rho^{-1}$ because of the symmetry of ρ.

To prove the transitivity of ρ^{-1}, assume that $(x, y) \in \rho^{-1}$ and $(y, z) \in \rho^{-1}$. This means that $(y, x), (z, y) \in \rho$, and because of the transitivity of ρ, we obtain $(z, x) \in \rho$; hence, $(x, z) \in \rho^{-1}$, which proves that ρ^{-1} is transitive. ∎

This proposition allows us to build a *new* poset closely related to the original.

Definition 3.5.2 *The* dual *of the poset* (M, ρ) *is the poset* (M, ρ^{-1}).

It is interesting to observe that certain concepts valid for a poset have a counterpart for their dual poset. For instance, x is an upper bound for the set K in the poset (M, ρ) if and only if x is a lower bound for K in the dual poset. Similarly, $t = \sup K$ in the poset (M, ρ) if and only if $t = \inf K$ in the dual poset. The following table lists the pairs of dual notions.

Concept	Dual Concept
upper bound	lower bound
lower bound	upper bound
minimal element	maximal element
maximal element	minimal element
least element	greatest element
greatest element	least element
infimum	supremum
supremum	infimum
noetherian poset	artinian poset
artinian poset	noetherian poset

If all concepts occurring in a statement about posets are replaced by their dual, we obtain the *dual statement*. Furthermore, if a statement holds for a poset (M, ρ), its dual holds for the dual poset (M, ρ^{-1}). This allows us to formulate the following principle.

The Duality Principle for Posets. If a statement is true for all posets, then its dual is also true for all posets.

The validity of this principle follows from the fact that any poset can be regarded as the dual of some other poset. The duality principle is used to avoid trivial repetitions of certain arguments made for posets. There are statements involving both a concept and its dual. For such cases we need to prove only half of the statement; the other half follows by applying the duality principle.

For instance, once we proved the statement "any subset of a poset can have at most one least upper bound," the dual statement "any subset of a poset can have at most one greatest lower bound" follows.

3.6   Constructing New Posets

We have studied constructions of sets using Cartesian products, disjoint sums, sets of functions, etc. In this section, we present several standard methods for endowing these sets with partial orders.

Let I be a set, (M, μ) be a poset. The partial order μ generates a partial order ρ on the set of functions $I \longrightarrow M$ using the following definition. If $u, v : I \longrightarrow M$, we have $(u, v) \in \rho$ if $(u(i), v(i)) \in \mu$ for every $i \in I$.

The relation ρ on $I \longrightarrow M$ is a partial order. We verify only the antisymmetry and leave for the reader the proof of the reflexivity and transitivity. Assume that $(u, v), (v, u) \in \rho$ for $u, v : I \longrightarrow M$. We have $(u(i), v(i)) \in \mu$ and $(v(i), u(i)) \in \mu$ for every $i \in I$. Therefore, taking into account the antisymmetry of μ, we obtain $u(i) = v(i)$ for all $i \in I$; hence, $u = v$, which

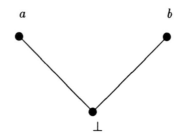

FIGURE 3.4. The Hasse diagram of the poset $\{\perp, a, b\}$.

proves the antisymmetry of ρ.

For a set of functions $F \subseteq I \longrightarrow M$, define the subset $F(i)$ of M as $F(i) = \{f(i) \mid f \in F\}$ for $i \in I$.

Theorem 3.6.1 *The subset F of the poset $(I \longrightarrow M, \rho)$ has a supremum if and only if $\sup F(i)$ exists for every $i \in I$ in the poset (M, μ).*

Proof. Suppose that $\sup F(i)$ exists for every $i \in I$ in the poset (M, μ). Define the mapping $g : I \longrightarrow M$ by $g(i) = \sup F(i)$ for every $i \in I$. We claim that g is $\sup F$.

If $f \in F$, then $(f(i), g(i)) \in \mu$ for every $i \in I$ because of the definition of g. This shows that $(f, g) \in \rho$; hence, g is an upper bound of F. Let h be an upper bound of F. For every $f \in F$, we have $(f(i), h(i)) \in \mu$ for $i \in I$. The definition of g implies $(g(i), h(i)) \in \mu$ for $i \in I$; hence, $g = \sup F$.

Conversely, assume that $k = \sup F$ exists in the poset $(I \longrightarrow M, \rho)$. We prove that $k(i)$ is $\sup F(i)$ for every $i \in I$ in the poset (M, μ).

The definition of k implies that for every $f \in F$ we have $(f, k) \in \rho$, that is, $(f(i), k(i)) \in \mu$ for every $i \in I$. Therefore, $k(i)$ is an upper bound of the set $F(i)$ for every $i \in I$.

Let l_i be an upper bound for $F(i)$ for $i \in I$. Define the function $l : I \longrightarrow M$ as $l(i) = l_i$ for $i \in I$. Clearly, l is an upper bound of the set F in the poset $(I \longrightarrow M, \rho)$, and therefore, $(k, l) \in \rho$. This, in turn, means that $(k(i), l(i)) = (k(i), l_i) \in \mu$, which shows that $\sup F(i)$ exists and it is equal to $k(i)$. ∎

Example 3.6.2 Consider the poset $(\{\perp, a, b\}, \rho)$, whose Hasse diagram is given in Figure 3.4. For $I = \{1, 2\}$, the set of mappings $I \longrightarrow \{\perp, a, b\}$ is given by the following table.

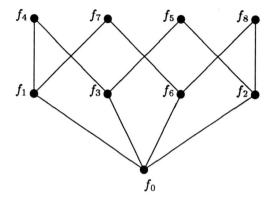

FIGURE 3.5. The Hasse diagram of the poset $\{0,1\} \longrightarrow \{\perp, a, b\}$.

	$f(1)$	$f(2)$
f_0	$\perp$	$\perp$
f_1	$\perp$	a
f_2	$\perp$	b
f_3	a	$\perp$
f_4	a	a
f_5	a	b
f_6	b	$\perp$
f_7	b	a
f_8	b	b

The Hasse diagram of $(I \longrightarrow M, \rho)$ is shown in Figure 3.5.

Another method for building partially ordered set starts from the disjoint sum of a family of sets.

Let $\{(D_i, \rho_i) \mid i \in I\}$ be a family of posets. Consider the disjoint sum

$$D = \bigcup \{D_i \times \{i\} \mid i \in I\}.$$

Definition 3.6.3 *The* sum *of the family* $\{(D_i, \rho_i) \mid i \in I\}$ *is the poset* (D, ρ), *where* $(x, y) \in \rho$ *if there exists* $i \in I$ *such that* $x = (u, i)$, $y = (v, i)$ *and* $(u, v) \in \rho_i$.

Example 3.6.4 The sum of two copies of the poset from Figure 3.4 is given in Figure 3.6.

Let $\{(D_i, \rho_i) \mid i \in I\}$ be a collection of posets. The product $\prod_{i \in I} D_i$ can be equipped with a partial order. Namely, we define the relation ρ on $\prod_{i \in I} D_i$ by $(t, s) \in \rho$ if $(p_i(t), p_i(s)) = (t(i), s(i)) \in \rho_i$ for every $i \in I$. It is

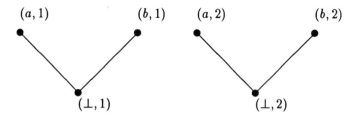

$(a,1)$ $(b,1)$ $(a,2)$ $(b,2)$

$(\perp,1)$ $(\perp,2)$

FIGURE 3.6. The diagram of the sum of two copies of $\{\perp, a, b\}$.

not difficult to verify that ρ is, indeed, a partial order, and we leave this verification to the reader. The relation ρ is denoted by $\prod_{i \in I} \rho_i$ or, when $I = \{1, \ldots, n\}$ by $\rho_1 \times \cdots \times \rho_n$.

Definition 3.6.5 *The* product *of the posets* $\{(D_i, \rho_i) \mid i \in I\}$ *is the poset* (D, ρ), *where* $D = \prod_{i \in I} D_i$. *This new poset will also be denoted by* $\prod_{i \in I}(D_i, \rho_i)$. *When* $I = \{1, \ldots, n\}$, *the product will be denoted by* $(D_1, \rho_1) \times \cdots \times (D_n, \rho_n)$.

Theorem 3.6.6 *Let* $\{(D_i, \rho_i) \mid i \in I\}$ *be a family of partially ordered sets. If* $H \subseteq \prod_{i \in I} D_i$, *then in the product poset,* $\sup H$ *exists if and only if* $\sup p_i(H)$ *exists for every* $i \in I$. *Moreover, if* $y = \sup H$, *then* $p_i(y) = \sup p_i(H)$ *for every* $i \in I$.

Proof. Assume that $y_i = \sup p_i(H)$ exists for every $i \in I$. We need to prove that the element y of $\prod_{i \in I} D_i$ defined by $p_i(y) = y_i$ is $\sup H$.

Consider an arbitrary element $z \in H$. Since $p_i(z) \in p_i(H)$, we have $(p_i(z), y_i) \in \rho_i$, that is, $(p_i(z), p_i(y)) \in \rho_i$ for every $i \in I$. This means that $(z, y) \in \rho$, which shows that y is an upper bound of H.

Suppose now that v is an arbitrary upper bound of H. To show that y is $\sup H$, we need to prove that y is *the least upper bound* of H, that is, $(y, v) \in \rho$ or, equivalently, that $(p_i(y), p_i(v)) \in \rho_i$ for every $i \in I$.

If v is an upper bound of H, then $p_i(v)$ is an upper bound of $p_i(H)$. Since $p_i(y) = y_i = \sup p_i(H)$, we obtain immediately $(p_i(y), p_i(v)) \in \rho_i$ for every $i \in I$.

Conversely, suppose that $\sup H$ exists. Let $y = \sup H$ and let $y_i = p_i(y)$ for every $i \in I$. We have $x_i \in p_i(H)$ if there is $x \in H$ such that $p_i(x) = x_i$. Since $(x, y) \in \rho$, it follows that $(x_i, p_i(y)) \in \rho_i$, which shows that $p_i(y)$ is an upper bound for $p_i(H)$.

Let w_i be an arbitrary upper bound of $p_i(H)$ for every $i \in I$. There is $w \in \prod_{i \in I} D_i$ such that $p_i(w) = w_i$, and we have $(y, w) \in \rho$ because of the fact that w is an upper bound for H. Consequently, $(p_i(y), p_i(w)) = (y_i, w_i) \in \rho_i$, and this means that $y_i = \sup p_i(H)$ for every $i \in I$. ∎

Now, let $\{D_i \mid 1 \leq i \leq n\}$ be a finite family of n sets. Another kind of order that can be introduced on $D_1 \times \cdots \times D_n$ is defined below.

Proposition 3.6.7 For $f, g \in D_1 \times \cdots \times D_n$ define $f \preceq g$ if $f = g$ or if there is k, $1 \leq k \leq n$, such that $f(k) \neq g(k)$, $f(i) = g(i)$ for $1 \leq i < k$ and $(f(k), g(k)) \in \rho_k$.
   The relation $\preceq$ is a partial order on $D_1 \times \cdots \times D_n$.

Proof. The relation $\preceq$ is obviously reflexive. Suppose now that $f \preceq g$ and $g \preceq f$ and that $f \neq g$. There are $k, h \in \mathbb{N}$ such that $f(i) = g(i)$ for $1 \leq i < k$, $(f(k), g(k)) \in \rho_k - \iota_{D_k}$, and $f(i) = g(i)$ for $1 \leq i < h$, $(f(h), g(h)) \in \rho_h - \iota_{D_h}$. If $k < h$, this leads to a contradiction since we cannot have $(f(k), g(k)) \in \rho_k - \iota_{D_k}$ and $f(k) = g(k)$. The case $h < k$ also results in a contradiction. For $k = h$, the previous supposition implies $(f(k), g(k)) \in \rho_k - \iota_{D_k}$ and $(g(k), f(k)) \in \rho_k - \iota_{D_k}$, which is contradictory because of the fact that $\rho_k - \iota_{D_k}$ is a strict partial order.
   Assume that $f \preceq g$ and $g \preceq l$ and that $f \neq g$, $g \neq l$. There are $k, h \in \mathbb{N}$ such that $f(i) = g(i)$ for $1 \leq i < k$, $(f(k), g(k)) \in \rho_k - \iota_{D_k}$, and $g(i) = l(i)$ for $1 \leq i < h$, $(g(h), l(h)) \in \rho_h - \iota_{D_h}$. Define p as being the least of the numbers k, h. For $1 \leq i < p$, we have $f(i) = g(i) = l(i)$. In addition, we have $(f(p), l(p)) \in \rho_p$. Three cases may arise now:

1. $f(p) = g(p)$ and $(g(p), l(p)) \in \rho_p - \iota_{D_p}$ (when $k > h$),

2. $(f(p), g(p)) \in \rho_p - \iota_{D_p}$ and $g(p) = l(p)$ (when $k < h$), and

3. $(f(p), g(p)), (g(p), l(p)) \in \rho_p - \iota_{D_p}$ (when $k = h$).

If $f = l$, then we have $f \preceq l$. Therefore, we can assume that $f \neq l$. In the first two of the cases mentioned above, this would imply immediately $f \preceq l$ because of the fact that $(f(p), l(p)) \in \rho_p - \iota_{D_p}$. The same conclusion can be reached in the third case because of the transitivity of the strict partial order $\rho_p - \iota_{D_p}$. ∎
   We refer to the partial order "$\preceq$" as the *lexicographic partial order* on $D_1 \times \cdots \times D_n$.

Example 3.6.8 Consider the poset $(\mathbb{N} \times \mathbb{N}, \preceq)$. We claim that this poset is well-founded. By Theorem 3.4.13, we have to prove that it is artinian. Observe that if $(m, n_0) \succ (m, n_1) \succ \ldots$ is a descending chain of pairs having the same first component, then $n_0 > n_1 > \ldots$ is a descending chain of natural numbers and such a chain is finite. Therefore, $(m, n_0) \succ (m, n_1) \succ \ldots$ must be a finite chain.
   Consider now an arbitrary descending chain

$$(p_0, q_0) \succ (p_1, q_1) \succ \ldots$$

in $(\mathbb{N} \times \mathbb{N}, \preceq)$. We have $p_0 \geq p_1 \geq \ldots$, and in this sequence we may have only finite "constant" fragments $p_k = p_{k+1} = \cdots = p_{k+l}$. Therefore, the chain of the first components of the pairs of the sequence $(p_0, q_0) \succ (p_1, q_1) \succ \ldots$ is ultimately decreasing, and this shows that the chain is finite.

3.7 Functions and Posets

Let (M_1, ρ_1) and (M_2, ρ_2) be two posets.

Definition 3.7.1 *A* morphism *between* (M_1, ρ_1) *and* (M_2, ρ_2) *or a mono-tonic mapping between* (M_1, ρ_1) *and* (M_2, ρ_2) *is a mapping* $f : M_1 \longrightarrow M_2$ *such that* $(u, v) \in \rho_1$ *implies* $(f(u), f(v)) \in \rho_2$ *for any* $u, v \in M_1$.
   The set of monotonic functions between (M_1, ρ_1) *and* (M_2, ρ_2) *will be denoted by* $M_1 \longrightarrow_m M_2$.

If f is a morphism and $(u, v)\rho_1$, $u \neq v$ implies $(f(u), f(v)) \in \rho_2$ and $f(u) \neq f(v)$, then we refer to f as a *strict morphism* or as a *strict monotonic function*.

Example 3.7.2 Consider a set M, the poset $(\mathcal{P}(M), \subseteq)$, and the functions $f, g : (\mathcal{P}(M))^2 \longrightarrow \mathcal{P}$, defined by $f(K, H) = K \cup H$ and $g(K, H) = K \cap H$, for $K, H \in \mathcal{P}(M)$. If the Cartesian product is equipped with the product partial order, then both f and g are monotonic. Indeed, if $(K_1, H_1) \subseteq (K_2, H_2)$, we have $K_1 \subseteq K_2$ and $H_1 \subseteq H_2$, which implies that

$$f(K_1, H_1) = K_1 \cup H_1 \subseteq K_2 \cup H_2 = f(K_2, H_2).$$

The argument for g is similar, and it is left to the reader.

Example 3.7.3 Let $\{(D_i, \rho_i) \mid i \in I\}$ be a collection of posets and let

$$(\prod_{i \in I} D_i, \rho)$$

be the product of these posets. The projections $p_i : \prod_{i \in I} D_i \longrightarrow D_i$ are monotonic mappings as the reader will easily verify.

Example 3.7.4 Let (M, ρ) be a poset. If I is an arbitrary set, we can consider the poset (I, ι_I). Any function $f : I \longrightarrow M$ is monotonic.

Theorem 3.7.5 *Let* (D_i, ρ_i), $1 \leq i \leq 3$ *be three posets and let* $f : D_1 \longrightarrow D_2$, $g : D_2 \longrightarrow D_3$ *be two monotonic mappings. The mapping* $g \circ f : D_1 \longrightarrow D_3$ *is also monotonic.*

   Proof. Let $x, y \in D_1$ be such that $(x, y) \in \rho_1$. In view of the monotonic-ity of f, we have $(f(x), f(y)) \in \rho_2$, and this implies $(g(f(x)), g(f(y))) \in \rho_3$, because of the monotonicity of g. Therefore, we obtain that $g \circ f$ is mono-tonic. ∎
   Let (D_1, ρ), (D_2, θ) be two posets. For a monotonic function $f : D_1 \longrightarrow D_2$, the quotient set, $D_1/\ker(f)$ can also be organized as a poset. Indeed, if $[x], [y] \in D_1/\ker(f)$, then we shall define $([x], [y]) \in \mu$ if $(f(x), g(x)) \in \theta$. The partial order μ is well defined because, if $x' \in [x]$ and $y' \in [y]$, we have $(f(x'), f(y')) = (f(x), f(y)) \in \theta$.

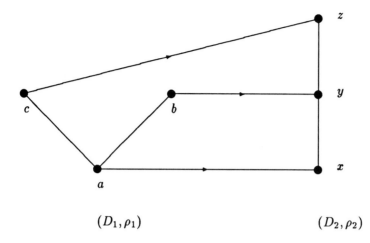

(D_1, ρ_1) (D_2, ρ_2)

FIGURE 3.7. The inverse of a monotonic bijection.

Proposition 3.7.6 *The mapping* $g : D_1 \longrightarrow D_1/\text{ker}(f)$ *considered between the posets* (D_1, ρ) *and* $(D_1/\text{ker}(f), \mu)$ *is monotonic.*

Proof. The argument is straightforward, and it is left to the reader as an exercise. ∎

Let $f : D_1 \longrightarrow D_2$ be a monotonic bijection between the posets (D_1, ρ_1) and (D_2, ρ_2). As we have seen in Chapter 2, the inverse f^{-1} is also a bijection. Nevertheless, the inverse *is not* necessarily monotonic, as follows from the next example.

Example 3.7.7 Let (D_1, ρ_1) and (D_2, ρ_2) be the posets whose Hasse diagrams are given in Figure 3.7 and consider the mapping $f : D_1 \longrightarrow D_2$ defined by $f(a) = x, f(b) = y$, and $f(c) = z$.

Let (D_1, ρ_1) and (D_2, ρ_2) be two posets. The previous considerations justify the following definition.

Definition 3.7.8 A poset isomorphism *is a monotonic bijective mapping* $f : D_1 \longrightarrow D_2$ *for which the inverse mapping* f^{-1} *is also monotonic.*
The posets (D_1, ρ_1) *and* (D_2, ρ_2) *are said to be* isomorphic.

Example 3.7.9 Let $\{p_1, p_2, \dots, p_n\}$ be the first n primes, $p_1 = 2$, $p_2 = 3$, $p_3 = 5$, etc. Let $m = p_1 \cdots p_n$ be their product and let D_m be the set of all divisors of m. Consider an arbitrary set $A = \{a_1, \dots, a_n\}$ having n elements.

We claim that the posets $(\mathcal{P}(A), \subseteq)$ and (D_m, δ) are isomorphic. Indeed, define the mapping $f : \mathcal{P}(A) \longrightarrow D_m$ by $f(\emptyset) = 1$ and $f(\{a_{i_1}, \dots, a_{i_k}\}) = p_{i_1} \cdots p_{i_k}$.

The mapping f is bijective. Indeed, for any divisor h of m, we have $h = p_{i_1} \cdots p_{i_k}$, and therefore, $h = f(\{a_{i_1}, \ldots, a_{i_k}\})$, which shows that f is surjective.

If $f(\{a_{i_1}, \ldots, a_{i_k}\}) = f(\{a_{j_1}, \ldots, a_{j_l}\})$, then $p_{i_1} \cdots p_{i_k} = p_{j_1} \cdots p_{j_l}$. This gives $k = l$ and $i_1 = j_1, \ldots, i_k = j_k$; hence, $\{a_{i_1}, \ldots, a_{i_k}\} = \{a_{j_1}, \ldots, a_{j_l}\}$, which proves that f is injective.

The mapping f is monotonic because, if $\{a_{i_1}, \ldots, a_{i_k}\} \subseteq \{a_{j_1}, \ldots, a_{j_l}\}$,

$$\{i_1, \ldots, i_k\} \subseteq \{j_1, \ldots, j_l\},$$

and this means that the number $p_{i_1} \cdots p_{i_k}$ divides $p_{j_1} \cdots p_{j_l}$.

The inverse mapping $g : D_m \longrightarrow \mathcal{P}(A)$ is also monotonic; we leave the argument to the reader.

Monotonic functions map chains to chains. We state this formally in the next theorem.

Theorem 3.7.10 *Let (M_1, ρ_1), (M_2, ρ_2) be two posets and $f : M_1 \longrightarrow_m M_2$ be a monotonic function. If $L \subseteq M_1$ is a chain in (M_1, ρ_1), then $f(L)$ is a chain in (M_2, ρ_2).*

Proof. Let $u, v \in f(L)$ be two elements of $f(L)$. There exist $x, y \in L$ such that $f(x) = u$ and $f(y) = v$. Since L is a chain, we have either $(x, y) \in \rho_1$ or $(y, x) \in \rho_1$. In the former case, the monotonicity of f implies $(u, v) \in \rho_2$; in the latter situation, we have $(v, u) \in \rho_2$. ∎

Partial functions from the set $A \rightsquigarrow B$ can be represented as monotonic functions between the flat posets $A_\perp$ and $B_\perp$. Let $\sqsubseteq_A, \sqsubseteq_B$ be the partial orders on the two sets.

Definition 3.7.11 *A function $f : A_\perp \longrightarrow B_\perp$ is strict if $f(\perp_A) = \perp_B$.*

We shall denote by $A \longrightarrow_s B$ the set of all strict functions between the posets $A_\perp$ and $B_\perp$.

Proposition 3.7.12 *A function between two flat posets is strict if and only if it is a strict monotonic function (between the flat posets $A_\perp$ and $B_\perp$) or it is the constant function $f(x) = \perp_B$ for every $x \in A$.*

Proof. Let f be a strict function and suppose that $x, y \in A_\perp$ with $(x, y) \in \rho_A$ and $x \neq y$; this implies $x = \perp_A$ and $y \in A$. If f is not the constant function mentioned above, we have $f(y) \in B$; hence, $f(y) \neq \perp_B = f(x)$; also,

$$(f(x), f(y)) = (\perp_B, f(y)) \in \sqsubseteq_B;$$

hence, f is a strict monotonic function.

Conversely, if f is the constant function defined by $f(x) = \perp_B$ for all $x \in A_\perp$, then f is clearly a strict function. Let us assume, therefore, that f is a strict monotonic function. Consider $x \in A$. Since $\perp_A \sqsubseteq_A x$ and

$\perp_A \neq x$, we have $f(\perp_A) \neq f(x)$ and $f(\perp_A) \sqsubseteq_B f(x)$. This can happen only if $f(\perp_A) = \perp_B$. The function f turns out to be strict in both cases. ∎

The representation of partial functions by strict functions between flat posets is formalized in the following proposition.

Proposition 3.7.13 *There exists a bijection* Ψ *between the sets* $A \rightsquigarrow B$ *and* $A \longrightarrow_s B$.

Proof. For $f : A \rightsquigarrow B$, define the mapping $\Psi(f) : A \longrightarrow_s B$ by

$$\Psi(f)(a) = \begin{cases} f(a) & \text{if } a \in \mathrm{Dom}(f) \\ \perp_B & \text{if } a \notin \mathrm{Dom}(f) \text{ or } a = \perp_A . \end{cases}$$

If is clear that $\Psi(f)$ is a strict function. The mapping Ψ is onto because, if $g : A_\perp \longrightarrow B_\perp$ is a strict function and $g : A \rightsquigarrow B$ is the partial function defined by $\mathrm{Dom}(g) = \{a \mid f(a) \neq \perp_B\}$ and $g(a) = f(a)$ if $f(a) \neq \perp_B$, we have $\Psi(g) = f$.

If $\Psi(f_1) = \Psi(f_2)$, where $f_1, f_2 : A \longrightarrow_s B$, then $\mathrm{Dom}(f_1) = \mathrm{Dom}(f_2)$ and $f_1(a) = f_2(a)$ for all $a \in \mathrm{Dom}(f_1) = \mathrm{Dom}(f_2)$, which proves that Ψ is injective. ∎

For a total function $f : A \longrightarrow B$ the corresponding strict partial function is given by

$$\Psi(f)(a) = \begin{cases} f(a) & \text{if } a \in A, \\ \perp_B & \text{if } a = \perp_A . \end{cases}$$

This is a special case of another interesting situation that occurs when we deal with a total function having n arguments $f : A^n \longrightarrow B$ and we seek to extend f to $(A_\perp)^n$.

Definition 3.7.14 *The natural extension of* f, $\hat{f} : (A_\perp)^n \longrightarrow B_\perp$ *is obtained by*

$$\hat{f}(a_1, \ldots, a_n) = \begin{cases} f(a_1, \ldots, a_n) & \text{if } a_i \neq \perp \text{ for } 1 \le i \le n, \\ \perp & \text{if } a_i = \perp \text{ for some } i, 1 \le i \le n. \end{cases}$$

Proposition 3.7.15 *The natural extension of a function* $f : A^n \longrightarrow B$ *is a monotonic function between the posets* $(A_\perp)^n$ *and* $B_\perp$.

Proof. The proof is left as an exercise to the reader. ∎

It is possible to consider extensions of total functions that are distinct from their natural extensions and yet obtain monotonic functions.

Example 3.7.16 Consider a set A and the function $f : \mathbf{Bool} \times A \times A \longrightarrow A$, where $\mathbf{Bool} = \{\mathbf{T}, \mathbf{F}\}$ and f is given by

$$f(x, y, z) = \begin{cases} y & \text{if } x = \mathbf{F} \\ z & \text{if } x = \mathbf{T}. \end{cases}$$

The definition of f justifies the notation $f(x, y, z) =$ if x then y else z.

We shall consider the extension of f to the product of flat posets $\mathbf{Bool}_\perp \times A_\perp \times A_\perp$ defined by

$$\text{if } x \text{ then } y \text{ else } z = \begin{cases} y & \text{if } x = \mathbf{F} \text{ for } z \in A_\perp \\ z & \text{if } x = \mathbf{T} \text{ for } y \in A_\perp \\ \perp & \text{if } x = \perp. \end{cases}$$

Let $\sqsubseteq$ be the order on the product $\mathbf{Bool}_\perp \times A_\perp \times A_\perp$. If $(x, y, z) \sqsubseteq (x', y', z')$, then, by examining all possible cases, we reach the conclusion that $f(x, y, z) \sqsubseteq f(x', y', z')$. We leave to the reader this verification.

3.8   Complete Partial Orders

Definition 3.8.1 *A complete poset or, in short, a cpo, is a poset (M, ρ) satisfying the following two conditions:*

1. *M has a least element, and*

2. *for every chain K of M, there is $\sup K$.*

In general, we shall denote the least element of a complete poset (M, ρ) by $\perp_M$ or, simpler, by $\perp$ if there is no risk of confusion.

Example 3.8.2 The simplest type of complete posets are the flat posets. Indeed, if (M, ρ) is such a poset then the first condition of Definition 3.8.1 is automatically satisfied. On the other hand, such a poset has only chains of length 2 having the form $\{\perp, x\}$, and clearly, any such chain has a supremum.

As an immediate generalization of flat posets, any poset that has a least element and whose chains are finite is also a complete poset.

Example 3.8.3 Let D be a set. The set $(\mathcal{P}(D), \subseteq)$ is a complete poset. Indeed, $\emptyset$ is the least element of the poset. For any chain of subsets of D, $\{D_i \mid i \in I\}$, the supremum is $\bigcup_{i \in I} D_i$.

Example 3.8.4 The poset $(A \rightsquigarrow B, \sqsubseteq)$ considered in Example 3.2.4 is a cpo. The least element of this poset is the empty partial function $\emptyset : A \rightsquigarrow B$. For a chain of partial functions, $L = \{f_i \mid i \in I\}$, we can define the function $f : A \rightsquigarrow B$ as having the domain $\text{Dom}(f) = \bigcup_{i \in I} \text{Dom}(f_i)$.

For $x \in \text{Dom}(f)$, there exists a function $f_i \in L$ such that $x \in \text{Dom}(f_i)$; we shall define $f(x)$ as $F(x) = f_i(x)$. This definition is correct because if $x \in \text{Dom}(f_j)$, then, taking into account that we have either $f_i \sqsubseteq f_j$ or $f_j \sqsubseteq f_i$, we obtain $f_i(x) = f_j(x)$.

We have $\sup L = f$. The function f is clearly an upper bound for L. On the other hand, if g is another partial function such that $f_i \sqsubseteq g$ for every

$i \in I$, we shall have $\bigcup_{i \in I} \mathrm{Dom}(f_i) \subseteq \mathrm{Dom}(g)$ and $f_i(x) = g(x)$ in all cases when $x \in \mathrm{Dom}(f_i)$. This allows us to infer that $f \sqsubseteq g$; hence, f is indeed the least upper bound for L.

In Section 3.6, we have considered three basic methods for constructing posets, namely, the formation of a set of functions, the products and the sum of posets. Now, we re-examine these methods in order to determine if, starting from complete posets, we can produce new complete posets. For the formation of the set of functions and for the product, the answer is affirmative.

Theorem 3.8.5 *Let (M, μ) be a cpo and let I be an arbitrary set. The poset $(I \longrightarrow M, \mu)$ is complete, where $(f, g) \in \mu$ if $(f(i), g(i)) \in \mu$ for all $i \in I$. Furthermore, if $\{f_j \mid j \in J\}$ is a chain in $(I \longrightarrow M, \mu)$, then, for $f = \sup\{f_j \mid j \in J\}$, we have $f(i) = \sup\{f_j(i) \mid j \in J\}$ for every $i \in I$.*

Proof. Note that we have used the same notation (μ) for the relation defined on the set of functions $I \longrightarrow M$ as for the relation defined on M.

The least element of the set of functions $I \longrightarrow M$ is the function f_0 defined by $f_0(x) = \bot_M$.

Consider now a chain of functions $\{f_j \mid j \in J\}$. For $i \in I$, the set $\{f_j(i) \mid j \in J\}$ is a chain in (M, μ), and therefore, we have the element $u_i = \sup\{f_j(i) \mid j \in J\}$. We shall define the function $f : I \longrightarrow M$ by $f(i) = u_i$ for every $i \in I$. It is straightforward to verify that $f = \sup\{f_j(i) \mid j \in J\}$; hence, $(I \longrightarrow M, \mu)$ is complete. ∎

Example 3.8.6 Consider a function $f : \mathbf{N}_\bot \longrightarrow \mathbf{N}_\bot$. We can approximate this function through a chain of functions $\{f_i \mid i \in \mathbf{N}\}$ defined by

$$f_i(n) = \begin{cases} f(n) & \text{if } n \leq i \\ \bot & \text{otherwise.} \end{cases}$$

It is straightforward to verify that $\{f_i \mid i \in \mathbf{N}\}$ is a chain. Furthermore, the supremum of this chain is the function f itself.

Theorem 3.8.7 *Let $\{(D_i, \rho_i) \mid i \in I\}$ be a family of complete partial orders. The product of this family $(\prod_{i \in I} D_i, \rho)$ is also a complete partial order.*

Proof. If $\bot_i$ is the least element of D_i, then the tuple t_0 defined by $t_0(i) = \bot_i$ for every $i \in I$, is the least element of $(\prod_{i \in I} D_i, \rho)$.

Let L be a chain in $(\prod_{i \in I} D_i, \rho)$. Because of Theorem 3.6.6, we need to prove only that each of the projections of L has a supremum in its own set. This follows immediately from the facts that a projection $p_i(L)$ of a chain L is also a chain in (D_i, ρ_i) and (D_i, ρ_i) is a complete poset. ∎

For the direct sum of posets, the problem is more complicated. As we can see from Example 3.6.4 the direct sum of a family of posets does not necessarily have a least element, even if each of the participating sets have one.

Let $\mathcal{C} = \{(D_i, \rho_i) \mid i \in I\}$ be a collection of posets. Let $\perp_i$ be the least element of the poset (D_i, ρ_i) for $i \in I$.

Definition 3.8.8 The separate sum *of $\mathcal{C}$ is the poset*

$$\left(\bigcup_{i \in I} D_i \times \{i\} \cup \{\perp\}, \rho_s \right),$$

where $\perp$ is a new element, $\perp \notin \bigcup_{i \in I} D_i$, and $(x, y) \in \rho_s$ if one of the following two things happen:

1. *there is $i \in I$ such that $x = (u, i)$, $y = (v, i)$ and $(u, v) \in \rho_i$, or*

2. *$x = \perp$ and $y \in \bigcup_{i \in I} D_i \times \{i\} \cup \{\perp\}$.*

There is another method for defining the sum of posets.

Definition 3.8.9 The coalesced sum *of $\mathcal{C}$ is the poset*

$$\left(\bigcup_{i \in I} ((D_i - \{\perp_{D_i}\}) \times \{i\}) \cup \{\perp\}, \sigma_s \right),$$

where $(x, y)\sigma_s$, if one of the following situations occur:

1. *there is $i \in I$ such that $x = (u, i)$, $y = (v, i)$, $u \neq \perp_{D_i}$, $v \neq \perp_{D_i}$, and $(u, v) \in \rho_i$ or,*

2. *$x = \perp$ and $y \in \bigcup_{i \in I} ((D_i - \{\perp_{D_i}\}) \times \{i\}) \cup \{\perp\}$.*

We leave to the reader, as an exercise, the proof of the following theorem.

Theorem 3.8.10 *If $\{(D_i, \rho_i) \mid i \in I\}$ are complete posets, then both their separate and their coalesced sums are complete posets.*

Let (M, ρ) be a complete poset and let H be a subset of M.

Definition 3.8.11 *H is a* complete subset *of the complete poset (M, ρ) or, simpler, a* sub-cpo *if (H, ρ) is a complete poset and, for each chain L such that $L \subseteq H$, the supremum of L in (H, ρ), denoted by $\sup_H L$, coincides with $\sup L$ in (M, ρ).*

The following theorem is a characterization of complete subsets.

Theorem 3.8.12 *H is a sub-cpo of the cpo (M, ρ) if H has a least element and for every chain $L \subseteq H$ we have $\sup L \in H$, where $\sup L$ is considered in the poset (M, ρ).*

Proof. The condition of the theorem is necessary. Indeed, let $L \subseteq H$ be a chain, where (H, ρ) is a sub-cpo. Clearly, $\sup_H L \in H$, and this implies $\sup L = \sup_H L \in H$.

In order to prove that the condition is sufficient, let us consider a chain L included in H. We need to prove that L has a supremum in H, knowing that L has a supremum in M and that $\sup L \in H$. Clearly, $\sup L$ is an upper bound for L with respect to the subset H; furthermore, if w is an upper bound of L in H, then, since $w \in M$, we also have $(\sup L, w) \in \rho$, and this proves that $\sup L$ is also the supremum of L in H and (H, ρ) is a complete poset for which $\sup_H L = \sup L$ for any chain L in H.

Example 3.8.13 We have seen that $(\mathcal{P}(\mathbf{N}), \subseteq)$ is a complete poset. Consider the collection $\mathcal{H} = \{\emptyset, \mathbf{N}\} \cup \{T \mid T \in \mathcal{P}(\mathbf{N}), T \text{ is finite }\}$.

$(\mathcal{H}, \subseteq)$ is itself a complete poset. Indeed, $\emptyset$ is its least element, and for a chain L we can have one of the following two cases

1. L is finite. In this case, the largest element of L is $\sup_H L$.

2. L is infinite. In this situation, $\sup_H L = \mathbf{N}$.

However, it is not difficult to see that $\mathcal{H}$ is not a sub-cpo of the cpo $(\mathcal{P}(\mathbf{N}), \subseteq)$. For instance, define $\mathcal{E}$ to be the chain $\mathcal{E} = \{E_n \mid n \geq 0\}$, where $E_n = \{0, 2, 4, \ldots, 2n\}$ for $n \geq 0$. We have $E_n \in \mathcal{H}$ for every $n \in \mathbf{N}$. On the other hand, $\sup_{\mathcal{H}} \mathcal{E} = \mathbf{N}$, while $\sup \mathcal{E}$ is the set E of all even numbers. Clearly, $E \notin H$; hence, $\mathcal{H}$ is not a complete subset.

Example 3.8.14 We have seen in Theorem 3.8.5 that if (M, μ) is a complete poset and I is an arbitrary set then the poset $(I \longrightarrow M, \rho)$ is complete. Assume now that (I, θ) is a poset and consider the set of monotonic functions $I \longrightarrow_m M$. We claim that $I \longrightarrow_m M$ is a closed subset of $(I \longrightarrow M, \rho)$. Note that the bottom function f_0 defined by $f(i) = \bot_M$ is monotonic. For a chain $\{f_j \mid j \in J\}$ of monotonic function in $(I \longrightarrow M, \rho)$, define the function $f : I \longrightarrow M$ by $f(i) = \sup\{f_j(i) \mid i \in I\}$. We need to prove that f is monotonic.

If $(i, i') \in \theta$, the monotonicity of a function f_j implies $(f_j(i), f_j(i')) \in \mu$. This, in turn, implies $(\sup\{f_j(i) \mid i \in I\}, \sup\{f_j(i') \mid i \in I\}) \in \mu$, that is, $(f(i), f(i')) \in \mu$ for all $i \in I$. Therefore, f is monotonic, and this concludes the argument.

Let (M_1, ρ_1) and (M_2, ρ_2) be two complete posets.

Definition 3.8.15 *A continuous function is a function* $f : M_1 \longrightarrow M_2$ *such that for every chain L of the cpo* (M_1, ρ_1) *the* $\sup f(L)$ *exists in* (M_2, ρ_2) *and* $f(\sup L) = \sup f(L)$.

In other words, a function $f : M_1 \longrightarrow M_2$ is continuous if it is compatible with the formation of suprema of chains. We shall denote the set of continuous functions between (M_1, ρ_1) and (M_2, ρ_2) by $M_1 \longrightarrow_c M_2$.

Let us remark that $M_1 \longrightarrow_c M_2 \subseteq M_1 \longrightarrow_m M_2$. Indeed, if $f : M_1 \longrightarrow_c M_2$ and $(x,y) \in \rho_1$, then $\{x,y\}$ is a chain, $\sup\{f(x), f(y)\}$ exists, and $f(\sup\{x,y\}) = f(y) = \sup\{f(x), f(y)\}$; this shows that $(f(x), f(y)) \in \rho_2$; hence, f is monotonic.

The above inclusion is, in general, a strict one.

Example 3.8.16 Consider the set **N** of natural numbers equipped with the usual partial order $(\mathbf{N}, \leq)$. In order to obtain a cpo, let us adjoin a new element ω to the set **N**. The partial order $\rho \subseteq (\mathbf{N} \cup \{\omega\}) \times (\mathbf{N} \cup \{\omega\})$ is defined by $(x,y) \in \rho$ if $x, y \in \mathbf{N}$ and $x \leq y$, or $x \in \mathbf{N} \cup \{\omega\}$ and $y = \omega$.

The set $(\mathbf{N} \cup \{\omega\}, \rho)$ is a cpo; its least element is 0, and for each chain L there exists $\sup L$. Indeed, if L is finite, then $\sup L$ is the greatest element of L; otherwise, $\sup L = \omega$.

Consider now the function $f : \mathbf{N} \cup \{\omega\} \longrightarrow \{0,1\}$, where $\{0,1\}$ is endowed with the usual order. The function f is defined by

$$f(n) = \begin{cases} 0 & \text{if } n \in \mathbf{N} \\ 1 & \text{if } n = \omega, \end{cases}$$

and it is obviously monotonic. For the chain **N**, we have $\sup \mathbf{N} = \omega$. However, $f(\mathbf{N}) = \{0\}$ and $\sup f(\mathbf{N}) = 0 \neq \omega = f(\sup \mathbf{N})$.

Continuous functions are characterized in the class of monotonic functions by the following theorem.

Theorem 3.8.17 *The monotonic function* $f : M_1 \longrightarrow_m M_2$ *between the complete posets* (M_1, ρ_1) *and* (M_2, ρ_2) *is continuous if for every chain* $L \subseteq M_1$ *we have* $(f(\sup L), \sup f(L)) \in \rho_2$.

Proof. As we have seen in Theorem 3.7.10, the set $f(L)$ is a chain. Therefore, $\sup f(L)$ exists in (M_2, ρ_2). Furthermore, for any $x \in L$, we have $(x, \sup L) \in \rho_1$, and applying the monotonicity of f, we obtain

$$(f(x), f(\sup L)) \in \rho_2.$$

This shows that $f(\sup L)$ is an upper bound for $f(L)$; hence,

$$(\sup f(L), f(\sup L)) \in \rho_2.$$

Combining this with the condition of the theorem, we have the equality $f(\sup L) = \sup f(L)$ because of the antisymmetry of ρ_2. This shows that the conditions of the theorem guarantee the continuity. ∎

There are many posets which present a great interest for computer science (e.g., flat posets and any finite products of such posets), whose chains have finite length. For functions defined on such posets, the monotonicity is equivalent to the continuity, as shown by the following theorem.

Theorem 3.8.18 *If* (M_1, ρ_1) *contains only finite chains, then*

$$M_1 \longrightarrow_c M_2 = M_1 \longrightarrow_m M_2.$$

Proof. Let $f : M_1 \longrightarrow_m M_2$ be a monotonic function and let $L = \{x_1, \ldots, x_n\}$ be a chain of the subset (M, ρ_1). The supremum of this chain coincides with its greatest element; hence, $f(\sup L) = f(x_n)$. The monotonicity of f implies that $f(L) = \{f(x_1), \ldots, f(x_n)\}$ is also a chain, and therefore, $\sup f(L) = f(x_n)$. The equality $f(\sup L) = \sup f(L)$ for every chain of (M_1, ρ_1) means that f is monotonic. ∎

Let $\{(D_i, \rho_i) \mid 1 \leq i \leq n\}$ be n flat posets and let (M, ρ) be a cpo.

Corollary 3.8.19 *Any monotonic function* $f : D_1 \times \cdots \times D_n \longrightarrow_m M$ *is continuous.*

Proof. Observe that in the product poset $\prod_{1 \leq i \leq n}(D_i, \rho_i)$ the maximal length of a chain is $n + 1$. The corollary follows immediately from the previous theorem. ∎

According to Theorem 3.8.5, the set of functions $M_1 \longrightarrow M_2$ between two cpos (M_1, ρ_1) and (M_2, ρ_2) is itself a cpo.

Theorem 3.8.20 *The set of continuous functions* $M_1 \longrightarrow_c M_2$ *between the cpos* (M_1, ρ_1) *and* (M_2, ρ_2) *is a sub-cpo of the cpo* $M_1 \longrightarrow M_2$.

Proof. The least element f_0 of $M_1 \longrightarrow M_2$ defined by $f_0(x) = \perp_{M_2}$ for all $x \in M_1$ is obviously a continuous function. Therefore, we need to prove that for any chain of continuous functions $L = \{f_i \mid i \in I\}$, $f = \sup L$ is also a continuous function.

Let $K = \{x_j \mid j \in J\}$ be a chain in the cpo (M_1, ρ_1). If we define the function $\phi : I \times J \longrightarrow M_2$ by $\phi(i, j) = f_i(x_j)$, then the four conditions of Corollary 3.3.15 are satisfied:

1. $u_j = \sup_{i \in I} \phi(i, j) = \sup_{i \in I} f_i(x_j)$ exists, since $\{f_i(x_j) \mid i \in I\}$ is a chain in M_2 for every $j \in J$. Furthermore, $u_j = f(x_j)$ for every $j \in J$, and the set $\{u_j \mid j \in J\}$ is a chain in M_2.

2. $v_i = \sup_{j \in J} \phi(i, j) = \sup_{j \in J} f_i(x_j) = f_i(\sup K)$ exists for every $i \in I$ since the image of any chain through a continuous function is a chain. The set $\{v_i \mid i \in I\}$ is a chain.

3. $\sup_{j \in J} \sup_{i \in I} \phi(i, j) = \sup_{j \in J} u_j = \sup_{j \in J} f(x_j)$ exists.

4. $\sup_{i \in I} \sup_{j \in J} \phi(i, j) = \sup_{i \in I} f_i(\sup K)$ exists.

Therefore, we have the equality

$$\sup_{j \in J} f(x_j) = \sup_{i \in I} f_i(\sup K),$$

which amounts to $\sup f(K) = f(\sup K)$. This proves that f is continuous; hence, $M_1 \longrightarrow_c M_2$ is a sub-cpo of $M_1 \longrightarrow M_2$. ∎

The rest of this section is dedicated to the study of some closure properties of the class of continuous functions.

Theorem 3.8.21 *Let* (M_i, ρ_i), $1 \le i \le 3$, *be three cpos and let* $f : M_1 \longrightarrow M_2$, $g : M_2 \longrightarrow M_3$ *be two continuous mappings. The mapping* $g \circ f : M_1 \longrightarrow M_3$ *is also continuous.*

Proof. Let $L = \{x_i \mid i \in I\}$ be a chain in (M_1, ρ_1). The functions f and g are continuous; hence, they are monotonic and so is $g \circ f$. Therefore, $f(L)$ and $g(f(L))$ are chains in (M_2, ρ_2) and (M_3, ρ_3), respectively.

Using the definition of $g \circ f$ and the continuity of f and g, we have $g \circ f(\sup L) = g(f(\sup L)) = g(\sup f(L)) = \sup g(f(L)) = \sup g \circ f(L)$, which proves that $g \circ f$ is continuous. ∎

Let us consider now n cpos (M_i, ρ_i) for $1 \le i \le n$ and their product $(M, \rho) = \prod_{1 \le i \le n}(M_i, \rho_i)$. We have seen that (M, ρ) is itself a cpo.

Theorem 3.8.22 *Any projection* $p_i : M \longrightarrow M_i$ *is a continuous mapping. Furthermore, a mapping* $f : S \longrightarrow M$, *where* (S, σ) *is a cpo, is continuous if and only if* $p_i \circ f$ *is continuous for every* i, $1 \le i \le n$.

Proof. Let $L = \{x_j \mid j \in J\}$ be a chain in (M, ρ). The members of this chain are n-tuples, and we shall write $x_j = (x_{j1}, \ldots, x_{jn})$ in order to introduce a notation for their components.

In view of the definition of ρ, we have the chains $L_i = \{x_{ji} \mid j \in J\}$ in (M_i, ρ_i) for $1 \le i \le n$, respectively. Also, $\sup L$ is $(\sup L_1, \ldots, \sup L_n)$. Since $L_i = p_i(L)$, we infer that $p_i(\sup L) = \sup L_i = \sup p_i(L)$, which amounts to the continuity of p_i.

The necessity of the condition contained by the second part of the theorem is straightforward. Indeed, if $f : S \longrightarrow M$ is continuous, applying Theorem 3.8.21, we obtain that every mapping $p_i \circ f$ is continuous.

Let $K = \{s_h \mid h \in H\}$ be a chain in S and let $s = \sup K$. We have

$$
\begin{aligned}
f(\sup K) &= f(s) = (p_1 \circ f(s), \ldots, p_n \circ f(s)) \\
&= (\sup p_1 \circ f(K), \ldots, \sup p_n \circ f(K)) \\
&= \sup f(K),
\end{aligned}
$$

because of the continuity of $p_1 \circ f, \ldots p_n \circ f$. This shows that f is continuous. ∎

Example 3.8.23 The previous theorem allows us to show that for any cpo (M, ρ) the function $f : M \longrightarrow M^n$ defined by

$$
f(x) = (\underbrace{x, \ldots, x}_{n})
$$

for $x \in M$ is continuous. This follows from the fact that for any projection $p_i : M^n \longrightarrow M$ the function $p_i \circ f$ is the identity 1_M, which is obviously continuous.

Corollary 3.8.24 *Consider n continuous functions $g_i : S_i \longrightarrow_c M_i$ for $1 \le i \le n$. The function*

$$g = g_1 \times \cdots \times g_n : S_1 \times \cdots \times S_n \longrightarrow M_1 \times \cdots \times M_n$$

defined by $g(s_1, \ldots, s_n) = (g_1(s_1), \ldots, g_n(s_n))$ for $(s_1, \ldots, s_n) \in S_1 \times \cdots \times S_n$ is continuous.

Proof. Note that $p_i \circ g(s_1, \ldots, s_n) = s_i = g_i(p_i(s_1, \ldots, s_n))$ for every $(s_1, \ldots, s_n) \in S_1 \times \cdots \times S_n$. Therefore, we have $p_i \circ g = g_i \circ p_i$. The functions $g_i \circ p_i$ are continuous and, because of Theorem 3.8.22, we obtain the continuity of g. ∎

In Exercise 25 of Chapter 2, we saw that there exists a bijection Φ between the sets $M_1 \longrightarrow (M_2 \longrightarrow M_3)$ and $M_1 \times M_2 \longrightarrow M_3$. Namely, if $f : M_1 \longrightarrow (M_2 \longrightarrow M_3)$, then $g = \Phi(f)$ is a function $g : M_1 \times M_2 \longrightarrow M_3$ defined by $g(x, y) = f(x)(y)$ for every $x \in M_1$ and $y \in M_2$. The function f is *said to be the curried form* of the function g, and the construction of f starting from g will be called *curryfication*.[3]

Naturally, the reverse process (that is, the the construction of g starting from f) will be called decurryfication.

We need a preliminary result.

Proposition 3.8.25 *Let (M_i, ρ_i) be three cpos, $1 \le i \le 3$. If g is a function, $g : M_1 \times M_2 \longrightarrow M_3$, and $f : M_1 \longrightarrow (M_2 \longrightarrow M_3)$ is the curried form of g, then g is monotonic if and only if f is monotonic and for every $x \in M_1$, $f(x)$ is a monotonic function between (M_1, ρ_1) and (M_2, ρ_2).*

Proof. Assume that g is monotonic and let $x, x' \in M_1$ such that $(x, x') \in \rho_1$. For any $y \in M_2$, we have $(g(x, y), g(x', y)) \in \rho_3$, and therefore,

$$(f(x)(y), f(x')(y)) \in \rho_3$$

for every $y \in M_2$. This shows that $(f(x), f(x')) \in \rho_3$; hence, f is monotonic. Taking into account the fact that for $(y, y') \in \rho_2$ we have $(g(x, y), g(x, y')) \in \rho_3$, we have $(f(x)(y), f(x)(y')) \in \rho_3$, which means that $f(x)$ is monotonic for every $x \in M_1$.

Conversely, suppose that f is monotonic and also that, $f(x)$ is monotonic for any $x \in M_1$. Consider $(x, y), (x', y')$, two pairs in $M_1 \times M_2$ such that $(x, x') \in \rho_1$ and $(y, y') \in \rho_2$. We have $(f(x), f(x')) \in \rho_3$, because of the monotonicity of f. In addition, both $f(x)$ and $f(x')$ are monotonic. These facts allow us to write $g(x, y) = f(x)(y)$, $(f(x)(y), f(x)(y')) \in \rho_3$ (because

---

[3]The American logician Haskell Brooks Curry was born on September 12, 1900 in Millis, Massachusetts and died on September 1, 1982. He taught at Pennsylvania State University for most of his career and then was briefly director of the Institute for Foundational Studies in the Philosophy of the Exact Sciences in Amsterdam, Holland. Curry was one of the founders of combinatory logic and a proponent of formalism as a philosophy of mathematics.

of the monotonicity of $f(x)$), $(f(x)(y'), f(x')(y')) \in \rho_3$ (because of the monotonicity of f) and, $f(x')(y') = g(x', y')$. Applying the transitivity of ρ_3, we obtain $(g(x, y), g(x', y')) \in \rho_3$; hence, g is monotonic. ∎

Theorem 3.8.26 *Consider three cpos, (M_i, ρ_i), $1 \leq i \leq 3$. Let g be a function, $g : M_1 \times M_2 \longrightarrow M_3$, and let $f : M_1 \longrightarrow (M_2 \longrightarrow M_3)$ be its curried form.*

The function g is continuous if and only if f is continuous, and $f(x)$ is a continuous function from (M_2, ρ_2) to (M_3, ρ_3) for every $x \in M_1$.

Proof. Suppose that g is a continuous function, $g : M_1 \times M_2 \longrightarrow_c M_3$. Let us prove, initially, that for $x \in M_1$ the mapping $f(x) : M_2 \longrightarrow M_3$ defined by $f(x)(y) = g(x, y)$ for every $y \in M_2$ is continuous.

Let $L = \{y_j \mid j \in J\}$ be a chain in (M_2, ρ_2). Observe that the set $\{(x, y_j) \mid j \in J\}$ is a chain in $M_1 \times M_2$ and $\sup\{(x, y_j) \mid j \in J\} = (x, \sup L)$. The continuity of g implies

$$\sup\{g(x, y_j) \mid j \in J\} = g(\sup\{(x, y_j) \mid j \in J\}),$$

that is, in terms of f, $\sup f(x)(L) = \sup\{f(x)(y_j) \mid j \in J\} = g(x, \sup L) = f(x)(\sup L)$, which proves that $f(x)$ is continuous for every $x \in M_1$.

Consider now a chain $K = \{x_i \mid i \in I\}$ in (M_1, ρ_1) and let $x = \sup K$. Also, since $f(K)$ is a chain in $M_2 \longrightarrow M_3$, we obtain the existence of $\sup f(K)$, and we need to prove that $f(x) = \sup f(K)$.

Since $(x_i, x) \in \rho_1$ for any $x_i \in K$, we have $(g(x_i, y), g(x, y)) \in \rho_3$ because of the fact that g is continuous and, therefore, monotonic. This implies $(f(x_i), f(x)) \in \rho_3$; hence, $f(x)$ is an upper bound of the chain $f(K)$. Suppose now that $(f(x_i), h) \in \rho_3$. This means that for any $y \in M_2$ we have $(f(x_i)(y), h(y)) \in \rho_3$ or, equivalently $(g(x_i, y), h(y)) \in \rho_3$. The continuity of g implies that $g(x, y) = \sup\{g(x_i, y) \mid x_i \in K\}$; hence, $(g(x, y), h(y)) \in \rho_3$. This shows that $(f(x), h) \in \rho_3$; hence, $f(x)$ is the supremum of $f(K)$.

Conversely, suppose that $f : M_1 \longrightarrow (M_2 \longrightarrow M_3)$ is continuous and $f(x)$ is continuous for every $x \in M_1$.

Because of the previous proposition, the function g is monotonic. The image of a chain $S = \{(x_i, y_i) \mid i \in I\} \subseteq M_1 \times M_2$ is a chain in M_3. Furthermore, $\sup S = (x, y)$, where $x = \sup\{x_i \mid i \in I\}$ and $y = \sup\{y_l \mid l \in I\}$). This allows us to write $g(\sup S) = g(x, y) = f(x)(y) = \sup_{l \in I} f(x)(y_l)$, because of the continuity of $f(x)$. Applying the continuity of f, we obtain $g(\sup S) = \sup_{l \in I} \sup_{i \in I} f(x_i)(y_l)$. By permuting the two suprema (which is possible according to Corollary 3.3.15), we have $g(\sup S) = \sup_{i \in I} \sup_{l \in I} g(x_i, y_l)$.

Since $(x_l, y_l), (x_i, y_i) \in S$, we have either $((x_l, y_l), (x_i, y_i)) \in \rho_1 \times \rho_2$ or $((x_i, y_i), (x_l, y_l)) \in \rho_1 \times \rho_2$. In the first case, we have $((x_i, y_i), (x_i, y_i)) \in \rho_1 \times \rho_2$; in the second case, $((x_i, y_l), (x_l, y_l)) \in \rho_1 \times \rho_2$. In both cases, we find pairs in the chain S "greater" than (x_i, y_l) in the sense of ρ_3. Therefore, $(g(\sup S), \sup g(S)) \in \rho_3$. This guarantees the continuity of g as we have seen in Theorem 3.8.17. ∎

Example 3.8.27 Let (M_1, ρ_1) and (M_2, ρ_2) be two cpos. Consider the identity mapping

$$1_{M_1 \longrightarrow_c M_2} : (M_1 \longrightarrow_c M_2) \longrightarrow (M_1 \longrightarrow_c M_2).$$

The mapping $1_{M_1 \longrightarrow_c M_2}$ is continuous. Also, for any function $f : M_1 \longrightarrow_c M_2$, we have $1_{M_1 \longrightarrow_c M_2}(f) = f$; hence, $1_{M_1 \longrightarrow_c M_2}(f)$ is continuous. Therefore, the mapping obtained by decurryfing $g : (M_1 \longrightarrow_c M_2) \times M_1 \longrightarrow_c M_2$ is continuous, because of Theorem 3.8.26. The mapping g is given by $g(h, x) = h(x)$ for $h : M_1 \longrightarrow_c M_2$ and $x \in M_1$. In other words, the mapping g embodies the *application* of the function f to the argument x. For this reason, we shall refer to $g(h, x)$ as apply(h, x).

3.9   The Axiom of Choice and Zorn's Lemma

Let $\mathcal{C}$ be a nonempty collection of pairwise disjoint, nonempty sets.

Definition 3.9.1 *A* selective set *for the collection $\mathcal{C}$ is a set K such that the intersection $K \cap M$ contains exactly one element for each set $M \in \mathcal{C}$.*

A notion similar to the notion of selective set is the notion of choice function.

Definition 3.9.2 *A* choice function *for a collection of sets $\mathcal{C}$ is a mapping f defined on $\mathcal{C}$ such that for each $M \in \mathcal{C}$ if $M \neq \emptyset$ then $f(M) \in M$.*
A selective function *for set P is a choice function for the collection $\mathcal{P}(P)$.*

The apparently innocuous question of whether for each such collection $\mathcal{C}$ a choice function exists turned out to be one of the central problems of modern mathematics. We hope to convince the reader that the answer to this question *does matter* for the concrete arguments required by current mathematical practice by providing several alternative formulations of the statement that asserts the existence of a selective set.

The axiom of choice was formulated for the first time by Ernst Zermello at the beginning of the 20th century. Numerous attempts at proving this axiom from other principles of set theory have failed; instead many equivalent formulations of the axiom of choice were obtained, and we present some of these formulations in this section.

Axiom 3.9.3 (Axiom of Choice) *Every collection of sets has a choice function.*

Note that the axiom of choice merely states the existence of a choice function. It is important to observe that the axiom of choice *states no criterion* for choosing the values of the choice function in each member M of the collection.

Example 3.9.4 Here is a nonmathematical example of the application of the axiom of choice. The reader has undoubtedly realized that socks are manufactured such that there is no difference between the socks that belong to the same pair. Suppose that we have an arbitrary set of socks. There is no criterion for deciding which socks are to be worn on the left feet. Yet, the axiom of choice gives us the possibility of stating that such a set exists. It suffices to use a choice function for the collection of all pairs of socks in order to pick up a left foot sock from each pair of socks.

Example 3.9.5 Consider the set of functions $\mathbf{R} \longrightarrow \mathbf{R}$. There exists a set of functions $S \subseteq \mathbf{R} \longrightarrow \mathbf{R}$ such that for any function $f : \mathbf{R} \longrightarrow \mathbf{R}$ either f or its opposite $-f$ belongs to S, but not both. Of course, the function $-f$ is defined by $(-f)(x) = -f(x)$ for every $x \in \mathbf{R}$.

We can show that the existence of the set S is a consequence of the axiom of choice. Indeed, consider the collection of sets:

$$\mathcal{C} = \{\{f, -f\} | f : \mathbf{R} \longrightarrow \mathbf{R}\}.$$

Any two distinct members of $\mathcal{C}$ are disjoint. The axiom of choice implies the existence of a choice function for $\mathcal{C}$. Let S be the set of values of the selective function. For each $f : \mathbf{R} \longrightarrow \mathbf{R}$ we have either $f \in S$ or $-f \in S$.

Theorem 3.9.6 *The axiom of choice is equivalent with the following statements:*

1. *each collection of nonempty, pairwise disjoint sets has a selective set;*

2. *for every set there exists a selective function;*

3. *the Cartesian product of every family of nonempty sets is nonempty.*

Proof. The axiom of choice implies (1). Let $\mathcal{C}$ be a collection of nonempty pairwise disjoint sets. The axiom of choice implies the existence of a selective function f for $\mathcal{C}$. Consider the set $K = \{f(M) | M \in \mathcal{C}\}$. We claim that K is a selective set for $\mathcal{C}$. Indeed, we have $f(M) \in K \cap M$ for every $M \in \mathcal{C}$, and since the sets of $\mathcal{C}$ are pairwise disjoint, it follows that $K \cap M$ consists of exactly one element $f(M)$.

(1) implies (2). If $M = \emptyset$, then the function $f : \mathcal{P}(\emptyset) \longrightarrow \{\emptyset\}$ defined by $f(\emptyset) = \emptyset$ can serve as a selective function. Therefore, we may assume that M is nonempty. Consider the collection $\mathcal{C}$ of all sets having the form $K \times \{K\}$ for $K \subseteq M$. Observe that if $K \neq H$ the sets $K \times \{K\}$ and $H \times \{H\}$ are disjoint. Indeed, if we would have $t \in (K \times \{K\}) \cap (H \times \{H\})$, then $t = (x, K) = (y, H)$, and this would imply $K = H$.

Let D be a selective set for $\mathcal{C}$. For every $K \subseteq M$, there is $x \in K$ such that $D \cap (K \times \{K\}) = \{(x, K)\}$. We can define the selective function on M by $f(K) = x$, where $(x, K) \in D$. Since $x \in K$, it follows that f is indeed a selective function for the set M.

(2) implies (3). Consider a collection of nonempty sets $C = \{M_i | i \in I\}$ and let $M = \bigcup \{M_i | i \in I\}$. If f is a selective function for M, then for every $M_i \in C$, we have $f(M_i) \in M_i$. Consider the mapping $t : I \longrightarrow M$ defined by $t(i) = f(M_i)$. From Definition 2.9.1, it follows that $t \in \prod_{i \in I} M_i$.

(3) implies the axiom of choice. Let $C = \{M_i | i \in I\}$ be a collection of sets. Consider the set of indices $J = \{i | i \in I, M_i \in C, M_i \neq \emptyset\}$ and the collection $D = \{M_i | M_i, i \in J\}$.

Let $t \in \prod D$. For every $i \in J$, we have $t(i) \in M_i$, which means that for every $M_i \in C$, $M_i \neq \emptyset$, we have $t(i) \in M_i$. This means that t is a choice function for C. ∎

It is interesting to remark that we have already used the axiom of choice. For instance, when proving that well-founded sets are the same as artinian sets (see Theorem 3.4.13), we have chosen the elements x_n of a poset (M, ρ) such that each x_n was a member of the set $\{x | x \in M, (x, x_{n-1}) \in \rho\}$ for $n \geq 1$. In other words, we have used a choice function for the collection of sets $\{\{x | x \in M, (x, x_{n-1}) \in \rho\} | n \geq 1\}$.

The following statement, also known as *Zorn's lemma*, is a formulation in terms of partially ordered sets of the axiom of choice.

Axiom 3.9.7 (Zorn's Lemma) *If (M, ρ) is a nonempty poset such that for any chain $L \subseteq M$ the set of upper bounds $L^s \neq \emptyset$, then M contains at least one maximal element.*

In this section, we discuss a variety of consequences of Zorn's lemma. One such consequence which is extremely important for various areas of computer science such as denotational semantics and formal language theory, is the Knaster-Tarski fixed point theorem. A less general version of this result can be obtained without using Zorn's lemma and will be presented later.

A *fixed point* of a mapping $f : M \longrightarrow M$ is an element of the set M that is mapped into itself by F, that is, for which we have $f(m) = m$. There are mappings that have no fixed point, while others may have one or several fixed points.

Example 3.9.8 The mapping $f : \mathbf{R} \longrightarrow \mathbf{R}$ defined by $f(x) = ax + b$ has one fixed point, namely, $\dfrac{b}{1-a}$ if $a \neq 1$; if $a = 1$ and $b \neq 0$, then f has no fixed points.

Theorem 3.9.9 (Knaster-Tarski Fixed Point Theorem.) *Let (M, ρ) be a complete partially ordered set. If $f : M \longrightarrow M$ is a monotonic function on M, then f has a least fixed point.*

Proof. Let C_f be the family of sub-cpos of the complete poset (M, ρ) that are closed under f. In other words, if $D \in C_f$, then D is a sub-cpo and $x \in D$ implies $f(x) \in D$. Note that $C_f \neq \emptyset$, since it contains at least the set M.

If $K \subseteq M$, denote by $\mathcal{C}_f(K)$ the collection of those sets from $\mathcal{C}_f$ that contain K. It is easy to see that $\bigcap\{D \mid D \in \mathcal{C}_f(K)\}$ is also a member of $\mathcal{C}_f(K)$, namely, the least such member.

Let $U \in \mathcal{C}_f(K)$ and let L be a chain contained in U. Since U is a complete subset, we have $\sup L \in U$. Using Zorn's lemma, we can infer the existence of a maximal element in U.

For $K = \{\bot\}$, we denote the set $\bigcap\{D \mid D \in \mathcal{C}_f(\{\bot\})\}$ by K_0. Let y_0 be a maximal element of K_0.

Consider the set $D_f = \{x \mid x \in M \text{ and } (x, f(x)) \in \rho\}$. This set is not empty since we have $\bot \in D_f$. Let $L = \{x_i \mid i \in I\}$ be a chain of (M, ρ) such that $L \subseteq D_f$ and let $x = \sup L$. The existence of x is guaranteed by the fact that (M, ρ) is a complete poset. We claim that $x \in L$. Indeed, notice that $(x_i, x) \in \rho$ for every $i \in I$. Therefore, from $(x_i, f(x_i)) \in \rho$ and $(f(x_i), f(x)) \in \rho$, we may infer $(x_i, f(x)) \in \rho$. Consequently, $(\sup\{x_i \mid i \in I\}, f(x)) \in \rho$, that is, $(x, f(x)) \in \rho$ which means $x \in D_f$. This shows that D_f is a closed subset of (M, ρ).

On the other hand, let $z \in D_f$, that is, $(z, f(z)) \in \rho$. Because of the monotonicity of f, we have $(f(z), f(f(z))) \in \rho$, which means that $f(z) \in D_f$. This allows us to conclude that $D_f \in \mathcal{C}_f(\{\bot\})$.

Clearly, we have $K_0 \subseteq D_f$. Therefore, $(y_0, f(y_0)) \in \rho$. The maximality of y_0 implies $y_0 = f(y_0)$; hence, y_0 is a fixed point.

Consider now w, an arbitrary fixed point of f. We claim that the set $E_w = \{u \mid u \in M, (u, w) \in \rho\}$ is a member of $\mathcal{C}_f(\{\bot\})$. Indeed, let L be a chain in E_w. Since w is an upper bound for L, we have $(\sup L, w) \in \rho$; hence, $\sup L \in E_w$. Therefore, E_w is a complete subset; in addition, $\bot \in E_w$.

E_w is also closed with respect to f, since, if $u \in E_w$, because of the monotonicity of f, we have $(f(u), f(w)) = (f(u), w) \in \rho$. Consequently, $E_w \in \mathcal{C}_f(\{\bot\})$, and this gives $K_0 \subseteq E_w$, which implies $(y_0, w) \in \rho$. This allows us to say that y_0 is, indeed, the least fixed point of f. ∎

The previous theorem can actually be proven without the use of Zorn's lemma; however, the proof is too involved to be given in this volume. The next theorem is a useful special case of Theorem 3.9.9, for which we can give a direct argument without the use of Zorn's lemma.

Theorem 3.9.10 *If $F : \mathcal{P}(A) \longrightarrow \mathcal{P}(A)$ is a monotonic function, then there is a subset K of A such that $F(K) = K$.*

Proof. Let $\mathcal{C}$ be the collection of subsets of A defined by

$$\mathcal{C} = \{L \in \mathcal{P}(A) \mid L \subseteq F(L)\}.$$

The collection $\mathcal{C}$ is not empty since $\emptyset \in \mathcal{C}$. Let $K_0 = \bigcup \mathcal{C}$. For any $L \in \mathcal{C}$, we have $F(L) \subseteq F(K_0)$, because of the monotonicity of F. This, in turn, implies $L \subseteq F(K_0)$, in view of the definition of $\mathcal{C}$; hence, $\bigcup\{L \mid L \in \mathcal{C}\} \subseteq F(K_0)$. We obtain in this manner $K_0 \subseteq F(K_0)$. Applying again F to $K_0 \subseteq F(K_0)$, we have $F(K_0) \subseteq F(F(K_0))$, and we can interpret this inclusion as $F(K_0) \in C$. Therefore, $F(K_0) \subseteq K_0$ by the definition of K_0, and we obtain $K_0 = F(K_0)$. ∎

Another important consequence of Zorn's lemma is the following theorem.

Theorem 3.9.11 *Let (M, ρ) be a poset. There exists a total order σ on M such that $\rho \subseteq \sigma$.*

Proof. Let $ORD_M(\rho)$ be the set of partial orders on the set M that contain the partial order ρ. We have the poset $(ORD_M(\rho), \subseteq)$. Let L be a chain of relations from $ORD_M(\rho)$.

In order to be able to apply Zorn's lemma to $(ORD_M(\rho), \subseteq)$, consider a chain $L = \{\rho_i | i \in I\}$ in this poset. We intend to show that the relation $\sigma = \bigcup\{\rho_i | i \in I\}$ is also a partial order relation.

Since $\rho \subseteq \sigma$, it is clear that σ is reflexive. Assume now that we have both $(x, y), (y, x) \in \sigma$. There are $\rho_i, \rho_j \in L$ such that $(x, y) \in \rho_i$ and $(y, x) \in \rho_j$. Since L is a chain, we can have either $\rho_i \subseteq \rho_j$ or $\rho_j \subseteq \rho_i$. In the first case, $(x, y), (y, x) \in \rho_j$, and because of the antisymmetry of ρ_j, we obtain $x = y$. The second case yields a similar conclusion, and we have the antisymmetry of σ.

To prove the transitivity of σ, consider $(x, y), (y, z) \in \sigma$. There are $\rho_i, \rho_j \in L$ such that $(x, y) \in \rho_i$ and $(y, z) \in \rho_j$. As before, we may have $\rho_i \subseteq \rho_j$ or $\rho_j \subseteq \rho_i$. In the first case, we have $(x, z) \in \rho_j \subseteq \sigma$ because of the transitivity of ρ_j. In the second case, we reach the same conclusion based on the transitivity of ρ_i. Therefore, σ is a partial order.

Applying Zorn's lemma, we obtain the existence of a maximal partial order θ on M that includes ρ. We claim that this is a total order on M.

Indeed, assume that we have $x, y \in M$ such that neither $(x, y) \in \theta$ nor $(y, x) \in \theta$. Define

$$\theta' = \theta \cup \{(x, y)\} \cup \{(u, y) | u \in M, (u, x) \in \theta\} \cup \{(x, v) | v \in M, (y, v) \in \theta\}.$$

The relation θ' is reflexive since $\rho \subseteq \theta \subset \theta'$. It is also easy to verify that θ' is transitive, and we leave this verification to the reader. We provide only the argument for antisymmetry.

Let $(r, s), (s, r) \in \theta'$. If both (r, s) and (s, r) belong to θ, then we have $r = s$ since θ is antisymmetric; otherwise, we should consider the following cases.

1. Suppose that $(r, s) = (u, y)$ (where $(u, x) \in \theta$) and $(s, r) = (y, u)$. This is impossible since $(u, x) \in \theta$ and $(y, u) \in \theta'$ imply $(y, x) \in \theta'$, which contradicts the definition of θ'.

2. If $(r, s) = (x, v)$ (where $(y, v) \in \theta$), we have $(s, r) = (v, x) \in \theta'$. This also implies the contradictory conclusion $(y, x) \in \theta'$.

This allows us to say that θ' is antisymmetric; hence, θ' is a partial order. On the other hand, we have $\theta \subset \theta'$, and θ is a *maximal* partial order including ρ. This contradiction shows that it is impossible to find in M two elements x, y such that $(x, y) \notin \theta$ and $(y, x) \notin \theta$. Therefore, we conclude that θ is a total order on M. ∎

Many applications of Zorn's lemma use the proof technique illustrated by the previous argument. Namely, a collection $\mathcal{C}$ of sets ordered by inclusion is considered such that the union of the sets of any given chain L belongs to $\mathcal{C}$. Then, application of Zorn's lemma allows us to infer the existence of a maximal element in $\mathcal{C}$, and we prove that this maximal element has certain properties. We illustrate again this method by proving yet another application of Zorn's lemma, namely the well-ordering axiom mentioned in Section 3.4.

Theorem 3.9.12 *Zorn's lemma implies the well-ordering axiom.*

Proof. Let M be a set and let $\mathcal{C}$ be the collection of posets defined by $(K, \sigma) \in \mathcal{C}$ if $K \subseteq M$ and (K, σ) is a well-ordered poset. Note that $\mathcal{C}$ is a nonempty collection since for any $x \in M$ we have $(\{x\}, \rho_x) \in \mathcal{C}$; here, $\rho_x = \{(x, x)\}$.

Define the relation $\preceq$ on $\mathcal{C}$ by $(K_1, \sigma_1) \preceq (K_2, \sigma_2)$ if the following three conditions are satisfied:

1. $K_1 \subseteq K_2$;

2. $\sigma_1 = \sigma_2|_{K_1}$; and

3. $(x, y) \in \sigma_2$ for every $x \in K_1$ and $y \in K_2 - K_1$.

It is easy to verify that $\preceq$ is a partial order on $\mathcal{C}$, and we leave this argument to the reader as an exercise. Moreover, if $(K_1, \sigma_1) \preceq (K_2, \sigma_2)$ and $(u, v) \in \sigma_2$, for $u \neq v$ and $u \in K_2$, $v \in K_1$, then $u \in K_1$.

Let L be a chain in $\mathcal{C}$ and let $U = \bigcup\{K \mid (K, \sigma) \in L\}$. Define the relation $\bar{\sigma}$ on U by

$$\bar{\sigma} = \{(x, y) | (x, y) \in \sigma, \text{ for some } (K, \sigma) \in L\}.$$

We leave to the reader to prove that $\bar{\sigma}$ is a total order on U.

We claim that $(U, \bar{\sigma})$ is a well-ordered poset. Indeed, let W be a nonempty subset of U. Take $w \in W$ and let $W_0 = \{x \in W \mid (x, w) \in \bar{\sigma}\}$. Observe that, since $w \in W_0$, we have $W_0 \neq \emptyset$.

The definition of U implies that there exists (H, ρ) in L such that $w \in H$. We claim that $W_0 \subseteq H$. Indeed, let $x \in W_0$. Since $x \in U$, there exists $(K, \theta) \in L$ such that $x \in K$. We can have either $(K, \theta) \preceq (H, \rho)$ or $(H, \rho) \preceq (K, \theta)$. In the first case, we have $K \subseteq H$; hence, $x \in H$. In the second case, from $x \in K$, $w \in H$, and $(x, w) \in \bar{\sigma}$, we infer again $x \in H$. Therefore, $W_0 \subseteq H$.

Since (H, ρ) is well-ordered, W_0 has a least element w_0. This element is also the least element of W. Let y be an element of W. If $y \in W_0$, then $(w_0, y) \in \bar{\sigma}$. If $y \notin W_0$, then $(y, w) \notin \bar{\sigma}$; hence, $(w, y) \in \bar{\sigma}$ (because of the fact that $(U, \bar{\sigma})$ is totally ordered). Since $(w_0, w) \in \bar{\sigma}$, we obtain $(w_0, y) \in \bar{\sigma}$, which shows that w_0 is, indeed, the least element of W. This justifies the claim made above, and we may conclude that $(U, \bar{\sigma})$ is an upper bound for L in $(\mathcal{C}, \preceq)$.

We proved that an arbitrary chain L in $\mathcal{C}$ has an upper bound; therefore, in view of Zorn's lemma, the collection $\mathcal{C}$ must contain some maximal element (G, γ) with respect to the partial order "$\preceq$". We have $G = M$, because otherwise there is $x \in M - G$, and we can define a well-order on $G \cup \{x\}$ by $\gamma_1 = \gamma \cup \{(t, x) \mid t \in G \cup \{x\}\}$. Applying the definition of $\preceq$, we have $(G, \gamma) \preceq (G \cup \{x\}, \gamma_1)$, which contradicts the maximality of (G, γ). Thus, $G = M$, and we have the well-ordered set (M, γ). ∎

Our next objective is to prove that the axiom of choice is equivalent to Zorn's lemma. We present initially a consequence of the axiom of choice in the realm of posets.

Let (M, ρ) be a poset.

Definition 3.9.13 *A chain L in (M, ρ) is* maximal *if, for every chain K of (M, ρ), $L \subseteq K$ implies $L = K$.*

Theorem 3.9.14 *The axiom of choice implies that every poset contains a maximal chain.*

Proof. Let (M, ρ) be a poset. The axiom of choice implies the existence of a selective function f on the set M.

For a chain L of (M, ρ), let $\hat{L}$ be the set of all elements $x \in M$ such that $L \cup \{x\}$ is also a chain.

Let $\mathcal{L}$ be the set of all chains of the poset (M, ρ). The selective function f generates a function $\varphi : \mathcal{L} \longrightarrow \mathcal{L}$ defined by

$$\varphi(L) = \begin{cases} L \cup \{f(\hat{L} - L)\} & \text{if } \hat{L} - L \neq \emptyset \\ L & \text{otherwise.} \end{cases}$$

Note that for any chain L the set $\varphi(L) - L$ contains at most one element.

A chain L is maximal if and only if $\hat{L} = L$, that is, if and only if $\hat{L} - L = \emptyset$. Therefore, we need to prove only the existence of a chain L such that $\varphi(L) = L$.

A *special family of chains* is a collection of chains $\mathcal{I} \subseteq \mathcal{L}$ that satisfies the following conditions:

1. $\emptyset \in \mathcal{I}$;

2. if $L \in \mathcal{I}$, then $\varphi(L) \in \mathcal{I}$, and

3. if $\mathcal{H} \subseteq \mathcal{I}$ and $(\mathcal{H}, \subseteq)$ is a totally ordered poset, then $\bigcup \{L \mid L \in \mathcal{H}\} \in \mathcal{I}$.

Observe that such special families of chains exist. For instance, $\mathcal{L}$ itself is a special family and, furthermore, the intersection of any collection of special families of chains is also a special family of chains.

Let $\mathcal{I}_0$ be the intersection of all special families of chains. We intend to show that $\mathcal{I}_0$ is *totally ordered* by the inclusion relation "$\subseteq$." If this is the case, condition 3) from the definition of special families of chains implies $\bar{L} = \bigcup\{L | L \in \mathcal{I}_0\} \in \mathcal{I}_0$, so $\varphi(\bar{L}) = \bar{L}$. This follows from $L \subseteq \varphi(L)$ for any chain L and from the fact that $\varphi(\bar{L}) \in \mathcal{I}_0$, which, in turn, means that $\varphi(\bar{L}) \subseteq \bar{L}$, because of the definition of $\bar{L}$.

A chain K of $\mathcal{I}_0$ is *comparable* if for any chain $L \in \mathcal{I}_0$ we have either $K \subseteq L$ or $L \subseteq K$. $\mathcal{I}_0$ contains comparable chains; for instance, the empty set $\emptyset$ is such a chain.

Let K be an arbitrary comparable chain. If $L \in \mathcal{I}_0$ and $L \subset K$, then $\varphi(L) \subseteq K$. Indeed, since $\varphi(L) \in \mathcal{I}_0$ and K is comparable, then $\varphi(L) \not\subseteq K$ implies $K \subset \varphi(L)$, and since $L \subset K$, this would contradict the fact that $\varphi(L) - L$ contains at most one element.

Let $\mathcal{K}$ be the collection of chains consisting of those chains $L \in \mathcal{I}_0$ for which $L \subseteq K$ or $\varphi(K) \subseteq L$. We prove that $\mathcal{K}$ is a special collection.

Since $\emptyset \in \mathcal{I}_0$ and $\emptyset \subseteq K$, we have $\emptyset \in \mathcal{K}$. Let $L \in \mathcal{K}$. Since $L \in \mathcal{I}_0$ we have $\varphi(L) \in \mathcal{I}_0$. We need to prove that $\varphi(L) \in \mathcal{K}$ (that is, either $\varphi(L) \subseteq K$ or $\varphi(K) \subseteq \varphi(L)$).

There are three possible cases: $L \subset K$, $L = K$, and $L \not\subseteq K$. If $L \subset K$, then $\varphi(L) \subseteq K$, in view of the previous argument. In the second case, we have, obviously, $\varphi(K) \subseteq \varphi(L)$. In the last situation, the definition of $\mathcal{K}$ implies $\varphi(K) \subseteq L$; hence, $\varphi(K) \subseteq \varphi(L)$.

In order to prove that $\mathcal{K}$ satisfies the third condition of the definition of a special collection of chains, let $\mathcal{H}$ be a totally ordered subset of $\mathcal{K}$ and let $H = \bigcup\{L | L \in \mathcal{H}\}$. Since $\mathcal{H} \subseteq \mathcal{K} \subseteq \mathcal{I}_0$, we have $H \in \mathcal{I}_0$. If there exists $L \in \mathcal{H}$ such that $\varphi(K) \subseteq L$, then $\varphi(K) \subseteq H$; if no such L exists, then $L \subseteq K$ for every $L \in \mathcal{H}$; hence, $H \subseteq K$.

We proved that $\mathcal{K}$ is a special collection included in $\mathcal{I}_0$, and since $\mathcal{I}_0$ is included in any special collection, we obtain $\mathcal{K} = \mathcal{I}_0$. In other words, for any chain $L \in \mathcal{I}_0$, we have $L \subseteq K$ or $\varphi(K) \subseteq L$. It follows that the chain $\varphi(K)$ is also comparable since for any chain $L \in \mathcal{I}_0$ we have $\varphi(K) \subseteq L$ or $L \subseteq K$, which implies $L \subseteq \varphi(K)$.

Since K was chosen to be an arbitrary comparable chain, it follows that the set of all comparable chain satisfies condition 2) of the definition of special collections of chains. The collection of comparable chains satisfies the first condition of the same definition since, as we observed before, $\emptyset$ is a comparable chain. Note that the third condition is also satisfied. Indeed, let $\mathcal{J}$ be a totally ordered collection of comparable chains, that is, a chain of comparable chains. If $J = \bigcup\{L | L \in \mathcal{J}\}$, then $J \in \mathcal{I}_0$ (since $\mathcal{I}_0$ satisfies the third condition). For every $L \in \mathcal{I}_0$, we have one of the following two situations:

1. there is $P \in \mathcal{J}$ such that $L \subseteq P$ (which implies $L \subseteq J$), or

2. $P \subseteq L$ for all $P \in \mathcal{J}$ (because of the comparability of any $P \in \mathcal{J}$); in this case, $J \subseteq L$.

We infer that the collection of all comparable chains is a special collection included in $\mathcal{I}_0$; therefore, it must coincide with $\mathcal{I}_0$, and this implies that every chain of $\mathcal{I}_0$ is comparable. This allows us to conclude that $(\mathcal{I}_0, \subseteq)$ is a totally ordered poset. ∎

Theorem 3.9.15 *The axiom of choice implies Zorn's lemma.*

Proof. Let (M, ρ) be a nonempty poset such that for every chain $L \subseteq M$ the set of upper bounds of L is nonempty. From the previous theorem, we infer the existence of a maximal chain L_1 in (M, ρ). Let x be an upper bound of L_1. We claim that x is a maximal element in (M, ρ). Suppose that there exists $y \in M$ such that $(x, y) \in \rho$ and $x \neq y$; in such a case, $L_2 = L_1 \cup \{y\}$ would be a chain strictly including L_1, and this contradicts the maximality of L_1. This shows that x is maximal. ∎

We already know that Zorn's lemma implies the well-ordering axiom. To come around full-circle, we need to show that the well-ordering axiom implies the axiom of choice.

Theorem 3.9.16 *The well-ordering axiom implies the axiom of choice.*

Proof. Let $\mathcal{C} = \{M_i | i \in I\}$ be a collection of nonempty pairwise disjoint sets and let $C = \bigcup\{M_i | i \in I\}$. The well-ordering axiom implies the existence of a relation $\rho \subseteq C \times C$ such that (C, ρ) is a well-ordered poset. Each member M_i of the collection $\mathcal{C}$ has a least element x_i, and the set $D = \{x_i | i \in I\}$ intersects each $M_i \in \mathcal{C}$ in exactly one element x_i. In other words, D is a selective set for $\mathcal{C}$. ∎

Now, we can group together the results described in this section.

Corollary 3.9.17 *The following are equivalent:*

1. *the axiom of choice;*

2. *the existence of a selective function for every nonempty set;*

3. *the nonemptiness of the Cartesian product of a family of nonempty sets;*

4. *the existence of maximal chains in every poset;*

5. *Zorn's lemma;*

6. *the well-ordering axiom.*

Proof. This corollary follows immediately from Theorems 3.9.12, 3.9.6, 3.9.14, 3.9.15 and 3.9.16. ∎

3.10   Exercises and Supplements

Partial Orders and Hasse Diagrams

1. Draw the Hasse diagram to represent the following posets:

    (a) $(\mathcal{P}(\{a, b, c, d\}), \subseteq)$;

    (b) $(\{n \mid n \in \mathbf{N}, 1 \leq n \leq 20\}, \delta)$, where δ is the divisibility relation;

    (c) $(\mathcal{P}(\{0, 1\} \times \{0, 1\}), \subseteq)$.

2. Determine all the possible Hasse diagrams for partial orders on a three-element set.

3. Let (M, ρ) be a poset. Prove that for any $K \subseteq M$ the pair (K, ρ_K) is also a poset, where $\rho_K = \rho \cap (K \times K)$ is *the trace of ρ on K*.

4. Let $\mathcal{P}(A \times A)$ be the set of relations on the set A. Define the mappings $\Phi : \mathcal{P}(A \times A) \longrightarrow \mathcal{P}(A \times A)$ and $\Psi : \mathcal{P}(A \times A) \longrightarrow \mathcal{P}(A \times A)$ as $\Phi(\rho) = \rho - \iota_A$ and $\Psi(\rho) = \rho \cup \iota_A$. Reformulate Proposition 3.2.6 in terms of the mappings Φ and Ψ and prove that for any partial order ρ we have $\Psi(\Phi(\rho)) = \rho$ and for any strict partial order ρ_1 we have $\Phi(\Psi(\rho_1)) = \rho_1$.

5. Let ρ_1 be the strict partial order corresponding to the partial order ρ. If θ is the transitive reduction of ρ show that for every relation σ such that $\sigma^+ = \rho_1$ we have $\theta \subseteq \sigma$. In other words, the transitive reduction is *the least* relation whose transitive closure is ρ_1.

6. Define a relation ρ to be antitransitive if for all $(a, b) \in \rho$ there is no path of length at least equal to 2 from a to b, i.e., for all $n \geq 2$, $\rho^n \cap \rho = \emptyset$. Note that an antitransitive relation is acyclic.

    Let M be a finite set.

    (a) If ρ is a strict partial order on M, then $\rho - \rho^2$ is an antitransitive relation on M.

    (b) If σ is an antitransitive relation on M, then σ^+ is a strict partial order on M.

    (c) The two processes outlined above are inverse to each other, i.e., if ρ is a strict partial order on M, then $(\rho - \rho^2)^+ = \rho$ and if σ is an antitransitive relation on M, then $\sigma^+ - (\sigma^+)^2 = \sigma$. Thus, when M is a finite set, there is a bijection between strict partial orders on M and antitransitive relations on M.

7. Taxonomies involving sets of entities are of special interest in databases and artificial intelligence. In order to define formally the notion of *taxonomy*, we introduce the notion of a *class of entities* as a pair

$\mathcal{K} = (K, A_K)$, where K is a set whose elements are referred to as *entities* and A_K is a collection of functions, called the *attributes* of the class $\mathcal{K}$, where $\text{Dom}(a) = K$ for every function $a \in A_K$.

If $\mathcal{H} = (H, A_H)$, $\mathcal{K} = (K, A_K)$ are classes of entities, and we write $\mathcal{H}$ **is-a** $\mathcal{K}$ if $H \subseteq K$, and for every $a \in A_K$ the restriction $a \restriction H$ belongs to A_H.

A *taxonomy* is a collection of classes of entities $\mathcal{T} = \{\mathcal{K}_t \mid t \in T\}$, such that if $\mathcal{H}, \mathcal{K} \in \mathcal{T}$, $\mathcal{H} = (H, A_H)$ and $\mathcal{K} = (K, A_K)$ then $H \subseteq K$ implies $\mathcal{H}$ **is-a** $\mathcal{K}$. This is the *inheritance property of taxonomies* .

(a) Prove that for any collection of classes of entities the relation is-a introduced above is a partial order.

(b) Let $\mathcal{C}$ be a collection of sets and let $\mathcal{F}$ be a collection of functions that is closed with respect to restriction (that is, if $a \in \mathcal{F}$, then every restriction of a is also a member of $\mathcal{F}$).

Prove that the collection $\mathcal{T} = \{(K, S_K) \mid K \in \mathcal{C}\}$ is a taxonomy, where $S_K = \{a \mid a \in \mathcal{F}, K = \text{Dom}(a)\}$.

Special Elements of Partially Ordered Sets

8. Determine all maximal and minimal elements of the posets mentioned under Exercise 1.

9. Give an example of a poset with *one* minimal element but without a least element.

10. Let K be a subset of the poset $(\mathbf{R}, \leq)$. Prove that $u = \sup K$ if and only if the following two conditions are satisfied:

   (a) for any $x \in K$, we have $x \leq u$, and

   (b) for any $\epsilon > 0$, there is $y \in K$ such that $u - \epsilon \leq y$.

   Formulate and prove a similar characterization for $\inf K$.

11. Let (M, ρ) be a subset.

   (a) Prove that if $K, H \subseteq M$ and $K \subseteq H$ then $H^s \subseteq K^s$ and $H^i \subseteq K^i$;

   (b) Describe the sets $\emptyset^s, \emptyset^i, M^s$, and M^i.

   (c) Show that there is $\sup \emptyset$ if and only if the set M has a least element and $\sup \emptyset$ is precisely the least element.

   (d) Show that there is $\inf \emptyset$ if and only if the set M has a greatest element and $\inf \emptyset$ is equal to the greatest element.

Chains

12. Let (D_1, ρ_1) and (D_2, ρ_2) be two *chains*. Prove that if $f : D_1 \longrightarrow D_2$ is a monotonic bijection then f^{-1} is also a monotonic bijection from D_2 to D_1.

13. Let ρ_1, ρ_2 be two total orders on a set M. Under which conditions is $\rho_1 \bullet \rho_2$ a total order on M?

14. Let (M, ρ) be a poset such that for every chain K of this poset sup K exists. Suppose that $f : M \longrightarrow M$ is a mapping for which $(t, f(t)) \in \rho_1$ for all $t \in M$, where ρ_1 is the strict partial order associated to ρ.

    For $x \in M$, define C_x as being the collection of all subsets K of M that have the following properties:

    1. $x \in K$,

    2. $y \in K$ implies $(x, y) \in \rho$,

    3. $y \in K$ implies $f(y) \in K$, and

    4. if H is a chain of M and $H \subseteq K$, then sup $H \in K$.

    In other words, C_x contains those subsets of K that contain x, all elements greater that x (in the sense of ρ), are closed with respect to f and with respect of suprema of chains.

    (a) Prove that for any $x \in M$ the collection C_x is not empty.

    (b) $P_x = \bigcap \{K \mid K \in C_x\}$ belongs to C_x.

    (c) Let $z \in P_x$ be such that $(u, z) \in \rho_1$ and $u \in P_x$ imply $(f(u), z) \in \rho$. The set

    $$R_z = \{v \mid v \in P_x, (v, z) \in \rho \text{ or } (f(z), v) \in \rho\}$$

    belongs to C_x.

    (d) Prove that the set

    $$Q_x = \{v \mid v \in P_x, u \in P_x, (u, v) \in \rho_1 \text{ imply } (f(u), v) \in \rho\}.$$

    is a member of C_x.

    (e) P_x is a chain in (M, ρ).

Solution:

(a) Observe that C_x is not empty since the set $\{y \mid y \in M, (x, y) \in \rho\}$ satisfies all four conditions mentioned above.

(b) The argument is straightforward.

(c) Let z be an element of P_x such that $u \in P_x$ and $(u, z) \in \rho_1$ imply $(f(u), z) \in \rho$.

We shall now prove that R_z satisfies the four conditions. Let us remark initially that $R_z \subseteq P_x$ for any $z \in Q_x$.

To prove that R_z satisfies the first condition, observe that since $z \in P_x$ we have $(x, z) \in \rho$ (because of the fact that P_x satisfies the second condition) and therefore $x \in R_z$.

The satisfaction of the second condition follows directly from $R_z \subseteq P_x$.

Let $y \in R_z$. We have $(y, f(y)) \in \rho_1$ by the definition of f. Also, we have $(y, z) \in \rho$ or $(f(z), y) \in \rho$.

In the first case, we have either $y = z$ or $(y, z) \in \rho_1$. If $y = z$, we have $(f(z), f(y)) \in \rho$; hence, $f(y) \in R_z$. If $(y, z) \in \rho_1$, we have $(f(y), z) \in \rho$ by the definition of z, which gives $f(y) \in R_z$.

In the second case, we have $(f(z), y) \in \rho$. In view of the fact that $(y, f(y)) \in \rho$, we obtain $(f(z), f(y)) \in \rho$, which proves that in any case $f(y) \in R_z$. Therefore, R_z satisfies the third condition.

Let H be a chain in R_z. Clearly, $\sup H \in P_x$ since $H \subseteq R_z \subseteq P_x$ and P_x satisfies the fourth condition. Because of the definition of R_z, we have for any $s \in H$ $(s, z) \in \rho$ or there is $s \in H$ such that $(f(z), s) \in \rho$. In the first case, $(\sup H, z) \in \rho$; in the second, $(f(z), \sup H) \in \rho$, and this shows that in any case $\sup H \in R_z$.

Note also that since $R_z \subseteq P_x$ we have $R_z = P_x$ *for any* $z \in P_x$.

(d) In view of the second condition which is satisfied by P_x, there is no $u \in P_x$ such that $(u, x) \in \rho_1$. Therefore, the implication of the definition of Q_x holds vacuously for x and Q_x satisfies the first condition.

Q_x satisfies the second condition since $Q_x \subseteq P_x$.

To prove that Q_x satisfies the third condition, let $y \in Q_x$. We proved above that $R_y = P_x$. Therefore, for any $u \in P_x$, we have either $(u, y) \in \rho$ or $(f(y), u) \in \rho$.

Let $r \in P_x$ such that $(r, f(y)) \in \rho_1$. We need to prove that this implies $(f(r), f(y)) \in \rho$ in order to show that $f(y) \in Q_x$.

Since we cannot have $(f(y), r) \in \rho$, we must have $(r, y) \in \rho$. If $r = y$, then we clearly have $(f(r), f(y)) \in \rho$. Otherwise, we have $(r, y) \in \rho_1$, and this gives $(f(r), y) \in \rho$ because of the fact that $y \in Q_x$. This, in turn, implies $(f(r), f(y)) \in \rho$; hence, Q_x satisfies the third condition.

Let K be a chain in Q_x. Since $Q_x \subseteq P_x$, we have $\sup K \in P_x$. Let $q \in Q_x$ such that $(q, \sup K) \in \rho_1$. Since $R_z = P_x$ for every $z \in Q_x$, we shall have $q \in R_z$ for all $z \in Q_x$; therefore, either $(q, z) \in \rho$ or $(f(z), q) \in \rho$ for every $z \in Q_x$. Hence, we have either $(f(z), q) \in \rho$ for every $z \in Q_x$ or there is some $z_1 \in Q_x$ such that $(q, z_1) \in \rho$. In the former case, $(z, q) \in \rho$ (since $(z, f(z)) \in \rho$) for all $z \in Q_x$. This also implies $(\sup K, q) \in \rho$, which contradicts our initial assumption about q.

Therefore, we can have only the latter case. If $z_1 \neq q$, then $(q, z_1) \in \rho_1$, and this implies $(f(q), z_1) \in \rho$ (because of the fact that $z_1 \in Q_x$) thus giving $(f(q), \sup K) \in \rho$.

If $z_1 = q$, then, since $\sup K \neq q$, we have $z_1 \neq \sup K$, and there exists $z_2 \in K$ with $(q, z_2) \in \rho_1$ (and, also $(z_2, \sup K) \in \rho$). This shows that $(f(q), z_2) \in \rho$; hence, $(f(q), \sup K) \in \rho$. We proved that $\sup K \in Q_x$, and therefore Q_x satisfies all conditions. This implies $Q_x = P_x$.

e) Let $u, v \in P_x$. In view of the fact that $P_x = Q_x = R_u$, we have $u \in Q_u$ and $v \in R_u$. Because of the definition of R_u, we have either $(v, u) \in \rho$ or $(f(u), v) \in \rho$. The second case implies $(u, v) \in \rho$ (since we have $(u, f(u)) \in \rho$). Therefore, we have either $(u, v) \in \rho$ or $(v, u) \in \rho$, which shows that P_x is indeed a chain.

15. For a poset (M, ρ), let $\mathcal{C}$ be the collection of all its chains. We have the poset $(\mathcal{C}, \subseteq)$, that is, the poset of all chains of the poset (M, ρ).

   (a) Prove that if $\mathcal{K}$ is a chain of $(\mathcal{C}, \subseteq)$, that is, if $\mathcal{K}$ is a chain of chains, then $\bigcup\{L \mid L \in \mathcal{K}\}$ is the supremum of $\mathcal{K}$.

   (b) Prove that if there exits a maximal element H of $\mathcal{C}$ such that $h = \sup H$ then h is a maximal element of the poset (M, ρ).

16. Prove that the poset $(\mathcal{P}(\mathbf{N}), \subseteq)$ is *not* well-founded.

   Hint: Consider the sequence of sets $\{K_n \mid n \geq 0\}$, where $K_n = \{p \mid p \in \mathbf{N}, p \geq n\}$.

17. Let $\mathcal{P}_f(\mathbf{N})$ be the set of all *finite* subsets of $\mathbf{N}$. Prove that the poset $(\mathcal{P}_f(\mathbf{N}), \subseteq)$ is well-founded.

18. A poset (M, ρ) has finite length if the set of the lengths of the chains of this poset has a greatest element n. The number n is the *length of the poset*.

   (a) Of course, every finite poset has finite length. The converse is not true; give an example of an infinite poset of length n.

   (b) Let (M, ρ) be a poset that has a finite length and the least element $\perp$. Define the height $h(x)$ of $x \in M$ as the greatest length of a chain joining $\perp$ to x.
   Prove that if $(x, y) \in \rho_1$ then $h(x) < h(y)$.

   (c) A poset (M, ρ) is *graded* by the function $g : M \longrightarrow \mathbf{Z}$ if the following two conditions are satisfied:

   i. $(x, y) \in \rho_1$ implies $g(x) < g(y)$, and
   ii. if y covers x (that is, $(x, y) \in \rho_1 - \rho_1^2$), then $g(y) = g(x) + 1$.

Prove that a poset (M, ρ) having the least element $\perp$ is graded by its own height function if and only if all maximal chains between the same endpoints have the same length.

19. For $n \in \mathbf{P}$, let $\varphi(n)$ be the number of distict prime divisors of n. Consider the binary relation

$$\rho = \{(m, n) | \varphi(m) < \varphi(n) \text{ or } \varphi(m) = \varphi(n) \text{ and } m < n\} \cup \iota_{\mathbf{P}}$$

on $\mathbf{P}$. Prove that $(\mathbf{P}, \rho)$ is a well-ordered set.

Constructing New Posets

20. Consider the chains $(\{0, 1, 2\}, \leq)$ and $(\{0, 1\}, \leq)$. Draw the Hasse diagram of their sum and product. Draw the Hasse diagram of the poset $(\{0, 1, 2\} \times \{0, 1\}, \preceq)$, where "$\preceq$" is the lexicographic partial order.

21. Consider the flat posets $\{a, b, c\}_{\perp}$ and $\{x, y\}_{\perp}$ and the set of functions $\{a, b, c\}_{\perp} \longrightarrow \{x, y\}_{\perp}$. Draw the Hasse diagram of the poset $(\{a, b, c\}_{\perp} \longrightarrow \{x, y\}_{\perp}, \sqsubseteq)$. Is this a flat poset?

22. Let $\{(D_i, \rho_i) \mid i \in I\}$ be a family of posets such that each poset has finite chains. Prove that both the sum and the product of the family enjoy the same property.

Functions and Posets

23. Let $f : M_1 \longrightarrow M_2$ be a function between the posets (M, ρ_1) and (M_2, ρ_2). Considering f as a binary relation $f \subseteq M_1 \times M_2$, prove that f is monotonic if and only if $f^{-1} \bullet \rho_1 \bullet f \subseteq \rho_2$.

24. A function $f : M_1 \longrightarrow M_2$ between the posets (M_1, ρ_1) and (M_2, ρ_2) is *dually monotonic* if f is a monotonic function between the poset (M_1, ρ_1) and the dual (M_2, ρ_2^{-1}) of the second poset.

Determine if the following functions are monotonic or dually monotonic relative to the posets specified below:

(a) $f : \mathbf{N} \longrightarrow \mathbf{N}$, where $f(n) = nk$ for some $k \in \mathbf{N}$, where $\mathbf{N}$ is equipped with the divisibility relation δ.

(b) $f, g, h : \mathbf{R} \longrightarrow \mathbf{R}$, defined respectively by $f(x) = x^2$, $g(x) = x^3$, and $h(x) = | x |$, for all $x \in \mathbf{R}$.

25. Let (M, ρ) be a poset. Suppose that $f : M \longrightarrow M$ is a function that is neither monotonic nor dually monotonic. Prove that there is a monotonic function $g : M \longrightarrow_m M$ such that the function $h = f \circ g$ is not dually monotonic.

Hint: Suppose that f is not monotonic. There are $x_0, y_0 \in M$ such that $(x_0, y_0) \in \rho$, but $(a, b) = (f(x_0), f(y_0)) \notin \rho$. Define the function

$g : M \longrightarrow M$ by $g(t) = x_0$ if $(t, b) \in \rho$ and $g(t) = y_0$, otherwise. It is easy to see that while g is monotonic the function $h = f \circ g$ is not dually monotonic. A similar argument works if f is not dually monotonic.

26. Consider two one-to-one functions $f, g : N \longrightarrow N$ such that any $m \in \text{Ran}(f)$ and $n \in \text{Ran}(g)$ are relatively prime and $0 \notin \text{Ran}(g)$.

Define the relation "$\preceq$" on $N \times N$ by $(a, b) \preceq (c, d)$ if

$$\frac{f(a)}{g(b)} \leq \frac{f(c)}{g(d)}.$$

(a) Prove that $(N \times N, \preceq)$ is a poset.

(b) Give examples of functions $f, g : N \longrightarrow N$ that satisfy the conditions imposed above.

(c) Prove that if for any $p \in N$ there is $q \in N$ such that $g(q) > p$ the poset $(N \times N, \preceq)$ is not a well-ordered set.

27. For $n \in N$, consider the set D_n of all natural numbers that divide n. Prove that if $n = p_1 \cdots p_k$, where $p_1, \ldots, p_k$ are k *distinct* prime numbers, then the poset (D_n, δ), where δ is the divisibility relation, is isomorphic to the poset $(\mathcal{P}(\{p_1, \ldots, p_k\}), \subseteq)$.

28. Prove that the poset $(V^*, \trianglelefteq)$, where $\mid V \mid \geq 2$, contains an infinite subset of pairwise incomparable elements.

Hint: Consider the set $\{a, ba, bba, bbba, \ldots\}$ in $\{a, b\}^*$.

29. Let "$\preceq$" be the lexicographic order on the finite alphabet $V = \{a_1, \ldots, a_n\}$. Show that if $y \preceq x \preceq yz$ then there exists $u \in V^*$ such that $x = yu$, where $u \preceq z$.

30. Let x, y be two words from V^*. We write $x \ll y$ if $x = x_1 \cdots x_n$ and $y = y_1 x_1 \cdots y_n x_n y_{n+1}$ for some $n \geq 0$, $x_i, y_j \in V^*$ for $1 \leq i \leq n$, and $1 \leq j \leq n + 1$. Prove that "$\ll$" is a partial order on V^*.

31. Let $(M, \leq)$ be a poset such that:

(a) for any $y \in M$, the set $M_y = \{x \mid x \in M, x \leq y\}$ is finite;

(b) any subset of M consisting of pairwise incomparable elements is finite.

Prove that every infinite subset K of M contains an infinite chain.

Hint: Assume that M contains a set K such that every chain in K is finite and prove that this assumption generates a contradiction.

32. Prove Proposition 3.7.15.

33. Consider a monotonic function $F : \mathcal{P}(A) \longrightarrow \mathcal{P}(A)$. Let $\mathcal{C} = \{L \in \mathcal{P}(A) \mid L \subseteq F(L)\}$ and $\mathcal{D} = \{L \in \mathcal{P}(A) \mid L \supseteq F(L)\}$ be two collections of sets defined by F.

   If $C = \bigcup \mathcal{C}$ and $D = \bigcap \mathcal{D}$, prove that if $F(X) = X$ for $X \in \mathcal{P}(A)$ then $D \subseteq X \subseteq C$.

34. Let $g, k, h : \mathbf{N}_\perp \longrightarrow_m \mathbf{N}_\perp$ and let $p : \mathbf{N}_\perp \longrightarrow_m \mathbf{Bool}_\perp$. Using the extension of the if-then-else function introduced in Example 3.7.16, consider the functions $f_1, f_2 : \mathbf{N}_\perp \longrightarrow_m \mathbf{N}_\perp$ defined by

$$f_1(n) = g(\text{ if } p(x) \text{ then } h(n) \text{ else } k(n)),$$
$$f_2(n) = \text{ if } p(x) \text{ then } g(h(n)) \text{ else } g(k(n)).$$

   Prove that

   (a) $f_1(n) \sqsubseteq f_2(n)$ for any $n \in \mathbf{N}_\perp$, and

   (b) if g is strict, then $f_1(n) = f_2(n)$ for every $n \in \mathbf{N}_\perp$.

35. Let (M, ρ) be a well-founded poset and let $f : M \longrightarrow_m M$ be a strict monotonic function on M. Prove that for every $x \in M$ we have $(x, f(x)) \in \rho$.

   Hint: Suppose that there is $x \in M$ such that $(x, f(x)) \notin \rho$. Since ρ is a total order, we must have $(f(x), x) \in \rho$ and $f(x) \neq x$. Using the strict monotonicity of f, we can build a descending chain, etc.

Complete Partial Orders

36. Prove Theorem 3.8.10.

37. Let C be a chain of the cpo (M, ρ). Suppose that C is the union of the chains C_1 and C_2. Prove that $\sup\{\sup C_1, \sup C_2\}$ exists and, furthermore, that we have

$$\sup C = \sup\{\sup C_1, \sup C_2\}.$$

   Solution: It is clear that $(\sup C_1, \sup C) \in \rho$ and $(\sup C_2, \sup C) \in \rho$ since $C_1, C_2 \subseteq C$. Suppose that $y \in M$ such that $(\sup C_1, y) \in \rho$ and $(\sup C_2, y) \in \rho$. For any $x \in C$, we have either $x \in C_1$ or $x \in C_2$. In the former case, $(x, y) \in \rho$ (since $(x, \sup C_1) \in \rho$ and $(\sup C_1, y)$); in the latter case, we reach the same conclusion. Therefore, y is an upper bound of C, and this implies $(\sup C, y) \in \rho$. This shows that $\sup\{\sup C_1, \sup C_2\}$ exists and that it equals $\sup C$.

38. Consider the cpo $(\mathcal{P}(\mathbf{N}), \subseteq)$ and a function $f : \mathbf{N} \longrightarrow \mathbf{N}$. Define the function $F : \mathcal{P}(\mathbf{N}) \longrightarrow \mathcal{P}(\mathbf{N})$ as $F(W) = f(W)$ for $W \in \mathcal{P}(\mathbf{N})$. Prove that F is continuous.

39. Consider the flat poset $\mathbf{N}_\perp$ and the strict functions $h, k : \mathbf{N}_\perp \longrightarrow_m \mathbf{N}_\perp$ and $g : \mathbf{N}_\perp \longrightarrow_m \mathbf{Bool}_\perp$, where $\mathbf{Bool} = \{\mathbf{T}, \mathbf{F}\}$.

    (a) Prove that every monotonic mapping from $\mathbf{N}_\perp \longrightarrow_m \mathbf{N}_\perp$ is continuous.

    (b) Define the mapping

$$F : (\mathbf{N}_\perp \longrightarrow_m \mathbf{N}_\perp) \longrightarrow (\mathbf{N}_\perp \longrightarrow \mathbf{N}_\perp)$$

    by $F(f)(x) = $ if $g(x)$ then $h(x)$ else $f(k(x))$. Prove that for any $f : \mathbf{N}_\perp \longrightarrow_m \mathbf{N}_\perp$ the mapping $F(f)$ is continuous. Furthermore, prove that F is a continuous function.

    Hint: Use the mapping *apply* introduced in Example 3.8.27.

40. Define the function $F : (\mathbf{N}_\perp \longrightarrow \mathbf{N}_\perp) \longrightarrow \mathbf{N}_\perp$ by

$$F(f) = \begin{cases} \perp & \text{if } f(n) = \perp \text{ for some } n \in \mathbf{N} \\ 0 & \text{otherwise.} \end{cases}$$

    In other words, F maps a partial function to $\perp$ and a total function to 0. Prove that F is monotonic but not continuous.

41. Let (M_i, ρ_i), $1 \le i \le 3$, be three cpos and let $f : M_1 \times M_2 \longrightarrow M_3$ be a function. For $x \in M_1$ and $y \in M_2$ define the functions $f_x : M_2 \longrightarrow M_3$ and $f_y : M_1 \longrightarrow M_3$ as $f_x(v) = f(x, v)$ for every $v \in M_2$ and $f_y(u) = f(u, y)$ for every $u \in M_1$. Prove that f is continuous if and only if both f_x and f_y are continuous for every $x \in M_1$ and $y \in M_2$.

42. *An isomorphism of cpos* is a bijection $f : M_1 \longrightarrow M_2$ such that both f and f^{-1} are continuous functions.

    Prove, by giving an example, that a continuous bijection is not necessarily an isomorphism of cpos.

Zorn's Lemma and the Axiom of Choice

43. Let $\rho \subseteq M \times M$ be a relation on the set M. Prove that there exists a maximal subset U of M (with respect to inclusion) such that $U \times U \subseteq \rho$.

44. Consider the function $f : \mathbf{R} \longrightarrow \mathbf{R}$ defined by $f(x) = ax^2 + bx + c$ for every $x \in \mathbf{R}$. Determine conditions involving the coefficients $a, b, c \in \mathbf{R}$ such that f has two fixed points, one or no fixed point.

45. Zorn's lemma is equivalent to the following apparently weaker statement:

    If (M, ρ) is a nonempty poset such that for any chain $K \subseteq M$ there exists $\sup K$, then M contains at least one maximal element.

46. Prove that the relation $\preceq$ considered in the proof of Theorem 3.9.12 is a partial order on $\mathcal{C}$.

3.11 Bibliographical Comments

There are a large number of books dealing with posets and with their applications to computer science. Hamilton's work [Ham82] is a very lucid presentation of fundamental aspects of set theory and is a useful general reference. Birkhoff's book [Bir73] is a basic source for partially ordered sets and for the algebraic structures that can be derived from partial orders (studied in subsequent chapters in this book). A standard reference for Zorn's axiom is [RR85].

An excellent source for applications in the area of program verification is [LS87].

4

Induction

4.1   Introduction
4.2   Induction on the Natural Numbers
4.3   Inductively Defined Sets
4.4   Proof by Structural Induction
4.5   Recursive Definitions of Functions
4.6   Constructors
4.7   Simultaneous Inductive Definitions
4.8   Propositional Logic
4.9   Primitive Recursive and Partial Recursive Functions
4.10 Grammars
4.11 Peano's Axioms
4.12 Well-Founded Sets and Induction
4.13 Fixed Points and Fixed Point Induction
4.14 Exercises and Supplements
4.15 Bibliographical Comments

4.1   Introduction

In this chapter, we discuss mathematical induction, a proof technique, and recursion, a method for defining functions. These two topics are of central importance in computer science and we will illustrate these subjects with many examples.

Philosophers distinguish between inductive reasoning, which attempts to establish general principles by examining particular cases, and deductive reasoning, which goes from general principles to other general principles or to particular cases. Mathematical reasoning is deductive, and mathematical induction is one particular way of doing mathematical reasoning. Thus, in spite of its name, mathematical induction is not a technique for carrying out reasoning that is "inductive" in the philosophical sense; it is, instead, one of many techniques that can be used to carry out the deductive reasoning that characterizes mathematics.

Most proofs in mathematics, when looked at abstractly enough, can be seen to show that every element of some set S has some property P. In non-inductive proofs ("inductive" from now on is meant in the mathematical sense and not the philosophical sense discussed above), one starts with an

177

arbitrary element x of S and gives a direct proof that x has property P. Sometimes this involves proof by cases, i.e., dividing S into subsets and then giving a separate argument for each subset to show that all the elements of the subset have property P. Nonetheless, in a noninductive proof, for each element x of S, the proof that x has property P is done without reference to any other elements of S.

An inductive proof that every element of S has property P has a different form. For some elements of S, it is possible to give a direct proof that the element has property P. For other elements x of S, however, one shows that $P(x)$ is true *if* $P(y)$ is true for some other element or elements y of S. A proof of this nature is not always successful. For example, suppose that we show that $P(x)$ is true *if* $P(x_1)$ is true. If x_1 was one of those elements of S for which we could show directly that P is true about the element, then we can conclude that $P(x)$ is true. If, on the other hand, we show that $P(x_1)$ is true *if* $P(x_2)$ is true, then we must look at x_2. If we can show directly that $P(x_2)$ is true, then we can conclude that $P(x)$ is true. Otherwise, we may get an x_3 and the process continues. If we eventually reach an x_i such that we can show directly that $P(x_i)$ is true, then we know that $P(x)$ is true. If not, then we have an infinite regress, and we cannot conclude that $P(x)$ is true.

In this chapter, we will give several principles of induction that say that certain proofs of the form described above do succeed in showing that every element of a set S has a property P. In most of these principles, we will have roughly the following situation. There is a way of measuring the complexity of elements of S; the elements x for which it is possible to show directly that $P(x)$ is true are the simplest elements according to this measure of complexity. For those elements x such that one can show that $P(x)$ is true *if* $P(y)$ is true for some element or elements y of S, the elements y are all simpler than x. If the notion of complexity is well enough behaved, the inductive proof works.

We will also consider in this chapter recursion, a method for defining functions. In a nonrecursive definition of a function f, one writes an equation $f(x) = \cdots$ where $\cdots$ denotes an expression involving already known quantities (and in particular, does not involve f). In such a definition, there is generally no difficulty in determining what the value of $f(x)$ is for any given x in the domain of f. In a recursive definition, on the other hand, the expression used to define $f(x)$ involves f. In such a "definition," it is not clear that a function is, in fact, defined. We will look at several versions of what it means for a recursive definition of a function to succeed, and then we will give general conditions under which such definitions do succeed.

4.2   Induction on the Natural Numbers

We begin our study of induction by looking at principles of induction that allow us to conclude that every natural number or, more generally, every integer greater than some fixed integer n_0, has some property. Each one of these principles can be justified using the following axiom.

Axiom 4.2.1 The well-ordering principle for the Natural Numbers *Every nonempty set of natural numbers has a least element.*

The well-ordering principle states that the usual ordering on the natural numbers is a well-ordering. We list it as an axiom rather than a theorem because we will not be giving a proof for it here. Instead, we choose to regard the principle as self-evident and use it to give proofs for other, perhaps not so self-evident, statements about the natural numbers.

The well-ordering principle can be used to show that every natural number has some property P. To do this, one considers the set S consisting of all natural numbers n for which $P(n)$ is false. If S can be shown to have no least element, then by the well-ordering principle, S is empty, and hence, every natural number has property P. We illustrate this proof technique in the following example.

Example 4.2.2 For each natural number n, let

$$S_n = 1 + 2 + \cdots + n$$

(with the usual convention that $S_0 = 0$). Note that for all n greater than or equal to 0 we have $S_{n+1} = S_n + (n+1)$. We will prove that, for all natural numbers n,

$$S_n = \frac{n(n+1)}{2}.$$

Let P be the property that the above formula is true for n and let S be the set of natural numbers for which $P(n)$ is false. If S is nonempty, then by the well-ordering principle, S has a least element m. Since $S_0 = 0$, $P(0)$ is true, and hence, m must be greater than 0. Thus, by definition of m, $P(m-1)$ must be true, i.e., $S_{m-1} = (m-1)m/2$. But, then we have

$$
\begin{aligned}
S_m &= S_{m-1} + m \\
&= \frac{(m-1)m}{2} + m \\
&= \frac{(m-1)m + 2m}{2} \\
&= \frac{(m-1+2)m}{2} \\
&= \frac{m(m+1)}{2}.
\end{aligned}
$$

Hence, $P(m)$ is true, a contradiction. It follows that S must be empty and, hence, that $P(n)$ is true for all natural numbers n, as desired.

Proofs that use the well-ordering principle in the style of the previous example are somewhat awkward and repetitious in that they use a proof by contradiction, which always has the same outward form. Various principles of induction are available that can be used to give less involved proofs. In essence, each principle of induction comes from a type of proof that involves the well-ordering principle to show that every integer greater than some fixed n_0 has a property P. The principle specifies exactly what must be true about P for the proof to work. A single appeal is made to the well-ordering principle to justify the induction principle, and from then on, one can work directly with the induction principle.

Before we can start giving the principles of induction, we need a slight strengthening of the well-ordering principle.

Theorem 4.2.3 *If n_0 is any integer, then every nonempty subset of $\{n \in \mathbf{N} \mid n \geq n_0\}$ has a least element.*

Proof: Let A be a nonempty subset of $\{n \mid n \geq n_0\}$. Define $B = \{n - n_0 \mid n \in A\}$. If $n \in A$, then $n \geq n_0$; so, $n - n_0 \geq 0$. Thus, B is a nonempty set of natural numbers, so by Axiom 4.2.1, it has a least element m. Since $m \in B$, there must be some $m' \in A$ with $m = m' - n_0$, i.e., $m' = m + n_0$. Now, if $n \in A$, then $n - n_0 \in B$; so, $m \leq n - n_0$, and hence, $m' = m + n_0 \leq n$. Thus, m' is the least element of A. ∎

Here is our first principle of induction.

Theorem 4.2.4 (Principle of Ordinary Induction) *Let n_0 be an integer and let P be a property of the integers greater than or equal to n_0. Suppose that*

1. *$P(n_0)$ is true;*

2. *for all $k \geq n_0$, if $P(k)$ is true, then $P(k+1)$ is true.*

Then, $P(n)$ is true for all integers n, $n \geq n_0$.

Proof: Suppose that P is a property of the integers greater than or equal to n_0 for which the two given hypotheses hold. Let S consist of those integers n greater than or equal to n_0 for which $P(n)$ is false. If S is nonempty, then by Theorem 4.2.3, S has a least element m. We cannot have $m = n_0$, since $P(n_0)$ is true, so m must be greater than n_0. Hence, by definition of m, $P(m-1)$ is true, and thus, by our second hypothesis, $P(m)$ is true. This contradiction shows that S is empty, and hence, that $P(n)$ is true for all integers n greater than or equal to n_0. ∎

In a proof that uses Theorem 4.2.4, the verification that $P(n_0)$ is true is called the *basis step*, and the verification that for all $k \geq n_0$, if $P(k)$ is true, then $P(k+1)$ is true, is called the *inductive step*. When, in the course of showing the inductive step, $P(k)$ is assumed, for the purpose of showing that $P(k+1)$ is true, this assumption is called the *inductive hypothesis*.

Although we have already proved Theorem 4.2.4 using the supposedly intuitively obvious well-ordering principle, it is useful to consider the intuitive justification for this theorem further. Suppose that P is a property of the natural numbers and that both the basis step and the inductive step in Theorem 4.2.4 have been carried out (with $n_0 = 0$). We can then see that $P(3)$ is true using a bottom-up approach illustrated by the following line-by-line proof.

Claim	Reason
1. $P(0)$ is true	Basis Step
2. $P(1)$ is true	Inductive Step with $k = 0$ and Line (1)
3. $P(2)$ is true	Inductive Step with $k = 1$ and Line (2)
4. $P(3)$ is true	Inductive Step with $k = 2$ and Line (3)

We can also give a top-down approach to showing that $P(3)$ is true which is very similar to the discussion of the introduction. First of all, the second hypothesis, with $k = 2$, says that $P(3)$ is true *if* $P(2)$ is true. Taking $k = 1$, we see that $P(2)$ is true *if* $P(1)$ is true. With $k = 0$, we see that $P(1)$ is true *if* $P(0)$ is true. But, by the first hypothesis, $P(0)$ is true. Hence, $P(3)$ (as well as $P(1)$ and $P(2)$) is true.

If we think of one integer n being "simpler" than another integer m if $n < m$, then the two steps needed to use the principle of ordinary induction are to show that $P(n)$ is true for the "simplest" integer n greater than or equal to n_0, namely, n_0, and to show that, for all other integers m greater than or equal to n_0, the truth of $P(m)$ follows from the truth of $P(m-1)$ (where $m-1$ is "simpler" than m). Hence, the principle of ordinary induction has the general form we described in the introduction.

To show how the principle of ordinary induction is used, we use it in the next example to reprove the formula already given in Example 4.2.2.

Example 4.2.5 Once again, we show that

$$S_n = \frac{n(n + 1)}{2}$$

for all natural numbers n, where

$$S_n = 1 + \cdots + n.$$

We let P be the property consisting of all n for which the formula is true and use Theorem 4.2.4 to show that $P(n)$ is true for all natural numbers n. The basis step is to show that $P(0)$ is true and this is immediate. For the inductive step, we make the inductive hypothesis that $k \geq 0$ and $P(k)$ is true and show that $P(k + 1)$ is true. Using some algebra and the induction hypothesis (in the second line), we get

$$
\begin{aligned}
S_{k+1} &= S_k + (k+1) \\
&= \tfrac{k(k+1)}{2} + (k+1) \\
&= \tfrac{k(k+1)+2(k+1)}{2} \\
&= \tfrac{(k+1)(k+2)}{2},
\end{aligned}
$$

which shows that $P(k+1)$ is true, if $P(k)$ is. Thus, by induction, $P(n)$ is true for all natural numbers n.

There is a vast number of formulas of the type given in the previous example that can be proven by induction. We give some more in the exercises, but turn now to examples involving finite sets and counting. Our first goal is to show a result that was promised in Section 2.4.

Theorem 4.2.6 *If $n \in \mathbf{N}$ and $f : \{0, \ldots, n-1\} \longrightarrow \{0, \ldots, n-1\}$ is an injection, then f is also a surjection.*

Proof. The proof is by induction on n. The basis step is to show that any injection from $\emptyset$ to itself is a surjection. This is immediate. For the inductive step, suppose that the result is true for $n \in \mathbf{N}$ and that $f : \{0, \ldots, n\} \longrightarrow \{0, \ldots, n\}$ is an injection. Let f' be the restriction of f to $\{0, \ldots, n-1\}$. We consider two cases. First, suppose that $f' : \{0, \ldots, n-1\} \longrightarrow \{0, \ldots, n-1\}$. Then, since f' is an injection, by the induction hypothesis, f' is a surjection. Since f is injective, we must then have $f(n) = n$, and thus, $f : \{0, \ldots, n\} \longrightarrow \{0, \ldots, n\}$ is a surjection.

If f' does not belong to $\{0, \ldots, n-1\} \longrightarrow \{0, \ldots, n-1\}$, then there must be an a, $0 \leq a < n$, such that $f(a) = n$, and this a is unique since f is injective. Also since f is injective, there must be a b, $0 \leq b < n$, with $f(n) = b$. Define a function $g : \{0, \ldots, n-1\} \longrightarrow \{0, \ldots, n-1\}$ by

$$
g(m) = \begin{cases} f(m) & \text{if } m \neq a \\ b & \text{if } m = a. \end{cases}
$$

Since f is an injection, it follows that g is an injection, and hence, by the induction hypothesis, $g : \{0, \ldots, n-1\} \longrightarrow \{0, \ldots, n-1\}$ is a surjection. Thus, all numbers j with $0 \leq j \leq n-1$ and $j \neq b$ are in the range of f. Since we already know that b and n are in the range of f, it follows that $f : \{0, \ldots, n\} \longrightarrow \{0, \ldots, n\}$ is a surjection. ∎

Corollary 4.2.7 *Let m, n be two natural numbers.*

1. *There is an injection from $\{0, \ldots, n-1\}$ to $\{0, \ldots, m-1\}$ if and only if $n \leq m$.*

2. *There is a surjection from $\{0, \ldots, n-1\}$ to $\{0, \ldots, m-1\}$ if and only if $n \geq m > 0$ or $n = m = 0$.*

3. *There is a bijection from* $\{0,\ldots,n-1\}$ *to* $\{0,\ldots,m-1\}$ *if and only if* $n = m$.

Proof. (1) If $n \leq m$, then the containment mapping is the desired injection from $\{0,\ldots,n-1\}$ to $\{0,\ldots,m-1\}$. Conversely, suppose that $f : \{0,\ldots,n-1\} \longrightarrow \{0,\ldots,m-1\}$ is an injection. If $n > m$, then we also have $f : \{0,\ldots,n-1\} \longrightarrow \{0,\ldots,n-1\}$. By Theorem 4.2.6, f must map onto $\{0,\ldots,n-1\}$, but this is impossible if we also have f mapping into $\{0,\ldots,m-1\}$ and $n > m$. Thus, we must have $n \leq m$.

(2) If $n = m = 0$, then the empty function is a surjection from $\{0,\ldots,n-1\}$ to $\{0,\ldots,m-1\}$. If $n \geq m > 0$, then we can define a surjection $f : \{0,\ldots,n-1\} \longrightarrow \{0,\ldots,m-1\}$ by defining

$$f(r) = \begin{cases} r & \text{if } 0 \leq r \leq m-1 \\ 0 & \text{if } m \leq r \leq n-1. \end{cases}$$

Conversely, suppose that $f : \{0,\ldots,n-1\} \longrightarrow \{0,\ldots,m-1\}$ is a surjection. Define a function $g : \{0,\ldots,m-1\} \longrightarrow \{0,\ldots,n-1\}$ by defining $g(r)$, for $0 \leq r \leq m-1$, to be the least t, $0 \leq t \leq n-1$, for which $f(t) = r$. (Such t exists since f is a surjection.) Then, g is an injection, and hence, by part (1), $m \leq n$. In addition, if $m = 0$, then we must also have $n = 0$, else the function f could not exist.

(3) If $n = m$, then the identity function is the desired bijection. Conversely, if there is a bijection from $\{0,\ldots,n-1\}$ to $\{0,\ldots,m-1\}$, then by part (1), $n \leq m$, while by part (2), $n \geq m$, so $n = m$. ∎

Corollary 4.2.8 *If A is a finite set, then there is a unique natural number n for which there exists a bijection from $\{0,\ldots,n-1\}$ to A.*

Proof. Suppose that $f : \{0,\ldots,n-1\} \longrightarrow A$ and $g : \{0,\ldots,m-1\} \longrightarrow A$ are both bijections. Then, $g^{-1}f : \{0,\ldots,n-1\} \longrightarrow \{0,\ldots,m-1\}$ is a bijection, so by Corollary 4.2.7, part (3), $n = m$. ∎

Now that we have seen that the cardinality of a finite set is well defined, we can extend Corollary 4.2.7 to arbitrary finite sets.

Corollary 4.2.9 *Let A and B be finite sets.*

1. *There is an injection from A to B if and only if $|A| \leq |B|$.*

2. *There is a surjection from A to B if and only if $|A| \geq |B|$.*

3. *There is a bijection from A to B if and only if $|A| = |B|$.*

Proof: Let $|A| = n$ and $|B| = m$ and let $f : \{0,\ldots,n-1\} \longrightarrow A$ and $g : \{0,\ldots,m-1\} \longrightarrow B$ be bijections. If $h : A \longrightarrow B$ is an injection, then $g^{-1}hf : \{0,\ldots,n-1\} \longrightarrow \{0,\ldots,m-1\}$ is an injection, so by Corollary 4.2.7, part (1), $n \leq m$, i.e., $|A| \leq |B|$. Conversely, if $n \leq m$, then there is an injection $k : \{0,\ldots,n-1\} \longrightarrow \{0,\ldots,m-1\}$, namely, the inclusion, and $gkf^{-1} : A \longrightarrow B$ is an injection.

The other parts are proven similarly. ∎

Corollary 4.2.10 *Let A and B be two finite sets with the same cardinality. If $h : A \longrightarrow B$, then the following are equivalent:*

1. h is an injection,

2. h is a surjection, and

3. h is a bijection.

Proof: Let $|A| = |B| = n$ and let $f : \{0, \ldots, n-1\} \longrightarrow A$ and $g : \{0, \ldots, m-1\} \longrightarrow B$ be bijections. If h is an injection, then $k = g^{-1}hf : \{0, \ldots, n-1\} \longrightarrow \{0, \ldots, n-1\}$ is also an injection, so by Theorem 4.2.6, k is a surjection. But, then $h = gkf^{-1}$ is also a surjection.

If h is a surjection, then, for each $b \in B$, $h^{-1}(\{b\})$ is nonempty. Thus, since f is a surjection, for each $b \in B$, there is some i, $0 \le i \le n - 1$, with $h(f(i)) = b$. We may thus define $k : B \longrightarrow A$ by defining $k(b)$ to be $f(i)$, where i is the least number with $h(f(i)) = b$. Then, k is a right inverse for h and k is an injection. By the first part of the proof, k is a surjection, hence, a bijection, and h, being a left inverse for k, must be k^{-1}, so h is also a bijection.

Finally, if h is a bijection, then h is an injection. ∎

Principles of induction for the integers can be used to prove facts about objects that are quite different from the integers. For example, to show that every finite sequence meets some condition $\mathcal{K}$, it suffices to use induction to show that every natural number n has the property P given by "every sequence of length n satisfies condition $\mathcal{K}$." Such a proof is called a proof by induction on the length of the sequence. Similarly, to show that every finite set meets some condition $\mathcal{K}$, it suffices to show by induction that every natural number n has the property P given by "every finite set with cardinality n satisfies condition $\mathcal{K}$." Such a proof is called a proof by induction on the cardinality of the finite set.

In a similar vein, suppose that P is a property of a set S and that $f : S \longrightarrow \mathbf{N}$. Let P' be the property of $\mathbf{N}$, where n has property P' if every element of $f^{-1}(\{n\})$ has property P. If we show (possibly by induction) that $P'(n)$ is true for all $n \in \mathbf{N}$, then $P(x)$ is true for $x \in S$.

Our first example of a proof by induction on the cardinality of a finite set shows a result that is rather obvious intuitively, but still deserves to be proved.

Theorem 4.2.11 *If A is a finite set and $C \subseteq A$, then C is finite and $|C| \le |A|$.*

Proof. We prove the result by induction on the cardinality of A. For the basis step, we note that if $|A| = 0$, then $A = \emptyset$, and the only subset of A is A, so the result holds.

For the inductive step, suppose that the result is true for sets of cardinality n and that $|A| = n + 1$, say, $f : \{0, \ldots, n\} \longrightarrow A$ is a bijection.

For $0 \leq i \leq n$, let $f(i) = a_i$. Let $B = \{a_0, \ldots, a_{n-1}\}$. Then, obviously, $|B| = n$. Let $C \subseteq A$. If $C \subseteq B$, then by the induction hypothesis, C is finite and $|C| \leq |B| = n < |A|$. Otherwise, we can write $C = C' \cup \{a_n\}$, where $C' \subseteq B$, so by the induction hypothesis, C' is finite and $|C'| \leq n$. But, then by Theorem 2.4.11, C is finite, and $|C| = |C'| + 1 \leq n + 1 = |A|$. ∎

In the next series of examples, we use induction to develop several formulas for calculating the sizes of finite sets.

Example 4.2.12 Let $n \in \mathbb{N}$ and let $\{A_i \mid 0 \leq i \leq n - 1\}$ be a family of pairwise disjoint finite sets. Then, we use induction on n to show that $\bigcup \{A_i \mid 0 \leq i \leq n - 1\}$ is finite and $|\bigcup \{A_i \mid 0 \leq i \leq n - 1\}| = |A_0| + |A_1| + \cdots + |A_{n-1}|$. (We use the convention that when $n = 0$ the sum on the right side of the equation equals 0.) For $n = 0$, the result is immediate since $\bigcup \emptyset = \emptyset$, which is finite and has cardinality 0.

Suppose that the result is true for some $n \in \mathbb{N}$ and that we have a family $\{A_i \mid 0 \leq i \leq n\}$ of pairwise disjoint finite sets. Then, we have

$$\bigcup \{A_i \mid 0 \leq i \leq n\} = \bigcup \{A_i \mid 0 \leq i \leq n - 1\} \cup A_n,$$

and $(\bigcup \{A_i \mid 0 \leq i \leq n - 1\}) \cap A_n = \emptyset$ since the A_i's are pairwise disjoint. Thus, by Theorem 2.4.11 and the induction hypothesis, we get $\bigcup \{A_i \mid 0 \leq i \leq n\}$ is finite and

$$
\begin{aligned}
\left|\bigcup \{A_i \mid 0 \leq i \leq n\}\right| &= \left|\bigcup \{A_i \mid 0 \leq i \leq n - 1\}\right| + |A_n| \\
&= |A_0| + \cdots + |A_{n-1}| + |A_n|,
\end{aligned}
$$

as desired.

Example 4.2.13 In this example, we present rigorously another useful counting principle, namely, we show that if A and B are two finite sets, then $A \cup B$ is finite and

$$|A \cup B| = |A| + |B| - |A \cap B|.$$

We let $P(n)$ be the property given by "if $|B| = n$, then for all finite sets A, $A \cup B$ is finite, and $|A \cup B| = |A| + |B| - |A \cap B|$," and use induction to show that $P(n)$ is true for all $n \in \mathbb{N}$. Note that if $B \subseteq A$, then $A \cup B = A$, and the desired formula becomes $|A| = |A| + |B| - |B|$, which is true, so the result is immediate in this case. This observation shows the basis step, since if $|B| = 0$, then $B \subseteq A$.

Suppose that $n \in \mathbb{N}$ and $P(n)$ is true. Let $|B| = n + 1$ and let A be a finite set. If $B \subseteq A$, then we have seen that the desired result is true, so suppose that $B \not\subseteq A$, say, $b \in B - A$. Let $B' = B - \{b\}$. Then, $|B'| = n$ and $A \cup B = (A \cup \{b\}) \cup B'$. Hence, by the induction hypothesis, $A \cup B$ is finite and

$$|A \cup B| = |A \cup \{b\}| + |B'| - |(A \cup \{b\}) \cap B'|.$$

Now, by Theorem 2.4.11, we have $|A \cup \{b\}| = |A| + 1$. We already know that $|B'| = |B| - 1$. Finally, we have $(A \cup \{b\}) \cap B' = (A \cap B') \cup (\{b\} \cap B') = (A \cap B') \cup \emptyset = A \cap B' = A \cap B$ (the last equality because $b \notin A$). Plugging these equalities into the displayed equation gives the desired result.

Example 4.2.14 We show that if A is a finite set, then $|\mathcal{P}(A)| = 2^{|A|}$. The proof is by induction on $|A|$; that is, we let P be the following property of n: "if $|A| = n$, then $|\mathcal{P}(A)| = 2^n$," and we prove that $P(n)$ holds for all natural numbers n.

For the basis step, we must show that if $|A| = 0$, then $|\mathcal{P}(A)| = 2^0 = 1$. But, if $|A| = 0$, then $A = \emptyset$, so $\mathcal{P}(A) = \{\emptyset\}$ and $|\mathcal{P}(A)| = 1$ as desired.

For the inductive step, we suppose that $P(k)$ is true for some $k \geq 0$, i.e., if $|A| = k$, then $|\mathcal{P}(A)| = 2^k$, and we try to show that $P(k + 1)$ is true using this induction hypothesis. Thus, suppose that $|A| = k + 1$. Let a be some element of A, and let $B = A - \{a\}$. Then, $|B| = k$, and hence, by the induction hypothesis, $|\mathcal{P}(B)| = 2^k$. Let D_1 consist of those subsets of A that do not contain a and let D_2 consist of those subsets of A that do contain a. Then, D_1 and D_2 are disjoint sets and $\mathcal{P}(A) = D_1 \cup D_2$.

Since D_1 is the same as $\mathcal{P}(B)$, we know that D_1 has 2^k elements. To count the elements in D_2, we define a function $\Psi : D_1 \longrightarrow D_2$ by letting $\Psi(S) = S \cup \{a\}$ for all $S \in D_1$. If $\Phi : D_2 \longrightarrow D_1$ is defined by $\Phi(T) = T - \{a\}$ for all $T \in D_2$, then it is easily verified that Ψ is both a left and a right inverse for Φ and, hence, that Ψ is a bijection. Thus, $|D_1| = |D_2| = 2^k$, so we get

$$|\mathcal{P}(A)| = |D_1| + |D_2| = 2^k + 2^k = 2^{k+1},$$

which is what we wanted.

We need to introduce some notation before giving our next example. If A is a set and $k \in \mathbf{N}$, then we let $\mathcal{P}_k(A)$ denote the set of all subsets of A that have cardinality k.

Example 4.2.15 If A is a finite set and $k > |A|$, then Theorem 4.2.11 (and common sense) tell us that $|\mathcal{P}_k(A)| = 0$. We will determine $|\mathcal{P}_k(A)|$ when A is finite and $0 \leq k \leq |A|$. In fact, we show by induction on n that if $|A| = n$ and $0 \leq k \leq n$, then

$$|\mathcal{P}_k(A)| = \tfrac{n!}{k!(n-k)!}.$$

Let $P(n)$ be the property that the given formula is true for sets of size n and all k, $0 \leq k \leq n$. When $n = 0$, the only k to consider is 0. We have

$$|\mathcal{P}_0(\emptyset)| = |\{\emptyset\}| = 1 = \tfrac{0!}{0!(0-0)!},$$

as desired.

Now, suppose that $n \in \mathbf{N}$ and $P(n)$ is true. Let $|A| = n + 1$ and let $0 \leq k \leq n+1$. If $k = 0$ or $k = n+1$, then $\frac{(n+1)!}{k!(n+1+k)!} = 1$. Since $\mathcal{P}_0(A) = \{\emptyset\}$

and $\mathcal{P}_{n+1}(A) = \{A\}$, this is the correct number. Thus, we may suppose that $1 \leq k \leq n$. Let a be some fixed element of A and let $B = A - \{a\}$, so $|B| = n$. We have

$$\mathcal{P}_k(A) = \mathcal{P}_k(B) \cup D,$$

where D consists of those k-element subsets of A that contain a. By the induction hypothesis (since $0 \leq k \leq n$), $|\mathcal{P}_k(B)| = \frac{n!}{k!(n-k)!}$. To count D, we note that the function $\Psi : \mathcal{P}_{k-1} \longrightarrow D$ given by $\Psi(S) = S \cup \{a\}$ is a bijection, and hence, by the induction hypothesis (since $0 \leq k - 1 \leq n$), we have $|D| = |\mathcal{P}_{k-1}(B)| = \frac{n!}{(k-1)!(n-(k-1))!}$. Since $\mathcal{P}_k(B)$ and D are disjoint, we get

$$|\mathcal{P}_k(A)| = \frac{n!}{k!(n-k)!} + \frac{n!}{(k-1)!(n-(k-1))!} = \frac{(n+1)!}{k!(n+1-k)!},$$

completing the induction.

Sometimes when you are trying an induction proof, you find that in order to show that $P(k+1)$ is true in the inductive step you need to know not just that $P(k)$ is true, but also that $P(m)$ is true, where m is some small number that is either equal to the basis value n_0 or else is slightly larger, and m is the same, no matter what value k has. For instance, in a proof by induction on the natural numbers, you may find that you can show that $P(k + 1)$ is true (for $k \geq 1$), if you use the assumption that $P(k)$ and $P(1)$ are both true. Occasionally, in the inductive step, you need to assume that $P(m)$ is true for several small values of m, but the values of m are the same no matter what k is. This situation can be covered by strong induction, which we discuss in the sequel, but it is really only a variation on the versions of induction already discussed, so we give a version of induction here that covers these situations. The idea is that we increase the number of basis steps, if needed, so that all of the values of m we need in the inductive step are covered in the basis steps, and then we include the truth of all of the basis steps as part of the induction hypothesis.

Theorem 4.2.16 *Let n_0 be an integer and let P be a property of the set $\{n \mid n \in \mathbf{N}, n \geq n_0\}$. Let ℓ be a natural number and suppose that*

1. *$P(j)$ is true for every j with $n_0 \leq j \leq n_0 + \ell$, and*

2. *for all $k \geq n_0 + \ell$, if $P(j)$ is true for every j with $n_0 \leq j \leq n_0 + \ell$ and $P(k)$ is true, then $P(k + 1)$ is true.*

Then, $P(n)$ is true for every $n \geq n_0$.

Proof. This theorem follows from Theorem 4.2.3 by a proof similar to that of Theorem 4.2.4. ∎

In the variation of induction just given, the basis steps are to show $P(n_0), P(n_0 + 1), \ldots, P(n_0 + \ell)$, so there are $\ell + 1$ basis steps. For example, if you find that in trying to do a proof by induction on the natural numbers you can show that $P(k+1)$ is true when $k \geq 1$ if you assume that

$P(k+1)$ and $P(1)$ are both true, then you would apply the above method with $\ell = 1$ and $n_0 = 0$. Step 1 would then require you to show that $P(0)$ and $P(1)$ are both true, while in Step 2, you would assume that $k \geq 1$ and $P(0), P(1)$, and $P(k)$ are all true, and you would use these assumptions to show that $P(k+1)$ is true.

Example 4.2.17 Let A and B be finite sets. We want to find a formula for $|A \times B|$ in terms of $|A|$ and $|B|$. A little experimentation suggests that

$$|A \times B| = |A||B|.$$

We will prove this equality by induction on $|B|$. Let $P(n)$ be the property "if $|B| = n$, then for any finite set A, $|A \times B| = |A||B|$." We use Theorem 4.2.16, with $n_0 = 0$ and $\ell = 1$, to show that $P(n)$ is true for all natural numbers n. The first basis step is for $n = 0$. If $|B| = 0$, then $B = \emptyset$, and we have

$$|A \times B| = |A \times \emptyset| = |\emptyset| = 0 = |A|0 = |A||B|,$$

establishing $P(0)$.

The second basis step is $|B| = 1$, say, $B = \{b\}$. Then, for any A there is a one-to-one correspondence between A and $A \times B$ given by $a \mapsto (a, b)$, so $|A \times B| = |A|$, and we have

$$|A \times B| = |A| = |A|1 = |A||B|,$$

so $P(1)$ is true.

For the inductive step, we assume that $k \geq 1$ and that $P(0), P(1)$, and $P(k)$ are all true, and we show that $P(k+1)$ is true. Suppose that $|B| = k + 1$. Let b be some particular element of B and let $B' = B - \{b\}$. Then, $|B'| = k$, so by $P(k)$ we know that $|A \times B'| = |A||B'| = |A|k$, and by $P(1)$, we know that $|A \times \{b\}| = |A|$. Now, $A \times B = A \times (B' \cup \{b\}) = (A \times B') \cup (A \times \{b\})$, and $A \times B'$ and $A \times \{b\}$ are disjoint sets, so

$$|A \times B| = |A \times B'| + |A \times \{b\}| = |A|k + |A| = |A|(k+1) = |A||B|,$$

so $P(k+1)$ holds.

Our next variation of induction is one that can be used when we want to show that every integer in some finite range of integers has a property.

Theorem 4.2.18 *Let n_0 and m_0 be two integers with $n_0 \leq m_0$. Let P be a property of the integers between n_0 and m_0. Suppose that*

1. *$P(n_0)$ is true, and*

2. *for all k with $n_0 \leq k < m_0$, if $P(k)$ is true, then $P(k+1)$ is true.*

Then, $P(n)$ is true for all n with $n_0 \leq n \leq m_0$.

Proof. This theorem follows from Theorem 4.2.3 by a proof similar to that of Theorem 4.2.4. ∎

Example 4.2.19 Consider the following program segment whose goal is to compute the product of two integer values x and y by adding x to 0 a total of y times:

$$p \leftarrow 0;$$
for $i \leftarrow 1$ **to** y **do**
$$\quad p \leftarrow p + x$$
endfor

Suppose that this program segment is run in a state where x is an integer and y is a natural number. Then, the body of the for loop will be executed y times. Let p_0 be the value of p before the first execution of the body of the for loop (so, $p_0 = 0$), and for each k with $1 \le k \le y$, let p_k be the value of p when the kth execution of the body of the for loop terminates. We will use Theorem 4.2.18 with $n_0 = 0$ and $m_0 = y$ to show that for each k with $0 \le k \le y$, $p_k = kx$. The basis step is immediate since $p_0 = 0 = 0x$. Suppose that $0 \le k < y$ and that $p_k = kx$. Then,

$$p_{k+1} = p_k + x = kx + x = (k+1)x,$$

as desired. Now, the value of p when execution of the program segment terminates is p_y. According to what we just proved, this means that the final value of p is yx, and hence, the program does work as we wanted it to.

The next version of induction that we consider in this section is one that can be applied in a situation where we do not know if the set of integers to which we are applying induction is a finite range or an infinite range. Such a situation can arise in the analysis of a computer program. For example, you may wish to show that some statement S about the program variables is true each time a loop in the program is executed. This suggests using induction to show that the property $P(n)$ given by "S is true after the loop is executed n times" holds for all n such that the loop is executed n times. If you do not know whether or not the loop terminates, then you do not know if the range of n for which you want to show that $P(n)$ is true is finite or infinite. It is also possible that for some input values the loop terminates, and for other input values, the loop does not terminate. It would be nice to be able to give one proof that covers both the terminating and nonterminating cases. The following version of induction, which is just a combination of Theorems 4.2.4 and 4.2.18, makes such a unified proof possible.

Theorem 4.2.20 *Let n_0 be an integer and let S be a set of integers that is either $\{n \mid n \in Z, n \ge n_0\}$ or $\{n \mid n \in Z, n_0 \le n \le m_0\}$ for some $m_0 \ge n_0$. Let P be a property of S. Suppose that*

1. $P(n_0)$ is true,

2. for all $k \geq n_0$, if $k+1 \in S$ and $P(k)$ is true, then $P(k+1)$ is true.

Then, $P(n)$ is true for every $n \in S$.

Proof. This theorem follows from Theorem 4.2.3 by a proof similar to that of Theorem 4.2.4. ∎

Example 4.2.21 Consider the following program segment whose goal is to calculate the product of two integers x, y by repeated addition of x.

```
p ← 0;
i ← 0;
while i ≠ y do
    p ← p + x; i ← i + 1
endwhile
```

Let p_0 and i_0 be the values of p and i just before the while loop is executed (so p_0 and i_0 both equal 0). For each $k \geq 1$, if the body of the while loop is executed at least k times, then let p_k and i_k be the values of p and i at the end of the kth execution of the body of the loop. Let S be the set of natural numbers k such that the body of the while loop is executed at least k times. Then, S is either $\{k \mid k \in Z, 0 \leq k \leq m_0\}$ for some natural number m_0 or S consists of all the natural numbers, and p_k and i_k are defined exactly for those k in S. We will use Theorem 4.2.20 to show that for each $k \in S$, $p_k = i_k x$. The base case is immediate since both p_0 and i_0 equal 0. Suppose that $k \geq 0, k+1 \in S$, and $p_k = i_k x$. Then,

$$p_{k+1} = p_k + x = i_k x + x = (i_k + 1)x = i_{k+1}x,$$

completing the induction.

We have just shown that the relationship "$p = ix$" holds before the while loop is executed and after each execution of the body of the loop. This is because our initialization made the relationship hold before the loop is executed, and because if the two assignment statements that constitute the body of the loop are executed with the relationship holding, then the relationship still holds after the statements are executed. A relationship such as $p = ix$, which holds before a loop is executed and after each execution of the loop, is called a *loop invariant*.

If execution of the while loop terminates, then the loop invariant $p = ix$ must be true when the loop terminates, and in addition, we must have $i = y$; so, if the loop terminates, we will have $p = yx$, which was our goal. What we have just shown is the *partial correctness* of our program segment; that is, *if* the program terminates, then it does what we wanted it to. The question of whether or not the program does terminate is handled separately. In fact, if the value of y is nonnegative, then, since the initial value of i is 0 and the value of i is incremented by 1 each time through the loop, we will eventually get $i = y$ and the loop will terminate, while if the value of y is negative, then the loop will never terminate. A program is called *totally correct* if the program is partially correct and it terminates for all input values.

(Actually, partial and total correctness of a program are defined with respect to some input and output specifications, called the *precondition* and *postcondition*. The postcondition states what should be true when the execution of the program terminates, and the precondition states what is assumed to be true before the program is executed. In our example, we could take the postcondition to be $p = xy$. If the precondition makes no restrictions on what is true before the program is executed (except for the implicit assumption that x and y are integers), then as we have seen, the program is partially correct, but not totally correct. If the precondition states that $y \geq 0$, then with respect to these pre- and postconditions, the program is totally correct.)

In all of the versions of induction that we have looked at so far, the induction step involves the proof that $P(k + 1)$ holds given the induction hypothesis that $P(k)$ is true (and, in one version, the additional assumption that all of the basis steps hold). In some proofs by induction, when you try to show that $P(k + 1)$ holds, you find that you can do this if you know that $P(j)$ holds for some value or values of j that are less or equal to k but not otherwise known. The principles of induction that we have so far do not appear to help in this situation, since they only allow you to assume that $P(k)$ (and possibly $P(m)$ for some basis values m that you know in advance) are true when you try to show that $P(k + 1)$ is true. The next principle of induction can be used in situations such as the one just described.

Theorem 4.2.22 (Principle of Strong Induction) *Let n_0 be an integer and let P be a property of the integers that are at least equal to n_0. Suppose that*

1. *$P(n_0)$ is true, and*

2. *for all $k \geq n_0$, if $P(j)$ is true for every j with $n_0 \leq j \leq k$, then $P(k + 1)$ is true.*

Then, $P(n)$ is true for every integer greater or equal to n_0.

Proof. This theorem follows from Theorem 4.2.3 by a proof similar to that of Theorem 4.2.4. ∎

To give an intuitive argument for the correctness of the principle of strong induction in another way, suppose that both the basis step and the inductive step of strong induction are true with $n_0 = 0$. We will show in a line-by-line fashion why $P(3)$ is true.

Claim	Reason
1. $P(0)$ is true	Basis Step
2. $P(1)$ is true	Inductive Step with $k = 0$ and Line (1)
3. $P(2)$ is true	Inductive Step with $k = 1$ and Lines (1),(2)
4. $P(3)$ is true	Inductive Step with $k = 2$ and Lines (1),(2),(3)

The only difference between this and the similar line-by-line argument we made for the first principle of induction we looked at is that each line here needs all of the previous lines as justification instead of just the immediately preceding line. There is nothing special about the number 3 in this argument. A similar argument could be made for any natural number n to show that $P(n)$ is true.

Example 4.2.23 For our first example of strong induction, we consider an algorithm called quicksort, which has as input a finite sequence **a** of integers and produces as output a rearrangement of **a** into increasing order. The quicksort algorithm makes use of the algorithm split. The input to split is a nonempty finite sequence of integers $\mathbf{a} = \langle a_1, \ldots, a_n \rangle$. The output is an integer a and two sequences $\mathbf{a}_L$ and $\mathbf{a}_R$ meeting the following specifications:

1. $\mathbf{a}_L \langle a \rangle \mathbf{a}_R$ is a rearrangement of **a**,

2. every entry in $\mathbf{a}_L$ is less or equal to a, and

3. every entry in $\mathbf{a}_R$ is greater or equal to a.

For example, if $\mathbf{a} = \langle 6, 10, 4, 3, 4, 14, 1, 4, 7 \rangle$, then one possible output for split is $\mathbf{a}_L = \langle 4, 3, 1 \rangle$, $a = 4$, and $\mathbf{a}_R = \langle 6, 10, 4, 14, 7 \rangle$. We do not care how the split algorithm works as long as it meets its specification.

Once the split algorithm is applied to a sequence **a** to get sequences $\mathbf{a}_L, \mathbf{a}_R$, and a, the problem of sorting **a** has been split into two independent subproblems: sorting $\mathbf{a}_L$ and sorting $\mathbf{a}_R$. This is the idea behind the quicksort algorithm that we now give.

Given the input sequence **a**, if **a** is the empty sequence, then **a** is already sorted, so the output for quicksort is the same as the input. Otherwise, apply the split algorithm to **a** to get $\mathbf{a}_L, \mathbf{a}_R$, and a. Recursively apply the quicksort algorithm to $\mathbf{a}_L, \mathbf{a}_R$ to produce outputs $\mathbf{b}_L, \mathbf{b}_R$, respectively. The output for quicksort applied to **a** is then $\mathbf{b}_L \langle a \rangle \mathbf{b}_R$.

We will use strong induction to show that no matter what finite sequence the quicksort algorithm is applied to, the algorithm terminates and produces the correct output. Let $P(n)$ be the property that if quicksort is applied to a sequence of length n, then the algorithm terminates and produces as output a rearrangement of the input into increasing order. Then, $P(0)$ is true since quicksort applied to the empty sequence produces the empty sequence as output. Now, suppose that $k \geq 0$ and that $P(j)$ is true for all j, $0 \leq j \leq k$. If quicksort is applied to a sequence **a** of length $k + 1$, then after split is called to produce $\mathbf{a}_L, \mathbf{a}_R$, and a, quicksort is recursively applied to a_L and a_R, each of which has length less or equal to k. Thus, by the induction hypothesis, quicksort, when applied to these sequences, produces sorted rearrangements. It follows that quicksort, when applied to a, terminates. Furthermore, since b_L and b_R are increasing rearrangements of a_L, a_R, the output $b_L \langle a \rangle b_R$ is a rearrangement of $a_L \langle a \rangle a_R$, which in turn (because of the way split works) is a rearrangement of a, and again, because of the way split works, the output is in increasing order. Hence, quicksort works when applied to a.

Note in the previous proof why we need strong induction. When split divides a sequence a into two parts a_L and a_R, we know nothing about the lengths of a_L and a_R except that they are less or equal to k. Hence, when we carry out the proof by induction, we need an induction hypothesis that applies to sequences of any length between 0 and k. This is exactly what strong induction gives.

Example 4.2.24 In this example, we show a fundamental property of the integers, namely, that every integer greater or equal to 2 has a prime factorization (i.e., it can be obtained as the product of one or more prime numbers). We use Theorem 4.2.22 with $n_0 = 2$. The basis step is immediate since 2 is itself prime and therefore is the product of one prime (itself). Suppose that $k \geq 2$, and that every natural number j with $2 \leq j \leq k$ has a prime factorization. We must show that $k + 1$ has a prime factorization. If $k + 1$ is prime, then this is certainly true. On the other hand, if $k + 1$ is not prime, then $k + 1$ must be evenly divisible by some number r bigger than 1 and less than $k + 1$, say, $k + 1 = rs$. Then, we have $2 \leq r, s \leq k$, so by the induction hypothesis, both r and s can be written as products of primes. Combining these prime factorizations, we get a prime factorization for $k + 1$.

Example 4.2.25 In this example, we consider an algorithm known as binary search. The input consists of a nonempty finite sequence a of integers in increasing order and an integer q. The output is "yes" if one of the entries in a is q, and otherwise, the output is "no." If a has length 1, then binary search proceeds by simply comparing q with the one entry in a and giving the appropriate yes or no answer. If a has length greater than 1, then divide a into two "halves" b_1 and b_2 of as close to the same lengths as possible. (If a has odd length, then let b_1 be the longer of the two halves. For instance, if $a = \langle 2, 5, 8, 9 \rangle$, then $b_1 = \langle 2, 5 \rangle$ and $b_2 = \langle 8, 9 \rangle$, while if $a = \langle 2, 5, 8, 9, 12 \rangle$, then $b_1 = \langle 2, 5, 8 \rangle$ and $b_2 = \langle 9, 12 \rangle$.) Compare q with the last element of b_1. If q is less than or equal to this element, then recursively apply binary search to b_1 and q and give as output for a and q the same output as is obtained from this recursive call; otherwise, recursively apply binary search to b_2 and q and give as output the output from this recursive call.

We wish to show that this algorithm is correct; that is, we want to show that if binary search is applied to a nonempty sequence of integers $\mathbf{a}$ in increasing order and an integer q, then the output is yes or no, according to whether or not q appears in $\mathbf{a}$. We use Theorem 4.2.22 with $n_0 = 1$ to show that if binary search is applied to an increasing sequence of length n and an integer q, then the algorithm terminates and output is correct. This is obvious for $n = 1$. Suppose that $k \geq 1$ and the algorithm works for sequences of length between 1 and k. If binary search is applied to an increasing sequence of length $k+1$, then the two halves $\mathbf{b}_1$ and $\mathbf{b}_2$ will both be nonempty increasing sequences of lengths less than or equal to k. (This is the key observation.) Thus, by the induction hypothesis, when binary search is applied to either half, the algorithm terminates with the correct answer for that half. If this answer is yes, then obviously q appears in $\mathbf{a}$. If the answer is no, then it follows from the way we chose which half to use together with the fact that $\mathbf{a}$ is increasing that q does not appear in the other half either and, hence, does not appear in $\mathbf{a}$. Thus, binary search works for sequences of length $k + 1$.

We will analyze binary search further by considering how many comparisons are needed between q and elements of $\mathbf{a}$ in the course of carrying out the algorithm. To keep things simple, we will only consider the case when $|\mathbf{a}|$ is a power of 2. We will show, using ordinary induction on the natural numbers, that if $|\mathbf{a}| = 2^n$, then binary search makes $n + 1$ comparisons when applied to $\mathbf{a}$ and any q. If $|\mathbf{a}| = 2^0 = 1$, then 1 comparison is made, so the result holds for $n = 0$. Suppose that the result is true for sequences of length 2^k and that binary search is applied to a sequence of length 2^{k+1}. Then, after 1 comparison is made, binary search is recursively applied to one of the halves of the original sequence, and this half will have length 2^k. By the induction hypothesis, the recursive application of binary search will make $k + 1$ comparisons, and hence, the total number of comparisons is $k + 2$, which is what we wanted to show.

Note that when $|\mathbf{a}| = 2^n$, $n + 1 = \log_2(|\mathbf{a}|) + 1$, and hence, the number of comparisons required is much smaller than the length of $\mathbf{a}$.

Another variation of strong induction is one that allows for several basis steps.

Theorem 4.2.26 *Let n_0 be an integer and let P be a property of the integers greater than or equal to n_0. Let ℓ be a natural number and suppose that*

1. *$P(j)$ is true for every j with $n_0 \leq j \leq n_0 + \ell$, and*

2. *for all $k \geq n_0 + \ell$, if $P(j)$ is true for every j with $n_0 \leq j \leq k$, then $P(k + 1)$ is true.*

Then, $P(n)$ is true for every integer $n \geq n_0$.

Proof. This theorem follows from Theorem 4.2.3 by a proof similar to that of Theorem 4.2.4. ∎

Example 4.2.27 The version of quicksort we gave in Example 4.2.23 is correct, but if you trace through what it does when the input is a sequence of length 1, you will see that it does unnecessary work. A slight improvement would be to rephrase the method this way: If the input has length 0 or 1, then it is already sorted, so let the output equal the input. Otherwise, proceed as in the previous version.

Proving that this version is correct is no harder than proving that the original version is correct, but it is now most natural to use Theorem 4.2.26 with $n_0 = 0$ and $\ell = 1$. The two basis values are then the two cases handled specially, namely, 0 and 1.

Example 4.2.28 Consider the sequence of integers $\langle F_n \rangle_{n \geq 1}$ defined by

$$\begin{aligned} F_1 &= 1, \\ F_2 &= 1, \\ F_n &= F_{n-1} + F_{n-2} \text{ for } n \geq 3. \end{aligned}$$

(It should be obvious that there is one and only one sequence of integers satisfying the above definition, but if you are sceptical, see Example 4.5.2.) The first few numbers in the sequence (referred to as the Fibonacci sequence) are $1, 1, 2, 3, 5, 8, 13, 21$. The sequence has many applications in mathematics and computer science.

We will use Theorem 4.2.26 to prove the important formula

$$F_n = \frac{1}{\sqrt{5}} \left(\left(\frac{1 + \sqrt{5}}{2} \right)^n - \left(\frac{1 - \sqrt{5}}{2} \right)^n \right).$$

Let $\phi = \frac{1+\sqrt{5}}{2}$ and $\hat{\phi} = \frac{1-\sqrt{5}}{2}$. The desired equality can now be expressed as $F_n = \frac{1}{\sqrt{5}}(\phi^n - \hat{\phi}^n)$. Note that ϕ and $\hat{\phi}$ are the two solutions of the equation $x^2 = x + 1$. We will use $n_0 = 1$ and $\ell = 1$ in Theorem 4.2.26. (Because of the way the Fibonacci sequence is defined, it is almost inevitable that an induction proof about the sequence will have two basis steps.)

The two basis steps are easy to verify; if you plug in $n = 1$ or $n = 2$ into the right side of the equation, you get 1, which is the right answer. Suppose that $k \geq 2$ and that the formula is correct for all j with $1 \leq j \leq k$. Then, $k + 1 \geq 3$; so, using the induction hypothesis, we get

$$\begin{aligned} F_{k+1} &= F_k + F_{k-1} \\ &= \tfrac{1}{\sqrt{5}}(\phi^k - \hat{\phi}^k) + \tfrac{1}{\sqrt{5}}(\phi^{k-1} - \hat{\phi}^{k-1}) \\ &= \tfrac{1}{\sqrt{5}}(\phi^{k-1}(\phi + 1) - \hat{\phi}^{k-1}(\hat{\phi} + 1)) \\ &= \tfrac{1}{\sqrt{5}}(\phi^{k-1}(\phi^2) - \hat{\phi}^{k-1}(\hat{\phi}^2)) \\ &= \tfrac{1}{\sqrt{5}}(\phi^{k+1} - \hat{\phi}^{k+1}), \end{aligned}$$

as desired.

Our next version of strong induction is one that can be applied to finite sets of integers.

Theorem 4.2.29 *Let n_0 and m_0 be two integers with $n_0 \leq m_0$. Let P be a property of the integers between n_0 and m_0. Suppose that*

1. *$P(n_0)$ is true, and*

2. *for all k with $n_0 \leq k < m_0$, if $P(j)$ is true for every j with $n_0 \leq j \leq k$, then $P(k+1)$ is true.*

Then, $P(n)$ is true for all n with $n_0 \leq n \leq m_0$.

Proof. This theorem follows from Theorem 4.2.3 by a proof similar to that of Theorem 4.2.4. ∎

Our next version of strong induction is sometimes called *course-of-values induction* or *complete induction*. This form of induction seems at first to be just like strong induction but with the basis step omitted. We will see that it is exactly equivalent to strong induction.

Theorem 4.2.30 *Let n_0 be an integer and let P be a property of the integers greater than or equal to n_0. Suppose that*

for all k with $n_0 \leq k$, if $P(j)$ is true for every j with $n_0 \leq j < k$, then $P(k)$ is true.

Then, $P(n)$ is true for every $n \geq n_0$.

Proof. This theorem follows from Theorem 4.2.3 by a proof similar to that of Theorem 4.2.4. ∎

Note that this version of induction has an inductive step, but appears to missing a basis step. Let $S(k)$ stand for the statement of the inductive step in course-of-values induction. Then, $S(n_0 + 1)$ is the statement

if $P(n_0)$ is true, then $P(n_0 + 1)$ is true,

while $S(n_0 + 2)$ is the statement

if $P(n_0)$ and $P(n_0 + 1)$ are true, then $P(n_0 + 2)$ is true,

and so on. Thus, proving the inductive step in course-of-values induction certainly establishes the inductive step for strong induction as well. To see where the basis step comes from, we must consider $S(n_0)$, which is the statement

if $P(j)$ is true for all j with $n_0 \leq j < n_0$, then $P(n_0)$ is true.

Now, there are, of course, no integers j with $n_0 \leq j < n_0$, so the statement "$P(j)$ is true for every j with $n_0 \leq j < n_0$" is true, and hence, if $S(n_0)$ is true, $P(n_0)$ must be true. Thus, showing the one step in course-of-values induction is equivalent to showing both steps in strong induction, and so the two principles are equivalent.

Although course-of-values induction makes it possible to state (in a perhaps confusing way) an induction principle with only a single step, it may appear that it cannot be of any practical value since the basis step still has to be carried out. In fact, although course-of-values induction is never essential, it can be used to slightly simplify the presentation of certain induction proofs. The typical situation where it can be useful is the following: in proving the inductive step of a strong induction argument, you have to consider at least two cases, one of which does not involve the inductive hypothesis and includes the basis value as a special case. Then, a separate argument for the basis step is not needed since it is taken care of while showing the inductive step. This is where an appeal to course-of-values induction can be useful.

As an example, consider our proof in Example 4.2.24 that every integer has a prime factorization. In the proof of the inductive step, we had to consider two possibilities. If $k + 1$ is prime, then the result for $k + 1$ is immediate and does not involve the inductive hypothesis. If $k + 1$ is not prime, then an argument that uses the inductive hypothesis is given. The basis step, with $k = 2$, is covered by the same argument that worked for the first case in the proof of the inductive step. Thus, we may use course-of-values induction to give a slightly simpler proof, as we do in the next example.

Example 4.2.31 Let $P(n)$ be the property that n is the product of one or more prime numbers. We use Theorem 4.2.30 with $n_0 = 2$ to show that $P(n)$ is true for all $n \geq 2$. Suppose that $k \geq 2$ and that $P(j)$ is true for all j with $2 \leq j < k$. If k is prime, then $P(k)$ is obviously true. If not, then we can write $k = rs$, where $2 \leq r, s < k$, and we can use the inductive hypothesis to finish the proof, as we did before.

In our next example, we again use course-of-values induction.

Example 4.2.32 Let Σ be an alphabet. We will show that if u, v are in Σ^* and $uv = vu$, then there are a string $z \in \Sigma^*$ and natural numbers n, m with $u = z^n$ and $v = z^m$. The proof will be by induction on $|u| + |v|$. That is, we let $P(p)$ be the property "if $|u| + |v| = p$ and $uv = vu$, then there are z, n, m with $u = z^n$ and $v = z^m$," and then we prove that $P(p)$ is true for all $p \geq 0$ using course-of-values induction. Suppose that $k \geq 0$ and that $P(j)$ is true for all j with $0 \leq j < k$. We must show that $P(k)$ is true, so suppose that $|u| + |v| = k$ and that $uv = vu$. We will consider the case where $|u| \leq |v|$. A similar argument will work if $|v| < |u|$. First, if $u = \lambda$, then we may set $z = v, n = 0, m = 1$ to get the desired result. If $u \neq \lambda$, then we use the fact that $uv = vu$ and $|u| \leq |v|$ to conclude that v must begin with u, say, $v = uv'$ for some string v', and since $u \neq \lambda$, $|v'| < |v|$. Substituting $v = uv'$ into $uv = vu$, we get $uuv' = uv'u$. Canceling the u's at the beginning of this equation, we get $uv' = v'u$. Now, since $|v'| < |v|$, we have $|u| + |v'| < |u| + |v|$, so we may apply the induction hypothesis to get z', n', m' with $u = z'^n$ and $v' = z'^{m'}$. Since $v = uv' = z'^n z'^{m'} = z'^{n' + m'}$, we may set $z = z', n = n', m = n' + m'$ to get the desired relationship.

Note that in the previous example, the basis step $P(0)$ is taken care of in the proof of the inductive step since if $|u| + |v| = 0$, then $u = \lambda$.

We conclude this section with an cautionary example that shows that you must be careful when you make a proof by induction.

Example 4.2.33 We will give a (false) proof by strong induction that every natural number is even. The basis step is to show that 0 is even. This is certainly true. Now, suppose that $k \geq 0$ and that every j with $0 \leq j \leq k$ is even. We must show that $k + 1$ is even. We have $k + 1 = (k - 1) + 2$. According to the induction hypothesis, $k - 1$ is even, and 2 is certainly even, so $k + 1$, the sum of these two numbers, is also even.

What is wrong with this proof? (You might want to think about this before reading further.) The problem is with the application of the induction hypothesis to the number $k - 1$. The induction hypothesis only applies to numbers between 0 and k. For any positive value of k, $k - 1$ will fall in this range. However, when $k = 0$, $k - 1 = -1$, and hence, it is not correct to apply the induction hypothesis to $k - 1$ when $k = 0$. In other words, when $k = 0$, the proof of the induction step begins by writing $1 = (-1) + 2$, which is correct, but the induction hypothesis cannot be applied to -1.

Note that the proof of the induction step is correct when $k > 0$. The moral of this example is that when you apply the induction hypothesis to a number, you must be sure that the number really is in the range of numbers to which the induction hypothesis applies.

In this section, we have discussed many principles of induction and shown how each one can be derived from Theorem 4.2.3, which in turn follows from the well-ordering principle. What is less obvious is that each of these induction principles can be used to derive the well-ordering principle.

Theorem 4.2.34 *The well-ordering principle and the principle of ordinary induction are equivalent.*

Proof. We have already seen that the principle of ordinary induction can be derived from the well-ordering principle. Conversely, assume that the principle of ordinary induction holds. We will show the contrapositive of the well-ordering principle. Let A be a set of natural numbers that has no least element. We will show that A is empty by using ordinary induction to show that, for every $n \in \mathbf{N}$, no natural number $m \leq n$ belongs to A. Let $P(n)$ be this property. For $n = 0$, $P(n)$ is true; otherwise, 0 would be the least element of A. Suppose that $n \in \mathbf{N}$ and $P(n)$ holds, i.e., no number less than or equal to n is in A. If $n + 1 \in A$, then $n + 1$ would be the least element of A. Therefore, $P(n + 1)$ holds. By ordinary induction, we may conclude that $P(n)$ is true for every n and so $A = \emptyset$. ∎

The importance of the previous theorem is not so much in the statement of the theorem (after all, both of the principles are true, so it is not surprising that we can prove that they are equivalent), but rather in the proof of the theorem, which shows how little we need to assume about the arithmetic of the natural numbers in order to prove the equivalence. (See Exercise 18 for more information.)

Slight variations of this proof can be used to show that all of the other induction principles we have looked at also imply the well-ordering principle, and, hence, are equivalent to it.

4.3 Inductively Defined Sets

Most of the time in mathematics, when you want to define a new set A, this set is defined to be all elements of some already known set B that satisfy some condition, where the condition only refers to already known ideas. For any given element of B then, to determine if the element belongs to A or not, you just check to see if it satisfies the condition, something that in principle at least is straightforward. Some definitions of sets, however, are most naturally given in a different form, one that involves specifying what the simplest elements of the set are and then giving rules for building up more complicated elements of the set out of simpler ones. We see such a definition in the next example.

Example 4.3.1 We define a set EXP of syntactically correct integer valued expressions. Our definition is similar to but simpler than definitions in many programming languages. We assume that a set of integer constants CONST and a set of valid variable names VAR has already been defined. The set of integer expressions is defined by the following rules.

1. Each integer constant is in EXP.

2. Each variable name is in EXP.

3. If E is in EXP, then so is (E).

4. If E is in EXP, then so is $+E$.

5. If E is in EXP, then so is $-E$.

6. If E_1 and E_2 are both in EXP, then so is $E_1 + E_2$.

7. If E_1 and E_2 are both in EXP, then so is $E_1 - E_2$.

8. If E_1 and E_2 are both in EXP, then so is $E_1 * E_2$.

9. If E_1 and E_2 are both in EXP, then so is E_1/E_2.

If you try to think of this definition as giving you the set EXP of integer expressions in one straightforward step, then the definition may seem circular since Rules (3)-(9) require you to know what the set of integer expressions is before these rules can be applied. If you think of the set EXP as being built up step by step through the use of these rules, then the circularity is gone. Rules (1) and (2) give you the simplest possible integer expressions. Rules (3)-(9) can then be used repeatedly to build up more and more complicated integer expressions. Something is an integer expression if it can be shown to be an integer expression by a finite number of applications of these rules.

Consider, for instance, the expression $-(a + b)/25$. If a and b are variable names and 25 is an integer constant, then $-(a + b)/25$ is an integer expression according to the definition of Example 4.3.1. We can see this by means of a line-by-line proof as follows.

	Claim	Reason
1.	a is in EXP	Rule (2)
2.	b is in EXP	Rule (2)
3.	$a + b$ is in EXP	Rule (6) and Lines (1),(2)
4.	$(a + b)$ is in EXP	Rule (3) and Line (3)
5.	$-(a + b)$ is in EXP	Rule (5) and Line (4)
6.	25 is in EXP	Rule (1)
7.	$-(a + b)/25$ is in EXP	Rule (9) and Lines (5),(6)

Figure 4.1 gives another way to see that $-(a + b)/25$ is an integer expression according to the definition in Example 4.3.1. This figure shows a labeled tree. Each node of the tree is labeled with an integer expression. The leaves are labeled with expressions that are integer expressions according to Rules (1) or (2). Each interior node is labeled with an integer expression that follows by one of Rules (3)-(9), applied to the labels of the children of the given interior node. The root is labeled with $-(a + b)/25$. The tree gives a graphic representation of how the integer expression $-(a + b)/25$ is built up using the rules of the definition.

With this example as background, we give a description of and some terminology for inductive definitions. We label this as a definition, but it is not mathematically precise. We give a more rigorous treatment of inductive definitions in Section 4.6.

Definition 4.3.2 *An* inductive definition *of a set S is a definition that consists of a collection of* rules. *The rules are of two types.* **Basis rules** *are ones that state unconditionally that certain elements are in the set.* **Inductive rules** *are ones that state that an element is in the set if certain other elements are in the set. The element that is put into the set by an inductive rule is called the* conclusion *of the rule. The elements that have to be in the set in order for the conclusion to be in the set are called the*

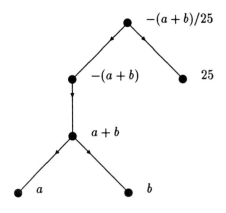

FIGURE 4.1. A tree for $-(a + b)/25$.

hypotheses *of the rule. The elements of S are those objects that can be shown to be in S by a finite number of applications of the rules.*

Whenever you have an inductively defined set S, you can use either of the two ways that we have already discussed to show that an object is an element of S. One way is with a line-by-line proof. Each line in the proof is either an element put into S by a basis rule or it is the conclusion of an inductive rule whose hypotheses appear as earlier lines in the proof. The last line of the proof is the element that the proof shows is in S.

The second way to show that an object is an element of S is with a labeled tree that shows how the object is built up using the rules. Each leaf is labeled with an element put into S by a basis rule. Each interior node is labeled with the conclusion of an inductive rule whose hypotheses are the labels of the children of the node. The label of the root of the tree is the element that is shown to be in S by the tree.

We consider some further examples of inductively defined sets. It is sometimes useful to give an inductive definition of a set that has already been defined in a noninductive way or to give several different inductive definitions for a set. This is because, as we shall see in the next few sections, each inductive definition of a set gives a way to show that every member of the set has a property. Also, this kind of definition of a set provides a way to define functions from the set to other sets. Of course, when we give multiple definitions for the same set, we must worry about whether or not the definitions are all equivalent.

Example 4.3.3 We define a subset S of the natural numbers by the following inductive definition.

    1. 0 is in S.

2. If $n \in S$, then $n + 1 \in S$.

The first rule is a basis rule, and the second one is an inductive one with one hypothesis. The fact that S is, in fact, all of **N** is simply a restatement of the principle of induction for the natural numbers.

Example 4.3.4 We once again define a subset S of **N**.

1. 0 is in S.

2. If $n \in$ **N** and every natural number less or equal to n is in S, then $n + 1 \in S$.

The first rule is a basis rule. The second rule is actually infinitely many different inductive rules. The n-th rule (for $n \in$ **N**) has $n + 1$ hypotheses (namely, the $n + 1$ natural numbers less than or equal to n) and it has conclusion $n + 1$. The fact that S is all of **N** is just a restatement of the principle of strong induction for the natural numbers.

Example 4.3.5 We define a subset S of the positive natural numbers.

1. 1 is in S.

2. 2 is in S.

3. If n and $n + 1$ are both in S, then so is $n + 2$.

We can use strong induction, with two basis steps, to show that every positive natural number is in S. The two basis rules in the definition of S show that 1 and 2 are in S. Now, suppose that $k \geq 2$ and that every natural number j with $1 \leq j \leq k$ is in S. Then, in particular (since $k \geq 2$), $k - 1$ and k are in S, so by the inductive rule, $k + 1 \in S$, completing the induction.

Example 4.3.6 We define a subset S of **N**.

1. 0 is in S.

2. If n is in S, then $n + 2$ is in S.

A little thought should convince you that S is equal to $E = \{2n \mid n \in N\}$, the set of even natural numbers. It is easy to show by induction that $E \subseteq S$; i.e., we show by induction that for every natural number n, $2n \in S$. The basis step is to show that $2 \cdot 0 = 0 \in S$. This is true by the basis rule in the definition of S. For the inductive step, suppose that $2k \in S$. Then, according to the inductive rule, $2k + 2 = 2(k + 1) \in S$, completing the induction.

The remaining step is to show that $S \subseteq E$, i.e., that every number in S is even. We do not have a technique to do this now, but it might be instructive for you to think about why this is true. In the next section, we will develop a technique that will allow us to finish the proof that $S = E$.

Example 4.3.7 If A is a nonempty set, then we have already defined
Seq(A) to be the set of all finite sequences of elements of A. This definition
is not inductive, but it is also possible to give an inductive definition of the
same set. To that end, we define inductively a subset S of Seq(A) and then
show that S is all of Seq(A).

1. λ is in S.

2. If $w \in S$ and $x \in A$, then $wx \in S$.

We must show that every element of Seq(A) is an element of S. To that end,
let $P(n)$ be the property of the natural numbers given by "if w is a sequence
of elements of A of length n, then $w \in S$." If we can show that $P(n)$ is true
for every natural number n, then we will have shown that every sequence
of elements of A is in S. We use induction on the natural numbers. Since
the only sequence of length 0 is λ, the first rule in the definition of S tells
us that $P(0)$ is true. Suppose that $P(k)$ is true for some $k \geq 0$. We show
that $P(k+1)$ is true. Let w' be an element of Seq(A) of length $k+1$. Let
w consist of the first k symbols in w' and let x be the last symbol in w'.
Then, by the induction hypothesis, $w \in S$, and hence, by the second part
of the definition of S, $wx = w' \in S$, which shows that $P(k+1)$ is true.

  In the inductive definition of Seq(A) just given, there is one basis rule. It
might seem that there is also one inductive rule, and the definition could
be interpreted this way, but it is better to interpret the definition as giving
one inductive rule for each element of A. For instance, if $A = \{0,1\}$, then
there are two inductive rules, one that says "if $w \in S$, then $w0 \in S$" and
the other that says "if $w \in S$, then $w1 \in S$."

Example 4.3.8 If A is a set, we give a slight variation on the previous
example by defining another subset S of Seq(A).

1. For each $x \in A$, $(x) \in S$.

2. If $w \in S$ and $x \in A$, then $wx \in S$.

By a proof similar to the one in the previous example, S is equal to Seq$^+(A)$,
the set of all nonnull sequences over A.

Example 4.3.9 Let $\Sigma = \{a,b\}$. We define a subset S of Σ^*.

1. λ is in S.

2. If $w \in S$, then $awb \in S$.

3. If $w \in S$, then $bwa \in S$.

4. If w and u are in S, then $wu \in S$.

The first rule is a basis rule. The next two rules are inductive with one
hypothesis. The last rule is inductive with two hypotheses. As an example
of using these rules, we show that $abbaba \in S$.

Claim	Reason
1. $\lambda \in S$	Rule (1)
2. $ab \in S$	Rule (2) and Line (1)
3. $baba \in S$	Rule (3) and Line (2)
4. $abbaba \in S$	Rule (4) and Lines (2) and (3)

A little experimentation should convince you that every string in S has the same number of a's as b's. We will be able to show that this is true in the next section. You might enjoy trying to figure out exactly what strings are in the set S before you go on to the next section.

Example 4.3.10 In this example, we define a set of programs that we call *while programs*. Although the language defined here will be much simpler than most real programming languages, study of this language will give insight into syntactic and semantic issues common to most procedural languages.

We assume that we have some alphabet Σ making up a character set for the language and that there are sets VAR of variable names, EXP of expressions, and COND of conditions, each of which is a language over Σ. In addition, we assume that **if, then, else, endif, while, do, endwhile**, $\leftarrow$, and ; are distinct symbols not in Σ. We now define WP, the set of while programs, to be a set of strings of symbols over the alphabet Σ augmented by the nine symbols just mentioned.

1. If V is a variable name and E is an expression, then

$$V \leftarrow E$$

is in WP.

2. If C is a condition and P_1 and P_2 are while programs, then

$$\textbf{if } C \textbf{ then } P_1 \textbf{ else } P_2 \textbf{ endif}$$

is a while program.

3. If C is a condition and P is a while program, then

$$\textbf{while } C \textbf{ do } P \textbf{ endwhile}$$

is a while program.

4. If P_1 and P_2 are while programs, then so is

$$P_1 ; P_2$$

In this definition, the first rule is a basis rule and the other rules are inductive. Rules (2) and (4) have two hypotheses while Rule (3) has one hypothesis. Rules (2) and (3) can be regarded as single rules or as giving a whole family of rules, one for each condition C. We call programs put into WP by Rules (1), (2), (3), and (4) *assignment statements*, *if statements*, *while statements*, and *compound statements*, respectively.

If x, y, and z are variable names, $x - y$ and $y - x$ are expressions, and $\min(x, y) > 0$ and $x \geq y$ are conditions, then, as is easily seen by several applications of the above rules, the following is a while program:

while $\min(x, y) > 0$ **do**
    if $x \geq y$
      then $x \leftarrow x - y$
      else $y \leftarrow y - x$
    endif
endwhile;
$z \leftarrow \max(x, y)$

In presenting this program, we have divided it into lines and added spacing to improve readability. This should not obscure the fact that a program is determined solely by the sequence of symbols of which it is made and not by how it is displayed.

Example 4.3.11 Let Σ be an alphabet. We will define the set REG_Σ of *regular expressions* over Σ. Let $($, $)$, $\cup$, $\circ$, $*$, and ϕ be distinct symbols not in Σ and let Σ_1 be the alphabet obtained from Σ by adding these symbols. REG_Σ is the subset of Σ_1^* defined by the following.

1. ϕ is in REG_Σ.

2. For each $x \in \Sigma$, x is in REG_Σ.

3. If R and S are in REG_Σ, then so is $(R \cup S)$.

4. If R and S are in REG_Σ, then so is $(R \circ S)$.

5. If R is in REG_Σ, then so is R^*.

We write the $*$ in R^* as a superscript just for notational convenience. It must be understood that R^* simply means the sequence consisting of those symbols in R followed by $*$.

If x and y are in Σ, then $(((x \circ y) \cup x)^* \cup x^*)$ is a regular expression over Σ. Regular expressions are meant to denote languages. It should be fairly obvious, in an intuitive way, how one can obtain a language from such an expression, but we will not be able to make this idea precise until we have studied inductively defined sets further.

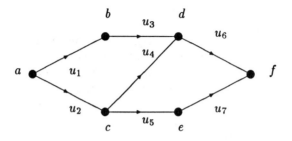

FIGURE 4.2. Graph $G = (\{a, g, c, d, e, f\}, U)$.

Example 4.3.12 Let $G = (A, U)$ be a directed graph. We have introduced in Definition 2.6.7 the notion of path in a directed graph. Now, we are interested in an inductive definition of $\mathrm{Path}(G)$, the set of paths in G as a subset of $\mathrm{Seq}(U)$.

  1. Every sequence u_1 of length 1 from $\mathrm{Seq}(U)$ belongs to $\mathrm{Path}(G)$.

  2. If $(u_1, \ldots, u_n)$ is a path in the graph G, $u \in U$, and $\omega(u_n) = \alpha(u)$, then $(u_1, \ldots, u_n, u)$ belongs to $\mathrm{Path}(G)$.

Consider, for instance the graph $G = (\{a, b, c, d, e, f\}, U)$ given in Figure 4.2. Let us prove that (u_2, u_4, u_6) is a path in the graph G.

	Claim	Reason
1.	$u_2 \in \mathrm{Path}(G)$	Rule (1)
2.	$(u_2, u_4) \in \mathrm{Path}(G)$	Rule (2), Line (1), and $\omega(u_2) = \alpha(u_4) = c$
3.	$(u_2, u_4, u_6) \in \mathrm{Path}(G)$	Rule (2), Line (2), and $\omega(u_4) = \alpha(u_6) = d$

4.4   Proof by Structural Induction

In this section, we discuss an inductive method to prove that every member of an inductively defined set has a property. The method is called structural induction because the induction is on the structure of the set, as given by the inductive definition of the set.

Theorem 4.4.1 (Principle of Structural Induction) *Let S be an inductively defined set and let P be a property of the elements of S. Suppose that*

1. *for every basis rule in the definition of* S, *if* x *is put into* S *by the rule, then* $P(x)$ *is true;*

2. *for every inductive rule in the definition of* S, *if* P *is true about all of the hypotheses of the rule, then* P *is true about the conclusion of the rule.*

Then, $P(x)$ *is true for all* x *in* S.

Proof: Our definition of an inductively defined set is somewhat imprecise, but we can still show why the principle of structural induction is true. This proof will be given again in a more precise setting in Section 4.6.

Let S be an inductively defined set and let P be a property of S that meets the two conditions of the theorem. Let x be an element of S. We must show that $P(x)$ is true. Since x is in S, there must be a line-by-line proof of this fact, say, $x_1, \ldots, x_m$ is a sequence of elements of S such that $x_m = x$ and each x_i is either put into S by a basis rule or else is the conclusion of an inductive rule all of whose hypotheses appear as previous lines in the proof. In order to show that $P(x)$ is true, we will show by strong induction that $P(x_n)$ is true for every n with $1 \leq n \leq m$. The basis step is $n = 1$. But, x_1 must be put into S by a basis rule, so by the first condition, $P(x_1)$ is true. Now, suppose that $1 \leq k < m$ and $P(x_j)$ is true for all j with $1 \leq j \leq k$. If x_{k+1} is put into S by a basis rule, then by the first condition, $P(x_{k+1})$ is true. The other possibility is that x_{k+1} is the conclusion of an inductive rule, all of whose hypotheses are among $x_1, \ldots, x_k$. But, then by the induction hypothesis, P is true about all of the hypotheses of the rule, so by the second condition, P is true about the conclusion of the rule, namely, x_{k+1}. This completes the induction. Since P is true about every line in the proof, it must be true about the last line of the proof, namely, $x_m = x$. ∎

Note that when you use proof by structural induction over an inductively defined set, each basis rule in the definition of the set gives rise to a basis step in the proof, and each inductive rule in the definition of the set gives rise to an inductive step in the proof. When you carry out an inductive step in the proof and assume that all of the hypotheses of an inductive rule have the given property, this assumption is called the induction hypothesis.

To illustrate the principle of structural induction, we apply it to the particular examples of inductively defined sets that we have seen previously.

Corollary 4.4.2 *Let* P *be a property of the set EXP of integer valued expressions defined in Example 4.3.1. Suppose that*

1. *every integer constant has property* P;

2. *every variable name has property* P;

3. *if the expression* E *has property* P, *then so do* $(E), +E,$ *and* $-E$;

4. if the expressions E_1 and E_2 both have property P, then so do $E_1 + E_2, E_1 - E_2, E_1 * E_2, E_1/E_2$.

Then, every integer expression has property P.

In the previous corollary, items (1) and (2) are basis steps; item (3) gives three different inductive steps, which, to save space, we give as a single item. Item (4) gives four inductive steps. The inductive steps in (3) come from inductive rules with one hypothesis. The steps in (4) come from inductive rules with two hypotheses.

If the principle of structural induction is applied to the inductive definition of example 4.3.3, then the result is just a restatement of the principle of induction for the natural numbers. If the principle of induction is applied to the inductive definition in Example 4.3.4, then the result is the principle of strong induction for the natural numbers. Thus, structural induction may be regarded as a generalization of the principles of induction we have studied for the natural numbers.

If we apply the principle of structural induction to the definition in Example 4.3.5 we get the following.

Corollary 4.4.3 *Let P be a property of the positive integers and suppose that*

1. $P(1)$ is true;

2. $P(2)$ is true;

3. if n is a positive integer and $P(n)$ and $P(n+1)$ are both true, then $P(n+2)$ is true.

Then, $P(n)$ is true for all positive integers n.

Applying the principle of structural induction to the remaining inductive definitions of the previous section, we get the following corollaries.

Corollary 4.4.4 *Let S be the set of natural numbers defined in Example 4.3.6 and let P be a property of S. Suppose that*

1. $P(0)$ is true;

2. if $P(n)$ is true, then $P(n+2)$ is true.

Then, $P(n)$ is true for all $n \in S$.

Corollary 4.4.5 *Let A be any set and let P be a property of $Seq(A)$. Suppose that*

1. $P(\lambda)$ is true;

2. for each $x \in A$, if $P(w)$ is true, then $P(wx)$ is true.

Then, $P(w)$ is true for all $w \in Seq(A)$.

Corollary 4.4.6 *Let A be any set and let P be a property of $Seq^+(A)$. Suppose that*

1. *for each $x \in A$, $P((x))$ is true;*

2. *for each $x \in A$, if $P(w)$ is true, then $P(wx)$ is true.*

Then, $P(w)$ is true for all $w \in Seq^+(A)$.

Corollary 4.4.7 *Let S be the set of strings of a's and b's defined in Example 4.3.9 and let P be a property of S. Suppose that*

1. *$P(\lambda)$ is true;*

2. *if $P(w)$ is true, then $P(awb)$ is true;*

3. *if $P(w)$ is true, then $P(bwa)$ is true;*

4. *if $P(w)$ and $P(u)$ are both true, then $P(wu)$ is true.*

Then, $P(w)$ is true for all $w \in S$.

Corollary 4.4.8 *Let WP be the set of while programs defined in Example 4.3.10 and let P' be a property of while programs. Suppose that*

1. *for all variables V and expressions E, $P'(V \leftarrow E)$ is true;*

2. *for all conditions C and while programs P_1, P_2, if $P'(P_1)$ and $P'(P_2)$ are true, then $P'(\text{if } C \text{ then } P_1 \text{ else } P_2 \text{ endif})$ is true;*

3. *for all conditions C and while programs P, if $P'(P)$ is true, then $P'(\text{while } C \text{ do } P \text{ endif})$ is true;*

4. *for all while programs P_1, P_2, if $P'(P_1)$ and $P'(P_2)$ are true, then $P'(P_1; P_2)$ is true.*

Then, $P'(P)$ is true for all while programs P.

Corollary 4.4.9 *Let Σ be an alphabet, let REG_Σ be the set of regular expressions over Σ as defined in Example 4.3.11 and let P be a property of regular expressions. Suppose that*

1. *$P(\phi)$ is true;*

2. *for all $x \in \Sigma$, $P(x)$ is true;*

3. *for all regular expressions R and S, if $P(R)$ and $P(S)$ are true, then $P((R \cup S))$ is true;*

4. *for all regular expressions R and S, if $P(R)$ and $P(S)$ are true, then $P((R \circ S))$ is true;*

5. *for all regular expressions R, if $P(R)$ is true, then $P(R^*)$ is true.*

Then, $P(R)$ is true for all regular expressions R over Σ.

Now, that we have all these methods of proof specified, we will give some examples of how they can be used.

Example 4.4.10 Let S be the set of natural numbers defined in Example 4.3.6. As promised in that example, we can now show that every member of S is even by using Corollary 4.4.4. Let P be the property of being even. Then, $P(0)$ is true since $0 = 2 \cdot 0$, so the basis step is complete. The inductive step is to show that if $P(n)$ is true, then $P(n+2)$ is true. If n is even, say, $n = 2k$, then $n + 2 = 2k + 2 = 2(k+1)$, so $n + 2$ is even. This completes the induction, and according to Corollary 4.4.4, every number in S is even. Combined with what we already showed in Example 4.3.6, we now see that S is equal to the set E of even natural numbers.

Example 4.4.11 Let S be the set of sequences over $\{a, b\}$ defined in Example 4.3.9. We will use Corollary 4.4.7 to show that every sequence in S has the same number of a's as b's. We will write $N_a(w)$ to be the number of a's in the string w, and similarly for $N_b(w)$. Let $P(w)$ be the property "$N_a(w) = N_b(w)$." We must show that $P(w)$ is true for all strings $w \in S$. The basis step is to show that $P(\lambda)$ is true. But, $N_a(\lambda) = 0 = N_b(\lambda)$, so $P(\lambda)$ is true. The first inductive step is to assume that $P(w)$ is true and show that $P(awb)$ is true. But, if $N_a(w) = N_b(w)$, then

$$N_a(awb) = N_a(w) + 1 = N_b(w) + 1 = N_b(awb),$$

so $P(awb)$ is true. The second inductive step is to assume that $P(w)$ is true and show that $P(bwa)$ is true. This is quite similar to the first inductive step. The final inductive step is to assume that $P(w)$ and $P(u)$ are both true and show that $P(wu)$ is true. But, if $N_a(w) = N_b(w)$ and $N_a(u) = N_b(u)$, then we get

$$N_a(wu) = N_a(w) + N_a(u) = N_b(w) + N_b(u) = N_b(wu),$$

which establishes the third inductive step.

Thus, every string in S has the same number of a's as b's. What is much less obvious is that every string with the same number of a's as b's belongs to S. We show this fact using strong induction on the natural numbers. Let $P(n)$ be the property that if w is a string of a's and b's that contains exactly n a's and n b's, then $w \in S$. We will show that $P(n)$ is true for all natural numbers n, and this will establish our claim. One key part of our proof will be left for the exercises.

$P(0)$ is true because if w is a string in $\{a, b\}^*$ with $N_a(w) = 0 = N_b(w)$, then $w = \lambda$, and hence, $w \in S$ by the basis rule in the definition of S. Now, suppose that $k \geq 0$ and that $P(j)$ is true for all j with $0 \leq j \leq k$. We must show that $P(k+1)$ is true, so suppose that $N_a(w) = k+1 = N_b(w)$. We consider several cases depending on what the first and last symbols in w are. (Note that $|w| = 2(k+1) > 1$.) If w begins with an a and ends with a b, then $w = aub$ for some string $u \in \{a,b\}^*$. Since $N_a(w) = k+1 = N_b(w)$, we must have $N_a(u) = k = N_b(u)$. Hence, by the induction hypothesis, $u \in S$, and thus by the first inductive rule in the definition of S, $aub = w \in S$. If w begins with a b and ends with an a, a similar proof, using the second inductive rule in the definition of S, can be carried out.

The remaining possibility is that w starts and ends with the same symbol. In this case, we claim that we can write $w = uv$, where $N_a(u) = N_b(u)$, $N_a(v) = N_b(v)$, and u and v are both shorter than w. We leave the proof of this claim to the exercises. (See Exercise 33.) Since u and v are both shorter than w, the induction hypothesis applies to them, and hence, they are both in S. Thus, by the third inductive rule in the definition of S, $uv = w \in S$. This completes the proof.

Example 4.4.12 In this example, we consider the set WP of while programs defined in Example 4.3.10. By a proof very similar to that in Example 4.8.5, it can be shown that every while program has the same number of **if**'s as **endif**'s and the same number of **while**'s as **endwhile**'s. We leave this to the reader and give a somewhat more complicated example of a proof by structural induction for WP. The fact that we prove may seem rather artificial and uninteresting, but it will be useful to us later on.

Define the *level* of a symbol in a while program to be the number of **if**'s and **while**'s in the program to the left of the symbol minus the number of **endif**'s and **endwhile**'s in the program to the left of the symbol. We will show that every occurrence of the semicolon symbol (;) in a while program has a nonnegative level.

Let $P'(P)$ be the property "every semicolon in P has a nonnegative level." We use Corollary 4.4.8 to show that $P'(P)$ is true for every while program P. The basis step is to show $P'(P)$ for every assignment statement P. This is immediate since assignment statements do not contain any semicolons. The first inductive step is to assume that $P'(P_1)$ and $P'(P_2)$ are true and that C is a condition and show that $P'(P)$ is true where $P = $ **if** C **then** P_1 **else** P_2 **endif**. Any occurrence of the semicolon symbol in P must be in P_1 or in P_2. If the occurrence is in P_1, then the level of the occurrence as a symbol in P is one more than the level of the occurrence as a symbol in P_1. By the inductive hypothesis then, the level of such an occurrence is positive. Similarly, since P_1 has the same number of **if**'s and **while**'s as **endif**'s and **endwhile**'s, the level of a semicolon that occurs in P_2, when considered as a symbol in P, is one more than its level as a symbol in P_2, and by inductive hypothesis, this latter level is nonnegative. Hence, $P'(P)$ is true.

Next, we suppose that $P'(P_1)$ is true and show that $P'(P)$ is true for $P = $ **while** C**do** P**endwhile**. Any occurrence of the semicolon symbol in

P must be part of P_1. The level of such an occurrence as a symbol in P is one more than its level as a symbol in P_1, and hence, is positive.

Finally, suppose that $P('P_1)$ and $P'(P_2)$ are both true. We show that $P'(P)$ is true for $P = P_1; P_2$. The semicolon between P_1 and P_2 has level 0. Any other semicolon in P is in either P_1 or P_2 and has the same level in P as it does in whichever of P_1 or P_2 it occurs, so by the induction hypothesis, we are done.

Example 4.4.13 Let Σ be an alphabet. By a proof similar to that of Example 4.8.5 but using Corollary 4.4.9, it is easy to show that every regular expression over Σ has the same number of left parentheses as right parentheses.

We define the *level* of a symbol in a regular expression to be the number of left parentheses in the expression to the left of the symbol minus the number of right parentheses in the expression to the left of the symbol. Using Corollary 4.4.9, one can give a proof similar to that of the previous example to show that in a regular expression every occurrence of the symbols $\cup$ and $\circ$ has a positive level.

Example 4.4.14 Let A be any set and let ρ be a binary relation on A. We define a new binary relation $\hat{\rho}$ on A by the following inductive definition.

1. Every pair in ρ is also in $\hat{\rho}$.

2. If (x, y) and (y, z) are in $\hat{\rho}$, then so is (x, z).

The first rule is a basis rule, and the second one is inductive. We claim that $\hat{\rho}$ is ρ^+, the transitive closure of ρ. To see this, first note that, by the basis rule, $\rho \subseteq \hat{\rho}$, and by the inductive rule, $\hat{\rho}$ is transitive. Thus, $\hat{\rho}$ is a transitive relation that contains ρ. Suppose that σ is a transitive relation that contains ρ. We must show that $\hat{\rho} \subseteq \sigma$. Let $P(x, y)$ be the property "$(x, y) \in \sigma$." We use structural induction on the definition of $\hat{\rho}$ to show that $P(x, y)$ is true for all (x, y) in $\hat{\rho}$. The basis step is to show that $P(x, y)$ is true for all $(x, y) \in \rho$, i.e., that if $(x, y) \in \rho$, then $(x, y) \in \sigma$. This is just a restatement of the fact that $\rho \subseteq \sigma$. The inductive step is to show that if $P(x, y)$ and $P(y, z)$ are both true, then $P(x, z)$ is true, i.e., that if $(x, y) \in \sigma$ and $(y, z) \in \sigma$, then $(x, z) \in \sigma$. This is just a restatement of the fact that σ is transitive. Thus, by structural induction, $P(x, y)$ is true for all $(x, y) \in \hat{\rho}$, i.e., $\hat{\rho} \subseteq \sigma$, as desired, and $\hat{\rho}$ is in fact equal to σ^+.

Since $\hat{\rho} = \rho^+$, we get the following principle of induction.

Let A be any set and ρ any binary relation on A. Let P be a property of $A \times A$. Suppose that

1. $P(x, y)$ is true for all $(x, y) \in \rho$;

2. if $P(x, y)$ and $P(y, z)$ are both true, then $P(x, z)$ is true.

Then, $P(x, y)$ is true for all $(x, y) \in \rho^*$.

Example 4.4.15 This example uses a slight variation on the last example to introduce another way of obtaining the transitive closure of a relation. Let ρ be a binary relation on a set A and define a new relation $\hat{\rho}$ by

1. every pair in ρ is also in $\hat{\rho}$;

2. if $(y, w) \in \hat{\rho}$ and $(w, z) \in \rho$, then $(y, z) \in \hat{\rho}$.

You should note the small difference between this definition of $\hat{\rho}$ and the one given in the previous example. We now show that the $\hat{\rho}$ we have just defined is also equal to the transitive closure of ρ.

First of all, $\hat{\rho}$ contains ρ because of the basis rule in the definition of $\hat{\rho}$. Next, we must show that $\hat{\rho}$ is transitive. The proof is short but tricky and understanding it requires a real understanding of structural induction. Let $P(y, z)$ be the property "for all x, if $(x, y) \in \hat{\rho}$, then $(x, z) \in \hat{\rho}$." We will use structural induction to show that $P(y, z)$ is true for all $(y, z) \in \hat{\rho}$. This will establish that $\hat{\rho}$ is transitive. The basis step is to show that $P(y, z)$ is true if $(y, z) \in \rho$, i.e., we must show that if $(y, z) \in \rho$ and $(x, y) \in \hat{\rho}$, then $(x, z) \in \hat{\rho}$. This is true by the inductive rule in the definition of $\hat{\rho}$. The inductive step is to assume that $P(y, w)$ is true and that $(w, z) \in \rho$ and show that $P(y, z)$ is true, i.e., that if $(x, y) \in \hat{\rho}$, then $(x, z) \in \hat{\rho}$. Now, if $(x, y) \in \hat{\rho}$, then by the inductive hypothesis $P(y, w)$, we must have $(x, w) \in \hat{\rho}$. But, since $(w, z) \in \rho$, by the inductive rule in the definition of $\hat{\rho}$, we have $(x, z) \in \hat{\rho}$ as desired. Thus, by structural induction, $P(y, z)$ is true for all (y, z) in $\hat{\rho}$, and hence, $\hat{\rho}$ is transitive.

Finally, we must show that if σ is any relation on A that contains ρ and is transitive, then $\hat{\rho} \subseteq \sigma$. Let $P(x, y)$ be the property "$(x, y) \in \sigma$." We show by structural induction that $P(x, y)$ is true for all $(x, y) \in \hat{\rho}$. The basis step is to show that $P(x, y)$ is true for all $(x, y) \in \rho$, i.e., that every pair in ρ is also in σ. This is given to us. The inductive step is to assume that $P(y, w)$ is true, i.e., that $(y, w) \in \sigma$, and that $(w, z) \in \rho$, and show that $P(y, z)$ is true. But, if $(w, z) \in \rho$, then $(w, z) \in \sigma$, so by transitivity of σ, $(y, z) \in \sigma$, which is what we wanted.

Since $\hat{\rho} = \rho^{+}$, we get another principle of induction for the transitive closure.

Let ρ be a relation on a set A and let P be a property of $A \times A$. Suppose that

1. $P(x, y)$ is true for all $(x, y) \in \rho$;

2. if $P(x, y)$ is true and $(y, z) \in \rho$, then $P(x, z)$ is true;

Then, $P(x, y)$ is true for all $(x, y) \in \rho^{+}$.

Example 4.4.16 Let A be a set and let ρ be a relation on A. For each element u of A, we define ρ_u^{+} to be $\{y \in A \mid (u, y) \in \rho^{+}\}$. It is sometimes useful, especially in the study of formal languages and other rewriting systems, to have an inductive characterization of the set ρ_u^{+}. To that end, we define inductively a set S_u and then show that it is equal to ρ_u^{+}.

1. If $(u, y) \in \rho$, then $y \in S_u$.

2. If $y \in S_u$ and $(y, z) \in \rho$, then $z \in S_u$.

We first show that $S_u \subseteq \rho_u^+$. Let $P(y)$ be the property "$y \in \rho_u^+$." We use structural induction to show that $P(y)$ is true for all $y \in S_u$. The basis step is to show that $P(y)$ is true if $(u, y) \in \rho$. But, if $(u, y) \in \rho$, then $(u, y) \in \rho^+$, so $y \in \rho_u^+$. The inductive step is to assume that $P(y)$ is true (i.e., $(u, y) \in \rho^+$) and that $(y, z) \in \rho$ and show that $P(z)$ is true. But, if $(y, z) \in \rho$, then $(y, z) \in \rho^+$, and since $(u, y) \in \rho^+$, we get by transitivity of ρ^+ that $(u, z) \in \rho^+$, which establishes $P(z)$.

To finish the proof, we must show that $\rho_u^+ \subseteq S_u$. Let $P(x, y)$ be the property "if $x = u$, then $y \in S_u$." If we can show that $P(x, y)$ is true for all (x, y) in ρ^+, then we will have shown that $\rho_u^+ \subseteq S_u$. We will use structural induction using the inductive definition of ρ^+ given in the previous example. The basis step is to show that $P(x, y)$ is true if $(x, y) \in \rho$. But, if $x = u$ and $(x, y) \in \rho$, then by the basis rule in the definition of S_u, we have $y \in S_u$, and hence, $P(x, y)$ is true. The inductive step is to assume that $P(y, w)$ is true and that $(w, z) \in \rho$ and show that $P(y, z)$ is true. But, if $y = u$, then since $P(y, w)$ is true, we have $w \in S_u$, and since $(w, z) \in \rho$, we get from the inductive rule in the definition of S_u that $z \in S_u$. This establishes $P(y, z)$ and finishes the proof.

Because $S_u = \rho_u^+$, we get the following principle.

Let ρ be a relation on A and let P be a property of $A \times A$. Suppose that

1. $P(y)$ is true for all y with $(u, y) \in \rho$;

2. if $P(y)$ is true and $(y, z) \in \rho$, then $P(z)$ is true.

Then, $P(y)$ is true for all y such that $(u, y) \in \rho^+$.

Example 4.4.17 Let A consist of all people who have ever lived and let ρ be the relation defined by $(x, y) \in \rho$ if and only if x is a parent of y. Then, (x, y) is in the transitive closure ρ^+ of ρ if and only if y is a descendant of x. Suppose that

1. all of Jane's children are interesting;

2. all of the children of an interesting person are also interesting.

Then, we can conclude immediately, using the inductive principle of the previous example, that all of Jane's descendants are interesting, since Jane's descendants are exactly those y such that $(\text{Jane}, y) \in \rho^+$.

4.5   Recursive Definitions of Functions

In mathematics, the usual way to define a function f is to write an equation $f(x) = \ldots$, where $\ldots$ is an expression that is supposed to give the value of the function f at the domain element x. In most definitions of this sort, the expression ... involves only already known quantities, and the meaning of the definition is clear – to evaluate $f(x)$, you calculate the value of the expression ... and assign its value to $f(x)$. Sometimes however, f appears as part of the expression .... Such a function "definition" is called *recursive*, and it is much less obvious how such a definition is supposed to work because of its seeming circularity. After all, in order to calculate the value of $f(x)$, it seems that you have to already know the values of f. In fact, not all recursive "definitions" do define functions. In this section, we will explain what it means for a recursive definition to succeed in defining a function, and we will give a general condition that allows us to show that a large class of inductive definitions succeeds.

Any purported recursive definition of a function f may be viewed as a condition that the function f is supposed obey. We will say that the definition succeeds if there is one and only one function satisfying the condition of the "definition." In this case, the unique function f satisfying the condition is the function defined by the definition. For example, suppose that we wish to define a function from the natural numbers to the natural numbers. If we try the recursive "definition"

$$f(n) = f(n),$$

then our definition fails because every function from $\mathbf{N}$ to $\mathbf{N}$ satisfies this condition. At the other extreme, consider the recursive definition

$$f(n) = f(n) + 1.$$

This definition also fails but for the opposite reason; there is no function from $\mathbf{N}$ to itself satisfying this condition. On the other hand, the recursive definition

$$f(n) = 2f(n)$$

does succeed. The unique function satisfying this condition is the constant function $f(n) = 0$.

The examples of the previous paragraph, although useful for clarifying what it means for a recursive definition to succeed, are not at all typical of the type of recursive definition usually encountered in mathematics. In our artificial examples, we tried to define $f(x)$ in terms of itself. Such a definition might be called an implicit definition of a function. We are more interested in explicit recursive definitions, i.e., ones that by their nature give a method for calculating the value of $f(x)$. We will consider definitions where we define $f(x)$ not in terms of itself but in terms of other values $f(y)$, where the y's are in some sense "simpler" than x. For the "simplest" values of x, the value of $f(x)$ is given outright (i.e., nonrecursively). For instance, consider the following recursive definition of a function from $\mathbf{N}$ to $\mathbf{N}$.

Example 4.5.1 We define a function fac : $\mathbf{N} \longrightarrow \mathbf{N}$ by

$$\mathrm{fac}(n) = \begin{cases} 1 & \text{if } n = 0 \\ n \cdot \mathrm{fac}(n-1) & \text{if } n > 0 \end{cases}$$

Note how the value of $\mathrm{fac}(n)$ is given nonrecursively when n equals 0, the "simplest" natural number and is given recursively in terms of $f(n-1)$ when $n > 0$, where $n - 1$ is a "simpler" natural number than n. The definition of fac gives an explicit method for computing $\mathrm{fac}(n)$. For example, to compute $\mathrm{fac}(3)$, we have

$$\begin{aligned} \mathrm{fac}(3) &= 3 \cdot \mathrm{fac}(2) \\ &= 3 \cdot 2\mathrm{fac}(1) \\ &= 3 \cdot 2 \cdot 1\mathrm{fac}(0) \\ &= 3 \cdot 2 \cdot 1 \cdot 1 \\ &= 6. \end{aligned}$$

You may have recognized the function fac as being the factorial function $\mathrm{fac}(n) = n!$. Note the connection between this recursive definition of the function fac and the inductive definition of the natural numbers given in Example 4.3.3. In this inductive definition, 0 is the value put into the set by the basis rule, and the value of $\mathrm{fac}(0)$ is given nonrecursively. If $n > 0$, then n is put into the set by the inductive rule with hypothesis $n - 1$, and we have defined the value of $f(n)$ recursively in terms of $f(n - 1)$. This is one instance of the type of recursive function definition we will be studying in this section.

We consider another recursively defined function in the next example.

Example 4.5.2 We define a function fib from **P** to **P** by

$$\mathrm{fib}(n) = \begin{cases} 1 & \text{if } n = 1 \text{ or } n = 2 \\ \mathrm{fib}(n-1) + \mathrm{fib}(n-2) & \text{if } n > 2. \end{cases}$$

The value $\mathrm{fib}(n)$ is the n-th Fibonacci number F_n, which was introduced in Example 4.2.28. It is easy to calculate $\mathrm{fib}(3) = \mathrm{fib}(2) + \mathrm{fib}(1) = 1 + 1 = 2$, $\mathrm{fib}(4) = \mathrm{fib}(3) + \mathrm{fib}(2) = 2 + 1 = 3$, and so on.

This definition exactly parallels the inductive definition of **P** given in Example 4.3.5. The value of $\mathrm{fib}(n)$ for the two values of n put into the set by the two basis rules is given nonrecursively, while the value of $\mathrm{fib}(n)$ if n is put into the set by the inductive rule is given by applying a known function (addition) to the values of fib on the two hypotheses ($n - 1$ and $n - 2$) which put n into the set.

So far, we have considered only functions defined on a subset of the integers. However, the style of recursive function definition we are considering can be used for other inductively defined sets as well.

So far, we have seen three recursive definitions of functions that are based on inductive definitions of sets, and all three of these definitions seem to succeed in the sense given at the beginning of the section, although we have no way of proving this now. You may be lead to think that recursive definitions of functions based on inductive definitions of sets always succeed, but this is not the case, as we will now show.

Example 4.5.3 Let EXP be the set of expressions introduced in Example 4.3.1. We will attempt to define a function EVAL that maps an expression to its value. If E is in EXP, then EVAL(E) will be either an integer or $\perp$. If EVAL(E) $=\perp$, this means that evaluation of E causes an error. We assume that we have two functions MEM and V such that MEM is a function from variable names to $\mathbf{Z} \cup \{\perp\}$ and V is a function from integer constants to $\mathbf{Z}$. If const is an integer constant, then V(const) is the value of the constant and if var is a variable name, then MEM(var) is the value stored in the variable if there is one, and otherwise, MEM(var)$=\perp$. We try to define EVAL by

$$
\begin{aligned}
\text{EVAL(const)} &= \text{V(const) for every constant const} \\
\text{EVAL(var)} &= \text{MEM(var) for every variable name var} \\
\text{EVAL}((E)) &= \text{EVAL}(E) \\
\text{EVAL}(+E) &= \text{EVAL}(E) \\
\text{EVAL}(-E) &= \begin{cases} \perp & \text{if EVAL}(E) =\perp \\ -\text{EVAL}(E) & \text{otherwise.} \end{cases}
\end{aligned}
$$

$$
\begin{aligned}
&\text{EVAL}(E_1 + E_2) \\
&= \begin{cases} \perp & \text{if EVAL}(E_1) =\perp \text{ or EVAL}(E_2) =\perp \\ \text{EVAL}(E_1)+\text{EVAL}(E_2) & \text{otherwise.} \end{cases} \\
&\text{EVAL}(E_1 - E_2) \\
&= \begin{cases} \perp & \text{if EVAL}(E_1) =\perp \text{ or EVAL}(E_2) =\perp \\ \text{EVAL}(E_1)-\text{EVAL}(E_2) & \text{otherwise.} \end{cases} \\
&\text{EVAL}(E_1 * E_2) \\
&= \begin{cases} \perp & \text{if EVAL}(E_1) =\perp \text{ or EVAL}(E_2) =\perp \\ \text{EVAL}(E_1)\text{EVAL}(E_2) & \text{otherwise.} \end{cases} \\
&\text{EVAL}(E_1/E_2) \\
&= \begin{cases} \perp & \begin{array}{l}\text{if EVAL}(E_1) =\perp \text{ or EVAL}(E_2) =\perp \\ \text{or EVAL}(E_2) = 0\end{array} \\ \lfloor\text{EVAL}(E_1)/\text{EVAL}(E_2)\rfloor & \text{otherwise.} \end{cases}
\end{aligned}
$$

Let us now try to use this definition to evaluate the expression $2 + 3 * 4$. Since this expression can be obtained by applying the $+$ rule to the two expressions 2 and $3*4$, we get EVAL($2+3*4$) = EVAL(2)+EVAL($3*4$) = V(2)+(EVAL(3)EVAL(4)) = 2+(V(3)V(4)) = 2+(3·4) = 14. On the other hand, $2+3*4$ can be obtained by applying the $*$ rule to $2+3$ and 4, so we get EVAL($2+3*4$) = EVAL($2 + 3$)EVAL(4) = 5 · 4 = 20. Since a function cannot have two different values for the same argument, we see that there is no function EVAL satisfying the above definition. The "definition" does not actually define anything.

(Of course, we know by the usual precedence rules that we really want the expression $2 + 3 * 4$ to evaluate to 14. However, there is nothing in our inductive definition of EXP that reflects these precedence rules. If we gave a more complicated inductive definition of EXP that reflected precedence rules correctly, then we could use that more complicated definition to successfully define EVAL.)

We have now seen two successful recursive definitions of functions based on inductive definitions of sets and one unsuccessful one. It would be nice to have some criterion to tell when a recursive function definition of the sort we have been looking at will work. If we look at the unsuccessful definition, we see that the problem is not so much with the recursive definition of the function, but rather with the inductive definition of the set EXP. The problem is that a given expression can be built up in two different ways. It turns out that when this cannot happen, recursive function definitions succeed.

If S is an inductively defined set and for each element of S there is only one way to build up that element using the rules in the definition of S, then the definition of S is said to meet the *unique readability condition*. When an inductive definition of a set meets the unique readability condition, recursive definitions of functions based on the inductive definition of the set succeed in the sense that given such a definition there is one and only one function that satisfies the definition. We now describe this situation in a little more detail.

First, we describe somewhat more exactly what a recursive definition of a function is. Full details will be given in Section 4.6.

Definition 4.5.4 *Let S be an inductively defined set and let T be any set. A recursive definition for a function f from S to T is one that has this form: For each basic rule in the definition of S, the value of f when applied to an element put into S by the basic rule is given nonrecursively by an already known function. For each inductive rule in the definition of S, the value of f when applied to the conclusion of the inductive rule is given by some known function applied to f's values on the hypotheses of the rule.*

Intuitively, the unique readability condition for an inductive definition of a set S states that every member of S can be built up in only one way using the rules in the inductive definition. A more exact version of this is given in the next definition. Full details are given in Section 4.6.

Definition 4.5.5 *An inductive definition of a set S is said to satisfy the unique readability condition if the following hold.*

  1. For each element of the set S, there is only one rule that puts the element into S.

2. *If an element is put into S by an inductive rule, then there is only one sequence of hypotheses that can be used with the rule to put the element into the set.*

Note that the unique readability condition is a condition on the inductive definition and not on the set itself. It is quite possible to have two different definitions of the same set, one of which meets the condition and the other of which does not meet the condition.

We now state the theorem which explains why the unique readability condition is important.

Theorem 4.5.6 *Let S be an inductively defined set and suppose that the inductive definition of S satisfies the unique readability condition. Then, any recursive function definition based on the inductive definition of S succeeds in the sense that there is one and only one function satisfying the condition given in the definition.*

This theorem is too vague to be proven now, but a more precise version will be proven in Section 4.6.

We now look at the inductive definitions of Section 4.3 and determine whether or not they meet the unique readability condition.

Example 4.5.7 The inductive definition of EXP given in Example 4.3.1 does not meet the unique readability condition. Consider, for example, the expression $2 + 3 * 4$. This expression can be obtained by applying the $+$ rule to the two expressions 2 and $3 * 4$, and it can also be obtained by applying the $*$ rule to the two expressions $2 + 3$ and 4. Thus, the same expression is put into EXP by two different rules. This violates the first part of the unique readability condition. For another example, consider the expression $2 + 3 + 4$. This expression can only be obtained by the $+$ rule. However, it can be obtained in two different ways, i.e., applied to $2 + 3$ and 4, or applied to 2 and $3 + 4$. Hence, this definition does not meet the second part of the unique readability condition either.

Example 4.5.8 The definition of the set S in Example 4.3.3 (which equals $\mathbf{N}$) satisfies the unique readability condition. This is because any number put into S by the inductive rule must be greater than 0, and hence, no element can be put into S by both the basis rule and the inductive rule. Thus, the first part of the condition is met. Furthermore, if n is put into S by the inductive rule, then the only possible hypothesis is $n - 1$, so the second part of the condition is met.

Thus, any recursive function definition based on the definition in Example 4.3.3 succeeds. In particular, there is a unique function fac satisfying the conditions given in Example 4.5.1.

Example 4.5.9 By proofs similar to that in the last example, each of the inductive definitions in Examples 4.3.4, 4.3.5, and 4.3.6 satisfies the unique readability condition. Thus, in particular, there is a unique function fib satisfying the conditions given in Example 4.5.2.

Example 4.5.10 We show that the definition of Seq(A) given in Example 4.3.7 meets the unique readability condition. We will call the one basic rule the λ-rule, and we will call the inductive rule that states that if w is in Seq(A) then wx is in Seq(A) the x-rule. (There is one x-rule for each x in A.)

If a string is put into Seq(A) by the x-rule, then the string ends in x and, hence, cannot equal λ. Thus, λ can only be put into Seq(A) by the λ-rule. Further, any nonnull string can only be put into Seq(A) by the x-rule, where x is the last symbol in the string. Thus, the first part of the unique readability condition is met by this definition. To see that the second part is met, note that if w' is put into Seq(A) by the x-rule, then the hypothesis w of the rule must consist of the string obtained from w' by removing the final x. Thus, w is uniquely determined by w', and the second part of the condition is also met.

Example 4.5.11 By a proof similar to the one in the preceeding example, the definition of Seq$^+(A)$ given in Example 4.3.8 meets the unique readability property.

Example 4.5.12 We show that the definition of the set S of sequences of a's and b's given in Example 4.3.9 does *not* meet the unique readability condition. To see this, first note that $\lambda \in S$ by the basis rule and hence by Rules (2) and (3) applied to λ both ab and ba are in S. Now, applying Rule (2) with ba as the hypothesis, we get $abab \in S$ and applying Rule (4) with both hypotheses equal to ab, we again get $abab \in S$. Thus, $abab$ can be put into S by both Rules (2) and (4), so the definition of S does not meet the unique readability condition.

For all of the examples we have looked at so far, it has been easy to tell whether or not the definition in question satisfied the unique readability condition. In the next example, we see a more difficult proof that a definition we have given meets the condition.

We will now look at some more examples of recursively defined functions. Sometimes it is useful to consider an already defined function and find a set of recursive equations that the function satisfies. We can then characterize the function as being the unique solution to the set of equations. Our next examples will illustrate these ideas.

Example 4.5.13 Let A be a set. We have already defined the length of a sequence, and we can consider length to be a function from Seq(A) to **N**. It is easy to see that $|\lambda| = 0$, and for any $w \in$ Seq(A) and $x \in A$, $|wx| = |w| + 1$. Since the definition of Seq(A) in Example 4.3.7 meets the unique readability condition, we can say that length is the unique function f from $\text{Seq}(A)$ to $\mathbf{N}$ that satisfies the equations

$$f(\lambda) = 0,$$
$$f(wx) = f(w) + 1 \text{ for each } w \in \text{Seq}(A) \text{ and } x \in A.$$

Example 4.5.14 Let A be any set. Then, concatenation is a binary operation on $\text{Seq}(A)$. We would like to give a recursive characterization of the concatenation function. Since concatenation is a function of two arguments and we only have means for giving recursive equations for functions of one argument, we must proceed indirectly . For each sequence w in $\text{Seq}(A)$, define a function $\text{cat}_w : \text{Seq}(A) \longrightarrow \text{Seq}(A)$ by $\text{cat}_w(w') = ww'$. We will give a recursive characterization of cat_w. For each $x \in A$, let $M_x : \text{Seq}(A) \longrightarrow \text{Seq}(A)$ be given by $M_x(w) = wx$. For any $w \in A$, we have

$$\text{cat}_w(\lambda) = w\lambda = w,$$

and for all $w' \in \text{Seq}(A), x \in A$,

$$\text{cat}_w(w'x) = w(w'x) = (ww')x = M_x(\text{cat}_w(w')).$$

According to Example 4.5.10, we can conclude that for each $w \in A$, cat_w is the unique function $f : \text{Seq}(A) \longrightarrow \text{Seq}(A)$ that satisfies

$$f(\lambda) = w,$$
$$f(w'x) = M_x(f(w')) \text{ for all } w' \in \text{Seq}(A) \text{ and } x \in A.$$

Example 4.5.15 Let A be a set. If w is a finite sequence of elements of A, then we would like to define the *reversal* of w, denoted by w^R, to be the sequence obtained from w by reversing the entries in w. For instance, $abcad^R = dacba$. We can do this nonrecursively by defining w^R to be the string with the same length as w and such that for each i with $1 \le i \le |w|$ we have $w^R(i) = w(|w|+1-i)$. We then have $\lambda^R = \lambda$, and for all $w \in \text{Seq}(A)$ and $x \in A$, we have $(wx)^R = xw^R$. To see the latter, note that $|(wx)^R| = |wx| = |w| + 1$, while $|xw^R| = |x| + |w^R| = 1 + |w|$, so the two sequences have the same length $|w| + 1$ and further, $(wx)^R(1) = wx(|wx| + 1 - 1) = wx(|wx|) = x(|wx| - |w|) = x(1) = x$, while $xw^R(1) = x(1) = x$, so the two sequences have the same first element. Finally, if $2 \le i \le |w| + 1$, then $(wx)^R(i) = wx(|wx| + 1 - i) = wx(|w| + 2 - i) = w(|w| + 2 - i)$, and so $xw^R(i) = w^R(i - |x|) = w^R(i - 1) = w(|w| + 1 - (i - 1)) = w(|w| + 2 - i) = (wx)^R(i)$.

Thus, we can give an equivalent recursive definition of the reversal function by

$$\lambda^R = \lambda,$$
$$(wx)^R = xw^R \text{ for } w \in \text{Seq}(A) \text{ and } x \in A.$$

To prove properties of the reversal, we can use either one of these equivalent definitions, whichever gives the easier proof. For example, to prove that for all sequences w, u in Seq(A), $(wu)^R = u^R w^R$, we use the recursive definition of reversal. Let $P(u)$ be the property that for all $w \in$ Seq(A), $(wu)^R = u^R w^R$. We will use structural induction as given in Corollary 4.4.5 to prove that $P(u)$ is true for all $u \in$ Seq(A). The basis step is to show $P(\lambda)$. For any $w \in$ Seq(A), we have

$$(w\lambda)^R = w^R = \lambda w^R = \lambda^R w^R,$$

so $P(\lambda)$ is true. To prove the inductive step, we assume that $P(u)$ is true, and we show that $P(ux)$ is true, where $u \in$ Seq(A) and $x \in A$. For any $w \in$ Seq(A), we have

$$
\begin{aligned}
(w(ux))^R &= ((wu)x)^R && \text{by associativity of concatenation} \\
&= x(wu)^R && \text{by definition of reversal} \\
&= x(u^R w^R) && \text{by induction hypothesis } P(u) \\
&= (xu^R)w^R && \text{by associativity} \\
&= (ux)^R w^R && \text{by definition of reversal}
\end{aligned}
$$

which establishes $P(ux)$.

In our next few examples, we look at how recursive definitions can be used to make precise ideas that we have been using already but without ever defining precisely. We first look at some recursive definitions of the arithmetic operations on the natural numbers.

Example 4.5.16 We take as given the set $\mathbf{N}$ of natural numbers and the *successor* function $S(n) = n + 1$. (Note that although we assume that we can find the successor of a number by adding 1 to it, we are not assuming that we have already defined the addition operation.) We will now show how to give a recursive definition of addition. Intuitively, adding n to a is the same as adding 1 to a n times, i.e., $a + n = a + 1 + 1 + \cdots + 1$, where the $\cdots$ indicates that the addition is to be carried out n times. The problem with this is the imprecise use of the $\cdots$. We use a recursive definition to remove this imprecision. Because addition is a binary operation, we must use the strategy of Example 4.5.14; that is, for each $a \in \mathbf{N}$, we will recursively define a function add_a with the idea that $\mathrm{add}_a(n) = a + n$. Once we are done, we can define addition by $n + m = \mathrm{add}_n(m)$. We define add_a by

$$
\begin{aligned}
\mathrm{add}_a(0) &= a, \\
\mathrm{add}_a(n + 1) &= S(\mathrm{add}_a(n)).
\end{aligned}
$$

A somewhat confusing but easier to read rephrasing of this definition is

$$
\begin{aligned}
a + 0 &= a, \\
a + (n + 1) &= (a + n) + 1.
\end{aligned}
$$

Example 4.5.17 Let us take the addition operation on the natural numbers as given and show how to use it to define recursively the multiplication operation. The intuition here is that $a \cdot n$ is $a + a + a + \cdots + a$, where the $\cdots$ stands for n copies of a being added together. We once again use a recursive definition to make this idea precise. For each natural number a, we define a function $\text{mult}_a : \mathbf{N} \longrightarrow \mathbf{N}$ by

$$\text{mult}_a(0) = 0,$$
$$\text{mult}_a(n + 1) = \text{mult}_a(n) + a.$$

We then define the product $n \cdot m$ to be $\text{mult}_n(m)$. We can reexpress this definition by

$$a \cdot 0 = 0,$$
$$a \cdot (n + 1) = a \cdot n + 1.$$

Example 4.5.18 In this example, we define exponentiation of natural numbers recursively. The intuition is that a^n is n copies of a multiplied together. To make this precise, for each $a \in \mathbf{N}$, we define $\exp_a : \mathbf{N} \longrightarrow \mathbf{N}$ by

$$\exp_a(0) = 1,$$
$$\exp_a(n + 1) = \exp_a(n) \cdot a.$$

We can then set $n^m = \exp_n(m)$. This definition can be rephrased as

$$a^0 = 1,$$
$$a^{n+1} = a^n \cdot a.$$

Example 4.5.19 It often happens that we wish to define a power a^n where n is a natural number and a is some sort of object such as a matrix or a sequence that can in some sense by multiplied by itself. By generalizing the previous example, we can provide a uniform way of making such definitions.

Let $(D, *, e)$ be a triple consisting of a set D, a binary operation $*$ on D, and an element e of D, which is an identity element for $*$. (This means that for all $d \in D$, $e * d = d = d * e$.) For each $d \in D$, we define recursively a function $\exp_d : \mathbf{N} \longrightarrow D$, given by

$$\exp_d(0) = e,$$
$$\exp_d(n + 1) = \exp_d(n) * d.$$

In most (but not all) applications of this definition, it is appropriate to denote $\exp_d(n)$ by d^n. In that case, the defining equations can be rewritten as

$$d^0 = e,$$
$$d^{n+1} = d^n * d.$$

In other words, d^n is n copies of d *'d together. Note that $\exp_d(1) = \exp_d(0 + 1) = \exp_d(0) * d = e * d = d$, so $d^1 = d$, as it should.

In the next few examples, we illustrate how this general definition of the previous example can be used.

Example 4.5.20 Let A be any set and let m be a positive integer. Then, if we let D be $M_{m \times m}(A)$, the set of all m by m matrices over A, $*$ be matrix multiplication, and e be I_m, the m by m identity matrix, we can apply Example 4.5.19 to define for each matrix $M \in D$ and natural number n the matrix M^n, which is the n copies of the matrix M multiplied together. The defining equations are

$$M^0 = I_m,$$
$$M^{n+1} = M^n M.$$

Example 4.5.21 Let A be a a set, let $D = \text{Seq}(A)$, let $*$ be concatenation, and let $e = \lambda$. Then, applying Example 4.5.19, we get a definition for w^n for each $w \in D$ and $n \in \mathbf{N}$. The sequence w^n is n copies of w concatenated. We have for every sequence w

$$w^0 = \lambda,$$
$$w^{n+1} = w^n w.$$

Example 4.5.22 Let $D = \mathbf{R}$, let $*$ be multiplication, and let $e = 1$. Then, applying Example 4.5.19, we get a definition of r^n for every real number r and natural number n with

$$r^0 = 1,$$
$$r^{n+1} = r^n r.$$

Example 4.5.23 Let A be a set and let $D = \mathcal{P}(A \times A)$, the set of all binary relations over A. Let $*$ be relation product and let $e = \iota_A$. Then, applying Example 4.5.19, we get for each relation R over A and natural number n a definition of R^n with

$$\rho^0 = \iota_A,$$
$$\rho^{n+1} = \rho^n \bullet \rho.$$

In Chapter 2, we have already defined ρ^n for $n \geq 1$ using paths, i.e., we defined ρ^n to consist of those pairs (a, b) such that there is a path of length n in ρ from a to b. Now, we have another definition of ρ^n. We must show that these two definitions are the same. For a fixed relation ρ over A, define $g_\rho : \mathbf{N} \longrightarrow D$ by $g_\rho(0) = \iota_A$, and for $n \geq 1$, $g_\rho(n) = \rho^n$, where ρ^n is as defined by paths of length n. Then, according to Proposition 2.6.10, we have $g_\rho(n+1) = g_\rho(n) \bullet g_\rho(1) = g_\rho(n) \bullet \rho$ for $n \geq 1$, and it is easy to see that we also have $g_\rho(n+1) = g_\rho(n) \bullet \rho$ when $n = 0$. Thus, g_ρ satisfies the same defining equations as the new definition of ρ^n. Because these equations have a unique solution, the new definition of ρ^n is the same as the old for all $n \geq 1$.

Example 4.5.24 Let A be a set, let $D = A \rightsquigarrow A$, let $*$ be function composition, and let $e = 1_A$. Using Example 4.5.19, we get a definition of the *n-iteration of the partial function f, f^n* for each partial function f from A to A and natural number n with

$$f^0 = 1_A,$$
$$f^{n+1} = f^n \circ f.$$

If f is considered to be a relation, then f^n as defined in this example is exactly the same as f^n as defined by the previous example.

Consider the function $f : \mathbf{R} \longrightarrow \mathbf{R}$, defined by $f(x) = ax + b$ for every $x \in \mathbf{R}$, where a, b are two real coefficients. We have $f^0(x) = x$ for $x \in \mathbf{R}$. On the other hand, we have

$$f^n(x) = a^n x + b(1 + a + \cdots + a^{n-1}),$$

for $n \geq 1$, as can be verified immediately. Indeed, for $n = 1$, we have $f^1(x) = f(f^0(x)) = f(x)$; also, $f^1(x) = ax + b$, which ends the base step.

For the induction step, assume that $f^n(x) = a^n x + b(1 + a + \cdots + a^{n-1})$. The next iteration is given by

$$\begin{aligned} f^{n+1}(x) &= f(f^n(x)) = a(a^n x + b(1 + a + \cdots + a^{n-1})) + b \\ &= a^{n+1}x + b(1 + a + \cdots + a^n), \end{aligned}$$

which concludes our proof.

The next example gives another illustration of how a recursive definition can be used to make precise an idea we have already been using, namely, the sum of a finite sequence of real numbers.

Example 4.5.25 Let $a = (a_1, \ldots, a_n)$ be a sequence of real numbers. Then, by $a_1 + \cdots + a_n$, we mean the sum of all the entries in a. Although it is intuitively clear what this means, we have not yet been able to define this idea precisely. Now, we can.

We define a function SUM : $\mathrm{Seq}(\mathbf{R}) \longrightarrow \mathbf{R}$ by the following recursive definition.

$$\mathrm{SUM}(\lambda) = 0,$$
$$\mathrm{SUM}(wx) = \mathrm{SUM}(w) + x \text{ for } w \in \mathrm{Seq}(\mathbf{R}) \text{ and } x \in \mathbf{R}.$$

The usual notation for SUM(a) where $a = (a_1, \ldots, a_n)$ is $\sum_{i=1}^n a_i$. We can have $n = 0$ in this last notation in which case we are talking about the sum of the null sequence, which is 0. Using this notation, we can rephrase our definition as

$$\sum_{i=1}^{0} a_i = 0,$$
$$\sum_{i=1}^{n+1} a_i = (\sum_{i=1}^{n} a_i) + a_{n+1}.$$

Using this definition, it is now possible to give precise proofs of various properties of sums of sequences. As an example, we will prove that for any sequence $(a_1, \ldots, a_n)$ of reals and any real c, $\sum_{i=1}^{n} ca_i = c \sum_{i=1}^{n} a_i$. The proof is by induction on n, the length of the sequence. For $n = 0$, we have

$$\sum_{i=1}^{0} ca_i = 0 = c0 = c \sum_{i=1}^{0} a_i.$$

Assuming the result true for n, we have

$$
\begin{aligned}
\sum_{i=1}^{n+1} ca_i &= (\sum_{i=1}^{n} ca_i) + ca_{n+1} && \text{by definition of SUM} \\
&= c(\sum_{i=0}^{n} a_i) + ca_{n+1} && \text{by induction hypothesis} \\
&= c((\sum_{i=1}^{n} a_i) + a_{n+1}) && \text{by distributive law for the reals} \\
&= c \sum_{i=1}^{n+1} a_i && \text{by definition of SUM,}
\end{aligned}
$$

which establishes the result for $n + 1$.

We now generalize the previous example in a way that has many applications.

Example 4.5.26 Let $(D, *, e)$ be any triple consisting of a set D, a binary operation $*$ on D, and an identity element e for $*$. We define a function $\Xi_* : \mathrm{Seq}(D) \longrightarrow D$. The idea is that $\Xi_*((a_1, \ldots, a_n)) = (\cdots ((a_1 *_2) * a_3) * \cdots * a_n)$, the product under $*$ of the elements of the sequence taken from left to right. Formally, we define

$$
\begin{aligned}
\Xi_*(\lambda) &= e, \\
\Xi_*(wx) &= \Xi_*(w) * x \text{ for each } w \in \mathrm{Seq}(D) \text{ and } x \in D.
\end{aligned}
$$

When $D = \mathbf{R}$, $*$ is $+$, and $e = 0$, then Ξ_+ is the same as SUM from the previous example.

In the next example, we give a slight variation on the definition in the previous example, which does not require that there be an identity element for $*$.

Example 4.5.27 Let D be a set and let $*$ be an operation on D. We define a function $\Xi_*^+ : \mathrm{Seq}^+(D) \longrightarrow D$ by

$$
\begin{aligned}
\Xi_*^+((x)) &= x \text{ for each } x \in D, \\
\Xi_*^+(wx) &= \Xi_*^+(w) * x \text{ for each } w \in \mathrm{Seq}^+(D) \text{ and } x \in D.
\end{aligned}
$$

Theorem 4.5.28 Let $(D, *, e)$ be as in Example 4.5.26. Then, for every $a \in Seq^+(D)$, $\Xi_*(a) = \Xi_*^+(a)$.

Proof: We use structural induction on $Seq^+(D)$ as in Corollary 4.4.6. For each $x \in D$, we have

$$\Xi_*((x)) = \Xi_*(\lambda x) = \Xi_*(\lambda) * x = e * x = x = \Xi_*^+((x)).$$

If $w \in Seq^+(D)$ and $x \in D$ and $\Xi_*(w) = \Xi_*^+(w)$, then

$$\Xi_*(wx) = \Xi_*(w) * x = \Xi_*^+(w) * x = \Xi_*^+(wx),$$

completing the induction. ∎

Because of the previous result, we will not distinguish between Ξ_* and Ξ_*^+ and will write Ξ_* for both of them. If $*$ has an identity element, then Ξ_* is defined for all finite sequences over D; otherwise, it is defined only for the nonnull sequences. In the next few examples, we give several applications of the Ξ_* function.

Example 4.5.29 Let $D = \mathbf{R}$, let $*$ be the multiplication operation $\cdot$, and let $e = 1$. Then, applying Example 4.5.26, we get $\Xi_\cdot : Seq(\mathbf{R}) \longrightarrow \mathbf{R}$, which just maps a sequence to the product of its elements. The usual notation for $\Xi_\cdot$ is $\prod\limits_{i=1}^{n}$, so we have

$$\prod_{i=1}^{0} a_i = 1,$$
$$\prod_{i=1}^{n+1} a_i = (\prod_{i=1}^{n} a_i)a_{n+1}.$$

Example 4.5.30 Let A be a set, let $D = \mathcal{P}(A)$, let $*$ be $\cup$, and let $e = \emptyset$. Then, $\Xi_\cup((S_1, \ldots, S_n))$ is the union of the S_i's taken from left to right and is usually written $\bigcup\limits_{i=0}^{n} S_i$. Note that according to this definition, the union of no sets is $\emptyset$. We have

$$\bigcup_{i=1}^{0} S_i = \emptyset,$$
$$\bigcup_{i=1}^{n+1} S_i = (\bigcup_{i=1}^{n} S_i) \cup S_{n+1}.$$

It is not hard to show (by induction on n) that for $n \geq 1$, $\bigcup\limits_{i=1}^{n} S_i$ is the same as $\bigcup\{S_i \mid 1 \leq i \leq n\}$, as defined in Chapter 1.

Example 4.5.31 Let A be a set, let $D = \mathcal{P}(A)$, let $*$ be $\cap$, and let $e = A$. Then, $\Xi_\cap((S_1, \ldots, S_n))$ is the intersection of the S_i's taken from left to right and is usually written $\bigcap\limits_{i=0}^{n} S_i$. Note that according to this definition, the intersection of no sets is the universal set A. We have

$$\bigcap_{i=1}^{0} S_i = A,$$
$$\bigcap_{i=1}^{n+1} S_i = (\bigcap_{i=1}^{n} S_i) \cap S_{n+1}.$$

It is not hard to show (by induction on n) that for $n \geq 1$, $\bigcap_{i=1}^{n} S_i$ is the same as $\bigcap \{S_i \mid 1 \leq i \leq n\}$, as defined in Chapter 1.

Example 4.5.32 Let A be a set, let $D = \text{Seq}(A)$, let $*$ be the concatenation operation $\cdot$, and let $e = \lambda$. Then, we will write CAT in place of Ξ_*. For any sequence $(w_1, \ldots, w_n)$ of sequences over A, $\text{CAT}((w_1, \ldots, w_n))$ is the concatenation of all these sequences taken left to right. We have

$$\text{CAT}(\lambda) = \lambda,$$
$$\text{CAT}((w_1, \ldots, w_{n+1})) = \text{CAT}((w_1, \ldots, w_n)) \cdot w_{n+1}.$$

If Σ is an alphabet and L is a language over Σ, then $L^* = \text{CAT}(\text{Seq}(L))$, that is, L^* consists of all strings of symbols from Σ that can be obtained by concatenating together 0 or more strings from L.

If $(D, *, e)$ are as in Example 4.5.26, and a is a finite sequence of elements of D of length greater or equal to 3, then $\Xi_*(a)$ gives only one way of combining together under $*$ the elements of a in the given order. For example, if $a = (a_1, a_2, a_3)$, then we can combine these elements as $(a_1 * a_2) * a_3$, which is $\Xi_*(a)$, or as $a_1 * (a_2 * a_3)$. These two quantities need not be the same in general, but they will be if $*$ is associative. When $a = (a_1, a_2, a_3, a_4)$, there are five different ways of parenthesizing the elements to combine them under $*$. If $*$ is associative, then all five of these ways of evaluating the product of the elements of a under $*$ are equal to $\Xi_*(a)$. For example, using associativity twice, we get $a_1 * ((a_2 * a_3) * a_4) = (a_1 * (a_2 * a_3)) * a_4 = ((a_1 * a_2) * a_3) * a_4$. It seems likely that if $*$ is associative, then for any sequence a of length greater or equal to 2, no matter how parentheses are put in, when the product of all the elements in a under $*$ is taken, the same result is obtained. This is true, and is often stated, but not so often proved. We are now able to give a proof of this fact. One way to proceed is to define formally what a proper parenthesizing of an expression is. This would take us too far afield. Instead, we take another approach.

We first prove a lemma about Ξ_*.

Lemma 4.5.33 *Let D be a set and let $*$ be an associative operation on D. Then, for all $a, b \in \text{Seq}^+(D)$, we have $\Xi_*(a) * \Xi_*(b) = \Xi_*(ab)$.*

Proof: We prove by induction on b that for all nonnull finite sequences a the desired equality holds. The basis step is to show this for $b = (x)$ for each element x of D. But, for any finite nonnull sequence a, we have

$$\Xi_*(a) * \Xi_*((x)) = \Xi_*(a) * x = \Xi_*(ax),$$

so the result holds for $b = (x)$. For the inductive step, suppose that the result holds for b and that $x \in D$. Then, for any finite nonnull sequence a, we have

$$
\begin{aligned}
&\Xi_*(a) * \Xi_*(bx) \\
&= \Xi_*(a) * (\Xi_*(b) * x) && \text{by definition of } \Xi_* \\
&= (\Xi_*(a) * \Xi_*(b)) * x && \text{by associativity of } * \\
&= \Xi_*(ab) * x && \text{by induction hypothesis} \\
&= \Xi_*((ab)x) && \text{by definition of } \Xi_* \\
&= \Xi_*(a(bx)) && \text{by associativity of concatenation,}
\end{aligned}
$$

and hence, the result is true for bx, completing the induction. ∎

Our next definition helps us to define precisely what it means to take the product of the elements in a sequence under some grouping.

Definition 4.5.34 *Let D be a set and let $*$ be a binary operation on D. We define a relation $\Rightarrow$ on $Seq(D)$ by $a \Rightarrow b$ if there are sequences a' and a'' and elements d_1, d_2 of D such that $a = a' d_1 d_2 a''$ and $b = a' d_1 * d_2 a''$. (In other words, b is a sequence of length one less than the length of a obtained from a by combining two adjacent entries under $*$.)*

As usual, we write $\overset{*}{\Rightarrow}$ for the reflexive, transitive closure of $\Rightarrow$. If $d \in D$, then $a \overset{*}{\Rightarrow} d$ means $a \overset{*}{\Rightarrow} (d)$.

If $a \overset{*}{\Rightarrow} d$, then this means that d can be obtained from the sequence a by combining the elements of a under $*$ in some grouping. For example, $(d_1, d_2, d_3, d_4, d_5) \Rightarrow (d_1 * d_2, d_3, d_4, d_5) \Rightarrow (d_1 * d_2, d_3, d_4 * d_5) \Rightarrow ((d_1 * d_2) * d_3, d_4 * d_5) \Rightarrow ((d_1 * d_2) * d_3) * (d_4 * d_5)$. The next lemma proves an intuitively obvious fact about $\overset{*}{\Rightarrow}$. The proof is somewhat intricate and could be skipped on a first reading.

Lemma 4.5.35 *Let $*$ be a binary operation on a set D. If $a \overset{*}{\Rightarrow} b$ and $|b| = 2$, then there are sequences $\hat{a}$ and $\hat{\hat{a}}$ with $a = \hat{a}\hat{\hat{a}}$, $\hat{a} \overset{*}{\Rightarrow} b_1$, and $\hat{\hat{a}} \overset{*}{\Rightarrow} b_2$.*

Proof: Let $P(a, b)$ be the property "if $|b| = 2$, then there are $\hat{a}, \hat{\hat{a}}$ with $a = \hat{a}\hat{\hat{a}}$, $\hat{a} \overset{*}{\Rightarrow} b_1$, and $\hat{\hat{a}} \overset{*}{\Rightarrow} b_2$." We use the method from the third part of Exercise 36 to show that $P(a, b)$ is true for all a, b with $a \overset{*}{\Rightarrow} b$. The basis step $P(b, b)$ is immediate. (If $|b| = 2$, let $\hat{a} = b_1$, $\hat{\hat{a}} = b_2$.) Suppose that $P(c, b)$ is true and that $a \Rightarrow c$. If $|b| = 2$, then by $P(c, b)$, there are $\hat{c}$ and $\hat{\hat{c}}$ with $c = \hat{c}\hat{\hat{c}}$, $\hat{c} \overset{*}{\Rightarrow} b_1$, $\hat{\hat{c}} \overset{*}{\Rightarrow} b_2$. Since $a \overset{*}{\Rightarrow} c$, there must be sequences a', a'' and elements d_1, d_2 of D with $a = a' d_1 d_2 a''$ and $c = a' d_1 * d_2 a''$. Hence, $\hat{c}\hat{\hat{c}} = c = a' d_1 * d_2 a''$. We consider two cases.

Case 1: $|\hat{c}| \leq |a'|$. Then, for some $\bar{c}$, $a' = \hat{c}\bar{c}$, so $\hat{c}\bar{c} d_1 * d_2 a'' = c = \hat{c}\hat{\hat{c}}$, and hence, $\bar{c} d_1 * d_2 a'' = \hat{\hat{c}}$. Let $\hat{a} = \hat{c}$ and $\hat{\hat{a}} = \bar{c} d_1 d_2 a''$. Then, $\hat{a}\hat{\hat{a}} = \hat{c}\bar{c} d_1 d_2 a'' = a' d_1 d_2 a'' = a$, $\hat{a} = \hat{c} \overset{*}{\Rightarrow} b_1$, and $\hat{\hat{a}} \Rightarrow \bar{c} d_1 * d_2 a'' = \hat{\hat{c}} \overset{*}{\Rightarrow} b_2$, so $\hat{\hat{a}} \overset{*}{\Rightarrow} b_2$, as desired.

Case 2: $|\hat{c}| > |a'|$. Then, for some $\bar{c}$, $a' d_1 * d_2 \bar{c} = \hat{c}$, so $a' d_1 * d_2 \bar{c}\hat{\hat{c}} = c = a' d_1 * d_2 a''$, and hence, $\bar{c}\hat{\hat{c}} = a''$. Let $\hat{a} = a' d_1 d_2 \bar{c}$ and $\hat{\hat{a}} = \hat{\hat{c}}$. Then,

$\hat{a}\hat{\bar{a}} = a'd_1d_2\bar{c}\hat{c} = a'd_1d_2a'' = a$, $\hat{a} = a'd_1d_2\bar{c} \Rightarrow a'd_1 * d_2\bar{c} = \hat{c} \stackrel{*}{\Rightarrow} b_1$ (so $\hat{a} \stackrel{*}{\Rightarrow} b_1$) and $\hat{\bar{a}} = \hat{c} \stackrel{*}{\Rightarrow} b_2$, as desired. ∎

Now, we are ready to prove our main result about associative operations.

Theorem 4.5.36 *Let $*$ be an associative operation on a set D. For every nonnull sequence a of elements of D, if $d \in D$ and $a \stackrel{*}{\Rightarrow} d$, we have $d = \Xi_*(a)$.*

Proof: The proof is by strong induction on the length of a. If $|a| = 1$, the result is obvious, since if $a \stackrel{*}{\Rightarrow} d$, then $a = (d)$. Suppose that $|a| > 1$ and that the result is true for sequences of length less than $|a|$. If $a \stackrel{*}{\Rightarrow} d$, then since $|a| > 1$, there must be a sequence b of length 2 with $a \stackrel{*}{\Rightarrow} b \Rightarrow (d)$. By Lemma 4.5.35, there are sequences a', a'' with $a = a'a''$, $a' \stackrel{*}{\Rightarrow} b_1$, and $a'' \stackrel{*}{\Rightarrow} b_2$. Neither a' nor a'' can be the null sequence, so both have length less than $|a|$. By the induction hypothesis, $b_1 = \Xi_*(a')$ and $b_2 = \Xi_*(a'')$. But, $d = b_1 * b_2$, so using Lemma 4.5.33, we get $d = b_1 * b_2 = \Xi_*(a') * \Xi_*(a'') = \Xi_*(a'a'') = \Xi_*(a)$, as desired. ∎

The last result says that if $*$ is an associative operation on D, then no matter what grouping is used to evaluate a product of three or more elements, the same result will be obtained. Thus, for such an operation, one can write expressions such as $d_1 * d_2 * d_3 * d_4$ without worrying about the locations of the parentheses.

4.6 Constructors

In this section, we give a framework that allows us to give a precise formulation of what an inductively defined set is and what the unique readability condition is. We prove the principle of structural induction, give a precise formulation of recursive definition of a function over an inductively defined set, and show that such definitions work when the unique readability condition is met.

We begin by considering the simplest possible example, and then afterward, we generalize.

Definition 4.6.1 *Let U be a set and let f be a function from U to U.*

1. *A subset C of U is called* closed under f *if for all $x \in C$, $f(x)$ is also in C (i.e., $f(C) \subseteq C$).*

2. *Let B be a subset of U. A subset C of U is called the* closure of B under f *if the following hold.*

   (a) *$B \subseteq C$.*

   (b) *C is closed under f.*

*(c) If C' is a subset of U that contains B and is closed under f,
then $C \subseteq C'$.*

The use of the word "the" in the definition of the closure of B under f
is justified by the following result.

Theorem 4.6.2 *Let U be a set, let f be a unary function on U, and let
B be a subset of U. If C_1 and C_2 are both closures of B under f, then
$C_1 = C_2$.*

Proof: Since C_1 is a closure of B under f, $B \subseteq C_1$ and C_1 is closed
under f. Thus, by the third condition in the definition of C_2 as a closure
for B under f, we must have $C_2 \subseteq C_1$. By a similar argument, $C_1 \subseteq C_2$,
and hence, $C_1 = C_2$. ∎

Example 4.6.3 Let $U = \mathbf{Z}$ and let $f(n) = n+1$. Then, for any integer n_0,
$\{n \in \mathbf{Z} \mid n \geq n_0\}$ is closed under f and in fact is the closure of $B = \{n_0\}$
under f. In particular, $\mathbf{N}$ is the closure of $\{0\}$ under f.

Example 4.6.4 Let $U = \mathbf{N}$ and let $f(n) = n + 2$. Then, the even natural
numbers are the closure of $\{0\}$ under f, the odd natural numbers are the
closure of $\{1\}$ under f, and $\mathbf{N}$ is the closure of $\{0,1\}$ under f.

Example 4.6.5 Let n be a positive integer, let $U = \{0,\ldots,n-1\}$, and
let $f(m) = (m+1) \bmod n$ for each $m \in U$. Then, for any $m \in U$, U is the
closure of $\{m\}$ under f.

Theorem 4.6.6 *Let U be a set and f be a unary function on U. Then,
for every subset B of U, the closure of B under f exists.*

Proof: Let B be a subset of U and let $\mathcal{S}$ be the collection of all subsets
of U that contain B and are closed under f, i.e., $\mathcal{S} = \{S \mid B \subseteq S \subseteq$
U and S is closed under $f\}$. Note that $\mathcal{S}$ is nonempty since $U \in \mathcal{S}$. Let
$C = \bigcap\{S \mid S \in \mathcal{S}\}$. We show that C is the closure of B under f. First
of all, every element of $\mathcal{S}$ contains B, so C, the intersection of all the
elements of $\mathcal{S}$, also contains B. Next, suppose that $x \in C$. Then, for all
$S \in \mathcal{S}$, $x \in S$. Since each $S \in \mathcal{S}$ is closed under f, for all $S \in \mathcal{S}$, $f(x) \in S$,
and thus, $f(x) \in C$. This shows that C is closed under f. Finally, suppose
that C' contains B and is closed under f. Then, $C' \in \mathcal{S}$, so $C \subseteq C'$. Thus,
C is the closure of B under f. ∎

We now prove the result that corresponds to structural induction for the
situation we are considering.

Theorem 4.6.7 *Let f be a unary function on a set U and let C be the
closure under f of a subset B of U. Suppose that P is a property of U such
that*

1. every element of B has property P;

2. *for all $x \in C$, if $P(x)$ is true, then $P(f(x))$ is true.*

Then, every element of C has property P.

Proof: Let $A = \{x \in C \mid P(x)$ is true$\}$. Then, according to the two given conditions, A is a subset of U that contains B and is closed under f. Thus, (by definition of C as the closure of B under f), $C \subseteq A$. (In fact, $A = C$ since A is a subset of C.) Thus, every $x \in C$ has property P. ∎

Note that in the previous theorem, if P is a property of C, then P is a property of U (since our official definition of a property of a set X is that it is a subset of X), so the previous theorem applies to properties of C as well.

We can now apply Theorem 4.6.7 to the examples of closures of sets under a single unary operation that we saw earlier.

Example 4.6.8 The principle of ordinary induction (Theorem 4.2.4) follows from Theorem 4.6.7 applied to Example 4.6.3.

Example 4.6.9 If we apply Theorem 4.6.7 to Example 4.6.4, one of the results we get is this one. Let P be a property of $\mathbf{N}$ such that 0 and 1 both have property P and such that whenever n is a natural number with property P, then $n + 2$ also has property P. Then, every natural number has property P.

One easy application of structural induction is the following.

Theorem 4.6.10 *Let U be a set, let $f : U \longrightarrow U$, and let C be the closure under f of a subset B of U. Then, $C = B \cup f(C)$.*

Proof: Since C is closed under f, $f(C) \subseteq C$, and by definition of C, $B \subseteq C$, so $B \cup f(C) \subseteq C$. For the other inclusion, let $P(x)$ be the property "$x \in B \cup f(C)$." We use Theorem 4.6.7 to show that $P(x)$ is true for all $x \in C$. The basis is to show that $P(x)$ is true for all $x \in B$, which is obvious. The inductive step is to assume that $x \in C$ and $P(x)$ is true and show that $P(f(x))$ is true; this is also obvious since if $x \in C$, then $f(x) \in f(C)$. ∎

We now wish to generalize our discussion so that we get more widely applicable results. First of all, we want to consider more than one function when taking the closure. Second, we want to allow more than one argument in our functions. Third, we need to consider families not just of functions, but of partial functions, or even more generally of arbitrary relations. When we make all three of these modifications, we get a very general approach with a wide variety of applications.

Definition 4.6.11 *Let U be a set.*

1. *If n is a natural number, then an n-ary constructor on U is a relation from U^n to U.*

2. *If R is an n-ary constructor on U and A is a subset of U^n, then $R(A)$ is the set $\{u \in U \mid$ there exists a sequence $a \in A$ with $aRu\}$.*

Note that an n-ary operation or partial operation on U is an n-ary constructor on U. We allow $n = 0$ in the definition of n-ary constructor. A 0-ary constructor is a relation from $U^0 = \{\lambda\}$ to U. Thus, a 0-ary constructor on U may be thought of as a subset of U. A 1-ary constructor is a relation from U^1 to U. Since we identify U^1 with U, a 1-ary constructor on U can be considered to be the same as a binary relation on U.

Definition 4.6.12   *1. Let R be an n-ary constructor on U and let C be a subset of U. Then, C is said to be* closed under R *if $R(C^n) \subseteq C$. (In other words, for every n-tuple $(x_1, \ldots, x_n)$ of elements of C and every element x of U, if $((x_1, \ldots, x_n), x) \in R$, then $x \in C$.)*

2. Let $\mathcal{R}$ be a family of constructors on U and let C be a subset of U. Then, C is called closed under $\mathcal{R}$ *if C is closed under each R in $\mathcal{R}$.*

3. Let $\mathcal{R}$ be a family of constructors on U and let B be a subset of U. Then, a subset C of U is called the closure *of B under $\mathcal{R}$ if the following conditions are met.*

(a) $B \subseteq C$.

(b) C is closed under $\mathcal{R}$.

(c) If C' is a subset of U that contains B and is closed under $\mathcal{R}$, then $C \subseteq C'$.

Theorem 4.6.13 *Let $\mathcal{R}$ be a family of constructors on the set U and let B be any subset of U. Then, the closure of B under $\mathcal{R}$ exists and is unique.*

Proof: The uniqueness of the closure of B under $\mathcal{R}$ is shown exactly as in Theorem 4.6.2. To see that the closure of B under $\mathcal{R}$ exists, we let $\mathcal{S}$ be the collection of all subsets of U that contain B and are closed under $\mathcal{R}$. Then, $\mathcal{S}$ is nonempty since $U \in \mathcal{S}$. Let $C = \bigcap \{S : S \in \mathcal{S}\}$. The verification that C meets the conditions in the definition of the closure of B under $\mathcal{R}$ is much as in Theorem 4.6.6 and is left to the reader. ∎

Theorem 4.6.14 *Let U be a set, let $\mathcal{R}$ be a family of constructors on U, let B be a subset of U, and let C be the closure of B under $\mathcal{R}$. Suppose that P is a property of U such that the following are met.*

1. Every element of B has property P.

2. For all $R \in \mathcal{R}$, $x_1, \ldots, x_n \in C$ (where R is n-ary), and $x \in U$, if $((x_1, \ldots, x_n), x) \in R$ and $P(x_i)$ is true for all i with $1 \leq i \leq n$, then $P(x)$ is true.

Then, $P(x)$ is true for all $x \in C$.

Proof: Let $A = \{x \in C \mid P(x) \text{ is true}\}$. Then, according to the two conditions given on P, A is a subset of U that contains B and is closed under $\mathcal{R}$, so $C \subseteq A$. Thus, every $x \in C$ has property P. ∎

Theorem 4.6.15 *Let U be a set, let $\mathcal{R}$ be a family of constructors on U, let B be a subset of U, and let C be the closure of B under $\mathcal{R}$. Then, $C = B \cup \bigcup_{R \in \mathcal{R}} R(C^{n_R})$, where for each $R \in \mathcal{R}$, R is n_R-ary.*

Proof: This theorem is a consequence of Theorem 4.6.14, just as Theorem 4.6.10 follows from Theorem 4.6.7. ∎

We now make the connection between this development and inductively defined sets. Let S be an inductively defined subset of a set U. For each inductive rule R in the definition of S, say, with n hypotheses, we define an n-ary constructor R' on U by letting $R' = \{((x_1, \ldots, x_n), x) \mid x \text{ is}$ a conclusion of rule R with hypotheses $x_1, \ldots, x_n\}$. Let B consist of the elements of U put into S by the basis rules in the definition of S. Then, S is the closure of B under the family of constructors $\{R' \mid R \text{ is an inductive rule}$ in the definition of $S\}$. In this way, we can make the idea of an inductively defined set precise. Suppose that we now apply Theorem 4.6.14. The first condition of the theorem, applied to this situation, says that every element put into S by a basic rule has property P, while the second condition states that for every inductive rule, if all of the hypotheses of the rule are in S and have property P, then the conclusion of the rule also has property P. These are the conditions of the principle of structural induction (Theorem 4.4.1). Thus, Theorem 4.6.14 provides a precise justification for the principle of structural induction.

We now look at the inductively defined sets we have considered and show how they can be obtained as closures of sets under families of constructors. In fact, in most cases, we will use families of operations, which is a special case of a family of constructors, since an n-ary operation on a set is an n-ary constructor on the set.

Example 4.6.16 We show how the set EXP of integer valued expressions, which was defined in Example 4.3.1, can be seen as the closure of a set under a family of operations. Let CONST be the set of integer constants, and let VAR be the set of valid variable names. Let $A = \text{CONST} \cup \text{VAR} \cup \{(,), +, -, *, /\}$, and let $U = \text{Seq}(A)$. Let $\mathcal{F} = \{f_i : 1 \leq i \leq 7\}$, where the f_i are defined by

$$f_1(\alpha) = (\alpha)$$
$$f_2(\alpha) = +\alpha$$
$$f_3(\alpha) = -\alpha$$
$$f_4(\alpha, \beta) = \alpha + \beta$$
$$f_5(\alpha, \beta) = \alpha - \beta$$
$$f_6(\alpha, \beta) = \alpha * \beta$$
$$f_7(\alpha, \beta) = \alpha/\beta$$

Let $B = \text{CONST} \cup \text{VAR}$. Then, EXP is the closure of B under $\mathcal{F}$.

Example 4.6.17 We have already seen in Example 4.6.3 that $\mathbf{N}$ is the closure of $B = \{0\}$ under the function $f(n) = n + 1$, and this shows how Example 4.3.3 can be interpreted in our current approach.

Example 4.6.18 Let $U = \mathbf{N}$, let $B = \{0\}$, and for each $n \in \mathbf{N}$, let R_n be the $n + 1$-ary constructor on U given by $R_n = \{((0, \ldots, n), n + 1)\}$. (Note that R_n is a partial $n + 1$-ary operation on U.) Then, Example 4.3.4 states that $\mathcal{N}$ is the closure of B under the family of partial operations $\{R_n \mid n \in \mathbf{N}\}$.

Example 4.6.19 Let $U = \mathbf{P}$, let $B = \{1, 2\}$, and let R be the constructor on U given by $R = \{((n, n + 1), n + 2) \mid n \in \mathbf{P}\}$. (Once again note that R is, in fact, a partial operation.) Then, Example 4.3.5 states that $\mathbf{P}$ is the closure of B under R.

Example 4.6.20 Let $U = \mathbf{N}$, let $B = \{0\}$, and let $f(n) = n + 2$. Then, Examples 4.3.6 and 4.4.10 show that the closure of B under f is the set of even natural numbers (as was already mentioned in Example 4.6.4).

Example 4.6.21 Let A be a set and let $U = \mathrm{Seq}(A)$. There are two ways to look at the inductive definition of U given in Example 4.3.7. One way is to regard U as being the closure of $\{\lambda\}$ under the single constructor $R = \{(w, wx) \mid w \in U \text{ and } x \in A\}$. Note that this constructor is not a partial function. Another way to regard the definition is to break up R into functions and define, for each $x \in A$, a function $f_x : U \longrightarrow U$ by $f_x(w) = wx$. Then, U is the closure of $\{\lambda\}$ under the family of functions $\mathcal{F} = \{f_x \mid x \in A\}$.

Example 4.6.22 Let A, U, and $\mathcal{F}$ be as in the previous example. Let $B = A$. (More precisely, let $B = \{(x) \mid x \in A\}$.) Then, according to Example 4.3.8, $\mathrm{Seq}^+(A)$ is the closure of B under the family of functions $\mathcal{F}$.

Example 4.6.23 Let $U = \{a, b\}^*$ and let $\mathcal{F} = \{f_1, f_2, f_3\}$, where the f_i's are given by

$$
\begin{aligned}
f_1(w) &= awb \\
f_2(w) &= bwa \\
f_3(w, u) &= wu
\end{aligned}
$$

Then, the set S defined in Example 4.3.9 is the closure of $\{\lambda\}$ under $\mathcal{F}$.

Example 4.6.24 Let A be any set, let $U = A \times A$, and let ρ be a binary relation on A. Define a binary constructor R on U by

$$
R = \{(((x, y), (y, z)), (x, z)) \mid x, y, z \in A\}.
$$

(Note that R is in fact a partial binary operation on U.) Then, according to Example 4.4.14, ρ^+, the transitive closure of ρ is the closure of ρ under the constructor R.

Example 4.6.25 Let A be any set, let $U = A \times A$, and let ρ be a binary relation on A. For each pair $(w, z) \in U$, we define a unary partial operation $f_{(w,z)}$ on U. The domain of $f_{(w,z)}$ is $\{(y, w) \mid y \in A\}$, and for each such (y, w), $f_{(w,z)}((y, w)) = (y, z)$. Then, according to Example 4.4.15, the transitive closure of ρ is the closure of ρ under the family of partial functions $\{f_{(w,z)} \mid (w, z) \in A \times A\}$.

Example 4.6.26 Let A be a set, $U = A \times A$, and let ρ be a binary relation on A. Then, ρ can be considered to be a unary constructor on R. According to Example 4.4.16, for each $u \in A$, $\{y \in A \mid (u, y) \in \rho^+\}$ is the closure of $\{y \mid (u, y) \in \rho\}$ under the constructor ρ.

Now, that we have made precise the ideas of inductively defined sets and structural induction, we must next analyze what the unique readability condition means in terms of the closure of a set under a family of constructors and then prove that recursive function definitions succeed when the condition is met. The precise analog of the unique readability condition is that the closure of a set under a family of constructors be "freely generated" in a sense to be defined shortly. We will once again begin with the simplest possible case and then generalize to an arbitrary family of constructors.

Definition 4.6.27 *Let U be a set, let $f : U \longrightarrow U$, let B be a subset of U, and let C be the closure of B under f. Then, C is said to be* freely *generated by f over B if the following conditions are met:*

1. $f(C) \cap B = \emptyset$ (i.e., for no $x \in C$ is $f(x) \in B$);

2. $f{\restriction}C$ is injective (i.e., if $x, y \in C$ and $x \neq y$, then $f(x) \neq f(y)$).

Example 4.6.28 Let $U = \mathbf{Z}$ and $f(n) = n + 1$. Then, as we saw in Example 4.6.3, for every $n_0 \in \mathbf{Z}$, $C_{n_0} = \{n \in \mathbf{Z} \mid n \geq n_0\}$ is the closure of $\{n_0\}$ under f. In fact, as we now show, C_{n_0} is freely generated by f over $\{n_0\}$. This is because if $n \in C_{n_0}$, then $f(n) = n + 1 > n_0$, and hence, $f(n) \notin \{n_0\}$, and because f is injective. In particular, $C_0 = \mathbf{N}$ is freely generated by f over $\{0\}$.

Note that in the previous example, $f(U)$, the range of f on all of U, is not disjoint from C_{n_0}. All that is required is that the range of f on C_{n_0} be disjoint from $\{n_0\}$, which it is. By the same token, although in this case f is injective on all of U, this is not required by the definition. All that is required is that the restriction of f to the closure be injective.

Example 4.6.29 Let $U = \mathbf{N}$ and $f(n) = n + 2$. Then, as stated in Example 4.6.4, $\mathbf{N}$ is the closure of $B = \{0, 1\}$ under f. In fact, $\mathbf{N}$ is freely generated by f over B because if $n \in \mathbf{N}$, then $f(n) = n + 2 \geq 2$, and hence, $f(n) \notin B$ and also, f is injective. By similar arguments, the even natural numbers are freely generated by f over $\{0\}$, and the odd natural numbers are freely generated by f over $\{1\}$.

Example 4.6.30 Let n be a positive integer, let $U = \{0, \ldots, n-1\}$, and let $f(m) = (m+1) \bmod n$ for each $m \in U$. Then, as stated in Example 4.6.5, for any $m \in U$, U is the closure of $\{m\}$ under f. In this case, U is not freely generated by f over $\{m\}$ because $f(U) = U$, and hence, $f(U)$ is not disjoint from $\{m\}$. (Note, however, that f is injective, so at least the second condition is met.)

We now prove three results, in order of increasing generality, each of which states that a certain type of recursive function definition based on a closure which is freely generated succeeds.

Theorem 4.6.31 Let U be a set, let $f : U \longrightarrow U$, let B be a subset of U, and let C be the closure of B under f. Suppose that C is freely generated by f over B. Then, for any set A and functions $G : B \longrightarrow A$, $H : A \longrightarrow A$, there is a unique function $g : C \longrightarrow A$ satisfying the recursive equations

$$g(x) = G(x) \text{ for all } x \in B,$$
$$g(f(x)) = H(g(x)) \text{ for all } x \in C.$$

Proof: We first show the uniqueness. Suppose that g_1 and g_2 are both solutions of the recursive equations. Let $P(x)$ be the property "$g_1(x) = g_2(x)$." We use structural induction (Theorem 4.6.7) to show that $P(x)$ is true for all $x \in C$. First of all, for $x \in B$, we have $g_1(x) = G(x) = g_2(x)$, so $P(x)$ is true. Now, suppose that $x \in C$ and that $P(x)$ is true. We must show that $P(f(x))$ is true. But, since g_1 and g_2 are both solutions of the recursive equations, we have $g_1(f(x)) = H(g_1(x)) = H(g_2(x)) = g_2(f(x))$, so $P(f(x))$ is true, completing the induction.

Now, we show that g exists. We define g by the inductive definition:

1. for all $x \in B$, $(x, G(x)) \in g$;

2. for all $x \in C, a \in A$, if $(x, a) \in g$, then $(f(x), H(a)) \in g$.

More precisely, define $f' : C \times A \longrightarrow C \times A$ by $f'(x, a) = (f(x), H(a))$. Then, g is the closure of $B' = \{(x, G(x)) \mid x \in B\}$ under f'.

This, of course, only defines g as a set of ordered pairs. We must first show that g is, in fact, a function. To this end, let $P(x)$ be the property "there is exactly one $a \in A$ with $(x, a) \in g$." We use induction to show that $P(x)$ is true for all $x \in C$. The basis step is to show that $P(x)$ is true if $x \in B$. But, if $x \in B$, then $(x, G(x)) \in B' \subseteq g$. Furthermore, if a is any element of A with $(x, a) \in g$, then by Theorem 4.6.10, either $(x, a) \in B'$, in which case $a = G(x)$, or for some $(x', a') \in C \times A$, $(x, a) = f'(x', a') = (f(x'), H(a'))$. But, this latter possibility gives $x = f(x')$ with $x' \in C$, contradicting the free generation of C by f over B. Thus, $G(x)$ is the only $a \in A$ with $(x, a) \in g$, and $P(x)$ is true.

For the inductive step, suppose that $x \in C$ and that $P(x)$ is true. Let a be the unique element of A with $(x, a) \in g$. Then, $f'(x, a) = (f(x), H(a)) \in g$.

Furthermore, if a' is some element of A with $(f(x), a') \in g$, then by Theorem 4.6.10, either $(f(x), a') \in B'$, which is impossible since that would mean $f(x) \in B$, which contradicts free generation, or $(f(x), a') = f'(x'', a'') = (f(x''), H(a''))$ for some $(x'', a'') \in g$. In the latter case we have $f(x) = f(x'')$ with both x and x'' in C. Since by free generation of C by f over B we have f injective on C, this means that $x = x''$, and hence, $(x, a'') = (x'', a'') \in g$, so by uniqueness of a, $a = a''$, so $a' = H(a'') = H(a)$. This shows that $H(a)$ is the unique $a' \in A$ with $(f(x), a') \in g$, so $P(f(x))$ is true.

Now, that we know that g is a function, it is easy to show that it is a solution of the given recursive equations. First of all, for each $x \in B$, we have $(x, G(x)) \in g$, so $g(x) = G(x)$ as desired. Now, if $x \in C$, then $(x, g(x)) \in g$, so $f'(x, g(x)) = (f(x), H(g(x)))$ is also in g, so $g(f(x)) = H(g(x))$, again as desired.∎

The previous theorem can be applied to any set freely generated by a function over a basis set. We are most interested in the application of the theorem to Example 4.6.3. Before we do this, we note that if B has only a single element b, then for any set A, there is a injective correspondence between functions from B to A and elements of A. We make use of this in the following corollary.

Corollary 4.6.32 *Let n_0 be an integer, let A be a set, let $a \in A$, and let $H : A \longrightarrow A$. Then, there is a unique function $g : \{n \in \mathbf{Z} : n \geq n_0\} \longrightarrow A$ that satisfies*

$$g(n_0) = a,$$
$$g(n + 1) = H(g(n)) \text{ for all } n \geq n_0.$$

Proof: This is just Theorem 4.6.31 applied to Example 4.6.28. ∎

Since an infinite sequence is just a function with domain $\mathbf{N}$, we can rephrase this result as a way of defining recursive sequences.

Corollary 4.6.33 *Let A be a set, let $a \in A$, and let $H : A \longrightarrow A$. Then, there is a unique infinite sequence b over A that satisfies*

$$b_0 = a,$$
$$b_{n+1} = H(b_n) \text{ for all } n \in \mathbf{P}.$$

We now show how some of the recursive function definitions we gave in Section 4.5 can be put into the form given in Corollary 4.6.32.

Example 4.6.34 In Example 4.5.16, we defined for each $a \in \mathbf{N}$, a function $\mathrm{add}_a : \mathbf{N} \longrightarrow \mathbf{N}$ by

$$\mathrm{add}_a(0) = a,$$
$$\mathrm{add}_a(n + 1) = S(\mathrm{add}_a(n)).$$

This definition is of the form given by Corollary 4.6.32 if we take $A = \mathbf{N}$, $n_0 = 0$, and $H = S$.

Example 4.6.35 In Example 4.5.17, we defined for each element a of $\mathbf{N}$ a function $\text{mult}_a : \mathbf{N} \longrightarrow \mathbf{N}$ by

$$\text{mult}_a(0) = 0,$$
$$\text{mult}_a(n+1) = \text{mult}_a(n) + a.$$

This definition has the form of Corollary 4.6.32 if we take $A = \mathbf{N}$, $n_0 = 0$, $a = 0$, and $H(x) = x + a$.

Example 4.6.36 In Example 4.5.19, we had a set D, a binary operation $*$ on D, and an identity element e for $*$. For each $d \in D$, we defined $\exp_d : \mathbf{N} \longrightarrow D$ by

$$\exp_d(0) = e,$$
$$\exp_d(n+1) = \exp_d(n) * d.$$

This definition has the form of Corollary 4.6.32 if we take $A = D$, $n_0 = 0$, $a = e$, and $H(x) = x * d$.

There are some fairly simple recursive definitions over the natural numbers that cannot be put into the form of Corollary 4.6.32. For example, the definition of the function fac in Example 4.5.1 cannot be put into this form. The problem is that the value of $\text{fac}(n+1)$ depends on n as well as $\text{fac}(n)$. The next theorem is a slight modification of the last one, which allows for this.

Theorem 4.6.37 *Let U be a set, let $f : U \longrightarrow U$, let B be a subset of U, and let C be the closure of B under f. Suppose that C is freely generated by f over B. Then, for any set A and functions $G : B \longrightarrow A$, $H : C \times A \longrightarrow A$, there is a unique function $g : C \longrightarrow A$ satisfying the recursive equations*

$$g(x) = G(x) \text{ for all } x \in B,$$
$$g(f(x)) = H(x, g(x)) \text{ for all } x \in C.$$

Proof: The uniqueness is as in Theorem 4.6.31. To see that g exists, define g by the inductive definition:

1. for all $x \in B$, $(x, G(x)) \in g$;

2. for all $x \in C, a \in A$, if $(x, a) \in g$, then $(f(x), H(x, a)) \in g$.

More precisely, define $f' : C \times A \longrightarrow C \times A$ by $f'(x, a) = (f(x), H(x, a))$. Then, g is the closure of $B' = \{(x, G(x)) \mid x \in B\}$ under f'. The fact that g is a function and that g satisfies the given recursive equations is as in Theorem 4.6.31. ∎

It is easy to show that our first version of recursive function definitions (Theorem 4.6.31) can be derived from the version just given. (You are asked to do this in Exercise 44.) What is less obvious is that the second version can be derived from the first, and hence, is no more powerful. (See Exercise 45.) Applying this theorem to Example 4.6.28, we get the following.

Corollary 4.6.38 *Let n_0 be an integer, let A be a set, let $a \in A$, and let $H : \{n \in \mathbf{Z} \mid n \geq n_0\} \times A \longrightarrow A$. Then, there is a unique function $g : \{n \in \mathbf{Z} : n \geq n_0\} \longrightarrow A$ that satisfies*

$$g(n_0) = a,$$
$$g(n+1) = H(n, g(n)) \text{ for all } n \geq n_0.$$

Expressing this in terms of sequences, we have the following.

Corollary 4.6.39 *Let A be a set, let $a \in A$, and let $H : \mathbf{N} \times A \longrightarrow A$. Then, there is a unique infinite sequence b over A that satisfies*

$$b_0 = a,$$
$$b_{n+1} = H(n, b_n) \text{ for all } n \in \mathbf{P}.$$

Example 4.6.40 In Example 4.5.1, we defined a function fac : $\mathbf{N} \longrightarrow \mathbf{N}$. Rephrasing this definition, we have

$$\text{fac}(0) = 1,$$
$$\text{fac}(n+1) = (n+1)\text{fac}(n) \text{ for all } n \in \mathbf{N}.$$

This definition has the form of Corollary 4.6.38 if we take $A = \mathbf{N}$, $n_0 = 0$, $a = 1$, and $H(x, y) = (x+1)y$.

We often want to define a function of several arguments using recursion on one of the arguments and having the other arguments be parameters not taking part in the recursion. The following version of recursive function definition allows that.

Theorem 4.6.41 *Let U be a set, let $f : U \longrightarrow U$, let B be a subset of U, and let C be the closure of B under f. Suppose that C is freely generated by f over B. Then, for any sets P and A and functions $G : P \times B \longrightarrow A$, $H : P \times C \times A \longrightarrow A$, there is a unique function $g : P \times C \longrightarrow A$ satisfying the recursive equations*

$$g(p, x) = G(p, x) \text{ for all } p \in P, \ x \in B,$$
$$g(p, f(x)) = H(p, x, g(p, x)) \text{ for all } x \in C.$$

Proof: To see uniqueness, suppose that g_1 and g_2 are both solutions of the recursive equations. Let $P'(x)$ be the property "for all $p \in P$, $g_1(p, x) = g_2(p, x)$." It is not hard to show by induction that $P'(x)$ is true for all $x \in C$.

To see that g exists, define g by the inductive definition:

1. for all $p \in P$ and $x \in B$, $((p, x), G(p, x)) \in g$;

2. for all $p \in P, x \in C, a \in A$, if $((p, x), a) \in g$, then we have

$$((p, f(x)), H(p, x, a)) \in g.$$

More precisely, define $f' : (P \times C) \times A \longrightarrow (P \times C) \times A$ by $f'((p,x),a) = ((p, f(x)), H(p,x,a))$. Then, g is the closure of $B' = \{((p,x), G(p,x)) \mid p \in P \text{ and } x \in B\}$ under f'.

Let $P'(x)$ be the property "for all $p \in P$, there is a unique $a \in A$ with $((p,x),a) \in g$." Then, much as in the Theorems 4.6.31 and 4.6.37, it is possible to show that $P'(x)$ is true for all $x \in C$ and, hence, that g is a function. The fact that g satisfies the recursive equations now follows easily. ∎

In Exercise 46, you are asked to show that our second version of recursive function definition can be derived from the one just given. It is also possible to derive the version just given from the second version in at least two different ways. (See Exercises 47 and 48.)

If we apply the preceding theorem to Example 4.6.28 we get the following.

Corollary 4.6.42 *Let n_0 be an integer, let P and A be sets, let $G : P \longrightarrow A$, and let $H : P \times \{n \in \mathbf{Z} \mid n \geq n_0\} \times A \longrightarrow A$. Then, there is a unique function $g : P \times \{n \in \mathbf{Z} : n \geq n_0\} \longrightarrow A$ that satisfies*

$$g(p, n_0) = G(p) \text{ for all } p \in P,$$
$$g(p, n+1) = H(p, n, g(p, n)) \text{ for all } n \geq n_0.$$

Note that in the previous corollary, since B has only one element, we replace a function $G : P \times B \longrightarrow A$ with a function $G : P \longrightarrow A$.

In many applications of Theorem 4.6.41, the parameter set P is C, so that we define a function $g : C \times C \longrightarrow A$. More generally, we can let $P = C^n$ and get a function $g : C^n \times C \longrightarrow A$. Because there is an obvious bijection between $C^n \times C$ and C^{n+1}, we can identify the two sets and consider the function g obtained to be a function from C^{n+1} to A.

With Theorem 4.6.41, we can define directly functions that we had to approach indirectly before, as we show in the next examples.

Example 4.6.43 Given the successor function S on the natural numbers, we define the addition function add by

$$\mathrm{add}(a, 0) = a \text{ for all natural numbers } a,$$
$$\mathrm{add}(a, n+1) = S(\mathrm{add}(a, n)) \text{ for all natural numbers } a, n.$$

This definition has the form of Corollary 4.6.42 if we take $n_0 = 0$, $P = A = \mathbf{N}$, $G(a) = a$, and $H(x, y, z) = S(z)$.

Example 4.6.44 We define the multiplication function mult on the natural numbers by

$$\mathrm{mult}(a, 0) = 0 \text{ for all } a \in \mathbf{N},$$
$$\mathrm{mult}(a, n+1) = \mathrm{mult}(a, n) + a \text{ for all } a, n \in \mathbf{N}.$$

This definition has the form of Corollary 4.6.42 if we take $n_0 = 0$, $P = A = \mathbf{N}$, $G(a) = 0$, and $H(x, y, z) = z + x$.

Example 4.6.45 Let D be a set, $*$ a binary operation on D, and e an identity element for $*$. Then, we define a function $\exp : D \times \mathbf{N} \longrightarrow D$ by

$$\exp(d, 0) = e \text{ for all } d \in D,$$
$$\exp(d, n + 1) = \exp(d, n) * d \text{ for all } d \in D, n \in \mathbf{N}.$$

We can put this definition into the form of Corollary 4.6.42 if we take $n_0 = 0$, $P = A = D$, $G(d) = e$, and $H(x, y, z) = z * x$.

So far, all of our theorems have dealt with defining total functions recursively. Sometimes we wish to define a partial function recursively. We will prove a version of Theorem 4.6.41 that can be used in such a situation.

Theorem 4.6.46 *Let U be a set, let $f : U \longrightarrow U$, let B be a subset of U, and let C be the closure of B under f. Suppose that C is freely generated by f over B. Then, for any sets P and A and partial functions $G : P \times B \rightsquigarrow A$, $H : P \times C \times A \rightsquigarrow A$, there is a unique partial function $g : P \times C \rightsquigarrow A$ satisfying the recursive equations*

$$g(p, x) = G(p, x) \text{ for all } p \in P, \, x \in B,$$
$$g(p, f(x)) = H(p, x, g(p, x)) \text{ for all } p \in P, x \in C,$$

where we interpret these equations to mean that

1. *for all $p \in P$ and $x \in B$, $g(p, x)$ is defined if and only if $G(p, x)$ is defined and if $g(p, x)$ is defined, then the given equality holds, and*

2. *for all $p \in P, x \in C$, $g(p, f(x))$ is defined if and only if $g(p, x)$ and $H(p, x, g(p, x))$ are both defined, and if $g(p, f(x))$ is defined, then the given equality holds.*

Proof: To see uniqueness, suppose that g_1, g_2 are both solutions of the recursive equations. Let $P'(x)$ be the property "for all $p \in P$, $g_1(p, x)$ is defined if and only if $g_2(p, x)$ is defined and if they are defined they are equal." It is not hard to show by structural induction that $P'(x)$ is true for all $x \in C$. To see the existence of g, choose ω with $\omega \notin A$ and define $G_\omega : P \times B \longrightarrow A_\omega$ by

$$G_\omega(p, b) = \begin{cases} G(p, b) & \text{if } G(p, b) \text{ is defined} \\ \omega & \text{otherwise.} \end{cases}$$

Similarly, define $H_\omega : P \times A_\omega \times C \longrightarrow A$ by

$$H_\omega(p, c, z) = \begin{cases} H(p, c, z) & \text{if } z \in A \text{ and } H(p, c, z) \text{ is defined} \\ \omega & \text{otherwise.} \end{cases}$$

Then, Theorem 4.6.41 gives us a function $g_\omega : P \times C \longrightarrow A_\omega$ satisfying

$$g_\omega(p, x) = G_\omega(p, x) \text{ for all } p \in P, \, x \in B,$$
$$g_\omega(p, f(x)) = H_\omega(p, x, g(p, x)) \text{ for all } p \in P, x \in C.$$

Define $g : P \times C \rightsquigarrow A$ by letting the domain of g be all (p, x) for which $g_\omega(p, x) \neq \omega$ and defining $g(p, x) = g_\omega(p, x)$ for these (p, x). Then, it is not hard to see that g is the desired function. ∎

The technique used in the previous theorem of converting a partial function to a total one by adding an element to the codomain can be used in any of the theorems we will prove about recursively defined total functions to get a corresponding theorem about recursively defined partial functions. Theorem 4.6.41 is not quite a special case of Theorem 4.6.46, because in Theorem 4.6.41 the function g produced is total while Theorem 4.6.41 only gives us a partial g even if G and H are total.

We now wish to generalize this discussion of recursively defined functions to the situation of a closure under an arbitrary family of constructors. Although the notation gets cumbersome, most of the ideas have already been introduced in the simplified setting we have discussed so far. Before we begin, let us agree that if R is a constructor on some set, then n_R will be the natural number n such that R is an n-ary constructor.

Definition 4.6.47 *Let U be a set, let $\mathcal{R}$ be a family of constructors on U, let B be a subset of U and let C be the closure of B under $\mathcal{R}$. Then, C is said to be freely generated by $\mathcal{R}$ over B if the following conditions are met.*

1. *For each $R \in \mathcal{R}$, $R(C^{n_R}) \cap B = \emptyset$.*

2. *For each $R, R' \in \mathcal{R}$ with $R \neq R'$, $R(C^{n_R}) \cap R'(C^{n_{R'}}) = \emptyset$.*

3. *For each $R \in \mathcal{R}$, R is injective on C^{n_R} (i.e., if $a, a' \in C^{n_R}$, $x \in C$, and $aRx, a'Rx$, then $a = a'$).*

The idea behind this definition is that each element of C gets into C in exactly one way.

Theorem 4.6.48 *Let U be a set, let $\mathcal{R}$ be a family of constructors on U, let $\{B_j \mid j \in J\}$ be a family of pairwise disjoint subsets of U, let $B = \bigcup\{B_j \mid j \in J\}$, and let C be the closure of B under $\mathcal{R}$. Suppose that C is freely generated by $\mathcal{R}$ over B. Suppose also that A is a set, for each $j \in J$, $G_j : B_j \longrightarrow A$, and for each $R \in \mathcal{R}$, $H_R : A^n \longrightarrow A$, where $n = n_R$. Then, there is a unique function $g : C \longrightarrow A$ satisfying*

$$
\begin{aligned}
g(x) &= G_j(x) && \text{for all } j \in J, x \in B_j, \\
g(x) &= H_R(g(x_1), \dots, g(x_n)) && \text{for all } x_1, \dots, x_n, x \in C \\
& && \text{with } ((x_1, \dots, x_n), x) \in R.
\end{aligned}
$$

Proof: The proof is similar to that of Theorem 4.6.31; only the notation is more complicated. First, we show uniqueness. Suppose that g_1 and g_2 are both solutions of the recursive equations. Let $P(x)$ be the property that $g_1(x) = g_2(x)$. We use induction on x (Theorem 4.6.14) to show that $P(x)$ is true for all $x \in C$. For $x \in B$, there must be a $j \in J$ with $x \in B_j$, and we have $g_1(x) = G_j(x) = g_2(x)$. Now, suppose that $R \in \mathcal{R}$ is n-ary, $x_1, \dots, x_n$

and x are in C, $((x_1, \ldots, x_n), x) \in R$, and $P(x_i)$ is true for all i, $1 \le i \le n$. Then, $g_1(x) = H_R(g_1(x_1), \ldots, g_1(x_n)) = H_R(g_2(x_1), \ldots, g_2(x_n)) = g_2(x)$, so $P(x)$ is true, completing the induction.

To see that g exists, define g (as a set of ordered pairs):

1. for all $j \in J$, $x \in B_j$, $(x, G_j(x)) \in g$;

2. for every $R \in \mathcal{R}$ (say, R is n-ary), $x_1, \ldots, x_n, x \in C$, with $((x_1, \ldots, x_n), x) \in R$ and $a_1, \ldots, a_n \in A$, if $(x_1, a_1), \ldots, (x_n, a_n)$ are all in g, then so is $(x, H_R(a_1, \ldots, a_n))$.

More precisely, for each $R \in \mathcal{R}$ with R n-ary, define an n-ary constructor R' on $C \times A$ by $R' = \{(((x_1, a_1), \ldots, (x_n, a_n)), (x, H_R(a_1, \ldots, a_n))) \mid x_1, \ldots, x_n, x \in C, a_1, \ldots, a_n \in A$, and $((x_1, \ldots, x_n), x) \in R\}$. Let $\mathcal{R}'$ be $\{R' \mid R \in \mathcal{R}\}$. Then, g is the closure of $B' = \{(x, G_j(x)) \mid j \in J, x \in B_j\}$ under $\mathcal{R}'$.

We first show that g is a function. Let $P(x)$ be the property given by "there is a unique $a \in A$ with $(x, a) \in g$." We use induction to prove that $P(x)$ is true for every $x \in C$, and hence, that g is a function. If $x \in B$, then, for some $j_0 \in J$, $x \in B_{j_0}$, and hence, $(x, G_{j_0}(x)) \in B' \subseteq g$. Now, suppose that a is any element of A with $(x, a) \in g$. If $(x, a) \in B'$, then, since the B_j's are pairwise disjoint, we must have $a = G_{j_0}(x)$. The other possibility is that, for some n-ary $R' \in \mathcal{R}'$, there is an n-tuple $((x_1, a_1), \ldots, (x_n, a_n))$ of elements of $C \times A$ with $(((x_1, a_1), \ldots, (x_n, a_n)), (x, a)) \in R'$. But, this means that $((x_1, \ldots, x_n), x) \in R$ and contradicts the free generation of C. Thus, $P(x)$ is true.

Now, suppose that $R \in \mathcal{R}$, $x_1, \ldots, x_n, x \in C$, $((x_1, \ldots, x_n), x) \in R$, and $P(x_i)$ is true for all i with $1 \le i \le n$. Then, for $1 \le i \le n$, let a_i be the unique element of A with $(x_i, a_i) \in g$. Then, since g is closed under R', $(x, H_R(a_1, \ldots, a_n)) \in g$.

Suppose that $(x, a) \in g$. We cannot have $(x, a) \in B'$, since this would imply $x \in B$ and thereby contradict the free generation of C. Thus, we must have some m-ary $R_1' \in \mathcal{R}'$ and an m-tuple $((y_1, a_1'), \ldots, (y_m, a_m'))$ of elements of g with $(((y_1, a_1'), \ldots, (y_m, a_m')), (x, a)) \in R_1'$. This means that $((y_1, \ldots, y_m), x) \in R_1$ and $a = H_{R_1}(a_1', \ldots, a_m')$. By free generation, we must have $R_1 = R$ (so $m = n$), and for all i with $1 \le i \le n$, $y_i = x_i$. Since $P(x_i)$ is true for all i, $1 \le i \le n$, we must have $a_i' = a_i$ for all such i. Thus, $a = H_R(a_1, \ldots, a_n)$. This shows that $P(x)$ is true and completes the induction, so g is a function.

Now, if $x \in B_j$, then $(x, G_j(x)) \in g$, so $g(x) = G_j(x)$. If $R \in \mathcal{R}$ is n-ary, $x_1, \ldots, x_n, x \in C$, and $((x_1, \ldots, x_n), x) \in R$, then $(x_1, g(x_1)), \ldots, (x_n, g(x_n))$ are all in g, so by definition of g, $(x, H_R(g(x_1), \ldots, g(x_n))) \in g$, so $g(x) = H_R(g(x_1), \ldots, g(x_n))$. ∎

Since we often use a family of constructors, which is, in fact, a family of partial operations, we restate Theorem 4.6.48 for that situation.

Corollary 4.6.49 *Let U be a set, let $\mathcal{F}$ be a family of partial operations on U, let $\{B_j \mid j \in J\}$ be a family of pairwise disjoint subsets of U, let $B = \bigcup\{B_j \mid j \in J\}$, and let C be the closure of B under $\mathcal{F}$. Suppose that C is freely generated by $\mathcal{F}$ over B. Suppose also that A is a set, for each $j \in J$, $G_j : B_j \longrightarrow A$, and for each $f \in \mathcal{F}$, $H_f : A^n \longrightarrow A$, where $n = n_f$. Then, there is a unique function $g : C \longrightarrow A$ satisfying*

$$
\begin{aligned}
g(x) &= G_j(x) \text{ for all } j \in J, x \in B_j, \\
g(f(x_1, \ldots, x_n)) &= H_f(g(x_1), \ldots, g(x_n)) \text{ for all } n\text{-ary } f \text{ in } \mathcal{F} \\
&\quad \text{and } (x_1, \ldots, x_n) \in C^n \cap dom(f).
\end{aligned}
$$

Theorem 4.6.50 *Let U be a set, let $\mathcal{R}$ be a family of constructors on U, let $\{B_j \mid j \in J\}$ be a family of pairwise disjoint subsets of U, let $B = \bigcup\{B_j \mid j \in J\}$, and let C be the closure of B under $\mathcal{R}$. Suppose that C is freely generated by $\mathcal{R}$ over B. Suppose also that A is a set, for each $j \in J$, $G_j : B_j \longrightarrow A$, and for each $R \in \mathcal{R}$, $H_R : A^n \times C^n \longrightarrow A$, where $n = n_R$. Then, there is a unique function $g : C \longrightarrow A$ satisfying*

$$
\begin{aligned}
g(x) &= G_j(x) \text{ for all } j \in J, x \in B_j, \\
g(x) &= H_R(g(x_1), \ldots, g(x_n), x_1, \ldots, x_n) \text{ for all } x_1, \ldots, x_n, x \in C, \\
&\quad \text{with } ((x_1, \ldots, x_n), x) \in R.
\end{aligned}
$$

Proof: The proof is obtained from the proof of Theorem 4.6.48, just as the proof of Theorem 4.6.37 was obtained from that of Theorem 4.6.31. ∎

Corollary 4.6.51 *Let U be a set, let $\mathcal{F}$ be a family of partial operations on U, let $\{B_j \mid j \in J\}$ be a family of pairwise disjoint subsets of U, let $B = \bigcup\{B_j \mid j \in J\}$, and let C be the closure of B under $\mathcal{F}$. Suppose that C is freely generated by $\mathcal{F}$ over B. Suppose also that A is a set, for each $j \in J$, $G_j : B_j \longrightarrow A$, and for each $f \in \mathcal{F}$, $H_f : A^n \times C^n \longrightarrow A$, where $n = n_f$. Then, there is a unique function $g : C \longrightarrow A$ satisfying the following conditions:*

1. for all $j \in J, x \in B_j$;

$$
g(x) = G_j(x).
$$

2. for all n-ary f in $\mathcal{F}$ and $(x_1, \ldots, x_n) \in C^n \cap dom(f)$,

$$
g(f(x_1, \ldots, x_n)) = H_f(g(x_1), \ldots, g(x_n), x_1, \ldots, x_n).
$$

Theorem 4.6.52 *Let U be a set, let $\mathcal{R}$ be a family of constructors on U, let $\{B_j \mid j \in J\}$ be a family of pairwise disjoint subsets of U, let $B = \bigcup\{B_j \mid j \in J\}$, and let C be the closure of B under $\mathcal{R}$. Suppose that C is freely generated by $\mathcal{R}$ over B. Suppose also that P and A are sets, for each $j \in J$, $G_j : P \times B_j \longrightarrow A$, and for each $R \in \mathcal{R}$, $H_R : P \times A^n \times C^n \longrightarrow A$ where $n = n_R$. Then, there is a unique function $g : C \longrightarrow A$ satisfying the following conditions:*

1. *for all* $j \in J, x \in B_j$, *and* $p \in P$,

$$g(p, x) = G_j(p, x);$$

2. *for all* $x_1, \ldots, x_n, x \in C$, *with* $((x_1, \ldots, x_n), x) \in R$, *and* $p \in P$,

$$g(p, x) = H_R(p, g(x_1), \ldots, g(x_n), x_1, \ldots, x_n).$$

Proof: The proof is obtained from the proof of Theorem 4.6.48, just as the proof of Theorem 4.6.41 was obtained from that of Theorem 4.6.31. ∎

Corollary 4.6.53 *Let U be a set, let $\mathcal{F}$ be a family of partial operations on U, let $\{B_j \mid j \in J\}$ be a family of pairwise disjoint subsets of U, let $B = \bigcup\{B_j \mid j \in J\}$, and let C be the closure of B under $\mathcal{F}$. Suppose that C is freely generated by $\mathcal{F}$ over B. Suppose also that P and A are sets, for each $j \in J$, $G_j : P \times B_j \longrightarrow A$, and for each $f \in \mathcal{F}$, $H_f : P \times A^n \times C^n \longrightarrow A$, where $n = n_f$. Then, there is a unique function $g : P \times C \longrightarrow A$ satisfying the following conditions:*

1. *for all* $j \in J, x \in B_j$, *and* $p \in P$,

$$g(p, x) = G_j(p, x);$$

2. *for all* n-*ary* f *in* $\mathcal{F}$, $p \in P$ *and* $(x_1, \ldots, x_n) \in C^n \cap dom(f)$,

$$g(p, f(x_1, \ldots, x_n)) = H_f(p, g(x_1), \ldots, g(x_n), x_1, \ldots, x_n).$$

Let S be an inductively defined subset of U. If we obtain S as the closure of B under a family of constructors $\mathcal{R}$, where B consists of the elements put into S by basis rules and $\mathcal{R}$ is obtained from the inductive rules in the definition of S in the way described before, and if we further divide B as $\bigcup\{B_j \mid j \in J\}$, where each B_j consists of the elements put into B by one particular basis rule, then the verification that the B_j's are pairwise disjoint and that S is freely generated by $\mathcal{R}$ over B is exactly the verification that the inductive definition of S satisfies the unique readability condition of Definition 4.5.5. Thus, the sequence of theorems culminating in Theorem 4.6.52 may be viewed as stating and proving precisely Theorem 4.5.6.

4.7 Simultaneous Inductive Definitions

In this section, we discuss simultaneous inductive definitions of several sets. We first discuss the subject from an informal viewpoint and then give the formal mathematical apparatus involved.

We begin with some examples.

Example 4.7.1 We define two sets of natural numbers D and E.

1. 0 is in E.

2. If n is in E, then $n + 1$ is in D.

3. If n is in D, then $n + 1$ is in E.

Note how this definition consists of rules, just as in an inductive definition of a single set. What is different here is that the rules that put an element into one set involve hypotheses from another set. One shows that an element belongs to one of the sets being constructed by using the rules repeatedly. For instance, to prove that $4 \in E$ we use the following argument.

	Claim	Reason
1.	$0 \in E$	by Rule (1)
2.	$1 \in D$	by Rule (2) and Line (1)
3.	$2 \in E$	by Rule (3) and Line (2)
4.	$3 \in D$	by Rule (2) and Line (3)
5.	$4 \in E$	by Rule (3) and Line (4)

Example 4.7.2 Let $T = \{a, b\}$. We define three subsets S, A, and B of T^* by the following inductive definition.

1. (a) λ is in S.

    (b) If u is in B, then au is in S.

    (c) If u is in A, then bu is in S.

2. (a) If u is in S, then au is in A.

    (b) If u and v are in A, then buv is in A.

3. (a) If u is in S, then bu is in B.

    (b) If u and v are in B, then auv is in B.

We can use this definition to show that $bbabaa \in S$.

	Claim	Reason
1.	$\lambda \in S$	by Rule (1a)
2.	$a \in A$	by Rule (2a) and Line (1)
3.	$baa \in A$	by Rule (2b) and Line (2)
4.	$babaa \in A$	by Rule (2b) and Lines (2) and (3)
5.	$bbabaa \in S$	by Rule (1b) and Line (4)

With these two examples as motivation, we now give a description of and terminology for simultaneous inductive definitions. Although we label this as a definition, it is not mathematically precise. Later in this section, we will see a more rigorous treatment.

Definition 4.7.3 *A simultaneous inductive definition of a family of sets is a definition that consists of a collection of* rules. *The rules are of two types.* Basis rules *are ones that state unconditionally that certain elements are in one of the sets being constructed.* Inductive rules *are ones that state that an element (called the* conclusion *of the rule) is in one of the sets being constructed if certain other elements (called the* hypotheses *of the rule) are in certain of the sets being defined. The elements of the sets being defined are those objects that can be shown to be in the sets using a finite number of applications of the rules.*

In the definition of Example 4.7.1, Rule (1) is a basis rule and the other two rules are inductive rules with one hypothesis. In the definition in Example 4.7.2, Rule (1a) is a basis rule and all of the other rules are inductive rules.

Just as for a single set defined inductively, there is a principle of structural induction for several sets defined by simultaneous induction.

Theorem 4.7.4 *Let S be a collection of sets defined by a simultaneous inductive definition, and for each $S \in S$, let P_S be a property of the elements of S. Suppose that*

1. *for every basis rule in the definition of S, if x is put into S by the rule, then $P_S(x)$ is true;*

2. *for every inductive rule in the definition of S, if each of the hypotheses of the rule has the property appropriate for the set of which it is a member, then the conclusion has the property appropriate for the set into which it is put.*

Then, for all $S \in S$ and $x \in S$, $P_S(x)$ is true.

We defer the proof of this principle until later in the section, where we give a more formal treatment.

Sometimes, given a family S of sets defined by simultaneous induction, one wishes to show that for some particular $S \in S$, every element of S satisfies some property, but does not wish to show anything about the elements of the other sets in S. Such a situation can be accomodated in the framework of the preceding theorem by letting $P_{S'}$ be the property that holds for all elements of S', for each $S' \in S$ with $S' \neq S$. In the proof, the steps involving rules that put elements into sets other than S become trivial.

Applying the structural induction principle to the two examples we gave earlier, we get the following corollaries.

Corollary 4.7.5 *Let D and E be the two sets defined in Example 4.7.1 and let P_D and P_E be properties of the elements of D and E, respectively. Suppose that*

1. $P_E(0)$ is true;

2. if $n \in E$ and $P_E(n)$ is true, then $P_D(n+1)$ is true;

3. if $n \in D$ and $P_D(n)$ is true, then $P_E(n+1)$ is true.

Then, for all $n \in E$, $P_E(n)$ is true, and for all $n \in D$, $P_D(n)$ is true.

Corollary 4.7.6 Let S, A, B be the sets constructed in Example 4.7.2 and let P_S, P_A, and P_B be properties of the elements of S, A, and B, respectively. Suppose that the following hold.

1. (a) $P_S(\lambda)$ is true.

   (b) If $u \in B$ and $P_B(u)$ is true, then $P_S(au)$ is true.

   (c) If $u \in A$ and $P_A(u)$ is true, then $P_S(bu)$ is true.

2. (a) If $u \in S$ and $P_S(u)$ is true, then $P_A(au)$ is true.

   (b) If $u, v \in A$ and $P_A(u)$ and $P_A(v)$ are both true, then $P_A(buv)$ is true.

3. (a) If $u \in S$ and $P_S(u)$ is true, then $P_B(bu)$ is true.

   (b) If $u, v \in B$ and $P_B(u)$ and $P_B(v)$ are both true, then $P_B(auv)$ is true.

Then, for all $u \in S$, $P_S(u)$ is true; for all $u \in A$, $P_A(u)$ is true; and for all $u in B$, $P_B(u)$ is true.

We next give examples illustrating the use of structural induction for simultaneous inductive definitions.

Example 4.7.7 Let D and E be the sets of natural numbers defined in Example 4.7.1. We use Corollary 4.7.5 to show that every element of E is even and every element of D is odd; that is, we let P_E be the property of being even and P_D be the property of being odd. The basis step is to show that $P_E(0)$ is true, i.e., to show that 0 is even. The first inductive step is to show that if $P_E(n)$ is true, then $P_D(n+1)$ is true, i.e., to show that if n is even, then $n+1$ is odd. The second inductive step amounts to showing that if n is odd, then $n+1$ is even. All of these steps are, of course, obvious.

In the converse direction, if we let $P(n)$ be the property given by "if n is even, then $n \in E$ and if n is odd, then $n \in D$," then it is easy to use ordinary induction to show that $P(n)$ is true for all natural numbers n. It follows from this and the result just shown that E is equal to the set of even natural numbers and D is equal to the set of odd natural numbers.

Example 4.7.8 Let S, A, and B be the sets defined in Example 4.7.2. We will show that S consists of all strings in $\{a, b\}^*$ with the same number of a's as b's, A consists of all strings of a's and b's with one more a than b, and B consists of all strings of a's and b's with one more b than a.

First, we let P_S be the property of having the same number of a's as b's, P_A be the property of having one more a than b, and P_B be the property of having one more b than a, and use Corollary 4.7.6 to show that $P_S(u)$ is true for every $u \in S$, $P_A(u)$ is true for every $u \in A$, and $P_B(u)$ is true for every $u \in B$. There are seven steps in the proof, but all are routine. For example, the first three steps amount to showing that λ has the same number of a's as b's; that if u has one more b than a, then au has the same number of a's as b's; and if u has one more a than b, then bu has the same number of a's as b's.

To show the converse, let $P(n)$ be the property given by "if $u \in \{a,b\}^*$ as length n, then if u has the same number of a's as b's, then $u \in S$; if u has one more a than b, then $u \in A$; and if u has one more b than a, then $u \in B$." We show that $P(n)$ is true for all natural numbers n by course-of-values induction. Suppose that n is a natural number and $P(n')$ is true for all natural numbers $n' < n$. Let u be a string of a's and b's of length n. Suppose first that u has the same number of a's as b's. If $u = \lambda$, then $u \in S$ by Rule (1a). Otherwise, u either starts with an a or a b. If $u = au'$, then u' has one more b than a, and hence, by the induction hypothesis, $u' \in B$, so by Rule (1b), $u = au' \in S$. A similar proof holds, using Rule (1c), if u starts with a b.

Now, suppose that u has one more a than b. If u starts with an a, say, $u = au'$, then u' has the same number of a's as b's, so by the induction hypothesis, $u' \in S$, and hence, by Rule (2a), $u \in A$. If $u = bu'$, then u' has two more a's than b's. We may write $u' = v'w'$, where v' and w' each have one more a than b. By the induction hypothesis, both v' and w' are in A, so $u = au' = av'w' \in A$ by Rule (2b).

A similar argument holds if u has one more b than a.

Next, we turn to recursive function definitions over a simultaneous inductive definition of a family of sets. It is quite easy to formulate the analog of the unique readability condition for simultaneous inductive definitions.

Definition 4.7.9 *A simultaneous inductive definition for a family S of sets is said to meet the* unique readability condition *if for every $s \in S$ and every $x \in S$ the following hold.*

1. *There is only one rule that puts x into S.*

2. *If the rule putting x into S is an inductive rule, then there is only one sequence of hypotheses that can be used with the rule to put x into S.*

Example 4.7.10 The definition of the sets E and D in Example 4.7.1 meets the unique readability condition. Indeed, 0 is put into E only by the first rule and a positive even number n is put into E only by the second rule with hypothesis $n - 1$. A positive odd number n is put into D only by the third rule with hypothesis $n - 1$.

Example 4.7.11 The inductive definition of S, A, and B given in Example 4.7.2 does meet the unique readability condition. Each element of each of these sets can be put into the set by only one rule, but the second part of the unique readability condition fails. For example, *babaa* is put into A by Rule (2b), applied to the hypotheses a and *baa* as well as the hypotheses *aba* and a.

In many cases, one starts with an inductive definition of a single set that does not meet the unique readability property and then, in order to obtain unique readability, gives a simultaneous inductive definition of several sets, including the original set, which does meet the unique readability condition. We look at examples of this now.

Example 4.7.12 In Example 4.5.7, we showed that the set EXP of integer valued expressions introduced in Example 4.3.1 does not meet the unique readability condition, which explains why the definition of the evaluation function EVAL given in Example 4.5.3 does not succeed. In this example, we use the simultaneous inductive definition to define the three sets EX, TERM, and FACT (whose elements are called expressions, terms, and factors, respectively) by a definition that meets the unique readability condition, and we show that EX is the same as EXP. Later, we will see how this allows us to successfully define the evaluation function.

1. (a) If E is an expression and T is a term, then $E+T$ is an expression.

    (b) If E is an expression and T is a term, then $E-T$ is an expression.

    (c) If T is a term, then T is an expression.

2. (a) If T is a term and F is a factor, then $T * F$ is a term.

    (b) If T is a term and F is a factor, then T/F is a term.

    (c) If F is a factor, then F is a term.

3. (a) Every variable name is a factor.

    (b) Every integer constant is a factor.

    (c) If E is an expression, then (E) is a factor.

    (d) If F is a factor, then $+F$ is a factor.

    (e) If F is a factor, then $-F$ is a factor.

We must first show that the set EX of expressions as defined by this definition is the same as the set EXP defined earlier. To show that EX is a subset of EXP, let P_{EX}, P_{TERM}, and P_{FACT} be the properties consisting of those expressions, terms, and factors, respectively, that are in EXP. It is then easy to use Theorem 4.7.4 and the rules in the definition of EXP to show that $P_{EX}(E)$ is true for every expression E, $P_{TERM}(T)$ is true for every term T, and $P_{FACT}(F)$ is true for every factor F, which shows that every expression in EX is also in EXP.

The converse is somewhat more difficult. Let P be the property of the elements of EXP consisting of those elements that are in EX. We will use induction on EXP to show that $P(E)$ is true for each E in EXP. The first basis step is to show that $P(c)$ is true if c is an integer constant. This follows from Rules (3b), (2c), and (1c) in the definition of EX, TERM, and FACT. Similarly, $P(v)$ is true for each variable name v.

The first inductive step is to assume that $P(E)$ is true, i.e., that E is in EX, and show that $+E$ is in EX. This requires a subinduction on the definition of EX and before that a subsubinduction on the terms to show that for any term T, $+T$ is also a term. Let $P'(T)$ be the property given by "$+T$ is a term." We use induction to show that $P'(T)$ is true for all terms T. The basis is to show that if F is a factor, then $+F$ is a term. This is true by Rules (3d) and (2c). Now, suppose that $+T$ is a term and F is a factor. Then, $+T * F$ is a term by the induction hypothesis and Rule (2a). The other inductive step is similar.

Now, we use induction to show that if E is in EX, then $+E$ is in EX. The basis step is to show that if T is a term, then $+T$ is an expression. As we have just shown, $+T$ is a term, and hence, by Rule (1c), $+T$ is an expression. For the inductive steps, assume that E is an expression such that $+E$ is in EX. Then, $+E + T$ and $+E - T$ are also in EX by the induction hypothesis and Rules (1a) and (1b).

Similarly, if E is in EX, then $-E$ is in EX.

Next, suppose that E_1 and E_2 are in EX. We must show that $E_1 + E_2$ is in EX. This requires a subinduction. Fix an expression E_1 and let P' be the property consisting of those expressions E_2 in EX such that $E_1 + E_2$ is also in EX. We show that $P'(E_2)$ is true for all E_2 in EX. We have $P'(T)$ for any term T by Rule (1a). Assume that $P'(E_2)$ is true and that T is a term. We must show that $P'(E_2 + T)$ is true, i.e., that $E_1 + E_2 + T$ is in EX. But, by the induction hypothesis, $E_1 + E_2$ is in EX, so by Rule (1a), we get what we want. The other inductive step is handled similarly.

By a similar proof, we can show that if E_1 and E_2 are both in EX, then so is $E_1 - E_2$.

Next, suppose again that E_1 and E_2 are both in EX. We must show that $E_1 * E_2$ and E_1/E_2 are both in EX. To this end, we first claim that if T_1 and T_2 are both terms, then $T_1 * T_2$ is also a term. This follows by induction on T_2, holding T_1 fixed. From this fact, it follows (not by induction, but simply by considering the three forms that an element of EX can have) that if E_1 is in EX and T_2 is a term, then $E_1 * T_2$ is in EX. Using this fact as the basis step, we have (by induction on E_2 with E_1 fixed) that if E_1 and E_2 are in EX, then $E_1 * E_2$ is in EX. A similar proof shows that if E_1 and E_2 are in EX, then $E_1 E_2$ is in EX.

We now have two equivalent definitions for a set of integer expressions, one more complicated than the other. The reason for considering the more

complicated definition is that it meets the unique readability condition, as
we now show.

We need to first make some reasonable assumptions; namely, we assume
that integer constants and variable names are strings over some alphabet
Σ, that no string over Σ is both an integer constant and a variable name,
and that $+$, $-$, $*$, $/$, $($, and $)$ are distinct symbols not in Σ.

We define the *level* of a symbol in an expression, term, or factor to be
the number of left parentheses to the left of the symbol minus the number
of right parentheses to the left of the symbol. We also define an occurrence
of a $+$ or $-$ in an expression, term, or factor to be a *unary* occurrence if it
is the first symbol in the word or else is preceded by one of the symbols $($,
$+$, $-$, $*$, or $/$. All other occurrences of $+$ and $-$ are called *binary*.

It is easy to show by induction that every expression, term, and factor is
a nonempty string whose final symbol is not $+$, $-$, $*$, $/$, or $($. Another easy
proof shows that every expression, term, or factor has the same number of
left and right parentheses.

Next, we show that every binary occurrence of a $+$ or $-$ in an expression
has a nonnegative level, and every binary occurrence of a $+$ of $-$ in a factor
or term has a positive level. We let P_{EX} be the property given by "every
binary occurrence of $+$ and $-$ has a nonnegative level" and P_{TERM} and
P_{FACT} be the property given by "every binary occurrence of $+$ and $-$ has
a positive level," and then we use simultaneous structural induction. Each
of the 11 rules in the definition gives rise to a step in the proof. We will
only give a few representative steps and leave the rest to the reader.

Corresponding to Rule (1a), we assume that $P_{EX}(E)$ and $P_{TERM}(T)$
are true and show that $P_{EX}(E + T)$ is true. A binary occurrence of $+$ or
$-$ in $E + T$ that is part of E must be a binary occurrence in E and thus
by the induction hypothesis, has a nonnegative level. The $+$ between E
and T has level 0 since E has the same number of left parentheses as right
parentheses. Any binary occurrence of $+$ or $-$ in $E + T$ that is part of T
must be a binary occurrence in T (it cannot be the first symbol in T since
this would be a unary occurrence in $E + T$), so by the induction hypothesis
it has a positive level in T. Since E has the same number of left and right
parentheses, the level of the binary $+$ or $-$ in $E + T$ is the same as its level
in T.

Corresponding to Rule (3a), note that if v is a variable name, then
$P_{FACT}(v)$ is vacuously true since there are no occurrences of $+$ or $-$ in v.

Corresponding to Rule (3c), suppose that P_E is true. Every binary occurrence of $+$ or $-$ in (E) is also binary as an occurrence in E and has a
level one more than its level in E, so it has a positive level.

Corresponding to Rule (3d), suppose that $P_{FACT}(F)$ is true. The initial
$+$ in $+F$ is unary, so every binary occurrence of a $+$ or $-$ in $+F$ is a binary
occurrence in F, and hence, by the induction hypothesis it has a positive
level.

By a similar proof, one can show that the level of every occurrence of $*$ and $/$ in an expression or term has a nonnegative level, and every occurrence of $*$ and $/$ in a factor has a positive level.

With these preliminary remarks behind us, we show unique readability for the definition of EX, TERM, and FACT. First, consider an expression E'. If E' is put into EX by Rule (1a), say, $E' = E + T$, then since E does not end with an operation symbol or left parenthesis, the $+$ between T and E is binary, and it is also at level 0. Further, since T has no binary $+$ or $-$ at level 0, this $+$ is the rightmost binary $+$ or $-$ in E' at level 0. Similarly, if E' is put into EX by Rule (1b) then the rightmost binary $+$ or $-$ at level 0 in E' is a $-$. If E' is put into EX by Rule (1c), then E' has no binary $+$ or $-$ at level 0. Thus, E' can only be put into EX by one rule. If that rule is (1c), then the hypothesis is uniquely determined as being E'. If the rule is (1a), then the hypothesis E is determined as consisting of exaclty those symbols to the left of the rightmost binary $+$ in E' at level 0, and hence, T is determined as well. Similarly, if E' is put into EX by Rule (1b), T and E are uniquely determined.

Let T' be a term. If T' is put into TERM by Rule (2a), then the rightmost $*$ or $/$ in T' at level 0 is a $*$. If T' is put into TERM by Rule (2b), then the rightmost $*$ or $/$ at level 0 in T' is a $/$. If T' is put into TERM by Rule (2c), then T' has no $*$ or $/$ at level 0. It follows from these remarks that T' is put into TERM by only one rule. In the case of Rule (2c), the hypothesis is obviously uniquely determined. If T' is put in by Rule (2a), then T consists of all symbols to the left of the rightmost $*$ in T' at level 0, and hence, F is determined as well. Similarly, if T' is put into TERM by Rule (2b), the hypotheses are uniquely determined.

Finally, let F be a factor. If F is put into FACT by Rule (3c), then F begins with (. If F is put into FACT by Rule (3d), then F begins with $+$, and if F is put in by Rule (3e), then F starts with $-$. If F is put in by one of the other two rules, F does not contain (, $+$, or $-$. Furthermore, we are assuming that no variable name is also an integer constant. It follows that F is put into FACT by only one rule, and if that rule is (3c), (3d), or (3e), then the hypothesis is obviously determined by F.

We now give the formal presentation of simultaneous inductive definitions. Let K be a nonempty set (which remains fixed throughout the rest of this section). Let $\mathcal{C} = \{C_k \mid k \in K\}$ and $\mathcal{D} = \{D_k \mid k \in K\}$ be two families of sets indexed by K. We write $\mathcal{C} \preceq \mathcal{D}$ if $C_k \subseteq D_k$ for every $k \in K$.

Definition 4.7.13 *A type over K is an element of $Seq(K) \times K$. If $\sigma = (\alpha, k)$ is a type over K and $|\alpha| = n$, then σ is called an n-ary type over K.*

Let $\mathcal{U} = \{U_k \mid k \in K\}$ be a family of sets indexed by K, fixed for the rest of this section. From this point on, whenever we write $\mathcal{C} \preceq \mathcal{D}$, we assume that both families of sets are indexed by K. We define $\hat{\mathcal{P}}(\mathcal{U})$ to be $\{\mathcal{C} \mid \mathcal{C} \preceq \mathcal{U}\}$.

If $\alpha \in \text{Seq}(K)$, say, $\alpha = (k_0, \ldots, k_{n-1})$, and $\mathcal{A} = \{A_k \mid k \in K\}$, then by A_α, we mean $A_{k_0} \times \cdots \times A_{k_{n-1}}$. (Note that $A_\lambda = \{\lambda\}$.) If $\mathcal{A} = \{A_k \mid k \in K\}$, $\mathcal{B} = \{B_k \mid k \in K\}$, and $\{g_k : A_k \longrightarrow B_k \mid k \in K\}$ is a collection of functions, we define $g_\alpha : A_\alpha \longrightarrow B_\alpha$ by

$$g_\alpha(u_0, \ldots, u_{n-1}) = (g_{k_0}(u_0), \ldots, g_{k_{n-1}}(u_{n-1}))$$

for every $(u_0, \ldots, u_{n-1}) \in A_\alpha$.

Definition 4.7.14 *A constructor over $\mathcal{U}$ is a pair $((\alpha, k,), R)$ where*

1. (α, k) is a type over K, and

2. R is a relation from U_α to U_k.

When we say "R is a constructor of type (α, k) over $\mathcal{U}$," what we mean is that $((\alpha, k), R)$ is a constructor over $\mathcal{U}$.

Often, we will abuse notation and refer to the constructor $((\alpha, k), R)$ as just R when the type (α, k) is clear from the context or irrelevant.

Definition 4.7.15 *Let R be a constructor of type (α, k) over $\mathcal{U}$ and let $\mathcal{A} = \{A_k \mid k \in K\}$ be a family of sets indexed by K such that $A_k \subseteq U_k$ for all $k \in K$. Then, $\mathcal{A}$ is* closed *under R if $R(A_\alpha) \subseteq A_k$.*

If $\mathfrak{R}$ is a family of constructors over $\mathcal{U}$, then we say that $\mathcal{A}$ is closed under $\mathfrak{R}$ if it is closed under each constructor in $\mathfrak{R}$.

Note that if $\mathcal{A} \preceq \mathcal{B} \preceq \mathcal{U}$ and R is a constructor of type (α, k) on $\mathcal{U}$, then $R(A_\alpha) \subseteq R(B_\alpha)$.

Definition 4.7.16 *Let $\mathfrak{R}$ be a family of constructors over $\mathcal{U}$ and let $\mathcal{B} \preceq \mathcal{U}$. A family of sets $\mathcal{C} \preceq \mathcal{U}$ is called the* closure *of $\mathcal{B}$ under $\mathfrak{R}$ if the following conditions are satisfied*

1. $\mathcal{B} \preceq \mathcal{C}$.

2. $\mathcal{C}$ is closed under $\mathfrak{R}$.

3. If $\mathcal{B} \preceq \mathcal{D} \preceq \mathcal{U}$ and $\mathcal{D}$ is closed under $\mathfrak{R}$, then $\mathcal{C} \preceq \mathcal{D}$.

The use of the word "the" in the definition of the closure of $\mathcal{B}$ under $\mathfrak{R}$ is justified by the following theorem.

Theorem 4.7.17 *Let $\mathfrak{R}$ be a family of constructors over $\mathcal{U}$ and let $\mathcal{B} \preceq \mathcal{U}$. If $\mathcal{C}$ and $\mathcal{D}$ are both closures of $\mathcal{B}$ under $\mathfrak{R}$, then $\mathcal{C} = \mathcal{D}$.*

Proof. The proof is similar to that of Theorem 4.6.2. ∎

Theorem 4.7.18 *Let $\mathfrak{R}$ be a family of constructors over $\mathcal{U}$. Then, for every $\mathcal{B} \preceq \mathcal{U}$, the closure of $\mathcal{B}$ under $\mathfrak{R}$ exists.*

Proof. Given $\mathcal{B} \preceq \mathcal{U}$, let $\{\mathcal{C}^i \mid i \in I\}$ be the collection of all families of sets such that $\mathcal{B} \preceq \mathcal{C}^i \preceq \mathcal{U}$ and $\mathcal{C}^i$ is closed under $\Re$. This collection is not empty since it contains at least $\mathcal{U}$.

If $\mathcal{C}^i = \{C_k^i \mid k \in K\}$, define $\mathcal{C} = \{C_k \mid k \in K\}$, where $C_k = \bigcap\{C_k^i \mid i \in I\}$. Then, $\mathcal{C}$ is the closure of $\mathcal{B}$ under $\Re$. The verifications of the first and third conditions are straightforward and left to the reader; we show only that the second condition is met. Let $R \in \Re$ be a constructor of type (α, k) on $\mathcal{U}$. We have $C_\alpha \subseteq C_\alpha^i$ for every $i \in I$, hence, $R(C_\alpha) \subseteq R(C_\alpha^i)$, which gives

$$R(C_\alpha) \subseteq \bigcap\{R(C_\alpha^i) \mid i \in I\}$$
$$\subseteq \bigcap\{C_k^i \mid i \in I\}$$
$$= C_k.\blacksquare$$

Theorem 4.7.19 *Let $\mathcal{B} \preceq \mathcal{U}$ and let $\mathcal{C} \preceq \mathcal{U}$ be the closure of $\mathcal{B}$ under the family of constructors $\Re$. Suppose that for each $k \in K$, P_k is a property of the elements of U_k, such that*

1. *for every $k \in K$, every element of B_k has property P_k;*

2. *for every constructor $R \in \Re$ of type $((k_0, \ldots, k_{n-1}), k)$, if $((u_0, \ldots, u_{n-1}), u) \in R$ and each u_j has property P_{k_j} for $0 \le j \le n-1$, then u has property P_k.*

Then, for all $k \in K$ and $u \in C_k$, u has property P_k.

Proof. Since a property of a set is a subset of that set, we have a family of sets $\mathcal{Q} = \{P_k \mid k \in K\} \preceq \mathcal{U}$. The hypotheses show that $\mathcal{Q}$ is closed under $\Re$ and $\mathcal{B} \preceq \mathcal{Q}$. Since $\mathcal{C}$ is the closure of $\mathcal{B}$ under $\Re$, we have $\mathcal{C} \preceq \mathcal{Q}$, which amounts to the desired conclusion. $\blacksquare$

Definition 4.7.20 *Let $\mathcal{B} \preceq \mathcal{U}$ and let $\mathcal{C} \preceq \mathcal{U}$ be the closure of $\mathcal{B}$ under the family of constructors $\Re$.*

We say that $\mathcal{C}$ is freely generated by $\Re$ over $\mathcal{B}$ if the following conditions are satisfied

1. *For each $R \in \Re$ of type (α, k), we have $R(C_\alpha) \cap B_k = \emptyset$.*

2. *If $((\alpha, k), R)$ and $((\alpha', k), R')$ are two distinct constructors in $\Re$, then $R(C_\alpha) \cap R'(C_{\alpha'}) = \emptyset$.*

3. *For each $R \in \Re$ of type (α, k), R restricted to C_α is a one-to-one relation.*

Theorem 4.7.21 *Let $\mathcal{A} = \{A_k \mid k \in K\}$ be a family of sets indexed by K. Let $\mathcal{B} \preceq \mathcal{C} \preceq \mathcal{U}$, where $\mathcal{C}$ is the closure of $\mathcal{B}$ under the family of constructors $\Re$ over $\mathcal{U}$. Suppose that for each $k \in K$ we have a function*

$G_k : B_k \longrightarrow A_k$ *and for each constructor* $((\alpha, k), R) \in \Re$ *we have a function* $H_R : A_\alpha \longrightarrow A_k$.

If $\mathcal{C}$ *is freely generated by* $\Re$ *over* $\mathcal{B}$, *then there is a unique family of functions* $\{g_k : C_k \longrightarrow A_k \mid k \in K\}$ *satisfying the following conditions:*

1. $g_k(x) = G_k(x)$ *for every* $x \in B_k$;

2. *for every* $R \in \Re$, *if* R *is of type* (α, k), $(\bar{u}, u) \in R$, *and* $\bar{u} \in C_\alpha$, *then*

$$g_k(u) = H_R(g_\alpha(\bar{u})).$$

Proof. Suppose that $\{g_k \mid k \in K\}$ and $\{g'_k \mid k \in K\}$ both satisfy the conditions of the theorem. We use Theorem 4.7.19 to show that, for all $k \in K$ and $x \in C_k$, $g_k(x) = g'_k(x)$. Let $P_k = \{x \in C_k \mid g_k(x) = g'_k(x)\}$ for each $k \in K$. We remind the reader that, for $\alpha = (k_0, \ldots, k_{n-1})$, P_α is the property $P_{k_0} \times \cdots \times P_{k_{n-1}}$.

For each $k \in K$ and $x \in B_k$, we have $P_k(x)$ holds because $g_k(x) = G_k(x) = g'_k(x)$. For the inductive step, assume that $((\alpha, k), R) \in \Re$, $(\bar{u}, u) \in R$, and $P_\alpha(\bar{u})$ is true, i.e., $g_\alpha(\bar{u}) = g'_\alpha(\bar{u})$. Then, $g_k(u) = H_R(g_k(\bar{u})) = H_R(g'_k(\bar{u})) = g'_k(u)$.

To show existence of $\{g_k \mid k \in K\}$, let $\mathcal{U}' = \{C_k \times A_k \mid k \in K\}$. Define $\mathcal{B}' = \{B'_k \mid k \in K\}$, where $B'_k = \{(x, G_k(x)) \mid x \in B_k\}$, and consider the closure $\mathcal{C}'$ of $\mathcal{B}'$ under the family of constructors $\Re' = \{R' \mid R \in \Re\}$, where

$$
\begin{aligned}
R' \;=\; & \{(((u_0, a_0), \ldots, (u_{n-1}, a_{n-1})), (u, H_R(a_0, \ldots, a_{n-1}))) \\
& \mid ((u_0, \ldots, u_{n-1}), u) \in R\}.
\end{aligned}
$$

If $\mathcal{C}' = \{g_k \mid k \in K\}$, then a proof similar to the one in Theorem 4.6.48 shows that each g_k is a function and that the conditions of the theorem are met. ∎

4.8   Propositional Logic

In this and the next few sections, we interrupt our development of induction in order to introduce several areas of great significance to computer science. In each of these areas, we will see that the techniques developed so far are essential for formulating and studying the basic concepts.

The subject we will look at in this section is an elementary area of mathematical logic known as propositional logic. Propositional logic provides a mathematical model for a certain type of reasoning. Although there is much reasoning, both mathematical and nonmathematical that cannot be captured by this model, it nonetheless has many applications in computer science, ranging from design of computer circuitry to programming languages.

The basic notion of propositional logic is that of a *statement* (also known as a *proposition*) which is an assertion that is either true or false. Statements are typically expressed as declarative sentences. For example, "$2 + 2 =$ 4" and "The earth is flat" express statements. The same statement can be expressed by different sentences. For instance, an alternative way of expressing "$2 + 2 = 4$" is "the sum of two and two is equal to four."

A *connective* is a way of combining statements into new statements such that the truth or falsity of the new statement depends only on the truth or falsity of the statements being combined. A connective that combines two statements into one is called a *binary connective* . Here, we introduce four binary connectives: conjunction, disjunction, conditional, and biconditional. A connective that produces a statement out of one statement is called a *unary connective,* although this is somewhat of a misnomer since no statements are being connected. We introduce one unary connective: negation. More generally, if a connective produces a statement starting from n statements, then we call the connective an *n-ary connective.*

The *conjunction* of two statements is true if and only if each of the statements is true. In English, the conjunction is usually expressed using the word "and," that is, if T_1 and T_2 are sentences expressing the statements S_1 and S_2, then "T_1 and T_2" expresses the conjunction of S_1 and S_2. For example, "$2 + 2 = 4$ and the earth is flat" expresses the conjunction of the two statements given above.

The *disjunction* of two statements is true if and only if at least one of the statements is true. In English, the disjunction is usually expressed using the word "or" in its inclusive (and/or) sense. Sometimes in English, the word "or" is used in an exclusive sense asserting that one or the other of the sentences connected by the word "or" is true, but not both. This use of the word "or" does not express disjunction, as the disjunction of two statements is true when both statements are true, as well as when exactly one of them is true.

The *conditional* of two statements is false if and only if the first statement (called the *hypothesis*) is true and the second statement (called the *conclusion*) is false. The conditional is intended to capture the idea of implication; that is, if T_1 and T_2 are sentences expressing the hypothesis S_1 and the conclusion S_2, then the conditional of S_1 and S_2 is meant to be the statement expressed by "if T_1, then T_2." Whether this is true in colloquial English is questionable. For example, there is no doubt that the assertion "if it rains tommorrow, then I will stay home" is true if its hypothesis and conclusion are both true and is false if its hypothesis is true and its conclusion is false. However, if the hypothesis is false, most people would say that it is impossible to determine whether or not the implication is true, while we have defined a conditional to be true if its hypothesis is false. From the mathematical point of view, the definition given here for the truth value of a conditional based on the truth values of its hypothesis and conclusion is the only appropriate one. To see why this is true, consider an assertion of the form "if A and B, then A." This assertion should be true no matter what the truth values of A and B are; in particular, if B is false, then the hypothesis is false, while the conclusion can be either true or false.

The *biconditional* of two statements is true if and only if the statements have the same truth value. If T_1 and T_2 are sentences expressing the statements S_1 and S_2, then the biconditional of S_1 and S_2 is expressed by the sentence "T_1 if and only if T_2."

The *negation* of a statement is true if and only if the statement is false. In English, the negation is usually expressed using the word "not". For example, "the sum of two and two is not equal to four" expresses the negation of "the sum of two and two is equal to four."

We wish to set up a mathematical model to allow us to study how the truth of complex statements depends on the truth of their component statements and the meanings of the connectives. In this model, we will be interested only in the forms that statements can have as combinations of simpler statements. For the simplest statements, we are only interested in whether the statement is true or false and ignore what they try to convey.

We start our mathematical model by selecting two objects and denoting them by **T** and **F**. **T** and **F** can be any two distinct objects; a common choice is **T** $= 1$ and **F** $= 0$, but any other choice would do for our model. We denote $\{\mathbf{T}, \mathbf{F}\}$ by **Bool**. Our idea, of course, is that **T** stands for true and **F** stands for false.

Next, we want to represent connectives in our model. This is done by truth functions, as introduced below.

Definition 4.8.1 *An n-ary truth function is a function from* **Bool**^{n} *to* **Bool***; 1-ary truth functions are called* unary *truth functions, and 2-ary truth functions are called* binary *truth functions.*

We use n-ary truth functions to represent n-ary connectives. In particular, we have the binary truth functions $f_\wedge$, $f_\vee$, $f_\to$, and $f_\leftrightarrow$ and the unary truth function $f_\sim$ to represent the connectives conjunction, disjunction, conditional, biconditional, and negation. These functions are given by the following tables.

x	y	$f_\wedge(x,y)$	$f_\vee(x,y)$	$f_\to(x,y)$	$f_\leftrightarrow(x,y)$
F	**F**	**F**	**F**	**T**	**T**
F	**T**	**F**	**T**	**T**	**F**
T	**F**	**F**	**T**	**F**	**F**
T	**T**	**T**	**T**	**T**	**T**

x	$f_\sim(x)$
F	**T**
T	**F**

Next, we need symbols to stand for simple statements. Let SV be some infinite set. We call the elements of SV *statement variables.* We use the letters p, q, r, s, t, with or without subscripts, to denote particular statement variables.

Compound statements are represented by the strings of symbols called formulas of propositional logic (or just formulas), which we now define.

Definition 4.8.2 *We define the set $\mathcal{F}_{prop}$ of formulas of propositional logic by the following inductive definition.*

   1. Every statement variable is a formula of propositional logic.

   2. If φ is a formula f propositional logic, then so is $(\sim \varphi)$.

   3. If φ and ψ are formulas of propositional logic, then so is $(\varphi \wedge \psi)$.

   4. If φ and ψ are formulas of propositional logic, then so is $(\varphi \vee \psi)$.

   5. If φ and ψ are formulas of propositional logic, then so is $(\varphi \rightarrow \psi)$.

   6. If φ and ψ are formulas o propositional logic, then so is $(\varphi \leftrightarrow \psi)$.

Example 4.8.3 If p, q, r are statement variables, then $((\sim (p \rightarrow q)) \wedge (r \vee p))$ is a formula of propositional logic. We can see this in a line-by-line proof as follows.

	Claim	Reason
1.	p is a formula	Rule (1)
2.	q is a formula	Rule (1)
3.	$(p \rightarrow q)$ is a formula	Rule (5) and Lines (1),(2)
4.	$(\sim (p \rightarrow q))$ is a formula	Rule (2) and Line (3)
5.	r is a formula	Rule (1)
6.	$(r \vee p)$ is a formula	Rule (4) and Lines (5),(1)
7.	$((\sim (p \rightarrow q)) \wedge (r \vee p))$ is a formula	Rule (3) and Lines (4),(6)

A tree illustration that shows that the expression is a formula of propositional logic is given in Figure 4.3.

The set of formulas of propositional logic can be seen as the closure of a set under a family of operations. Let $A = SV \cup \{(,), \sim, \wedge, \vee, \rightarrow, \leftrightarrow\}$, and let $U = \text{Seq}(A)$. We assume that $SV \cap \{(,), \sim, \wedge, \vee, \rightarrow, \leftrightarrow\} = \emptyset$. Let $\mathcal{G} = \{g_i \mid i \in \{\sim, \wedge, \vee, \rightarrow, \leftrightarrow\}\}$, where the g_i are given by

$$
\begin{aligned}
g_{\sim}(\alpha) &= (\sim \alpha), \\
g_{\wedge}(\alpha, \beta) &= (\alpha \wedge \beta), \\
g_{\vee}(\alpha, \beta) &= (\alpha \vee \beta), \\
g_{\rightarrow}(\alpha, \beta) &= (\alpha \rightarrow \beta), \\
g_{\leftrightarrow}(\alpha, \beta) &= (\alpha \leftrightarrow \beta).
\end{aligned}
$$

If $B = SV$, then the set of formulas of propositional logic is the closure of B under $\mathcal{G}$.

Structural induction applied to Definition 4.8.2 gives the following.

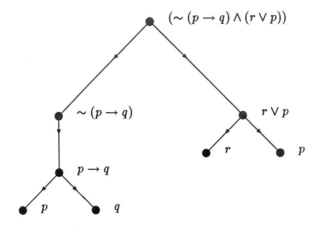

$$(\sim (p \to q) \land (r \lor p))$$

$\sim (p \to q)$

$r \lor p$

r p

$p \to q$

p q

FIGURE 4.3. A tree for$((\sim (p \to q)) \land (r \lor p))$.

Corollary 4.8.4 *Let P be a property of the formulas of propositional logic. Suppose that*

1. *$P(q)$ is true for every statement variable q;*

2. *if $P(\varphi)$ is true for some formula φ, then $P((\sim \varphi))$ is true;*

3. *if $P(\varphi)$ and $P(\psi)$ are both true, then P is also true about $(\varphi \land \psi), (\varphi \lor \psi), (\varphi \to \psi),$ and $(\varphi \leftrightarrow \psi).$*

Then, $P(\varphi)$ is true for every formula φ of propositional logic.

Example 4.8.5 We show that for every formula φ of propositional logic, the number of occurrences of left parentheses in φ equals the number of occurrences of right parentheses in φ. For any formula φ of propositional logic, let $L(\varphi)$ be the number of left parentheses in φ and let $R(\varphi)$ be the number of right parentheses in φ. Let $P(\varphi)$ be the property that $L(\varphi) = R(\varphi)$. We want to show that $P(\varphi)$ is true for all formulas φ of propositional logic. We do this by structural induction using Corollary 4.8.4.

The basis step is to show that $P(q)$ holds for any statement variable q. But, $L(q) = 0 = R(q)$ for any statement variable, establishing the basis step.

The first inductive step is to assume that $P(\varphi)$ is true and show that $P((\sim \varphi))$ is true. Since $P(\varphi)$ is true, $L(\varphi) = R(\varphi)$. But, then, we get

$$L((\sim \varphi)) = L(\varphi) + 1 = R(\varphi) + 1 = R((\sim \varphi)),$$

so $P((\sim \varphi))$ is true.

The next inductive step is to assume that $P(\varphi)$ and $P(\psi)$ are both true, that is, $L(\varphi) = R(\varphi)$ and $L(\psi) = R(\psi)$, and to show that $P((\varphi \wedge \psi))$ is true. We have

$$L((\varphi \wedge \psi)) = L(\varphi) + L(\psi) + 1 + R(\varphi) + R(\psi) + 1 = R((\varphi \wedge \psi)),$$

so $P((\varphi \wedge \psi))$ is true.

The other inductive steps are similar to the one just done.

Theorem 4.8.6 *The definition of the set $\mathcal{F}_{prop}$ of formulas of propositional logic given in Definition 4.8.2 satisfies the unique readability condition.*

Proof: We will use the result from Exercise 60 that every proper initial segment of a formula of propositional logic contains more left parentheses than right parentheses.

If a formula is put into S by the basic rule, then the formula is a single statement variable. If the formula is put into S by an inductive rule, then the formula begins with a (, so no formula is put into S by the basic rule and by one of the inductive rules. If a formula were put into S by the $\sim$-rule and by the C-rule, where C is a binary connective, then we would have $(\sim \varphi) = (\alpha C \beta)$ for some formulas φ, α, β, and this would mean that α begins with $\sim$, which is impossible since every formula is either a statement letter or begins with a (. If a formula is put into S by the C_1-rule and by the C_2-rule, where C_1 and C_2 are both binary connectives, then $(\varphi C_1 \psi) = (\alpha C_2 \beta)$ for some formulas $\varphi, \psi, \alpha, \beta$. By Example 4.8.5, α has the same number of left and right parentheses and, hence, cannot be a proper initial segment of φ. By similar reasoning, φ cannot be a proper initial segment of α. Thus, $\varphi = \alpha$. But, then, $C_1 = C_2$ and $\psi = \beta$. This shows that if $C_1 \neq C_2$, then no formula can be put into S by both the C_1- and C_2-rules. Thus, the first part of the definition of the unique readability condition is met. The second part is also met, because as we just showed, if $(\varphi C \psi) = (\alpha C \beta)$, where $\varphi, \psi, \alpha.\beta$ are formulas and C is a binary connective, then $\varphi = \alpha$ and $\psi = \beta$, while if $(\sim \varphi) = (\sim \psi)$, then it is obvious that $\varphi = \psi$. ∎

Definition 4.8.7 *A truth assignment is a function $v : SV \longrightarrow$ **Bool**.*

In an "obvious" way, a truth assignment v determines a *truth valuation* $\bar{v}$, which is a function from $\mathcal{F}_{prop}$ to **Bool**. For example, if v is a truth assignment with $v(p_1) = \mathbf{T}, v(p_2) = \mathbf{F}, v(p_3) = \mathbf{F}$, and $\bar{v}$ is the truth valuation determined by v, then $\bar{v}(((p_1 \rightarrow p_2) \vee p_3)) = \mathbf{F}$. Although it seems clear how $\bar{v}$ works, a precise definition should still be given and this seems clear how $\bar{v}$ works, a precise definition should still be given and this involves recursion.

$$\bar{v}(q) \;=\; v(q) \text{ for each statement variable } q,$$
$$\bar{v}((\sim \varphi)) \;=\; f_\sim(\bar{v}(\varphi)),$$
$$\bar{v}((\varphi \wedge \psi)) \;=\; f_\wedge(\bar{v}(\varphi), \bar{v}(\psi)),$$
$$\bar{v}((\varphi \vee \psi)) \;=\; f_\vee(\bar{v}(\varphi), \bar{v}(\psi)),$$
$$\bar{v}((\varphi \to \psi)) \;=\; f_\to(\bar{v}(\varphi), \bar{v}(\psi)),$$
$$\bar{v}((\varphi \leftrightarrow \psi)) \;=\; f_\leftrightarrow(\bar{v}(\varphi), \bar{v}(\psi)).$$

Note how the value of $\bar{v}(\varphi)$ for those formulas φ put into $\mathcal{F}_{prop}$ by the basis rule (i.e., the statement variables) is given nonrecursively, while the value of $\bar{v}(\varphi)$ for those formulas φ put into $\mathcal{F}_{prop}$ by inductive rules is given recursively in terms of the values of $\bar{v}$ when applied to the hypotheses of the rule. This definition can now be used in a step by step manner to evaluate $\bar{v}(\varphi)$ for any formula φ. For example, if $v(p_1) = \mathbf{T}, v(p_2) = \mathbf{F}, v(p_3) = \mathbf{F}$, then, according to the first clause of the definition, $\bar{v}(p_1) = \mathbf{T}, \bar{v}(p_2) = \mathbf{F}, \bar{v}(p_3) = \mathbf{F}$ and hence, according to the fifth clause of the definition, $\bar{v}((p_1 \to p_2)) = \mathbf{F}$, and thus, by the third clause of the definition, $\bar{v}(((p_1 \to p_2) \vee p_3)) = \mathbf{F}$.

The existence of the function $\bar{v}$ follows from the fact that the definition of $\mathcal{F}_{prop}$ meets the unique readability condition and Theorem 4.5.6.

Definition 4.8.8 *A formula φ is a* tautology *if for every truth assignment v, $\bar{v}(\varphi) = \mathbf{T}$.*

A formula φ is a contradiction *if for every truth assignment v, $\bar{v}(\varphi) = \mathbf{F}$.*

Example 4.8.9   1. The formula $(p \vee (\sim p))$, where p is a statement variable, is a tautology. To see this, let v be any truth assignment. Then, according to Definition 4.8.7,

$$\bar{v}((p \vee (\sim p))) \;=\; f_\vee(\bar{v}(p), \bar{v}((\sim p)))$$
$$= f_\vee(v(p), f_\sim(\bar{v}(p)))$$
$$= f_\vee(v(p), f_\sim(v(p))).$$

Thus, if $v(p) = \mathbf{T}$, we have

$$\bar{v}((p \vee (\sim p))) \;=\; f_\vee(\mathbf{T}, f_\sim(\mathbf{T}))$$
$$= f_\vee(\mathbf{T}, \mathbf{F})$$
$$= \mathbf{T},$$

and, if $v(p) = \mathbf{F}$, we have

$$\bar{v}((p \vee (\sim p))) \;=\; f_\vee(\mathbf{F}, f_\sim(\mathbf{F}))$$
$$= f_\vee(\mathbf{F}, \mathbf{T})$$
$$= \mathbf{T}.$$

2. More generally, for any formula φ, $(\varphi \vee (\sim \varphi))$ is a tautology. Indeed, for any truth assignment v,

$$\bar{v}((\varphi \vee (\sim \varphi))) = f_\vee(\bar{v}(\varphi), f_\sim(\bar{v}(\varphi))),$$

and one of the truth values $\bar{v}(\varphi)$ and $f_\sim(\bar{v}(\varphi))$ is $\mathbf{T}$, which implies that $\bar{v}((\varphi \vee (\sim \varphi))) = \mathbf{T}$.

3. For any formula φ, $(\varphi \rightarrow \varphi)$ is a tautology. Indeed, for any truth assignment v, we have

$$\bar{v}((\varphi \rightarrow \varphi)) = f_\rightarrow(\bar{v}(\varphi), \bar{v}(\varphi)) = \mathbf{T},$$

as shown by the table defining $f_\rightarrow$.

Example 4.8.10 For any formula φ, $(\varphi \wedge (\sim \varphi))$ is a contradiction, since for every truth assignment v, we have

$$\bar{v}((\varphi \wedge (\sim \varphi))) = f_\wedge(\bar{v}(\varphi), f_\sim(\bar{v}(\varphi))) = \mathbf{F},$$

because of the fact that for any v only one of $\bar{v}(\varphi)$ and $f_\sim(\varphi)$ is $\mathbf{T}$.

4.9 Primitive Recursive and Partial Recursive Functions

In this short section, we will define two very important classes of functions, the primitive recursive functions and the partial recursive functions. The careful study of these functions is a topic for the theory of computation. Our goal is simply to present the definitions using the results of the previous section.

First, we give a corollary that says that the type of recursive definition we are interested in always succeeds.

Corollary 4.9.1 *Let m be a natural number and suppose that $g : \mathbf{N}^m \rightarrow \mathbf{N}$ and $h : \mathbf{N}^{m+2} \rightarrow \mathbf{N}$. Then, there is a unique function $f : \mathbf{N}^{m+1} \rightarrow \mathbf{N}$ that satisfies*

$$f(x_1, \ldots, x_m, 0) = g(x_1, \ldots, x_m) \text{ for all } x_1, \ldots, x_m \text{ in } \mathbf{N},$$
$$f(x_1, \ldots, x_m, n+1) = h(x_1, \ldots, x_m, n, f(x_1, \ldots, x_m, n))$$
$$\text{for all } x_1, \ldots, x_m, n \text{ in } \mathbf{N}.$$

Proof: This follows from Corollary 4.6.42 with $n_0 = 0$, $P = \mathbf{N}^m$, $G = g$, and $H = h$ if the obvious identifications are made between $\mathbf{N}^m \times \mathbf{N} \times \mathbf{N}$ and $\mathbf{N}^{m+2}$ and between $\mathbf{N}^m \times \mathbf{N}$ and $\mathbf{N}^{m+1}$. ∎

The function f of the previous corollary is said to be obtained by *primitive recursion* from the functions g and h.

Suppose that $k \geq 1$, $n \in \mathbf{N}$, and that $g : \mathbf{N}^k \to \mathbf{N}$ and for each i with $1 \leq i \leq k$ we have $h_i : \mathbf{N}^n \to \mathbf{N}$. Then, recall from Chapter 2 that we have a function $g \circ (h_1, \ldots, h_k) : \mathbf{N}^n \to \mathbf{N}$ (also written $g(h_1, \ldots, h_k)$) defined by

$$g \circ (h_1, \ldots, h_k)(x_1, \ldots, x_n) = g(h_1(x_1, \ldots, x_n), \ldots, h_k(x_1, \ldots, x_n)).$$

We will say that the function $g \circ (h_1, \ldots, h_k)$ is obtained by composition from $g, h_1, \ldots, h_k$.

Definition 4.9.2   *1. The* zero function *is the function* $Z : \mathbf{N}^0 \to \mathbf{N}$ *given by* $Z() = 0$.

2. For each n, i with $1 \leq i \leq n$, we define the projection function p_i^n : $\mathbf{N}^n \to \mathbf{N}$ *by* $p_i^n(x_1, \ldots, x_n) = x_i$.

3. A function is called an initial function *if it is the zero function, one of the projection functions, or the successor function* $S(x) = x + 1$ *introduced earlier.*

Definition 4.9.3 *Let* $T = \bigcup_{n \geq 0}(\mathbf{N}^n \longrightarrow \mathbf{N})$. *The set of* primitive recursive functions *is the subset of T defined by the following.*

1. Every initial function is primitive recursive.

2. If g and $h_1, \ldots, h_k$ are all primitive recursive and f is obtained from these functions by composition, then f is primitive recursive.

3. If g and h are primitive recursive and f is obtained from g and h by primitive recursion, then f is primitive recursive.

Another way of phrasing this definition is that the primitive recursive functions are the closure of the initial functions under the operations of composition and primitive recursion; i.e., they are those functions that can be obtained from the initial functions by a finite number of applications of composition and primitive recursion.

We can, of course, rephrase this definition in terms of constructors. Let T be as in the previous definition and let B be the subset of T consisting of the initial functions. For each $k \geq 1$, define a $k + 1$-ary constructor $R_{C,k}$ on T by

$$R_{C,k} = \{((g, h_1, \ldots, h_k), g(h_1, \ldots, h_k)) \mid g(h_1, \ldots, h_k) \text{ is defined}\}$$

and define a binary constructor R_{PR} on T by

$$R_{PR} = \{((g, h), f) \mid f \text{ is obtained from } g, h \text{ by primitive recursion}\}.$$

(Note that $R_{C,k}$ is a partial $(k + 1)$-ary operation on T whose domain is the set of $(k + 1)$-tuples of functions such that the first entry is k-ary and there is some $n \in \mathbf{N}$ such that the remaining k functions are n-ary; and R_{PR} is a partial binary operation on T whose domain is the set of pairs (g, h) such that, for some $n \in \mathbf{N}$, g is n-ary and h is $(n+2)$-ary.) Then, the set of primitive recursive functions is the closure of B under the family of constructors consisting of R_{PR} and all of the $R_{C,k}$'s.

We now build up a small collection of primitive recursive functions that consists of those functions we will need later. We leave to the exercises further examples and closure properties of the set of primitive recursive functions.

Example 4.9.4 For $m, n \in \mathbf{N}$, we let K_m^n denote the constant function with n arguments whose value is m. We will show that all of these functions are primitive recursive. Note that K_0^0 is Z and

$$K_0^{n+1}(x_1, \ldots, x_n, 0) = K_0^n(x_1, \ldots, x_n),$$
$$K_0^{n+1}(x_1, \ldots, x_n, y+1) = p_{n+2}^{n+2}(x_1, \ldots, x_n, y, K_0^{n+1}(x_1, \ldots, x_n, y)).$$

So, an easy induction argument shows that K_0^n is primitive recursive for all $n \in \mathbf{N}$. Now, for a fixed n, we can show by induction on m that K_m^n is primitive recursive. The basis $m = 0$ has just been done. For the inductive step, observe that $K_{m+1}^n = S \circ K_m^n$.

Example 4.9.5 The recursive definition of the addition function add given in Example 4.6.43 can be used to show that add is primitive recursive since we can rewrite this definition as follows:

$$\mathrm{add}(x, 0) = p_1^1(x),$$
$$\mathrm{add}(x, y+1) = S \circ p_3^3(x, y, \mathrm{add}(x, y)).$$

Similarly, the definitions of the functions mult and exp given in Examples 4.6.44 and 4.6.45 can be used to show that these functions are primitive recursive.

Example 4.9.6 The *signum function* is the function sgn : $\mathbf{N} \longrightarrow \mathbf{N}$ given by

$$\mathrm{sgn}(x) = \begin{cases} 0 & \text{if } x = 0 \\ 1 & \text{if } x > 0. \end{cases}$$

We can show that sgn is primitive recursive by considering the following equivalent recursive definition:

$$\mathrm{sgn}(0) = Z(),$$
$$\mathrm{sgn}(y+1) = K_1^2(y, \mathrm{sgn}(y)).$$

In order to perform logical computations (i.e., computations with truth values) using primitive recursive functions, we need to represent truth values as numbers. In this section, we adopt the following convention shared by many programming languages (such as C and PL/1): 0 stands for **F** and every number greater than 0 stands for **T**.

Example 4.9.7 We define the functions f_{and}, $f_{\text{or}} : \mathbf{N}^2 \longrightarrow \mathbf{N}$, and $f_{\text{not}} : \mathbf{N} \longrightarrow \mathbf{N}$ by

$$f_{\text{and}}(x, y) = \begin{cases} 1 & \text{if } x > 0 \text{ and } y > 0 \\ 0 & \text{otherwise}, \end{cases}$$

$$f_{\text{or}}(x, y) = \begin{cases} 1 & \text{if } x > 0 \text{ or } y > 0 \\ 0 & \text{otherwise}, \end{cases}$$

$$f_{\text{not}}(x) = \begin{cases} 1 & \text{if } x = 0 \\ 0 & \text{otherwise}. \end{cases}$$

These functions are the analogs of the truth functions $f_\wedge$, $f_\vee$, and $f_\sim$ introduced in Section 4.8.

We can easily show that these three functions are primitive recursive since $f_{\text{and}} = \text{sgn} \circ \text{mult}$, and $f_{\text{or}} = \text{sgn} \circ \text{add}$, and for f_{not}, we have the following definition by primitive recursion:

$$f_{\text{not}}(0) = K_1^0()$$
$$f_{\text{not}}(y + 1) = K_0^2(y, f_{\text{not}}(y)).$$

To simplify notation, we will write x **and** y for $f_{\text{and}}(x, y)$, x **or** y for $f_{\text{or}}(x, y)$, and **not** x for $f_{\text{not}}(x)$.

It is easy to see that we have the following elementary properties:

$$(x \text{ and } y) \text{ and } z = x \text{ and } (y \text{ and } z)$$
$$\text{(associativity of and)},$$
$$x \text{ and } y = y \text{ and } x$$
$$\text{(commutativity of and)},$$
$$(x \text{ or } y) \text{ or } z = x \text{ or } (y \text{ or } z)$$
$$\text{(associativity of or)},$$
$$x \text{ or } y = y \text{ or } x,$$
$$\text{(commutativity of or)},$$

for every $x, y, z \in \mathbf{N}$.

The associativity of **and** and **or** allows us to avoid using parantheses; we write x **and** y **and** z instead of $(x$ **and** $y)$ **and** z or x **and** $(y$ **and** $z)$.

We define the *test functions* to be those functions that can be obtained from the projection functions and the functions f_{and}, f_{or}, and f_{not} by using composition; i.e., the set of test functions is the set of functions defined by the following inductive definition:

1. f_{and}, f_{or}, and f_{not} are test functions,

2. the projection functions are test functions, and

3. if g and $h_1, \ldots, h_k$ are all test functions, then $g(h_1, \ldots, h_k)$ is a test function.

The other class of functions we wish to define, the partial recursive functions, is a collection of partial functions, so we need a result analogous to Corollary 4.9.1 that applies to partial functions. We can get such a result as an immediate corollary of Theorem 4.6.46.

Corollary 4.9.8 *Let* m *be a natural number and suppose that* $g : \mathbf{N}^m \rightsquigarrow \mathbf{N}$ *and* $h : \mathbf{N}^{m+2} \rightsquigarrow \mathbf{N}$. *Then, there is a unique function* $f : \mathbf{N}^{m+1} \rightsquigarrow \mathbf{N}$ *that satisfies*

$$f(x_1, \ldots, x_m, 0) = g(x_1, \ldots, x_m) \text{ for all } x_1, \ldots, x_m \text{ in } \mathbf{N},$$
$$f(x_1, \ldots, x_m, n+1) = h(x_1, \ldots, x_m, n, f(x_1, \ldots, x_m, n))$$
$$\text{for all } x_1, \ldots, x_m, n \text{ in } \mathbf{N},$$

where the recursion equations are taken to mean that

1. *for all* $x_1, \ldots, x_m$ *in* $\mathbf{N}$, $f(x_1, \ldots, x_m, 0)$ *is defined if and only if* $g(x_1, \ldots, x_m)$ *is defined, and if* $f(x_1, \ldots, x_m, 0)$ *is defined, then the given equality holds, and*

2. *for all* $x_1, \ldots, x_m, n$ *in* $\mathbf{N}$, $f(x_1, \ldots, x_m, n+1)$ *is defined if and only if* $f(x_1, \ldots, x_m, n)$ *and* $h(x_1, \ldots, x_m, n, f(x_1, \ldots, x_m, n))$ *are both defined, and if* $f(x_1, \ldots, x_m, n+1)$ *is defined, then the given equality holds.*

Once again, the partial function f in the previous corollary is said to be obtained by primitive recursion from the partial functions g and h. Using this idea, we will define another important class of functions called the partial recursive functions after some preliminaries.

Suppose that $k \geq 1$, $n \in \mathbf{N}$, and $g : \mathbf{N}^k \rightsquigarrow \mathbf{N}$, and for each i with $1 \leq i \leq k$, we have $h_i : \mathbf{N}^n \rightsquigarrow \mathbf{N}$. Then, recall from Chapter 2 that we have a function $g \circ (h_1, \ldots, h_k) : \mathbf{N}^n \rightsquigarrow \mathbf{N}$ (also written $g(h_1, \ldots, h_k)$) defined by making the domain of $g \circ (h_1, \ldots, h_k)$ be the set of $x_1, \ldots, x_n$, which are in the domains of all the h_i's and are such that

$$(h_1(x_1, \ldots, x_n), \ldots, h_k(x_1, \ldots, x_n))$$

is in the domain of g and defining

$$g \circ (h_1, \ldots, h_k)(x_1, \ldots, x_n) = g(h_1(x_1, \ldots, x_n), \ldots, h_k(x_1, \ldots, x_n))$$

for such $x_1, \ldots, x_n$. We will say that the function $g \circ (h_1, \ldots, h_k)$ is obtained by composition from $g, h_1, \ldots, h_k$.

Another way of building new functions from old, which we need, is given in the following definition.

Definition 4.9.9 *Suppose that* $n \in \mathbf{N}$ *and* $f : \mathbf{N}^{n+1} \rightsquigarrow \mathbf{N}$. *Then the function* $\mu f : \mathbf{N}^n \rightsquigarrow \mathbf{N}$ *is defined by setting* $\mu f(x_1, \ldots, x_n)$ *to be the least natural number* y *such that* $(x_1, \ldots, x_n, y')$ *is in the domain of* f *for all* y' *with* $0 \leq y' \leq y$ *and* $f(x_1, \ldots, x_n, y) = 0$, *if such a* y *exists, and to be undefined otherwise.*

In order to indicate that $g = \mu f$, we sometimes write

$$g(x_1, \ldots, x_n) = \mu y[f(x_1, \ldots, x_n, y) = 0]$$

and read μy as "the least y." If $g = \mu f$, then we say that g is obtained from f by *minimization*.

Definition 4.9.10 *Let* $T = \bigcup_{n>0} \mathbf{N}^n \rightsquigarrow \mathbf{N}$. *Then, the* partial recursive functions *are the subset of* T *defined by the following.*

1. *Every initial function is partial recursive.*

2. *If* g *and* $h_1, \ldots, h_k$ *are all partial recursive, and* f *is obtained from these functions by composition, then* f *is partial recursive.*

3. *If* g *and* h *are partial recursive and* f *is obtained from* g *and* h *by primitive recursion, then* f *is partial recursive.*

4. *If* g *is partial recursive and* f *is obtained from* g *by minimization, then* f *is partial recursive.*

Another way of expressing this definition is that the partial recursive functions are the closure of the initial functions under the operations of composition, primitive recursion, and minimization; i.e., they are those functions that can be obtained from the initial functions by a finite number of applications of these operations. Just as for the primitive recursive functions, this definition can be expressed in terms of constructors.

A function that is total and belongs to the set of partial recursive functions is called a *total recursive function* or just a *recursive function*. It is immediate form the definitions that every primitive recursive function is recursive. Since there are partial recursive functions that are not total (as is easily seen), not every partial recursive function is primitive recursive. It is also true, but less easily shown, that there are (total) recursive functions that are not primitive recursive.

4.10   Grammars

In this section, we introduce grammars; they provide one method for generating languages formally. The subject of formal languages, of which this definition is one of the foundations, is a large one that we will not go into deeply. Our aim in presenting grammars is to give another example of an area in computer science in which induction is a vital tool and to draw a connection between a certain kind of grammar and simultaneous inductive definitions of sets of strings.

A grammar provides rules for generating strings of symbols. A rule is just an ordered pair of strings and the meaning of the rule is that any occurrence of the first component of the rule may be replaced by the second component of the rule. For instance, if we wish to generate $\{a^n b^n \mid n \geq 0\}$, then two rules suffice, namely, (S, aSb) and (S, λ). Given any string $a^n b^n$, we can generate it by starting with the start symbol S, applying the first rule n times, each time replacing S with aSb to obtain successively S, aSb, $aaSbb, \ldots, a^n Sb^n$, and then applying the second rule to obtain $a^n b^n$. Symbols in the target alphabet are called *terminals*. Auxiliary symbols used in giving the rules are called *nonterminals*, and one of them is the *start symbol*. Rules (called *productions*) are pairs of strings consisting of terminals and nonterminals with the restriction that the first component of each rule must contain at least one nonterminal.

We make these ideas precise in the following definitions.

Definition 4.10.1 *A* grammar *(also called a* phrase-structure grammar *or* type 0 grammar*) is a 4-tuple $G = (N, T, S, P)$, where N and T are disjoint alphabets, S is an element of N, and P is a subset of $(N \cup T)^* \times (N \cup T)^*$ such that the first component of each element of P contains at least one element of N.*

The elements of N are called the nonterminals *of G, the elements of T are called the* terminals *of G, S is called the* start symbol *of G, and the elements of P are called the* productions *of G. If $(\alpha, \beta) \in P$, then we write $\alpha \underset{G}{\rightarrow} \beta$ or just $\alpha \rightarrow \beta$ if G is understood.*

If $G = (N, T, S, P)$ is a grammar, we will generally use capital roman letters (except X), with or without subscripts, to denote single nonterminals, small roman letters early in the alphabet (a, b, c, d, e) to denote single terminals, small roman letters late in the alphabet (u, v, w, x, y, z) to denote elements of T^*, and greek letters to denote elements of $(N \cup T)^*$. Also, we use $X, X_1, \ldots$ to denote both nonterminal and terminal symbols of G.

Definition 4.10.2 *If $G = (N, T, S, P)$ is a grammar, then we define a relation $\underset{G}{\Rightarrow}$ on $(N \cup T)^*$ by saying that $\alpha \underset{G}{\Rightarrow} \beta$ if β can be obtained from α by replacing the left side of a production in G with the right side of that production or, more formally, if there are $\alpha_L, \alpha_M, \alpha_R, \beta_M$ in $(N \cup T)^*$ such that $\alpha = \alpha_L \alpha_M \alpha_R$, $\beta = \alpha_L \beta_M \alpha_R$, and $\alpha_M \underset{G}{\rightarrow} \beta_M$. If $\alpha \underset{G}{\Rightarrow} \beta$, then we say that β is directly derivable from α.*

Productions of the form $\alpha \rightarrow \lambda$ are called *erasure productions*. The effect of $\alpha \rightarrow \lambda$ is to erase the word α. A grammar without erasure productions is said to be *λ-free*.

Definition 4.10.3 *A word $\alpha \in (N \cup T)^*$ generates a word $\beta \in (N \cup T)^*$ in the grammar $G = (N, T, S, P)$ if there is a finite sequence of words over $N \cup T$: $d = (\gamma_0, \gamma_1, \ldots, \gamma_n)$ such that $\gamma_0 = \alpha$, $\gamma_n = \beta$, and $\gamma_k \underset{G}{\Rightarrow} \gamma_{k+1}$ for $0 \leq k \leq n - 1$.*

The sequence $(\gamma_0, \gamma_1, \ldots, \gamma_n)$ *is referred to as a* derivation *of* β *from* α in G, *where* n *is the* length *of the derivation.*

We write $\overset{n}{\underset{G}{\Rightarrow}}$ for the n-th power of $\underset{G}{\Rightarrow}$, $\overset{+}{\underset{G}{\Rightarrow}}$ for the transitive closure of $\underset{G}{\Rightarrow}$, and $\overset{*}{\underset{G}{\Rightarrow}}$ for the reflexive and transitive closure of $\underset{G}{\Rightarrow}$. (Hence, $\alpha \overset{+}{\underset{G}{\Rightarrow}} \beta$ if and only if there is an $n \geq 1$ such that $\alpha \overset{n}{\underset{G}{\Rightarrow}} \beta$, and $\alpha \overset{*}{\underset{G}{\Rightarrow}} \beta$ if and only if there is an $n \geq 0$ such that $\alpha \overset{n}{\underset{G}{\Rightarrow}} \beta$.) If there is no possibility of confusion, we will drop the subscript G in these notations. By the definition of the n-th power of a relation, there is a derivation of length n of β from α in the grammar G if and only if $\alpha \overset{n}{\underset{G}{\Rightarrow}} \beta$.

Definition 4.10.4 *Let* $G = (N, T, S, P)$ *be a grammar. The* language generated *by* G *is the set of words* $L(G)$ *given by*

$$L(G) = \{u \in T^* \mid S \overset{*}{\underset{G}{\Rightarrow}} u\}.$$

Example 4.10.5 Consider the grammar $G = (\{S, U, V\}, \{a, b, c\}, S, P)$, where P is given by

$$P = \{S \to abc, S \to aUbc, Ub \to bU, Uc \to Vbcc),$$
$$bV \to Vb, aV \to aaU, aV \to aa\}.$$

We prove that $L(G) = \{a^n b^n c^n \mid n \geq 1\}$.

Claim 1. For every $p \geq 1$, we have $S \overset{*}{\underset{G}{\Rightarrow}} a^p U b^p c^p$ and $S \overset{*}{\underset{G}{\Rightarrow}} a^p b^p c^p$. The argument is by induction on p. For $p = 1$, we have $S \underset{G}{\Rightarrow} aUbc$ since we have the production $S \underset{G}{\to} aUbc$ and $S \underset{G}{\Rightarrow} abc$ since we have the production $S \underset{G}{\to} abc$.

Suppose now that we have $S \overset{*}{\underset{G}{\Rightarrow}} a^p U b^p c^p$ for $p \geq 1$. Applying the production $Ub \to bU$, we have

$$S \overset{*}{\underset{G}{\Rightarrow}} a^p U b^p c^p \overset{*}{\underset{G}{\Rightarrow}} a^p b^p U c^p,$$

and since $p \geq 1$, we can apply the production $Uc \to Vbcc$. This allows us to write

$$S \overset{*}{\underset{G}{\Rightarrow}} a^p b^p V b c^{p+1},$$

and using the production $bV \to Vb$, we can "move" V toward the left, thus obtaining

$$S \overset{*}{\underset{G}{\Rightarrow}} a^p V b^{p+1} c^{p+1} \overset{*}{\underset{G}{\Rightarrow}} a^{p+1} U b^{p+1} c^{p+1}.$$

Note that we used in the last step the production $aV \to aaU$. If, instead, we use the production $aV \to aa$, we obtain $S \overset{*}{\underset{G}{\Rightarrow}} a^{p+1}b^{p+1}c^{p+1}$, which justifies the claim made above.

This shows that

$$\{a^n b^n c^n | n \geq 1\} \subseteq L(G).$$

Conversely, suppose that $u \in L(G)$. We have $u \in \{a, b, c\}^*$ and $S \overset{*}{\underset{G}{\Rightarrow}} u$. In order to prove the converse inclusion, we need to justify the next claim.

Claim 2. For every $n \in \mathbf{N}, n \geq 1$, if $S \overset{n}{\underset{G}{\Rightarrow}} \theta$, then one of the following is true:

1. $\theta = a^p b^p c^p$ for some $p \geq 1$;

2. $\theta = a^p b^q U b^r c^p$, where $q \geq 0$, $r \geq 1$, and $q + r = p$;

3. $\theta = a^p b^p U c^p$;

4. $\theta = a^p b^q V b^r c^{p+1}$, where $q \geq 1$, $r \geq 0$, and $q + r = p + 1$;

5. $\theta = a^p V b^{p+1} c^{p+1}$.

For $n = 1$, we have either $\theta = abc$ (case 1) or $\theta = aUbc$ (case 2). Suppose now that the claim is true for n and consider θ such that $S \overset{n+1}{\underset{G}{\Rightarrow}} \theta$. If $S \overset{n}{\underset{G}{\Rightarrow}} \sigma \underset{G}{\Rightarrow} \theta$, then σ is in one of the cases (2.-5.), by the inductive hypothesis. If $\sigma = a^p b^q U b^r c^p$ with $q \geq 0$, $r \geq 1$, and $q + r = p$, then we can apply only the production $Ub \to bU$; hence, $\theta = a^p b^{q+1} U b^{r-1} c^p$ and, therefore, θ is either in case (2) (if $r > 1$) or, if $r = 1$, in case (3).

If $\sigma = a^p b^p U c^p$, we can apply only the production $Uc \to Vbcc$, which means that $\theta = a^p b^p V b c^{p+1}$, that is, θ is in case (4).

When $\sigma = a^p b^q V b^r c^{p+1}$ with $q + r = p + 1$ and $q > 1$, we can apply only the production $bV \to Vb$, and this allows us to write $\theta = a^p b^{q-1} V b^{r+1} c^{p+1}$. If $q = 1$, θ is in case (5); otherwise, θ is in case (4).

Finally, if $\sigma = a^p V b^{p+1} c^{p+1}$, we can apply either production $aV \to aa$ or production $aV \to aaU$. In the first situation, $\theta = a^{p+1} b^{p+1} c^{p+1}$ (case 1); in the latter case, $\theta = a^{p+1} U b^{p+1} c^{p+1}$, and this corresponds to case (2), which proves our claim.

Observe now that if θ consists exclusively from terminal symbols then θ must be in case (1) and this shows that $L(G) \subseteq \{a^n b^n c^n | n \geq 1\}$.

In computer science we are interested in languages that can be generated by grammars satisfying certain restrictions. These classes of grammars constitute the *Chomsky hierarchy* and are introduced by the following.

Definition 4.10.6 *A grammar* $G = (N, T, S, P)$ *is of type i*, $0 \leq i \leq 3$ *if it satisfies the corresponding restriction (i) from the following list.*

(0) No restriction on productions.

(1) Each production from P is either of the form $\alpha Y\beta \rightarrow \alpha\theta\beta$, where $\alpha,\beta \in (N\cup T)^*$, $\theta \in (N\cup T)^+$, and $Y \in N$, or of the form (S,λ). In the latter case, S may not occur in the right side of any production of G.

(2) Each production in P is of the form $Y \rightarrow \theta$, where $Y \in N$ and $\theta \in (N\cup T)^*$.

(3) Each production in P is either of the form $Y \rightarrow u$ or of the form $Y \rightarrow uZ$, where $Y, Z \in N$ and $u \in T^*$.

These types of grammars allow us to define corresponding classes of languages.

Definition 4.10.7 *A language of type i is a language that can be generated by a grammar of type i, $0 \leq i \leq 3$.*

The class of languages of type i will be denoted by $\mathcal{L}_i$ for $0 \leq i \leq 3$. Grammars of type 1 are also known as *context-sensitive grammars* because of the fact that a rule of the form $(\alpha Y\beta, \alpha\theta\beta)$ allows the replacement of the nonterminal symbol Y by the word θ when Y occurs in the context (α, β). In the case of type 2 grammars, a nonterminal symbol Y can be replaced by a word θ in any context. For this reason, type 2 grammars are also known as *context-free grammars*. Type 3 grammars are also called *regular grammars*. The languages that belong to the classes $\mathcal{L}_3$, $\mathcal{L}_2$, and $\mathcal{L}_1$, are referred to as *regular, context-free,* and *context-sensitive languages*, respectively.

It is obvious from Definition 4.10.6 that $\mathcal{L}_3 \subseteq \mathcal{L}_2$ and $\mathcal{L}_1 \subseteq \mathcal{L}_0$. The inclusion $\mathcal{L}_2 \subseteq \mathcal{L}_1$ is also valid, but it is not an immediate consequence of the definition.

Example 4.10.8 The language $\{a^n b^n | n \in \mathbf{N}\}$ is of type 2 since the grammar $G = (\{S\}, \{a, b\}, S, \{(S, aSb), (S, \lambda)\})$, mentioned above, is of type 2.

A generalization of context-sensitive grammars is introduced in the following.

Definition 4.10.9 *A grammar $G = (N, T, S, P)$ is length-increasing if for every production $\alpha \underset{G}{\rightarrow} \beta$ we have $|\alpha| \leq |\beta|$ with the possible exception of the production $S \underset{G}{\rightarrow} \lambda$. If $S \underset{G}{\rightarrow} \lambda$, then S may not occur on the right side of any production of G.*

Clearly, every context-sensitive grammar is also length-increasing. It is interesting to note that every language generated by a length-increasing grammar also belongs to $\mathcal{L}_1$, as we shall prove in the sequel (see Supplement 75).

Definition 4.10.10 *Grammars* G, G' *with the same terminal alphabet are equivalent if they generate the same language, that is, if* $L(G) = L(G')$.

We present now a technical result for derivations in context-free grammars.

Theorem 4.10.11 *Let* $G = (N, T, S, P)$ *be a context-free grammar. If* $X_1 \cdots X_k \overset{n}{\underset{G}{\Rightarrow}} \alpha$, *where* $X_1, \ldots, X_k \in N \cup T$ *and* $\alpha \in (N \cup T)^*$, *then we can write* $\alpha = \alpha_1 \cdots \alpha_k$, *where* $X_i \overset{n_i}{\underset{G}{\Rightarrow}} \alpha_i$ *and* $\sum_{1 \le i \le k} n_i = n$.

Proof. We use an argument by induction on n, $n \ge 0$. For $n = 0$, we have $\alpha_i = X_i$ for $1 \le i \le k$, and the statement is obviously true; in this case, $n_1 = \cdots = n_k = 0$.

Assume that the statement is true for derivations of length n and let $X_1 \cdots X_k \overset{n+1}{\underset{G}{\Rightarrow}} \alpha$. If $X_1 \cdots X_k \overset{n}{\underset{G}{\Rightarrow}} \gamma \Rightarrow \alpha$, by the inductive hypothesis, we can write $\gamma = \gamma_1 \cdots \gamma_k$, where $X_i \overset{n_i}{\underset{G}{\Rightarrow}} \gamma_i$ for $1 \le i \le k$ and $\sum_{1 \le i \le k} n_i = n$.

Let (Y, β) be the production applied in the last step $\gamma \underset{G}{\Rightarrow} \alpha$. Y occurs in one of the words $\gamma_1, \ldots, \gamma_k$, say, γ_j. In this case, we can write $\gamma_j = \gamma'_j Y \gamma''_j$ and α can be written as $\alpha = \alpha_1 \cdots \alpha_k$, where $\alpha_i = \gamma_i$ for $1 \le i \le j - 1$, and $j + 1 \le i \le k$,

$$X_j \overset{n_j}{\underset{G}{\Rightarrow}} \gamma_j \underset{G}{\Rightarrow} \gamma'_j \beta \gamma''_j = \alpha_j,$$

which proves the statement. ∎

Example 4.10.12 Consider the context-free grammar $G = (N, T, S_0, P)$, where $N = \{S_0, S_1, S_2\}$, $T = \{a, b\}$, and P contains the following productions:

$$(S_0, aS_2), (S_0, bS_1),$$
$$(S_1, a), (S_1, aS_0), (S_1, bS_1S_1),$$
$$(S_2, b), (S_2, bS_0), (S_2, aS_2S_2).$$

We prove that $L(G)$ consists of all words over $\{a, b\}$ that contain an equal number of a's and b's. Let $n_X(\alpha)$ be the number of occurrences of symbol X in the word α.

Claim 1. We prove by strong induction on p, $p \ge 1$, that

1. if $n_a(u) = n_b(u) = p$, then $S_0 \overset{*}{\underset{G}{\Rightarrow}} u$;

2. if $n_a(u) = n_b(u) + 1 = p$, then $S_1 \overset{*}{\underset{G}{\Rightarrow}} u$;

3. if $n_b(u) = n_a(u) + 1 = p$, then $S_2 \overset{*}{\underset{G}{\Rightarrow}} u$.

In the first case, for $p = 1$, we have either $u = ab$ or $u = ba$; hence, we have either

$$S_0 \underset{G}{\Rightarrow} aS_2 \underset{G}{\Rightarrow} ab$$

or

$$S_0 \underset{G}{\Rightarrow} bS_1 \underset{G}{\Rightarrow} ba.$$

For the second case, $u = a$, and we have $S_1 \underset{G}{\Rightarrow} a$; the third case, for $u = b$, is similar.

Suppose that the claim holds for $p \leq n$. Again, we consider three cases for the word u:

1. $n_a(u) = n_b(u) = n + 1$;

2. if $n_a(u) = n_b(u) + 1 = n + 1$;

3. if $n_b(u) = n_a(u) + 1 = n + 1$.

In the first case, we may have four situations:

$1_1.$ $u = abt$, where $t \in \{a, b\}^*$ and $n_a(t) = n_b(t) = n$,

$1_2.$ $u = bat$, where $t \in \{a, b\}^*$ and $n_a(t) = n_b(t) = n$,

$1_3.$ $u = aav$ with $n_b(v) = n + 1$ and $n_a(v) = n - 1$, or

$1_4.$ $u = bbw$ with $n_a(w) = n - 1$ and $n_b(w) = n + 1$.

By the inductive hypothesis, we have $S_0 \overset{*}{\underset{G}{\Rightarrow}} t$, and therefore, we obtain one of the following derivations:

$$S_0 \underset{G}{\Rightarrow} aS_2 \underset{G}{\Rightarrow} abS_0 \overset{*}{\underset{G}{\Rightarrow}} abt = u,$$
$$S_0 \underset{G}{\Rightarrow} bS_1 \underset{G}{\Rightarrow} baS_0 \overset{*}{\underset{G}{\Rightarrow}} bat = u,$$

for the cases (1_1) and (1_2), respectively.

On other hand, if $u = aav$, we can write $v = v'v''$, where v' is the shortest prefix of v, where the number of b exceeds the number of a. Clearly, we must have $n_b(v') = n_a(a) + 1 = n$, and therefore, $n_b(v'') = n_a(v'') + 1 = n$. By the inductive hypothesis, we have $S_2 \overset{*}{\underset{G}{\Rightarrow}} v'$, $S_2 \overset{*}{\underset{G}{\Rightarrow}} v''$; hence,

$$S_0 \underset{G}{\Rightarrow} aS_2 \underset{G}{\Rightarrow} aaS_2S_2 \overset{*}{\underset{a}{\Rightarrow}} av'v'' = u,$$

which concludes the argument for (1_3). We leave to the reader the similar arguments for the remaining cases. This allows us to conclude that every word that contains an equal number of a's and b's belongs to $L(G)$.

In order to prove the reverse inclusion, we justify the following.

Claim 2. The following implications hold.

1. If $S_0 \overset{n}{\underset{G}{\Rightarrow}} \alpha$, then $n_a(\alpha) + n_{S_1}(\alpha) = n_b(\alpha) + n_{S_2}(\alpha)$.

2. If $S_1 \overset{n}{\underset{G}{\Rightarrow}} \alpha$, then $n_a(\alpha) + n_{S_1}(\alpha) = n_b(\alpha) + n_{S_2}(\alpha) + 1$.

3. If $S_2 \overset{n}{\underset{G}{\Rightarrow}} \alpha$, then $n_a(\alpha) + n_{S_1}(\alpha) + 1 = n_b(\alpha) + n_{S_2}(\alpha)$.

The proof is by strong induction on n, where $n \geq 1$. For $n = 1$, the verification is immediate. For instance, if $S_1 \underset{G}{\Rightarrow} \alpha$, we have $\alpha = a$, $\alpha = aS_0$, or $\alpha = bS_1S_1$; in every case, the equality is satisfied.

Suppose that the implications of Claim 2 hold for derivations no longer than n.

If $S_0 \overset{n+1}{\underset{G}{\Rightarrow}} \alpha$, the first production applied in the derivation is $S_0 \to aS_2$ or $S_0 \to bS_1$. In the first case, by Theorem 4.10.11, we have $\alpha = a\beta$, where $S_2 \overset{n}{\underset{G}{\Rightarrow}} \beta$, and by the inductive hypothesis, we have $n_a(\beta) + n_{S_1}(\beta) + 1 = n_b(\beta) + n_{S_2}(\beta)$, so

$$
\begin{aligned}
n_a(\alpha) + n_{S_1}(\alpha) &= n_a(\beta) + 1 + n_{S_1}(\beta) \\
&= n_b(\beta) + n_{S_2}(\beta) \\
&= n_b(\alpha) + n_{S_2}(\alpha).
\end{aligned}
$$

The second case has a similar treatment.

If $S_1 \overset{n+1}{\underset{G}{\Rightarrow}} \alpha$, we have three possibilities.

(a) If the first production of the derivation is $S_1 \to a$, then $\alpha = a$ and the equality corresponding to this case is obviously satisfied.

(b) If the first production is $S_1 \to aS_0$, we can write $\alpha = a\beta$, where $S_0 \overset{n}{\underset{G}{\Rightarrow}} \beta$; hence, $n_a(\beta) + n_{S_1}(\beta) = n_b(\beta) + n_{S_2}(\beta)$, so

$$
\begin{aligned}
n_a(\alpha) + n_{S_1}(\alpha) &= n_a(\beta) + 1 + n_{S_1}(\beta) \\
&= n_b(\beta) + n_{S_2}(\beta) + 1 \\
&= n_b(\alpha) + n_{S_2}(\alpha) + 1.
\end{aligned}
$$

(c) If the derivation begins with (S_1, bS_1S_1), we can write $\alpha = b\beta\gamma$, where $S_1 \overset{p}{\underset{G}{\Rightarrow}} \beta$ and $S_1 \overset{q}{\underset{G}{\Rightarrow}} \gamma$, where $p, q \leq n$. By the inductive hypothesis, $n_a(\beta) + n_{S_1}(\beta) = n_b(\beta) + n_{S_2}(\beta) + 1$, and $n_a(\gamma) + n_{S_1}(\gamma) = n_b(\gamma) + n_{S_2}(\gamma) + 1$. Consequently,

$$
\begin{aligned}
n_a(\alpha) + n_{S_1}(\alpha) &= n_a(\beta) + n_a(\gamma) + n_{S_1}(\beta) + n_{S_1}(\gamma) \\
&= n_b(\beta) + n_{S_2}(\beta) + 1 + n_b(\gamma) + n_{S_2}(\gamma) + 1 \\
&= n_b(\alpha) + n_{S_1}(\alpha) + 1.
\end{aligned}
$$

The case of the derivation $S_2 \overset{*}{\underset{G}{\Rightarrow}} \alpha$ can be treated in a similar manner, and it is left to the reader.

Let $u \in L(G)$. From the existence of the derivation $S_0 \overset{*}{\underset{G}{\Rightarrow}} u$ and Claim 2, we obtain $n_a(u) = n_b(u)$, which shows the reverse inclusion.

Theorem 4.10.13 *For every context-free grammar G, there is a context-free, λ-free grammar G' such that $L(G') = L(G) - \{\lambda\}$.*

Proof. Let G be $G = (N, T, S_0, P)$, where $N = \{S_0, \ldots, S_{n-1}\}$. Consider the sequence $\{U_0, \ldots, U_m, \ldots\}$ of subsets of N defined by

$$
\begin{aligned}
U_0 &= \{S_i | S_i \in N \text{ and } S_i \underset{G}{\rightarrow} \lambda\}, \\
U_{m+1} &= U_m \cup \{S_i | S_i \in N \text{ and } S_i \underset{G}{\rightarrow} \alpha \text{ for some } \alpha \in (U_m)^*\},
\end{aligned}
$$

for $m \in \mathbf{N}$.

Since $U_0 \subseteq U_1 \subseteq \cdots \subseteq N$, there is $k \in \mathbf{N}$ such that $U_k = U_{k+1} = \ldots$. We claim that $S_i \overset{+}{\underset{G}{\Rightarrow}} \lambda$ if and only if $S_i \in U_k$.

We prove by strong induction on $p \geq 1$ that if $S_i \overset{p}{\underset{G}{\Rightarrow}} \lambda$ then $S_i \in U_k$. For $p = 1$, if $S_i \underset{G}{\Rightarrow} \lambda$, then $S_i \in U_0$ and $U_0 \subseteq U_k$.

Suppose that the statement is true for derivations $S_i \overset{*}{\underset{G}{\Rightarrow}} \lambda$ of length no greater than p and let $S_i \overset{p+1}{\underset{G}{\Rightarrow}} \lambda$. The first production applied in this derivation must have the form $S_i \underset{G}{\rightarrow} S_{i_1} \cdots S_{i_q}$; therefore, we have

$$
S_{i_1} \cdots S_{i_q} \overset{p}{\underset{G}{\Rightarrow}} \lambda.
$$

Hence, $S_{i_l} \overset{p_l}{\underset{G}{\Rightarrow}} \lambda$, where $p_l \leq p$ for $1 \leq l \leq q$, by Theorem 4.10.11. By the inductive hypothesis, we have $S_l \in U_k$, so $S_{i_1} \cdots S_{i_q} \in (U_k)^*$, which implies $S_i \in U_{k+1} = U_k$.

Conversely, it is easy to prove (by induction on p) that for every $S \in U_p$ we have $S \overset{*}{\underset{G}{\Rightarrow}} \lambda$. We leave this argument to the reader.

Consider now the set of productions P', where

$$
\begin{aligned}
P' = \{(S_i, \alpha') | \alpha' &\neq \lambda, \text{ there is } S_i \underset{G}{\rightarrow} \alpha \text{ and} \\
\alpha' &\text{ is obtained from } \alpha \text{ by erasing 0 or more} \\
&\text{symbols from } U_k\}.
\end{aligned}
$$

If G' is the context-free grammar $G' = (N, T, S_0, P')$, then $L(G') = L(G) - \{\lambda\}$. Indeed, suppose that $S_i \overset{p}{\underset{G'}{\Rightarrow}} \gamma$. Clearly, $\gamma \neq \lambda$ since G' has no erasure productions. We prove, by strong induction on p, that we have $S_i \overset{*}{\underset{G}{\Rightarrow}} \gamma$.

For $p = 0$, the statement is trivially true. Assume that it holds for derivations of length less than or equal to p and let $S_i \overset{p+1}{\underset{G'}{\Rightarrow}} \gamma$. If the first production applied in this derivation is $S_i \xrightarrow[G']{} X_1 \cdots X_h$, then, by Theorem 4.10.11, we can write $\gamma = \gamma_1 \ldots \gamma_h$, where $X_j \xRightarrow[G']{p_j} \gamma_j$, $p_j \leq p$, for $1 \leq j \leq h$. By the inductive hypothesis we have $X_j \xRightarrow[G]{*} \gamma_j$ for $1 \leq j \leq h$. Furthermore, assume that the production $S_i \xrightarrow[G']{} X_1 \cdots X_h$ was obtained from the production $S_i \xrightarrow[G]{} \theta_0 X_1 \theta_1 \cdots X_h \theta_h$ from P, where $\theta_0, \ldots, \theta_h \in (U_k)^*$. Our previous discussion allows us to infer the existence of the derivations $\theta_q \xRightarrow[G]{*} \lambda$ for $0 \leq q \leq h$. By combining the derivations obtained above, we have

$$
\begin{aligned}
S_i \quad & \xRightarrow[G]{} \quad \theta_0 X_1 \theta_1 \cdots X_h \theta_h \\
& \xRightarrow[G]{*} \quad X_1 \cdots X_h \\
& \xRightarrow[G]{*} \quad \gamma_1 \cdots \gamma_k = \gamma.
\end{aligned}
$$

This allows us to infer that $L(G') \subseteq L(G) - \{\lambda\}$.

In order to prove the converse inclusion, consider a derivation $S_i \xRightarrow[G]{p} \gamma$, where $\gamma \neq \lambda$. We claim that $S_i \xRightarrow[G']{*} \gamma$. Again, the argument is by strong induction on $p \geq 0$.

The case $p = 0$ is trivially true. Assume that the statement holds for derivations of length of no more than p and let $S_i \xRightarrow[G]{p+1} \gamma$, where $\gamma \neq \lambda$. Let $\beta = X_1 \cdots X_l$ be the word that folllows S_i in the previous derivation, that is,

$$
S_i \xRightarrow[G]{} \beta \xRightarrow[G]{p} \gamma.
$$

By Theorem 4.10.11, we can write $\gamma = \gamma_1 \cdots \gamma_l$, where $X_j \xRightarrow[G']{p_j} \gamma_j$, where $p_j \leq p$ for $1 \leq j \leq l$. If $\gamma_j \neq \lambda$, by the inductive hypothesis, we have $X_j \xRightarrow[G']{*} \gamma_j$. On the other hand, if $\gamma_j = \lambda$, we have $X_j \in U_k$. Let

$$
\{j_1, \ldots, j_q\} = \{j \,|\, 1 \leq j \leq l \text{ and } \gamma_j \neq \lambda\},
$$

where $j_1 < \cdots < j_q$. The definition of P' implies that we have the production $S_i \xrightarrow[G']{} X_{j_1} \cdots X_{j_q}$ in P'. Therefore, we can write

$$
\begin{aligned}
S_i \quad & \xRightarrow[G']{} \quad X_{j_1} \cdots X_{j_q} \\
& \xRightarrow[G']{*} \quad \gamma_{j_1} \cdots \gamma_{j_q} \\
& = \quad \gamma.
\end{aligned}
$$

This implies that $L(G) - \{\lambda\} \subseteq L(G')$. ∎

Corollary 4.10.14 *For every context-free grammar $G = (N, T, S, P)$, there exists an equivalent context-free grammar G' that is also context-sensitive.*

Proof. Note that every context-free grammar without erasure productions is also context-sensitive. Suppose that $G = (N, T, S, P)$ is a grammar with erasure productions. We distinguish two cases:

Case 1: If $\lambda \notin L(G)$, construct the grammar G' as in the proof of Theorem 4.10.13. In this situation, $L(G') = L(G) - \{\lambda\} = L(G)$. The grammar G' is context-sensitive since it has no erasure productions.

Case 2: If $\lambda \in L(G)$, consider again the grammar G'. Let $S_0' \notin N \cup T$ be a new symbol and let G'' be the grammar

$$G'' = (N \cup \{S_0'\}, T, S_0', P' \cup \{(S_0', S_0), (S_0', \lambda)\}).$$

It is easy to see that G'' is context-sensitive and equivalent to G. ∎

Corollary 4.10.15 *We have $\mathcal{L}_2 \subseteq \mathcal{L}_1$.*

Proof. The inclusion follows from of Corollary 4.10.14. ∎

Let $G = (N, T, S_0, P)$ be a context-free grammar. Consider the family of sets $\mathcal{U} = \{U_{S_i} | S_i \in N\}$, where $U_{S_i} = T^*$ for every $S_i \in N$.

Definition 4.10.16 *The constructor that corresponds to the production $r = (S_i, w_0 S_{i_1} w_1 \ldots S_{i_k} w_k)$, $k > 0$, is $(((S_{i_1}, \ldots, S_{i_k}), S_i), R_r)$, given by*

$$R_r = \{((u_1, \ldots, u_k), w_0 u_1 w_1 \cdots u_k w_k) | u_i \in T^*, for 1 \leq i \leq k\}.$$

The family of constructors defined by the context-free grammar G is

$$\Re_G = \{R_r | r \in P\}.$$

Define the family of languages over T, $\mathcal{B}_G$, by

$$\mathcal{B}_G = \{B_{S_i} | S_i \in N\},$$

where $B_{S_i} = \{u | u \in T^*, S_i \underset{G}{\Rightarrow} u\}$ for $0 \leq i \leq n - 1$. Clearly, $\mathcal{B}_G \preceq \mathcal{U}$.

Example 4.10.17 For the context-free grammar G introduced in Example 4.10.12, the family $\mathcal{B}_G$ consists of $\{B_{S_0}, B_{S_1}, S_{S_2}\}$, where $B_{S_0} = \emptyset$, $B_{S_1} = \{a\}$, and $B_{S_2} = \{b\}$.

The constructors or $\Re_G$ are given by the following table

Type	Relation
$((S_2), S_0)$	$((w), aw)$
$((S_1), S_0)$	$((v), bv)$
$((S_0), S_1)$	$((u), au)$
$((S_1 S_1), S_1)$	$((v_1, v_2), bv_1 v_2)$
$((S_0), S_2)$	$((w), bw)$
$((S_2 S_2), S_2)$	$((w_1, w_2), aw_1 w_2)$

for every $u \in L_{S_0}$, $v, v_1, v_2 \in L_{S_1}$, and $w, w_1, w_2 \in L_{S_2}$.

Theorem 4.10.18 *The closure of* $\mathcal{B}_G$ *under* $\Re_G$ *is the family*

$$\mathcal{L}_G = \{L_{S_i} | 0 \le i \le n - 1\},$$

where $L_{S_i} = \{u | u \in T^*, S_i \overset{*}{\underset{G}{\Rightarrow}} u\}$ *for* $0 \le i \le n - 1$.

Proof. Let P_{S_i} be the property of the elements of U_{S_i} that consists of all words $u \in T^*$ such that $S_i \overset{*}{\underset{G}{\Rightarrow}} u$. We apply Theorem 4.7.19 to prove that $L_{S_i} \subseteq \{u | u \in T^*, S_i \overset{*}{\underset{G}{\Rightarrow}} u\}$ for $0 \le i \le n - 1$, that is, that every $u \in L_{S_i}$ has property P_{S_i} for $0 \le i \le n - 1$.

Note that every element w of B_{S_i} has the property P_{S_i}, that is, $S_i \overset{*}{\underset{G}{\Rightarrow}} w$, because of the definition of B_{S_i}, for every $S_i \in N$.

Now, let R_r be a constructor from $\Re_G$, where $r = (S_i, w_0 S_{i_1} w_1 \cdots S_{i_k} w_k)$ and the type of R_r is $((S_{i_1}, \ldots, S_{i_k}), S_i)$. Assume that $((u_1, \ldots, u_k), u) \in R_r$ (which means that $u = w_0 u_1 w_1 \cdots u_k w_k$) and that each u_j has property P_j for $1 \le j \le k$, that is, $S_j \overset{*}{\underset{G}{\Rightarrow}} u_j$, for $1 \le j \le k$. From the existence of the production $r = (S_i, w_0 S_{i_1} w_1 \cdots S_{i_k} w_k)$, we infer that

$$
\begin{aligned}
S_i &\overset{*}{\underset{G}{\Rightarrow}} && w_0 S_{i_1} w_1 \cdots S_{i_k} w_k \\
&\overset{*}{\underset{G}{\Rightarrow}} && w_0 u_1 w_1 \cdots u_k w_k \\
&= && u,
\end{aligned}
$$

which shows that $L_{S_i} \subseteq \{u | u \in T^*, S_i \overset{*}{\underset{G}{\Rightarrow}} u\}$ for $0 \le i \le n - 1$.

Conversely, we can prove by strong induction on $n \ge 1$ that if $S_i \overset{n}{\underset{G}{\Rightarrow}} u$ then $u \in L_{S_i}$. Note that for $n = 1$ we have $u \in B_{S_i}$; hence, $u \in L_{S_i}$ because $\mathcal{L}_G$ is the closure of $\mathcal{B}_G$ under $\Re_G$.

Assume now that our statement is true for derivations of length not greater than n and let $S_i \overset{n+1}{\underset{G}{\Rightarrow}} u$. Suppose that the first production used in this derivation is $r = (S_i, w_0 S_{i_1} w_1 \ldots S_{i_k} w_k)$. In this case, the word u can be written as

$$u = w_0 u_1 w_1 \cdots u_k w_k,$$

where $S_{i_j} \overset{m_j}{\underset{G}{\Rightarrow}} u_j$ and $1 \le m_j \le n$ for $1 \le j \le k$. By the inductive hypothesis, we have $u_j \in L_{S_{i_j}}$ for $1 \le j \le k$. Since $\mathcal{L}_G$ is closed under $\Re_G$ and we have the constructor $R_r \in \Re_G$, we infer that $u \in L_{S_i}$. ∎

Definition 4.10.19 *Let* $G = (N, T, S, P)$ *be a context-free grammar. A leftmost derivation is a derivation* $\gamma_0, \gamma_1, \ldots, \gamma_n$ *in* G *such that, for every* k, $0 \le k \le n - 1$, *if the production applied in deriving* γ_{k+1} *from* γ_k *is* (S_i, β) *then* $\gamma_k = \gamma_{kL} S_i \gamma_{kR}$, $\gamma_{k+1} = \gamma_{kL} \beta \gamma_{kR}$ *and* $\gamma_{kL} \in T^*$.

In other words, if $\gamma_0, \gamma_1, \ldots, \gamma_n$ *is a leftmost derivation then, at every step of this derivation, we always rewrite the leftmost nonterminal symbol.*

Example 4.10.20 Let G be the grammar introduced in Example 4.10.12. The derivation

$$
\begin{aligned}
S_0 &\underset{G}{\Rightarrow} bS_1 & &\underset{G}{\Rightarrow} bbS_1S_1 \\
&\underset{G}{\Rightarrow} bbS_1aS_0 & &\underset{G}{\Rightarrow} bbS_1aaS_2 \\
&\underset{G}{\Rightarrow} bbaaaS_2 & &\underset{G}{\Rightarrow} bbaaab
\end{aligned}
$$

is not leftmost since in deriving bbS_1aaS_2 from bbS_1aS_0 we do not replace the leftmost nonterminal S_1. It is not difficult to see that we can transform this derivation into a leftmost derivation by changing the order in which nonterminals are replaced. Namely, in grammar G, we have the leftmost derivation

$$S_0, bS_1, bbS_1S_1, bbaS_1, bbaaS_0, bbaaaS_0, bbaaab.$$

Theorem 4.10.21 Let $G = (N, T, S, P)$ be a context-free grammar. For every derivation of length n in G,

$$d = (S_i, \gamma_1, \ldots, \gamma_n),$$

where $\gamma_n \in T^*$, there is a leftmost derivation of length n, using the same productions as d, that allows us to derive γ_n from S_i.

Proof. The argument is by strong induction on $n \geq 1$. For $n = 1$, the statement is trivially true since any derivation $S_i \underset{G}{\Rightarrow} w_1$ is a leftmost derivation.

Suppose that the statement holds for derivations whose length is no more than n and let

$$d = (S_i, \gamma_1, \ldots, \gamma_{n+1})$$

be a derivation of length $n+1$. If the first production used in this derivation is $S_i \to w_0 S_{i_1} w_1 \cdots S_{i_k} w_k$, then we can write $\gamma_{n+1} = w_0 u_1 w_1 \cdots u_k w_k$, where $S_{i_j} \underset{G}{\overset{*}{\Rightarrow}} u_j$ for $1 \leq j \leq k$ are derivations of length no greater than n. By the inductive hypothesis, for each of these derivations, we obtain the existence of the leftmost derivations $d_j = (S_{i_j}, \ldots, u_j)$ for $1 \leq j \leq k$, which uses the same set of productions. Now, we obtain the existence of the leftmost derivation:

$$
\begin{aligned}
S_i &\underset{G}{\Rightarrow} w_0 S_{i_1} w_1 S_{i_2} \ldots S_{i_k} w_k \\
&\underset{G}{\overset{*}{\Rightarrow}} w_0 u_1 w_1 S_{i_2} \ldots S_{i_k} w_k \\
&\qquad \text{(using derivation } d_1) \\
&\underset{G}{\overset{*}{\Rightarrow}} w_0 u_1 w_1 u_2 \ldots S_{i_k} w_k \\
&\qquad \text{(using derivation } d_2) \\
&\quad \vdots \\
&\underset{G}{\overset{*}{\Rightarrow}} w_0 u_1 w_1 u_2 \ldots u_k w_k \\
&\qquad \text{(using derivation } d_k),
\end{aligned}
$$

which concludes our argument. ∎

Definition 4.10.22 *A context-free grammar* $G = (N, T, S, P)$ *is ambiguous if there is* $S_i \in N$ *and* $w \in T^*$ *such that there are two leftmost derivations from* S_i *to* w *in* G. *Otherwise,* G *is unambiguous.*

Example 4.10.23 Consider the context-free grammars

$$G_1 = (\{S_0\}, \{a\}, S_0, \{(S_0, S_0 S_0), (S_0, a)\})$$

and

$$G_2 = (\{S_0\}, \{a\}, S_0, \{(S_0, a S_0), (S_0, a)\}).$$

They both generate the language $\{a^n | n \geq 1\}$. Note that in G_1 we have distinct leftmost derivations:

$$S_0 \quad \underset{G_1}{\Rightarrow} S_0 S_0 \quad \underset{G_1}{\Rightarrow} S_0 S_0 S_0 \quad \underset{G_1}{\Rightarrow} a S_0 S_0$$
$$\underset{G_1}{\Rightarrow} a a S_0 \quad \underset{G_1}{\Rightarrow} a a a$$

and

$$S_0 \quad \underset{G_1}{\Rightarrow} S_0 S_0 \quad \underset{G_1}{\Rightarrow} a S_0 \quad \underset{G_1}{\Rightarrow} a S_0 S_0$$
$$\underset{G_1}{\Rightarrow} a a S_0 \quad \underset{G_1}{\Rightarrow} a a a,$$

which prove that G_1 is an ambiguous grammar.

On other hand, the equivalent grammar G_2 is unambiguous since for every a^n, $n \leq 1$, we have exactly one derivation:

$$S_0 \underset{G_2}{\Rightarrow} a S_0 \underset{G_2}{\Rightarrow} a^2 S_0 \cdots \underset{G_2}{\Rightarrow} a^n.$$

Theorem 4.10.24 *Let* $G = (N, T, S, P)$ *be a context-free grammar. Consider the families of languages* $\mathcal{B}_G$ *and* $\mathcal{L}_G$ *introduced in Theorem 4.10.18. $\mathcal{L}_G$ *is freely generated by* $\Re_G$ *over* $\mathcal{B}_G$ *if and only if the grammar* G *is unambiguous.*

Proof. Suppose that the grammar G is unambiguous. We need to prove the following.

1. For every production $r = (S_i, w_0 S_{i_1} w_1 \cdots S_{i_k} w_k) \in P$ with $k > 0$ and the corresponding constructor R_r of type $((S_{i_1}, \ldots, S_{i_k}), S_i)$, we have

$$R_r(L_{S_{i_1}} \times \ldots \times L_{S_{i_k}}) \cap B_{S_i} = \emptyset.$$

2. If r, r' are two different productions, $r = (S_i, w_0 S_{i_1} w_1 \cdots S_{i_k} w_k)$, and $r' = (S_i, w_0' S_{j_1} w_1' \cdots S_{j_l} w_l')$, then

$$R_r(L_{S_{i_1}} \times \cdots \times L_{S_{i_k}}) \cap R_{r'}(L_{S_{j_1}} \times \cdots \times L_{S_{j_l}}) = \emptyset.$$

3. For every production $r = (S_i, w_0 S_{i_1} w_1 \ldots S_{i_k} w_k) \in P$, the restriction of R_r to $L_{S_{i_1}} \times \ldots \times L_{S_{i_k}}$ is injective.

Suppose that the first condition is not satisfied. There are k terminal words $u_j \in L_{S_{i_j}}$ for $1 \leq j \leq k$ such that $u = w_0 u_1 w_1 \cdots u_k w_k \in B_{S_i}$. This means that $S_i \underset{G}{\Rightarrow} u$ and also that we have the leftmost derivation

$$
\begin{aligned}
S_i \quad &\underset{G}{\Rightarrow} \quad w_0 S_{i_1} w_1 \cdots S_{i_k} w_k \\
&\underset{G}{\overset{*}{\Rightarrow}} \quad w_0 u_1 w_1 \cdots S_{i_k} w_k \\
&\vdots \\
&\underset{G}{\overset{*}{\Rightarrow}} \quad w_0 u_1 w_1 \cdots u_k w_k = u,
\end{aligned}
$$

which contradicts the unambiguity of G.

Assume now that the second condition is not satisfied and let r, r' be two different productions,

$$
r = (S_i, w_0 S_{i_1} w_1 \cdots S_{i_k} w_k) \text{ and } r' = (S_i, w_0' S_{j_1} w_1' \cdots S_{j_l} w_l')
$$

such that there is $u \in R_r(L_{S_{i_1}} \times \cdots \times L_{S_{i_k}}) \cap R_{r'}(L_{S_{j_1}} \times \cdots \times L_{S_{j_l}})$. We obtain the existence of the words $u_1, \ldots, u_k, u_1', \ldots, u_l' \in T^*$ and of the leftmost derivations

$$
S_{i_g} \overset{*}{\underset{G}{\Rightarrow}} u_g \text{ and } S_{j_h} \overset{*}{\underset{G}{\Rightarrow}} u_h'
$$

for $1 \leq g \leq k$ and $1 \leq h \leq l$ such that

$$
u = w_0 u_1 w_1 \cdots u_k w_k = w_0' u_1' w_1' \cdots u_l' w_l'.
$$

This allows us to infer the existence of two distinct leftmost derivations:

$$
S_i \underset{G}{\Rightarrow} w_0 S_{i_1} w_1 \cdots S_{i_k} w_k \overset{*}{\underset{G}{\Rightarrow}} u,
$$

and

$$
S_i \underset{G}{\Rightarrow} w_0' S_{j_1} w_1' \cdots S_{j_l} w_l' \overset{*}{\underset{G}{\Rightarrow}} u,
$$

which contradicts the fact that G is unambiguous.

Finally, assume that the third condition is violated. There is a production $r = (S_i, w_0 S_{i_1} w_1 \cdots S_{i_k} w_k) \in P$ such that we have $(u_1, \ldots, u_k) \neq (u_1', \ldots, u_k')$ in $L_{S_{i_1}} \times \cdots \times L_{S_{i_k}}$ and

$$
u = w_0 u_1 w_1 \cdots u_k w_k = w_0 u_1' w_1 \cdots u_l' w_l.
$$

We have the leftmost derivations $S_{i_j} \overset{*}{\underset{G}{\Rightarrow}} u_j$ and $S_{i_j} \overset{*}{\underset{G}{\Rightarrow}} u_j'$ for $1 \leq j \leq k$, and there is at least one j such that $u_j \neq u_j'$. In turn, these derivations can be used to generate two distinct leftmost derivations for the word u, namely,

$$S_i \underset{G}{\Rightarrow} w_0 S_{i_1} w_1 \cdots S_{i_k} w_k$$

$$\underset{G}{\overset{*}{\Rightarrow}} w_0 u_1 w_1 \cdots S_{i_k} w_k$$

$$\vdots$$

$$\underset{G}{\overset{*}{\Rightarrow}} w_0 u_1 w_1 \cdots u_k w_k = u,$$

and

$$S_i \underset{G}{\Rightarrow} w_0 S_{i_1} w_1 \cdots S_{i_k} w_k$$

$$\underset{G}{\overset{*}{\Rightarrow}} w_0 u_1' w_1 \cdots S_{i_k} w_k$$

$$\vdots$$

$$\underset{G}{\overset{*}{\Rightarrow}} w_0 u_1' w_1 \cdots u_k' w_k = u,$$

which is, again, a contradiction.

Conversely, assume that the three conditions required for free generation are satisfied and let $u \in T^*$ be a word such that there are two distinct leftmost derivations $S_i \underset{G}{\overset{m}{\Rightarrow}} u$ and $S_i \underset{G}{\overset{n}{\Rightarrow}} u$. If $m = 1$ and the first production of the second derivation is $r' = (S_i, w_0' S_{j_1} w_1' \cdots S_{j_l} w_l')$ with $l > 0$, then we have $u \in B_{S_i}$, $u = w_0' u_1' \cdots u_l' w_l'$, and

$$u \in R_{r'}(\{(u_1', \ldots, u_l')\}) \subseteq R_{r'}(L_{S_{j_1} \cdots S_{j_l}}),$$

which contradicts the first condition. Therefore, we may assume that we have both $m \geq 2$ and $n \geq 2$.

Suppose that the initial productions r and r', applied in the above derivations, are distinct, where

$$r = (S_i, w_0 S_{i_1} w_1 \cdots S_{i_k} w_k) \text{ and } r' = (S_i, w_0' S_{j_1} w_1' \cdots S_{j_l} w_l').$$

The word u can be factored as

$$u = w_0 u_1 w_1 \cdots u_k w_k = w_0' u_1' w_1' \cdots u_l' w_l',$$

where $(u_1, \ldots, u_k) \in L_{S_{i_1} \cdots S_{i_k}}$ and $(u_1', \ldots, u_l') \in L_{S_{j_1} \cdots S_{j_l}}$. Also,

$$((u_1, \ldots, u_k), u) \in R_r \text{ and } ((u_1', \ldots, u_l'), u) \in R_{r'},$$

which means that

$$R_r(L_{S_{i_1} \cdots S_{i_k}}) \cap R_{r'}(L_{S_{j_1} \cdots S_{j_l}}) \neq \emptyset,$$

which contradicts the second condition.

If both derivations begin with the application of r, then we can write u as

$$u = w_0 u_{i_1} w_1 \cdots u_{i_k} w_k = w_0 u_{i_1}' w_1 \cdots u_{i_k}' w_k,$$

where

$$(u_{i_1}, \ldots, u_{i_k}), (u'_{i_1}, \ldots, u'_{i_k}) \in L_{S_{i_1} \cdots S_{i_k}},$$

which contradicts the third condition. We conclude that there are no two such distinct leftmost derivations. ∎

4.11   Peano's Axioms

So far, we have taken the natural numbers as given and have used elementary facts about the arithmetic of the natural numbers whenever we needed them without worrying about how one can prove that these properties are true. Because the natural numbers are so familiar to us, this situation is probably not too disturbing. However, it is quite natural at some point to want to identify a certain minimal set of assumptions about the natural numbers such that all further properties of interest about them can be derived from these minimal assumptions. In other words, we seek a set of axioms about the natural numbers (preferably few in number) such that each axiom is obviously true based on our intuitive understanding of the natural numbers and such that the set of axioms taken together is powerful enough to let us show all the usual properties of arithmetic on the natural numbers that we need. In this section, we will present such a set of axioms, first given by the Italian mathematician Giuseppe Peano in 1889.

Peano gave his axioms in terms of the number 1 and the successor function S, where $S(n)$ is the next largest integer after n. (Intuitively, $S(n) = n + 1$. However, $+$ is not among the primitive notions of Peano's system.) We now give a slightly modernized version of Peano's axioms. Because we have been working more with the natural numbers than the positive integers, we give axioms for the natural numbers.

Definition 4.11.1 (Peano's Axioms) *The set of natural numbers* N, *the element 0 of* N, *and the function* S *from* N *to itself satisfy the following axioms:*

Axiom 4.11.2 *For no n in N do we have $S(n) = 0$ (i.e., 0 is not in the range of S).*

Axiom 4.11.3 *If $n, m \in N$ and $S(n) = S(m)$, then $n = m$, (i.e., S is injective).*

Axiom 4.11.4 *If A is a subset of N, $0 \in A$, and $S(n) \in A$ for every $n \in A$, then $A = N$.*

The first axiom says that 0 is not the successor of any natural number. The second axiom says that if two numbers have the same successor, then they are the same. We have run into both these statements before (written with $n + 1$ in place of $S(n)$) and took them as being so obvious as to not require any explanation. The third axiom is just a rephrasing of the principle of induction for the natural numbers using sets instead of properties. Of the three equivalent principles of Theorem 4.2.34, the one that we have chosen as an axiom is the only one that we could choose because the other two mention the $\leq$ relation on N, which we do not have available to us yet in our axiomatic approach. Before we can assert in our axiomatic development that the principle of strong induction and the well-ordering principle hold for N, we will have to define $\leq$ and show those facts about arithmetic on N that were needed to prove Theorem 4.2.34.

Peano's axioms have an interesting interpretation from the point of view of inductively defined sets and recursively defined functions which we have developed in the past few sections. The third axiom states that the closure of 0 under the unary operation S is all of N, while the first two axioms imply that N is freely generated by S over 0. Thus, from our perspective, we see that Peano's axioms are exactly what is needed to justify recursive function definitions over the natural numbers. (It is doubtful that this is why Peano chose his axioms; in his work, he introduced recursively defined functions, but never justified the definitions by showing that they actually determined a unique function.) Although it is easy to see why Peano's axioms are important, it may be hard to believe that they are all that is needed to prove the usual properties of arithmetic. We now indicate why this is so.

Definition 4.11.5 *A* Peano structure *is a triple* $\mathcal{S} = (N, 0, S)$, *consisting of a set N, an element 0 of N, and a unary function S on N that satisfies all three of the Peano axioms.*

There are many Peano structures other than the natural numbers.

Example 4.11.6 Let $\mathcal{S}_1 = (N_1, 0_1, S_1)$, where N_1 is the odd natural numbers, $0_1 = 1$, and $S_1(n) = n + 2$. Then, $\mathcal{S}_1$ is a Peano structure.

Example 4.11.7 Let $\mathcal{S}_2 = (N_2, 0_2, S_2)$, where $N_2 = \Sigma^+$ for some one element alphabet $\Sigma = \{a\}$, $0_2 = a$, and $S_2(w) = wa$. Then, $\mathcal{S}_2$ is a Peano structure.

In the previous examples, we rely on intuitive knowledge about the natural numbers in asserting that we have Peano structures. (The natural numbers come into the second example because they are involved in the definition of sequence.) An axiomatic proof that these are Peano structures would have to wait until properties of arithmetic on the natural numbers have been developed formally.

If $\mathcal{S} = (N, 0, S)$ is any Peano structure, then N is freely generated by S over 0. Thus, we may apply both Theorems 4.6.37 and 4.6.41 to get the following two results.

Theorem 4.11.8 *If $S = (N, 0, S)$ is a Peano structure, A is a set, $a \in A$, and $h : A \times N \to A$, then there is a unique function $f : N \to A$ satisfying the following two equations:*

$$f(0) = a,$$
$$f(S(n)) = h(f(n), n) \text{ for all } n \in N.$$

Theorem 4.11.9 *If $S = (N, 0, S)$ is a Peano structure, A and P are sets, and $g : P \to A$, $h : P \times A \times N \to A$, then there is a unique function $f : P \times N \to A$ satisfying the following two equations:*

$$f(p, 0) = g(p) \text{ for all } p \in P,$$
$$f(p, S(n)) = h(p, f(p, n), n) \text{ for all } p \in P, n \in N.$$

These results are just Corollaries 4.6.38 and 4.6.42 with $S(n)$ in place of $n + 1$ and generalized to an arbitrary Peano structure.

Definition 4.11.10 *Let $S_1 = (N_1, 0_1, S_1)$ and $S_2 = (N_2, 0_2, S_2)$ be two structures with $0_i \in N_i$ and S_i a unary operation on N_i for $i = 1, 2$. A function $\varphi : N_1 \to N_2$ is called a* homomorphism *from S_1 to S_2 if $\varphi(0_1) = 0_2$ and $\varphi(S_1(n)) = S_2(\varphi(n))$ for all $n \in N_1$.*

An isomorphism *from S_1 to S_2 is a homomorphism that is also a bijection. If there is an isomorphism from S_1 to S_2, then S_1 and S_2 are called* isomorphic.

The idea is that two isomorphic structures are the same except for a renaming of the elements and that the isomorphism gives a renaming that transforms one structure into the other.

Theorem 4.11.11 *If $S_1 = (N_1, 0_1, S_1)$ is a Peano structure and $S_2 = (N_2, 0_2, S_2)$ is any structure with $0_2 \in N_2$ and S_2 a unary operation on N_2, then there is a unique homomorphism from S_1 to S_2.*

Proof: This follows immediately from Theorem 4.11.8 with $A = N_2, a = 0_2, h(a, n) = S_2(a)$. ∎

Theorem 4.11.12 *Any two Peano structures are isomorphic.*

Proof: Let $S_1 = (N_1, 0_1, S_1)$ and $S_2 = (N_2, 0_2, S_2)$ be two Peano structures. According to Theorem 4.11.11, there are homomorphisms φ from S_1 to S_2 and ψ from S_2 to S_1. Since $\psi\varphi(0_1) = \psi(0_2) = 0_1$ and for any $n \in N_1$, $\psi\varphi(S_1(n)) = \psi(S_2(\varphi(n))) = S_1(\psi\varphi(n))$, $\psi\varphi$ is a homomorphism from S_1 to itself. But, 1_{N_1} is also a homomorphism from S_1 to itself, so by the uniqueness in Theorem 4.11.11, $\psi\varphi = 1_{N_1}$. Similarly, $\varphi\psi = 1_{N_2}$. Thus, φ is a bijection and S_1 and S_2 are isomorphic. ∎

In the terminology we have introduced, our axiomatic approach to the natural numbers consists of assuming that the natural numbers N, together with 0 and the successor function, form a Peano structure. We then base all further proofs about the arithmetic of the natural numbers on this assumption alone (plus general set-theoretic facts). The previous theorem shows that the assumption that the natural numbers form a Peano structure completely characterizes the natural numbers, and this suggests that Peano's axioms are, in fact, strong enough to allow us to derive the usual properties of arithmetic on the natural numbers. We now indicate how this is done.

Definition 4.11.13 *Let $S = (N, 0, S)$ be any Peano structure. Then, the addition, multiplication, and exponentiation operations for S are the unique binary operations $f_+, f_\times, f_{exp}$ satisfying the following equations for all $n, m \in N$:*

$$f_+(n, 0) = n,$$
$$f_+(n, S(m)) = S(f_+(n, m)),$$

$$f_\times(n, 0) = 0,$$
$$f_\times(n, S(m)) = f_+(f_\times(n, m), n),$$

and

$$f_{exp}(n, 0) = S(0),$$
$$f_{exp}(n, S(m)) = f_\times(f_{exp}(n, m), n).$$

The $\leq$ relation on S is the binary relation on N defined by

$$n \leq m \text{ iff for some } p \in N, n + p = m.$$

The existence and uniqueness of the functions $f_+, f_\times$, and f_{exp} follow from Theorem 4.11.9 with $P = N$. (For example, to get the definition of $f_\times$ into the form of the theorem, take g to be the constant function 0 and define $h(x, y, z) = f_+(y, x)$. Note how we need to know that f_+ exists before this makes sense.) We will write $n + m$ for $f_+(n, m)$, nm for $f_\times(n, m)$, and n^m for $f_{exp}(n, m)$. In this notation, and writing $n + 1$ for $S(n)$, the definitions of the functions become

$$n + 0 = n,$$
$$n + (m + 1) = (n + m) + 1,$$

$$n0 = 0,$$
$$n(m + 1) = nm + 1,$$

and

$$n^0 = 1,$$
$$n^{m+1} = n^m n.$$

Note that we have defined 0^0 to be 1.

With these definitions and the third Peano axiom, it is now possible to prove the usual facts we have been assuming about ordinary arithmetic. We give one sample and leave some of the other verifications for the exercises.

Theorem 4.11.14 (Commutativity of +.) *Let $S = (N, 0, S)$ be any Peano structure (such as the natural numbers). Then, for any $n, m \in N$, we have $n + m = m + n$.*

Proof: Let $A = \{n \in N : n + m = m + n \text{ for all } m \in N\}$. We wish to show that $A = N$. We will use the third Peano axiom (i.e., the principle of induction). To show that $0 \in A$, we must show that $0 + m = m + 0$ for all $m \in N$. Since, by definition, $m + 0 = m$, this amounts to showing that $0 + m = m$ for all $m \in N$. Let $B = \{m \in N : 0 + m = m\}$. Then, $0 \in B$ since $0 + 0 = 0$, and if $m \in B$, then $m = 0 + m$, so $S(m) = S(0 + m) = 0 + S(m)$ (by definition of $+$), so $S(m) \in B$. By induction, this shows that $B = N$ and, hence, that $0 \in A$.

Now, suppose that $n \in A$. We must show that $S(n) \in A$, i.e., that $S(n) + m = m + S(n)$ for every $m \in N$. To this end, let $C = \{m \in N : S(n) + m = m + S(n)\}$. If we show that $C = N$, then we are done. The fact that $0 \in C$ follows from the fact that $0 \in A$. Suppose that $m \in C$. Then, we have the following calculation:

$$
\begin{aligned}
S(n) + S(m) &= S(S(n) + m) \text{ (definition of } +) \\
&= S(m + S(n)) \text{ (since } m \in C) \\
&= S(S(m + n)) \text{ (definition of } +) \\
&= S(S(n + m)) \text{ (since } n \in A) \\
&= S(n + S(m)) \text{ (definition of } +) \\
&= S(S(m) + n) \text{ (since } n \in A) \\
&= S(m) + S(n) \text{ (definition of } +).
\end{aligned}
$$

Thus, by induction $C = N$, completing the proof. ∎

Note that the previous proof only used the existence of the $+$ operation satisfying the conditions of Definition 4.11.13 and the third Peano axiom. Similar proofs can be given for all of the usual commutative, associative, and distributive laws for arithmetic in the natural numbers (or any other Peano structure). The usual facts about $\leq$, such as the fact that $\leq$ is a linear order, and all of the other arithmetic facts about the natural numbers that we have used can be shown as well. In particular, the arithmetic assumptions used in the reasoning in Theorem 4.2.34 can be justified, and hence, we can conclude that the principle of strong induction and the well-ordering principle hold in any Peano structure. We leave this development for the exercises.

Although Peano's axioms provide a satisfactory way of developing the arithmetic of the natural numbers axiomatically, there is still the obvious question of whether one can do no better than just assume that the natural numbers exist and form a Peano structure or whether one can actually construct the natural numbers using set-theoretic techniques and then prove that they form a Peano structure. In fact, it is possible to construct the natural numbers using set-theoretic techniques, but one must assume an axiom for set theory whose main purpose is to make the construction work. We indicate how this is done in the exercises.

4.12   Well-Founded Sets and Induction

In this section, we will look at a very general principle of induction. Many of the versions of induction that we have seen so far will turn out to be special cases of this one principle.

We need to consider a generalization of the notion of the well-founded poset introduced in Definition 3.4.12.

Let $\rho \subseteq S \times S$ be a binary relation on the set S and consider the relation $\rho_1 = \rho - \iota_S$. We have $(x, y) \in \rho_1$ if and only if $(x, y) \in \rho$ and $x \neq y$.

Definition 4.12.1 *An* infinite ρ-sequence *is a sequence* $s : \mathbf{P} \longrightarrow S$ *such that* $(s(n), s(n + 1)) \in \rho_1$.

The notion of infinite ρ-sequence is a generalization of the notion of an infinite ascending (or descending) sequence for posets. Namely, if (S, ρ) is a poset, then every infinite ρ-sequence is an infinite ascending sequence of the poset (see Definition 3.4.9). On the other hand, every infinite ρ^{-1}-sequence is an infinite descending sequence in the poset.

The next definition introduces a generalization of artinian and noetherian posets.

Definition 4.12.2 *A* well-founded set *is a pair* (S, ρ), *where* ρ *is a binary relation on* S *such that there are no infinite* ρ^{-1}-sequences in S.

The poset (S, ρ) is well-founded in the sense of Definition 4.12.2 if and only if the pair (S, ρ) is a well-founded poset (due to Theorem 3.4.13) or, equivalently, if the poset (S, ρ) is artinian.

Let S be a set and let $\rho \subseteq S \times S$ be a binary relation on S. We shall denote by $Pred_\rho(x)$ the set $\{y \mid y \in S, (y, x) \in \rho_1\}$, where $\rho_1 = \rho - \iota_S$. When there is no risk of confusion, we shall omit the subscript ρ, and we shall write $Pred(x)$ rather than $Pred_\rho(x)$. It is quite transparent that this notation is inspired by posets. Indeed, if (S, ρ) is a poset, the set $Pred_\rho(x)$ is the set of all elements of S that precede x with respect to the partial order ρ.

Theorem 4.12.3 Principle of Induction for Well-Founded Sets *Let* (S, ρ) *be a well-founded set. Let* P *be a property of* S *such that*

> *For every* $x \in S$, *if* $P(y)$ *is true for every* $y \in Pred(x)$, *then* $P(x)$ *is true.*

Then, $P(x)$ *is true for every* $x \in S$.

Proof: Suppose that P is a property of S that meets the condition of the theorem, but that $P(x)$ is false for some $x \in S$. We will define an infinite ρ^{-1}-sequence $\{x_n \mid n \geq 1\}$ such that for every n, $P(x_n)$ is false. We start by setting $x_1 = x$, where x is the element such that $P(x)$ is false. If we have defined x_n so that $P(x_n)$ is false, then, by our assumption about P, there must be a $y \in Pred(x_n)$ such that $P(y)$ is false. Choose such a y and set $x_{n+1} = y$. Then, $(x_{n+1}, x_n) \in \rho$ and $x_{n+1} \neq x_n$, since $x_{n+1} \in Pred(x_n)$, and by choice of y, $P(x_{n+1})$ is false. More precisely, define a function $f : S \to S$ by letting $f(y) = x$ if $P(y)$ is true and by defining $f(y)$ to be some $z \in Pred(y)$ with $P(z)$ false if $P(y)$ is false. (Such a z exists because of our assumption about P. The existence of such a function f may seem obvious, but to show that it exists, one needs the axiom of choice.) Then, $x_1 = x$ and $x_{n+1} = f(x_n)$.

Thus, starting with the existence of a single x such that $P(x)$ is false, we get a whole infinite ρ^{-1}-sequence of elements of S for which P fails to hold. Since the poset has no infinite ρ^{-1}-sequences, there must not be any x such that $P(x)$ is false. ∎

As we have noticed, every well-ordered poset is a well-founded poset and, consequently, a well-founded set. Thus, as a corollary to the previous theorem, we get the following.

Theorem 4.12.4 Principle of Induction for Well-Ordered Posets
Let (S, ρ) be a well-ordered poset. Let P be a property of S such that

> *for every $x \in S$, if $P(y)$ is true for every $y \in Pred(x)$, then $P(x)$ is true.*

Then, $P(x)$ is true for every $x \in S$.

We will show how many of the versions of induction we have seen so far are special cases of the principle of induction for well-founded sets. One observation will be useful when we do this. Suppose that $Pred(x)$ is empty. Then, the statement "for every $y \in Pred(x)$, $P(y)$ is true" is vacuously true. Consequently, for such an x, the statement "if $P(y)$ is true for every $y \in Pred(x)$, then $P(x)$ is true" is equivalent to the statement "$P(x)$ is true."

Every version of induction that we have given for a subset of the integers is a special case of the principle of induction for well-founded posets.

Example 4.12.5 Let n_0 be an integer and let $S = \{n \mid n \in \mathbf{N}, n \geq n_0\}$. Define the relation $\rho = \{(n, n+1) \mid n \geq n_0\}$. Then, (S, ρ) is a well-founded set, and Theorem 4.12.3, when applied to (S, ρ), gives Theorem 4.2.4, since $Pred(n_0) = \emptyset$ and $Pred(n+1) = \{n\}$ for $n \geq n_0$. Taking $n_0 = 0$ here gives the principle of induction for the natural numbers. (Note that ρ is not a well ordering in this example, so the principle of induction for well-ordered sets does not generalize this version of induction.)

Example 4.12.6 Let n_0 be an integer and let $S = \{n \mid n \in \mathbf{N}, n \geq n_0\}$. The pair $(S, \leq)$ is a well-founded partial order on S. (In fact, "$\leq$" is a well ordering on S.) Theorem 4.12.3 applied to this R gives Theorem 4.2.22. Taking $n_0 = 0$ here gives the principle of strong induction for the natural numbers.

Example 4.12.7 Let n_0 be an integer and let $S = \{n \mid n \in \mathbf{N}, n \geq n_0\}$. Let ℓ be a natural number. Define a relation ρ on S by $(n, m) \in \rho$ if $n \leq m$ and $m > n_0 + \ell$. Then, (S, ρ) is a well-founded set and Theorem 4.12.3 yields Theorem 4.2.26.

All the other versions of induction over a subset of the integers that we have given are also special cases of Theorem 4.12.3. We leave the verification of this fact to the exercises.

Example 4.12.8 Let $\mathcal{R}$ be a family of constructors on a set U, let B be a subset of U, and let C be the closure of B under $\mathcal{R}$. We define a binary relation θ on C as follows. For each $R \in \mathcal{R}$, if R is n-ary, $(x_1, \ldots, x_n)$ is an n-tuple of elements of C, x is an element of C, and $((x_1, \ldots, x_n), x) \in R$, then we put all of the pairs (x_i, x) into θ. In other words, $(x, y) \in \theta$ if and only if there is some n-ary R in $\mathcal{R}$ and some n-tuple $(x_1, \ldots, x_n)$ of elements of C that includes x such that $((x_1, \ldots, x_n), y) \in R$.

The pair (C, θ) might not be well-founded; for example, there might be a unary $R \in \mathcal{R}$ and $x \neq y$ in C with xRy and yRx. However, in many cases (C, θ) will be a well-founded set. For instance, if $U = \mathrm{Seq}(A)$ for some set A and $(x, y) \in \theta$ implies that x is a substring of y, then (C, θ) is well-founded. When (C, θ) is a well-founded set, Theorem 4.12.3 gives a way of showing that every element of C has some property P. If C is freely generated by $\mathcal{R}$ over B, then $Pred_\theta(x) = \emptyset$ for $x \in B$, while if $x \in R(C^n)$ for some n-ary $R \in \mathcal{R}$, say, $((x_1, \ldots, x_n), x) \in R$, with $(x_1, \ldots, x_n) \in C^n$, then $Pred(x) = \{x_1, \ldots, x_n\}$. Furthermore, when C is freely generated by $\mathcal{R}$ over B, then (C, θ) will be well-founded. To see this, let $P(x)$ be the property of C given by "there are no infinite θ^{-1}-sequences in (C, θ) starting with x." If $x \in B$, then, by the definition of free generation, there are no θ^{-1}-sequences of any length starting with x, so $P(x)$ is true. Now, suppose that $((x_1, \ldots, x_n), x) \in R$ for some $R \in \mathcal{R}$ and that $P(x_i)$ is true for all i, $1 \leq i \leq n$. Then, by the free generation of C, any θ^{-1}-sequence in (C, θ) starting with x must have one of the x_i's as its second element, and hence, by the induction hypothesis, there are no infinite θ^{-1}-sequences starting with x, so $P(x)$ is true. Thus, when C is freely generated by $\mathcal{R}$ over B, proof by induction over θ is identical to Theorem 4.6.14. Hence, many instances of proof by structural induction are special cases of proof by induction over a well-founded set.

Example 4.12.9 Let $S = \{n \mid n \in \mathbf{N}, n \geq 2\}$. Define a relation ρ on S by $(n, m) \in \rho$ if n divides m evenly and $n < m$. Then, (S, ρ) is well-founded since $(n, m) \in \rho$ implies that $n < m$. We use induction over (S, ρ)

Example 4.12.9 Let $S = \{n \mid n \in \mathbf{N}, n \geq 2\}$. Define a relation ρ on S by $(n, m) \in \rho$ if n divides m evenly and $n < m$. Then, (S, ρ) is well-founded since $(n, m) \in \rho$ implies that $n < m$. We use induction over (S, ρ) to show that every element of S can be written as the product of one or more primes. Suppose that $n \in S$ and that every number in $Pred_\rho(n)$ can be written as a product of one or more primes. If n is prime, then it is the product of a single prime, namely, itself. If n is not prime, then it can be written as $n = mp$, where m and p are both in $Pred_\rho(n)$. By the induction hypothesis, m and p are both products of one or more primes, so combining these prime factorizations, we see that n is the product of primes.

We proved this same result in Examples 4.2.24 and 4.2.31 using strong induction and course of values induction. It is instructive to compare the proofs.

4.13 Fixed Points and Fixed Point Induction

We have seen in Section 3.9 that, as a consequence of Zorn's lemma, each monotonic function over a cpo has a least fixed point. The reader will notice that the proof of this fact is existential: we have shown only that such a fixed point exists but not how to obtain it.

Theorem 4.13.1 Let (M, ρ) be a poset and let $f : M \longrightarrow_c M$ be a monotonic function on M. If $\perp$ is the least element of the poset (M, ρ), then the set $C_f = \{\perp, f(\perp), \ldots, f^n(\perp), \ldots\}$ is a chain.

Proof. We prove, by induction on $n \geq 0$, that $(f^n(\perp), f^{n+1}(\perp)) \in \rho$.

For $n = 0$, we have $(\perp, f(\perp)) \in \rho$, which is justified by the fact that $\perp$ is the least element of the poset (M, ρ).

Suppose that $(f^n(\perp), f^{n+1}(\perp)) \in \rho$. Using the monotonicity of f, we obtain immediately $(f^{n+1}(\perp), f^{n+2}(\perp)) \in \rho$.

We can show now that $(f^n(\perp), f^{n+p}(\perp)) \in \rho$ for any $p \geq 0$, using the fact that ρ is a partial order. This argument can be made by induction on p.

For $p = 0$, we have $(f^n(\perp), f^n(\perp)) \in \rho$ because of the reflexivity of ρ.

Assume now that $(f^n(\perp), f^{n+p}(\perp)) \in \rho$. Applying the result obtained in the first part of the proof, we have $(f^{n+p}(\perp), f^{n+p+1}(\perp)) \in \rho$, and the transitivity of ρ gives $(f^n(\perp), f^{n+p+1}(\perp)) \in \rho$.

We just proved that for $m, n \in N$, $m \geq n$, we have $(f^n(x), f^m(x)) \in \rho$. This shows that the set $\{\perp, f(\perp), \ldots, f^n(\perp), \ldots\}$ is a chain in (M, ρ). ∎

Let (M, ρ) be a cpo. We give a new proof of the Knaster-Tarski theorem for the case of continuous functions. The continuity gives us the possibility of giving an explicit description of the least fixed point of the function.

Theorem 4.13.2 Any continuous function $f : M \longrightarrow_c M$ has a least fixed point that is the least upper bound of the chain $\{\perp, f(\perp), f^2(\perp), \ldots\}$.

Note that $f(C_f) = \{f(\bot), \ldots, f^n(\bot), \ldots\}$ and this is also a chain; also, $\sup f(C_f) = x_0$. The continuity of f means that $\sup f(C_f) = f(\sup C_f)$, which means that $x_0 = f(x_0)$; hence, x_0 is a fixed point of f.

On the other hand, let y be an arbitrary fixed point of f; we have $y = f(y)$. We can prove, by induction on n, that $(f^n(\bot), y) \in \rho$. For $n = 0$, this amounts to $(\bot, y) \in \rho$, which is true in view of the definition of $\bot$. Suppose that $(f^n(\bot), y) \in \rho$. Using the monotonicity of f and the fact that y is a fixed point of f, we have $(f^{n+1}(\bot), f(y)) = (f^{n+1}(\bot), y) \in \rho$, which concludes the inductive argument. Since y is an upper bound for C_f, we have $(x_0, y) \in \rho$; hence, x_0 is the least fixed point of f. ∎

The least fixed point of f will be denoted by fix (f). This notation suggests the consideration of a function fix $: (M \longrightarrow_c M) \longrightarrow M$, which maps a continuous function $f : M \longrightarrow_c M$ to its least fixed point.

Theorem 4.13.3 *Let (M, ρ) be a cpo. The function fix $: (M \longrightarrow_c M) \longrightarrow M$ is continuous.*

Proof. Let us consider the functions $F_i : (M \longrightarrow_c M) \longrightarrow M$, which are defined by $F_i(f) = f^i(\bot)$ for $i \in \mathbf{N}$. As we proved in Theorem 4.13.1, $m < n$ implies $(F_m(f), F_n(f)) \in \rho$, which means that $\{F_i \mid i \in \mathbf{N}\}$ is a chain.

We prove, by induction on i, that each function F_i is continuous. For $i = 0$, $F_0(f) = f^0(\bot) = \bot$; therefore, F_0 is a constant function; hence, it is continuous.

Suppose that F_i is continuous. We have

$$F_{i+1}(f) = f^{i+1}(\bot) = \text{apply } (f, f^i(\bot)) = \text{apply } (f, F_i(f)),$$

which implies the continuity of F_{i+1}.

From the definition of fix(f), we obtain fix$(f) = \sup\{f^i(\bot) \mid i \in \mathbf{N}\} = \sup\{F_i(f) \mid i \in \mathbf{N}\} = \sup\{F_i \mid i \in \mathbf{N}\}(f)$. This allows us to infer (from Theorem 3.8.5) that fix is a continuous function as the supremum of a chain of continuous functions. ∎

An immediate consequence of Theorem 4.13.2 is given below.

Corollary 4.13.4 *Let (M, ρ) be a cpo. If $f : M \longrightarrow_c M$, then $(f(z), z) \in \rho$ implies $(fix(f), z) \in \rho$.*

Proof. We prove, by induction on i, that for any $i \in \mathbf{N}$ and $z \in M$ such that $(f(z), z) \in \rho$ we have $(f^i(\bot), z) \in \rho$. This is clearly true for $i = 0$. Suppose that this holds for i. Then, using the monotonicity of f, we have $(f^{i+1}(\bot), f(z)) \in \rho$, and this, combined with the fact that $(f(z), z) \in \rho$, implies $(f^{i+1}(\bot), z) \in \rho$. Consequently, z is an upper bound of the chain $\{\bot, f(\bot), \ldots, f^i(\bot), \ldots\}$, and this implies $(\text{fix}(f), z) \in \rho$. ∎

Corollary 4.13.5 *Let (M, ρ) be a cpo such that for every u and v in M, $\sup\{u, v\}$ exists and let $f : M \longrightarrow_c M$. Define $g : M \longrightarrow M$ by $g(x) = \sup\{x, f(x)\}$ for every $x \in M$. Then, g is continuous and $fix(g) = fix(f)$.*

Proof. The fact that g is continuous is not hard to show; we leave the argument to the reader. Let $x_0 = \text{fix}(f)$ and $y_0 = \text{fix}(g)$. We have $g(x_0) = \sup\{x_0, f(x_0)\} = \sup\{x_0, x_0\} = x_0$. Therefore, x_0 is a fixed point for g and $(y_0, x_0) \in \rho$. Conversely, since $\sup\{y_0, f(y_0)\} = y_0$, we have $(f(y_0), y_0) \in \rho$. By Corollary 4.13.4, we have $(x_0, y_0) \in \rho$, so $x_0 = y_0$. ∎

Let M be a set and let $\textbf{Bool} = \{\textbf{F}, \textbf{T}\}$. We remind the reader that for a predicate p on a set M we introduced in Section 2.3 the set $T_p = \{x \in M \mid p(x) = \textbf{T}\}$.

Consider now a cpo (M, ρ).

Definition 4.13.6 *A predicate* $p : M \longrightarrow \textbf{Bool}$ *is* admissible *if for every chain* $L \subseteq M$, $p(x) = \textbf{T}$ *for every* $x \in L$ *implies* $p(\sup L) = \textbf{T}$.

It is not difficult to see that if $\textbf{Bool}$ is equipped with the partial order defined by $\rho_{\textbf{Bool}} = \{(\textbf{F}, \textbf{F}), (\textbf{F}, \textbf{T}), (\textbf{T}, \textbf{T})\}$ then any admissible predicate is also a monotonic function between the posets (M, ρ) and $(\textbf{Bool}, \rho_{\textbf{Bool}})$.

Example 4.13.7 Let (M, ρ), (M', ρ') be two cpos. Consider two families of continuous functions $\{f_i \mid 1 \le i \le n$, where $f_i : M \longrightarrow_c M',\}$ and $\{g_i \mid 1 \le i \le n$, where $g_i : M \longrightarrow_c M',\}$.

Define the predicate $p : M \longrightarrow \textbf{Bool}$ by $p(x)$ if and only if $(f_i(x), g_i(x)) \in \rho'$ for all i, $1 \le i \le n$.

Let L be a chain in (M, ρ) such that $p(x) = \textbf{T}$ for all $x \in L$. We have $(f_i(x), g_i(x)) \in \rho'$ for $1 \le i \le n$ and $x \in L$. This means that $(f_i(x), \sup_{t \in L} g_i(t)) \in \rho'$ for all $x \in L$ or, equivalently, $(f_i(x), g_i(\sup L)) \in \rho'$ for $x \in L$, $1 \le i \le n$. This implies $(\sup_{x \in L} f_i(x), g_i(\sup L)) \in \rho'$, and therefore, $(f_i(\sup L), g_i(\sup L)) \in \rho'$ for $1 \le i \le n$. Consequently, $p(\sup L) = \textbf{T}$, which proves the admissibility of the predicate p.

Example 4.13.8 Again, let (M, ρ), (M', ρ') be two cpos and consider the functions $f, g : M \longrightarrow_c M'$. The predicate p defined by $p(x) = \textbf{T}$ if $f(x) = g(x)$ is admissible in view of the previous example. It suffices to remark that we can choose the families introduced above as $f_1 = f, f_2 = g$ and $g_1 = g, g_2 = f$. The predicate defined by these families of functions will be given by the conditions $(f(x), g(x)) \in \rho'$, and $(g(x), f(x)) \in \rho'$ which imply $f(x) = g(x)$.

For any subset K of a cpo (M, ρ), we define the characteristic predicate p_K of K by

$$p_K(x) = \begin{cases} \textbf{T} & \text{if } x \in K \\ \textbf{F} & \text{otherwise.} \end{cases}$$

Using the notation introduced above, we have $T_{p_K} = K$.

It is clear that if K is a sub-cpo of (M, ρ) then p_K is an admissible predicate. Indeed, if L is a chain in (M, ρ) such that $p_K(x) = \textbf{T}$ for all $x \in L$, then $L \subseteq K$. Therefore, the completeness of K implies $\sup L \in K$; hence, $p_K(\sup L) = \textbf{T}$.

Of course, if K contains a chain whose supremum is not in K, then p_K is not going to be admissible.

Example 4.13.9 Let (M_i, ρ_i), $i = 1, 2$ be two cpos. The set of all functions $M_1 \longrightarrow M_2$ is also a cpo. Since the set of monotonic functions is a sub-cpo of $M_1 \longrightarrow M_2$, the predicate p_{mon} defined by $p_{mon}(f) = \mathbf{T}$ if and only if f is monotonic is clearly admissible.

Example 4.13.10 In Example 3.8.6, we show that for every function $f :$ $\mathbf{N}_\perp \longrightarrow \mathbf{N}_\perp$ we can construct a sequence of approximations $\{f_i \mid i \in \mathbf{N}\}$ such that $f_i(x) = \perp$ for $x \geq i$.

Define the predicate p on $\mathbf{N}_\perp \longrightarrow \mathbf{N}_\perp$ by $p(g) = \mathbf{T}$ if and only if $g(x) = \perp$ for some $x \in \mathbf{N}$. In other words, p is in this case the characteristic predicate of the representations of the partial function. This predicate is not admissible because, if we choose f as the strict extension of a function defined on $\mathbf{N}$, then, for the approximating chain $\{f_i \mid i \in \mathbf{N}\}$, we have $p(f_i) = \mathbf{T}$, while $p(f) = p(\sup\{f_i \mid i \in \mathbf{N}\}) = \mathbf{F}$.

The next theorem offers a technique that can be used to prove that the least fixed point of a continuous function has properties that can be described by admissible predicates.

Theorem 4.13.11 (The Fixed point Induction Principle) *Consider a cpo (M, ρ). Assume that p is an admissible predicate on (M, ρ) and that $f : M \longrightarrow_c M$ is a continuous function.*

If $p(\perp) = \mathbf{T}$ and $p(x) = \mathbf{T}$ implies $p(f(x)) = \mathbf{T}$ for every $x \in M$, then

$$p(\mathit{fix}\ (f)) = \mathbf{T}.$$

Proof. We have seen in Theorem 4.13.1 that

$$C_f = \{\perp, f(\perp), \ldots, f^n(\perp), \ldots\}$$

is a chain in (M, ρ). Furthermore, fix $(f) = \sup C_f$.

We intend to prove that $p(u) = \mathbf{T}$ for every $u \in C_f$. Since any $u \in C_f$ is of the form $u = f^n(\perp)$ for $n \in \mathbf{N}$, we can perform this proof by induction on n. For $n = 0$, we have $p(\perp) = \mathbf{T}$, by the hypothesis. If we assume that $p(f^n(\perp)) = \mathbf{T}$, then, by applying the property of p given by the hypothesis, we obtain $p(f^{n+1}(\perp)) = \mathbf{T}$. Therefore, we may conclude that $p(u) = \mathbf{T}$ for every $u \in C_f$. The admissibility of p implies $p(\sup C_f) = \mathbf{T}$; hence, $p(\text{fix}\ (f)) = \mathbf{T}$. ∎

Fixed points of functions defined on complete partial orders can be used to give another way of understanding recursive function definitions. If we wish to define a function $f : A \longrightarrow B$ by a recursive definition $f(x) = \ldots$, where $\ldots$ is an expression that involves f, then we do not know automatically that there is such a function f. What we can say is that if $f : A \longrightarrow B$ is a given function, then we can use the expression $\ldots$ to define a new function $\hat{f} : A \longrightarrow B$ by $\hat{f}(x) = \ldots$. We thus get a functional $F : (A \longrightarrow B) \longrightarrow (A \longrightarrow B)$ by defining $F(f) = \hat{f}$, that is, $F(f)(x) = \ldots$. The function f we are seeking to define by the recursive definition is a fixed point of the functional F, that is, we are looking for a function $f : A \longrightarrow B$ with $F(f) = f$. The set $A \longrightarrow B$ is not a cpo, but $(A \rightsquigarrow B, \subseteq)$ is, and for the typical recursive definition, there is no problem in giving a meaning to the right-hand side expression ... even if f is only a partial function, so we can regard F as being a function from $A \rightsquigarrow B$ to itself and look for fixed points. In almost any conceivable recursive definition, a short argument shows that the functional F is monotonic, and hence, F has a least fixed point. In most cases, a slightly more involved argument will show that F is continuous, in which case F not only has a least fixed point, but we know how to calculate it. When F does have a least fixed point f, we can say that this f is the function defined by the recursive definition. From this point of view, any recursive definition whose corresponding functional has a least fixed point (and this covers just about all reasonable recursive definitions) defines some function; however, this function could be partial and in fact could be $\emptyset$, the partial function with empty domain. If the least fixed point is a total function, then, since the total functions are maximal elements of $A \rightsquigarrow B$, there is in fact only one fixed point, and hence, the recursive definition succeeds in the sense that we have given before, namely, that there is only one $f : A \longrightarrow B$ that satisfies the definition. The converse of this statement is not true, however; it is possible for a recursive definition to have a unique solution among the total functions and yet have its least fixed point be a partial function. We will see an example of this later.

When we put these ideas into practice, we will use $A_\perp \longrightarrow_s B_\perp$ in place of the isomorphic cpo $A \rightsquigarrow B$. Since many recursive definitions have similar forms, it will be useful to have some general results that say that if a recursive definition has a certain form, then the corresponding functional is continuous. In Exercise 39(a) of the previous chapter, we have seen that the function

$$F : (\mathbf{N}_\perp \longrightarrow_c \mathbf{N}_\perp) \longrightarrow (\mathbf{N}_\perp \longrightarrow_c \mathbf{N}_\perp),$$

defined by $F(f)(x) =$ if $g(x)$ then $h(x)$ else $f(k(x))$, is continuous, where $g : \mathbf{N}_\perp \longrightarrow \mathbf{Bool}_\perp$ and $h, k : \mathbf{N}_\perp \longrightarrow \mathbf{N}_\perp$ are strict extensions of functions defined on $\mathbf{N}$. Since h, k are monotonic, they are continuous (because of the fact that we deal here with flat posets). Likewise, if we restrict g to $\mathbf{N}$, we have $g(\mathbf{N}) \subseteq \mathbf{Bool}$, which means that g restricted to $\mathbf{N}$ is a predicate. A predicate such as

$$g(x) = \begin{cases} \mathbf{T} & \text{if } x = 0 \\ \mathbf{F} & \text{otherwise,} \end{cases}$$

will be simply denoted as "$x = 0$."

Many variations are possible on the above example. It is not difficult, for instance, to prove that functions such as

$$G(f)(x) = \text{if } g(x) \text{ then } h(x) \text{ else } q(x, f(k(x))),$$
$$H(f)(x) = \text{if } g(x) \text{ then } h(x) \text{ else } q(f(l(x)), f(k(x))),$$
$$K(f)(x) = \text{if } g(x) \text{ then } h(x) \text{ else } l(f(k(x))),$$

where g, h, k, l, are strict extensions of functions defined on $\mathbf{N}$ and $q :$ $\mathbf{N}_\perp \times \mathbf{N}_\perp \longrightarrow_m \mathbf{N}_\perp$.

In this context, we obtain such continuous functions (from $M \longrightarrow_c M$ to $M \longrightarrow_c M$) as $G(f)(x) = \text{if } x = 0 \text{ then } 1 \text{ else } x \cdot f(x-1)$, or $H(f)(x) = \text{if } x < 2 \text{ then } 1 \text{ else } f(x-1) + f(x-2)$. To obtain H, we made the obvious choices:

$$g(x) = \begin{cases} \mathbf{T} & \text{if } x < 2 \\ \mathbf{F} & \text{otherwise,} \end{cases}$$

$h(x) = 1$, and $q(x, y) = x + y$ for every $x, y \in \mathbf{N}$. In addition, we used the monotonic functions l, k defined by

$$l(x) = \begin{cases} \perp & \text{if } x = 0 \text{ or } x = \perp \\ x - 1 & \text{otherwise ,} \end{cases}$$

and

$$k(x) = \begin{cases} \perp & \text{if } x \leq 1 \text{ or } x = \perp \\ x - 2 & \text{otherwise.} \end{cases}$$

Example 4.13.12 Suppose that we try to define a function $f : \mathbf{N} \longrightarrow \mathbf{N}$ by $f(n) = f(n)$. From our previous perspective, we said that this definition fails because every function from $\mathbf{N}$ to $\mathbf{N}$ satisfies the equation. From our new perspective, let $F : (\mathbf{N}_\perp \longrightarrow_s \mathbf{N}_\perp) \longrightarrow (\mathbf{N}_\perp \longrightarrow_s \mathbf{N}_\perp)$ be the identity function. Then, we are looking for the least fixed point of F, and this is obviously f_0, the least element of $\mathbf{N}_\perp \longrightarrow_s \mathbf{N}_\perp$, which corresponds to the completely undefined partial function from $\mathbf{N}$ to $\mathbf{N}$.

Example 4.13.13 Suppose that we try to define a function $f : \mathbf{N} \longrightarrow \mathbf{N}$ by $f(n) = f(n) + 1$. From our old perspective, this definition fails because no function from $\mathbf{N}$ to $\mathbf{N}$ satisfies the equation. From our new perspective, let $F : (\mathbf{N}_\perp \longrightarrow_s \mathbf{N}_\perp) \longrightarrow (\mathbf{N}_\perp \longrightarrow_s \mathbf{N}_\perp)$ be defined by

$$F(f)(n) = \begin{cases} \perp & \text{if } f(n) = \perp \\ f(n) + 1 & \text{otherwise.} \end{cases}$$

Note that if f is strict, then $F(f)(\perp) = \perp$, so $F(f)$ is strict, as it should be. It is easy to see that if f_0 is the least element of $\mathbf{N}_\perp \longrightarrow_s \mathbf{N}_\perp$ (i.e., $f_0(n) = \perp$ for all $n \in \mathbf{N}_\perp$), then $F(f_0) = f_0$, and hence, this definition also has the completely undefined function as its least fixed point.

In the previous example, we could give the definition of F by $F(f)(n) = f(n) + 1$ if we use the usual convention that when addition is extended to a function from $\mathbf{N}_\perp \times \mathbf{N}_\perp$ to $\mathbf{N}_\perp$, it's value is $\perp$ when one or both of its arguments is $\perp$. We will use similar conventions without further mention.

Example 4.13.14 If we try to define a function from $\mathbf{N}$ to $\mathbf{N}$ by $f(n) = 2f(n)$, then from our previous perspective, this definition succeeds since the constant 0 is its only solution. If we define the corresponding functional F by $F(f)(n) = 2f(n)$, then this functional has $F(f_0) = f_0$, and hence, the least fixed point is again the function corresponding to the totally undefined function even though the functional has a unique solution among the functions that are strict extensions of total functions on $\mathbf{N}$.

Example 4.13.15 The usual definition of the factorial function can be expressed as

$$f(x) := \text{ if } x = 0 \text{ then } 1 \text{ else } x \cdot f(x-1). \tag{4.1}$$

If we take $G : (\mathbf{N}_\perp \longrightarrow_s \mathbf{N}_\perp) \longrightarrow (\mathbf{N}_\perp \longrightarrow_c \mathbf{N}_\perp)$ to be the corresponding functional, then as discussed above, G is continuous, and we want to see what the least fixed point of G is.

In order to determine the least fixed point of G, we need to compute the elements of the chain C_G. The least element of $\mathbf{N}_\perp \longrightarrow_c \mathbf{N}_\perp$ is the function f_0 defined by $f_0(x) = \perp$ for all $x \in \mathbf{N}_\perp$, and this is the least element of the chain C_G. Further members of C_G are computed by using the rule $f_i = G^i(f_0)$.

We claim that

$$f_i(n) = \begin{cases} n! & \text{if } n < i \\ \perp & \text{otherwise.} \end{cases}$$

For $i = 0$, the above equality is obviously satisfied. Suppose that it is valid for i and let us compute f_{i+1}. Since $f_{i+1} = G(f_i)$, we have

$$f_{i+1}(n) = G(f_i)(n) = \begin{cases} 1 & \text{if } n = 0 \\ n f_i(n-1) & \text{otherwise.} \end{cases}$$

Because of the inductive hypothesis, we have $f_i(n-1) = (n-1)!$ if $n-1 < i$, which amounts to $n < i+1$. Therefore, we obtain

$$f_{i+1}(n) = \begin{cases} 1 & \text{if } n = 0 \\ n \cdot (n-1)! & \text{if } 1 \le n < i+1 \\ \perp & \text{otherwise,} \end{cases}$$

which shows that $f_{i+1}(n) = n!$ if $n < i+1$ and $f_{i+1}(n) = \perp$ otherwise; this concludes our inductive argument concerning the sequence of functions C_G. It is clear that the members of the sequence $\{f_i \mid i \in \mathbf{N}\}$ approximate, with an increasing degree of definiteness, the factorial function.

Example 4.13.16 We have seen in Example 3.6.8 that the set $(N \times N, \preceq)$ is well-founded.

Consider now the following recursive definition of a function $f : N \times N \longrightarrow N$:

$$f(x, y) = \begin{cases} y+1 & \text{if } x = 0 \\ f(x-1, 1) & \text{if } x > 0 \text{ and } y = 0 \\ f(x-1, f(x, y-1)) & \text{otherwise.} \end{cases}$$

This function, known as *Ackermann's function*, is very important for many studies in complexity aspects of computer science, and we review in the exercises a number of its elementary properties. We wish to show that there is indeed a unique function f satisfying the given equation.

Ackermann's function is to be the least fixed point of the function

$$F : ((N \times N)_\perp \longrightarrow_s N_\perp) \longrightarrow ((N \times N)_\perp \longrightarrow_s N_\perp),$$

given by $F(f)(\perp) = \perp$, and for $(x, y) \in N \times N$,

$$F(f)(x, y) = \begin{cases} y+1 & \text{if } x = 0 \\ f(x-1, 1) & \text{if } x > 0 \text{ and } y = 0 \\ f(x-1, f(x, y-1)) & \text{otherwise.} \end{cases}$$

In the last clause of this definition, it is to be understood that if $f(x, y-1) = \perp$, then $f(x-1, f(x, y-1)) = \perp$. It is immediate that F is monotonic, and hence, that F has a least fixed point. The reader who has not read the theorem that states that every monotonic function on a cpo has a least fixed point or who does not accept the argument because it uses Zorn's lemma can prove that F is in fact continuous. Either way, F has a least fixed point f that we regard as being Ackermann's function.

In this example, we intend to prove, using the principle of the induction for well-founded posets, that for all $(x, y) \in N \times N$, $f(x, y) \neq \perp$, so that Ackermann's function is a total function on $N \times N$.

Let P be the property of $N \times N$ defined by "$f(x, y) \neq \perp$." Assume that $P(u, v)$ is true for all (u, v) such that $(u, v) \preceq (x, y)$. We prove that $P(x, y)$ is true. There are several cases to consider.

1. If $x = 0$, then since f is a fixed point of F, we must have $f(x, y) = F(f)(x, y) = y + 1 \neq \perp$.

2. If $x > 0, y = 0$, then $(x-1, 1) \prec (x, y)$, so by the induction hypothesis, $f(x-1, 1) \neq \perp$, so $f(x, y) = F(f)(x, y) = f(x-1, 1) \neq \perp$.

3. If $x > 0, y > 0$, then $(x, y-1) \prec (x, y)$, so by the induction hypothesis, $f(x, y-1) = m$ for some natural number m and $(x-1, m) \prec (x, y)$. So again, by the induction hypothesis, $f(x-1, m) \neq \perp$. Thus, $f(x, y) = F(f)(x, y) = f(x-1, f(x, y-1)) = f(x-1, m) \neq \perp$.

Let us now consider a few applications of fixed point induction.

Example 4.13.17 Consider the function

$$K : (\mathbf{N}_\perp \longrightarrow_s \mathbf{N}_\perp) \longrightarrow_c (\mathbf{N}_\perp \longrightarrow_s \mathbf{N}_\perp),$$

defined by

$$K(f)(n) = \text{ if } n = 0 \text{ then } 1 \text{ else } 2f(n-1) - 1.$$

Of course, it is not difficult to see that the least fixed point of K is the strict extension of the constant function $f(n) = 1$ for all $n \in \mathbf{N}$. However, the point of this example is the illustration of fixed point induction. Consider the predicate $p : (\mathbf{N}_\perp \longrightarrow \mathbf{N}_\perp) \longrightarrow \mathbf{Bool}$ defined by

$$p(h) = \mathbf{T} \text{ if and only if } \text{Ran}(h) \subseteq \{\perp, 1\}.$$

The predicate is clearly admissible. Using fixed point induction, we intend to prove that $\text{Ran}(k) \subseteq \{\perp, 1\}$, where $k = \text{fix}(K)$, is the least fixed point of K.

For the least element f_0 of $\mathbf{N}_\perp \longrightarrow \mathbf{N}_\perp$, we have $\text{Ran}(f_0) = \{\perp\}$; hence, $p(f_0) = \mathbf{T}$.

Suppose that $p(h) = \mathbf{T}$ and let us prove that $p(K(h)) = \mathbf{T}$. Since $\text{Ran}(h) \subseteq \{\perp, 1\}$ and

$$K(h)(n) = \text{ if } n = 0 \text{ then } 1 \text{ else } 2h(n-1) - 1,$$

there are two cases: when $n = 0$, $K(h)(0) = 1$; for $n > 0$, we have $K(h)(n) = 2h(n-1) - 1$. From the inductive assumption, we have $K(h)(n) = \perp$ if $h(n-1) = \perp$ and $K(h)(n) = 2 \cdot 1 - 1 = 1$, when $h(n-1) = 1$. Consequently, $K(h)(n) \in \{\perp, 1\}$; hence, $\text{Ran}(K(h)) \subseteq \{\perp, 1\}$.

Now, applying the fixed point induction, we may conclude that $\text{Ran}(k) \subseteq \{\perp, 1\}$.

Example 4.13.18 Consider the predicate p on $\mathbf{P}_\perp \longrightarrow \mathbf{P}_\perp$ defined by $p(f) = \mathbf{T}$ if the following two conditions are satisfied:

1. For every number $n \in \mathbf{P}$, if $f(n) \neq \perp$ then $f(k) \neq \perp$ for $1 \leq k \leq n$;

2. $f(n)$ is an even positive integer if and only if $f(n) \neq \perp$ and n is a multiple of 3.

The predicate p is admissible. Indeed, consider a chain of functions $L = \{f_i \mid i \in I\}$ in the cpo $(\mathbf{P}_\perp \longrightarrow \mathbf{P}_\perp, \sqsubseteq)$ such that each function f_i satisfies both conditions mentioned above.

Let f be the supremum of the chain L. Since we have

$$\{m \mid m \in \mathbf{P} \text{ and } f(m) \neq \perp\} = \bigcup_{i \in I} \{m \mid m \in \mathbf{P} \text{ and } f_i(m) \neq \perp\},$$

then, if $f(n) \neq \perp$, there is a function f_i in the chain such that $f_i(n) \neq \perp$. Consequently, $f_i(k) \neq \perp$ for $1 \leq k \leq n$, and since $f_i \sqsubseteq f$, this implies that $f(k) \neq \perp$.

Also, $f(n)$ is an even positive integer if and only if $f_i(n)$ is an even positive integer for some function $f_i \in L$. This happens if and only if $f_i(n) \neq \perp$ and n is a multiple of 3 for some $f_i \in L$, which in turn happens if and only if $f(n) \neq \perp$ and n is a multiple of 3. This shows that p is admissible.

The recursive definition of the Fibonacci sequence $f : \mathbf{P} \longrightarrow \mathbf{P}$ as the least fixed point of the continuous function

$$F : (\mathbf{P}_\perp \longrightarrow_c \mathbf{P}_\perp) \longrightarrow_c (\mathbf{P}_\perp \longrightarrow_c \mathbf{P}_\perp),$$

given by

$$F(f)(n) = \text{ if } n \leq 2 \text{ then } 1 \text{ else } f(n-1) + f(n-2),$$

allows us to prove that $f(n)$ is even if and only if n is a multiple of 3.

Indeed, the function $f_0 : \mathbf{P}_\perp \longrightarrow_s \mathbf{P}_\perp$, given by $f(n) = \perp$ for all $n \in \mathbf{P}$, satisfies vacuously the predicate p since there is no n such that $f(n) \neq \perp$.

On the other hand, suppose that the function $h : \mathbf{P}_\perp \longrightarrow_s \mathbf{P}_\perp$ satisfies the predicate p and let $g = F(h)$. We have

$$g(n) = \text{ if } n \leq 2 \text{ then } 1 \text{ else } h(n-1) + h(n-2).$$

Suppose that $g(n) \neq \perp$. If $n \leq 2$, the first defining condition of the predicate p is satisfied since $g(1) = g(2) = 1$. Suppose now that $n \geq 3$. Since $g(n) = h(n-1) + h(n-2)$, we must have $h(l) \neq \perp$ for every l, $1 \leq l \leq n-1$. Therefore, if $2 < k < n$, we have $g(k) = h(k-1) + h(k-2)$, and this implies that $g(k) \in \mathbf{P}$.

Assume that $g(n)$ is an even positive number. We must have $n \geq 3$ (since both $g(1)$ and $g(2)$ are not even). This means that $g(n) = h(n-1) + h(n-2)$. Either both $h(n-1)$ and $h(n-2)$ are even or they are both odd. Since at most one of two consecutive numbers can be a multiple of 3, in view of the fact that $p(h) = \mathbf{T}$, we may exclude the former case. Therefore, both $h(n-1)$ and $h(n-2)$ must be odd numbers. Again, since h satisfies the predicate p, $n-1$ and $n-2$ are not multiples of 3, and this implies that n is such a multiple.

Conversely, suppose that $g(n) \neq \perp$ and n is a multiple of 3. In view of the inductive hypothesis, $h(n-1)$ and $h(n-2)$ are both odd numbers, and because of the fact that $g(n) = h(n-1) + h(n-2)$, we obtain that $g(n)$ is even.

There is an interesting connection between least fixed points and inductively defined sets. In fact, we will show that if a set is defined by an inductive definition or a family of sets is defined by a simultaneous inductive definition, then in a natural way the set or family of sets can be regarded as being the least fixed point of a continuous function.

We begin by considering the simplest case. Suppose that U is a set, $f : U \longrightarrow U$, and $B \subseteq U$. Then, we have seen that $(\mathcal{P}(U), \subseteq)$ is a cpo and the least upper bound of a chain in this cpo is the union of the elements of the chain. Define $\Psi : \mathcal{P}(U) \longrightarrow \mathcal{P}(U)$ by $\Psi(D) = D \cup B \cup f(D)$. Then, it is immediate that for any $D \in \mathcal{P}(U)$, D contains B and is closed under f if and only if D is a fixed point of Ψ. Thus, the closure of B under f is the least fixed point of Ψ.

Define $\Phi : \mathcal{P}(U) \longrightarrow \mathcal{P}(U)$ by $\Phi(D) = B \cup f(D)$. We now show that Φ is continuous. To see this, let $\mathcal{C}$ be a chain in $(\mathcal{P}(U), \subseteq)$. Then, we have

$$\begin{aligned}
\Phi(\sup \mathcal{C}) &= \Psi \left(\bigcup \{ C \mid C \in \mathcal{C} \} \right) \\
&= B \cup f \left(\bigcup \{ C \mid C \in \mathcal{C} \} \right) \\
&= B \cup \bigcup \{ f(C) \mid C \in \mathcal{C} \},
\end{aligned}$$

while

$$\sup \{ \Phi(C) \mid C \in \mathcal{C} \} = \bigcup \{ B \cup f(C) \mid C \in \mathcal{C} \}.$$

It is easy to see that these two sets are the same; hence, Φ is continuous.

Thus, by Corollary 4.13.5, Ψ is continuous and Ψ and Φ have the same least fixed point. Thus, the closure of B under f is the least fixed point of either of the continuous functions Ψ and Φ. The point here is not so much to give another proof of the existence of the closure of a set under a function (after all, the proof we gave for Theorem 4.6.6 was very simple), but rather to give further evidence for the power of the least fixed point idea by showing how the closure of a set under a function is a special case of this concept.

We can generalize the discussion just given to the closure of a set under a family of constructors. Once again, let U be a set, and let B be a subset of U. Suppose that $\mathcal{R}$ is a family of constructors on U. Define $\Psi, \Phi : \mathcal{P}(U) \longrightarrow \mathcal{P}(U)$ by

$$\begin{aligned}
\Psi(D) &= D \cup B \cup \bigcup \{ R(D^{n_R}) \mid R \in \mathcal{R} \}, \\
\Phi(D) &= B \cup \bigcup \{ R(D^{n_R}) \mid R \in \mathcal{R} \},
\end{aligned}$$

where each $R \in \mathcal{R}$ is n_R-ary. If $D \in \mathcal{P}(U)$, then D is a fixed point of Ψ if and only if D contains B and is closed under $\mathcal{R}$. It is not hard to show that Φ, and hence Ψ, is continuous, and hence, that the closure of B under $\mathcal{R}$ is the least fixed point of a continuous function (either Ψ or Φ).

We can generalize once more, this time to simultaneous inductive definitions. Let K be a nonempty set, let $\mathcal{U} = \{ U_k \mid k \in K \}$ and let $\mathcal{B} \preceq \mathcal{U}$. (Recall that $\mathcal{B} \preceq \mathcal{U}$ means that $\mathcal{B} = \{ B_k \mid k \in K \}$ and $B_k \subseteq U_k$ for all $k \in K$.) We have defined $\hat{\mathcal{P}}(\mathcal{U})$ to be $\{ \mathcal{D} \mid \mathcal{D} \preceq \mathcal{U} \}$. It is easy to see that $(\hat{\mathcal{P}}(\mathcal{U}), \preceq)$ is a cpo. (In fact, $(\hat{\mathcal{P}}(\mathcal{U}), \preceq)$ is the product of the cpos $(U_k, \subseteq)$ with $k \in K$.) Note that for any $\mathcal{D}, \mathcal{E} \in \hat{\mathcal{P}}(U)$, $\sup \{ \mathcal{D}, \mathcal{E} \}$ is $\{ D_k \cup E_k \mid k \in K \}$.

Let $\mathfrak{R}$ be a family of constructors on $\mathcal{U}$. Define $\Psi, \Phi : \hat{\mathcal{P}}(\mathcal{U}) \longrightarrow \hat{\mathcal{P}}(\mathcal{U})$ by letting $\Psi(\mathcal{D})_k$ and $\Phi(\mathcal{D})_k$, the k-th components of $\Psi(\mathcal{D})$ and $\Phi(\mathcal{D})$, respectively, be given by

$$\begin{aligned}\Psi(\mathcal{D})_k &= D_k \cup B_k \cup \bigcup\{R(D_\alpha) \mid (\alpha, k, R) \in \Re\}, \\ \Phi(\mathcal{D})_k &= B_k \cup \bigcup\{R(D_\alpha) \mid (\alpha, k, R) \in \Re\}.\end{aligned}$$

Then, Ψ and Φ are continuous, and the closure of $\mathcal{B}$ under $\Re$ is the least fixed point of both Ψ and Φ.

Example 4.13.19 Let $G = (N, T, S_0, P)$ be a context-free grammar, where $N = \{S_0, \ldots, S_{n-1}\}$, and consider the family of constructors $\Re_G$ on the family of sets $\mathcal{U}_G = \{U_{S_i} \mid 0 \le i \le n-1\}$, where $U_{S_i} = T^*$ for $0 \le i \le n-1$.

If $\mathcal{B}_G = \{B_{S_i} \mid 0 \le i \le n-1\}$, then the closure of $\mathcal{B}_G$ under $\Re_G$ is the least fixed point of the continuous function $\Phi_G : \hat{\mathcal{P}}(\mathcal{U}) \longrightarrow_c \hat{\mathcal{P}}(\mathcal{U})$, whose component corresponding to S_k is given by

$$(\Phi_G(\{V_{S_i} \mid S_i \in N\}))_{S_k} = B_{S_k} \cup \bigcup\{R_r(V_\alpha) \mid ((\alpha, S_k), R_r) \in \Re_G\}.$$

For example, consider the grammar $G = (\{S_0, S_1, S_2\}, \{a, b\}, S_0, P)$ introduced in Example 4.10.12. The function Ψ_G is defined by

$$\begin{aligned}(\Phi(V_{S_0}, V_{S_1}, V_{S_2}))_{S_0} &= \{\lambda\} \cup a V_{S_2} \cup b V_{S_1}, \\ (\Phi(V_{S_0}, V_{S_1}, V_{S_2}))_{S_1} &= a V_{S_0} \cup b V_{S_1}, V_{S_1}, \\ (\Phi(V_{S_0}, V_{S_1}, V_{S_2}))_{S_2} &= b V_{S_0} \cup a V_{S_2}, V_{S_2}.\end{aligned}$$

4.14 Exercises and Supplements

Induction on Natural Numbers

1. For each $n \ge 0$, let $U_n = 0^3 + 1^3 + 2^3 + \ldots + n^3$. Prove that for all $n \ge 0$, $U_n = n^2(n+1)^2/4$.

2. Prove by induction on n that

    (a)
    $$1 + x + \cdots x^n = \frac{x^{n+1} - 1}{x - 1}$$

    for $x \ne 1$ and $n \ge 0$.

    (b)
    $$\sqrt{0} + \sqrt{1} + \cdots + \sqrt{n} < \frac{2}{3}(n+1)\sqrt{n+1}$$

    for $n \ge 0$.

3. Let M be a finite set and let $\mathcal{C} \subseteq \mathcal{P}(M)$ be a family of subsets of M such that $A \cap B \ne \emptyset$ for any $A, B \in \mathcal{C}$. Prove that $\mid \mathcal{C} \mid \le 2^{n-1}$, where $\mid M \mid = n$.

4. (a) Prove that for $n \geq 1$ we have

$$\frac{1}{n+1} + \frac{1}{n+2} + \cdots + \frac{1}{2n} > \frac{1}{2}.$$

   (b) Using the inequality proven in (a), show by induction that for $k \geq 0$ we have

$$1 + \frac{1}{2} + \cdots + \frac{1}{2^k} > \frac{k}{2}.$$

5. Prove that for $n > 3$ we have

$$\frac{n!}{n^n} < \frac{2}{n^2}.$$

6. Show that for $n \geq 1$

$$1!1 + 2!2 + \cdots + n!n = (n+1)! - 1.$$

7. Prove that the number of diagonals of an n-sided convex polygon is

$$d_n = \frac{n(n-3)}{2}$$

   for $n \geq 4$.

8. Prove by induction on n:

   (a)

$$1^2 + 3^2 + \cdots + (2n-1)^2 = \frac{4n^3 - n}{3}$$

   for $n \geq 1$;

   (b)

$$1 + \frac{1}{\sqrt{2}} + \cdots + \frac{1}{\sqrt{n}} > 2(\sqrt{n+1} - 1)$$

   for $n \geq 2$.

9. Prove by induction on n that for any real numbers $r_1, \ldots, r_n$ we have

$$| r_1 + \cdots + r_n | \leq | r_1 | + \cdots + | r_n |.$$

10. Assume that for all $n \geq 1$ we have the constant numbers a, b, c, d such that

$$1 \cdot 2 + 2 \cdot 3 + \cdots + n(n+1) = an^3 + bn^2 + cn + d.$$

   Determine a, b, c, d using some particular values for n. Prove the resulting formula by induction on n.

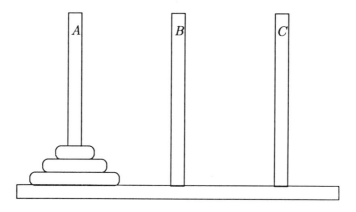

FIGURE 4.4. The Towers of Hanoi

11. Let p_n be the n-th prime natural number; for instance, we have $p_1 = 2$, $p_2 = 3$, etc.

   (a) Show that for $n \geq 1$ we have

   $$p_{n+1} < p_1 p_2 \cdots p_n + 1.$$

   (b) Prove by induction on $n \geq 1$ that

   $$p_n \leq 2^{2^{n-1}}.$$

   (c) Prove that for $n \geq 4$ we also have

   $$p_{n+1}^2 < p_1 p_2 \cdots p_n.$$

12. Consider the following game, known as the *Towers of Hanoi*. Three rods A, B, C are arranged in a row (see Figure 4.4), and on one rod is a pile of n different sized disks arranged top to bottom from smallest to largest. The player is supposed to move the disks, *one at a time* to any one of the rods, subjected to the condition, which held also at the beginning, that no disk be placed above a smaller disk. The aim of the game is to transfer the disks from rod A to rod C.

   (a) Prove by induction on the number of disks n that the transfer can be made in $2^n - 1$ moves.

   (b) If we add the extra condition that a disk may be transferred only between adjacent rods, then we need $3^n - 1$ moves.

13. What is wrong with the following "proof?"

    We show that for every $n \geq 3$, F_n, the n-th Fibonacci number, is even. We use strong induction (Theorem 4.2.22) with $n_0 = 3$. Since $F_3 = F_1 + F_2 = 1 + 1 = 2$, the basis step is established. For the inductive step, suppose that $k \geq 3$ and that the result is true for all j with $3 \leq j \leq k$. Then, $F_{k+1} = F_{k-1} + F_k$, and by the induction hypothesis, both F_{k-1} and F_k are even. Since the sum of two even numbers is even, F_{k+1} is even, completing the proof.

14. (a) Give a version of strong induction that allows for several basis steps and applies to a finite subset of the integers.

    (b) Give a version of strong induction similar to Theorem 4.2.20 that applies to a set of integers that might be finite or infinite.

15. (a) Use Theorem 4.2.3 to prove Theorem 4.2.29.

    (b) Use Theorem 4.2.3 to prove Theorem 4.2.30.

16. Derive the principle of strong induction directly from the principle of ordinary induction.

    Hint. Let n_0 be an integer and let P be a property of the integers greater than or equal to n_0 such that $P(n_0)$ is true, and for all $n \geq n_0$, if $P(k)$ is true for all k with $n_0 \leq k \leq n$, then $P(n + 1)$ is true. Define $P'(n)$ to be the property that $P(k)$ is true for every k with $n_0 \leq k \leq n$. Apply ordinary induction to show that $P'(n)$ is true for all $n \geq n_0$.

17. (a) Derive the well-ordering principle from the principle of strong induction.

    (b) Derive the well-ordering principle from Theorem 4.2.18.

18. (a) Let $\mathcal{S} = (N, 0, S, \leq)$ be a structure with $0 \in N$, S a unary function on N, and $\leq$ a binary relation on N such that the following four properties hold:

    C1. $\leq$ is a linear order on N,

    C2. 0 is the least element of N under $\leq$,

    C3. if $n \in N$ is $\neq 0$, then there is an $m \in N$ with $n = S(m)$,

    C4. for all $n \in N$, $\{m \in N \mid m < S(n)\} = \{m \in N \mid m \leq n\}$.

    Show that if any one of the following statements is true about $\mathcal{S}$, then they all are.

    I1. Well-ordering principle for $\mathcal{S}$: $\leq$ is a well-ordering on N (i.e., every nonempty subset of N has a least element under $\leq$).

    I2. Principle of strong induction for $\mathcal{S}$: If P is a property of N such that

i. $P(0)$ is true, and

ii. for all $k \in N$, if $P(j)$ is true for all $j \leq k$, then $P(S(k))$ is true,

then $P(n)$ is true for all $n \in N$.

I3. Principle of induction for $\mathcal{S}$: If P is a property of N such that

i. $P(0)$ is true, and

ii. for all $k \in N$, if $P(k)$ is true, then $P(S(k))$ is true,

then $P(n)$ is true for all $n \in N$.

(b) The natural numbers are an example of a structure that satisfies the conditions C1-C4 and also satisfies the principles I1-I3. Give an example of a structure that satisfies all of C1-C4 and does not satisfy any of the equivalent principles I1-I3.

(c) Show that if $\leq$ is a linear order, then condition C4 is equivalent to the condition

for all $n \in N$, $n < S(n)$, and there is no $m \in N$ with $n < m < S(n)$.

19. Using the well-ordering principle for $\mathbf{N}$, prove that if $m, n \in \mathbf{N}$ and $n > 0$ then there exists a unique pair of numbers (q, r) such that $m = nq + r$ and $0 \leq r < n$ (the Euclidean division algorithm).

Solution: Consider the set

$$T_{m,n} = \{k \mid k = m - nq, k \in \mathbf{N}, \text{ for some } q \in \mathbf{N}\}.$$

The set $T_{m,n}$ is nonempty since $m \in T_{m,n}$. From the well-ordering principle, it follows that it contains a least element r. We claim that $r < n$. Indeed, suppose that $r \geq n$, that is, $r = n+u$, where $r > u > 0$. This allows us to write $u = r - n = m - n(q + 1) \in T_{m,n}$, and this contradicts the definition of r.

We need to prove now that the pair (q, r) is unique. Suppose that $m = nq' + r'$, with $0 \leq r' < n$ and $(q, r) \neq (q', r')$. Since $nq + r = nq' + r'$, we have $n(q - q') = r' - r$. If $r \neq r'$, we also have $q \neq q'$. Consequently, $|r' - r| > n$, and this is absurd in view of the fact that $0 \leq r, r' < n$. On the other hand, if $r = r'$, we also have $q = q'$.

20. For $m, n \in \mathbf{N}$, consider the set

$$S_{m,n} = \{k \mid k = um + vn \text{ for some } u, v \in \mathbf{Z}\}.$$

(a) Prove that there is number $d \in \mathbf{N}$ such that $S_{m,n}$ is the set of multiples of d.

(b) Prove that if m, n are relatively prime then $1 \in S_{m,n}$.

Solution: Note that for any $m, n \in \mathbb{N}$ the set $S_{m,n}$ is nonempty. Therefore, from the well-ordering principle, we infer the existence of a positive natural number d that is the least element of $S_{m,n}$; we have $d = um + vn$ for some $u, v \in \mathbb{Z}$.

We claim that every member of $S_{m,n}$ is a multiple of d. Indeed, consider a number $h \in S_{m,n}$. There exist $s, t \in \mathbb{Z}$ such that $h = sm + tn$. From Exercise 19 it follows that we can write $h = dq + r$, where $0 \le r < d$. This implies

$$r = h - dq = (s - uq)m + (t - vq)n,$$

which shows that $r = 0$ (since, otherwise, it would contradict the definition of d). This proves that every member of $S_{m,n}$ is a multiple of d. Conversely, it is clear that every multiple of d is a member of $S_{m,n}$. Furthermore, since $m, n \in S_{m,n}$, it follows that d is a common divisor of m and n. Note that no divisor of d may belong to $S_{m,n}$, since this would imply the existence in $S_{m,n}$ of numbers that are not multiples of d. This proves that d is the greatest common divisor of m, n, and the second part of the statement follows immediately.

21. Prove, by induction on $k \ge 1$, that if $m_1, \ldots, m_k$ are pairwise relatively prime numbers and $a_1, \ldots, a_k \in \mathbb{Z}$ then there is a number x such that

$$x \equiv a_1 (\text{mod } m_1), \ldots, x \equiv a_k (\text{mod } m_k)$$

Solution: For $k = 1$, we can choose $x = a_1$.

Suppose that the statement is true for k and consider the existence of x such that

$$x \equiv a_1 (\text{mod } m_1), \ldots, x \equiv a_k (\text{mod } m_k), x \equiv a_{k+1} (\text{mod } m_{k+1}).$$

The inductive hypothesis implies the existence of y such that

$$y \equiv a_1 (\text{mod } m_1), \ldots, y \equiv a_k (\text{mod } m_k)$$

It is easy to see that the numbers $m_1 \cdots m_k$ and m_{k+1} are relatively prime. Therefore, there are $s, t \in \mathbb{Z}$ such that $1 = sm_1 \cdots m_k + tm_{k+1}$ because of Exercise 20. If $q = s(a_{k+1} - y)$, we have

$$
\begin{aligned}
y + qm_1 \cdots m_k &= y + sm_1 \cdots m_k (a_{k+1} - y) \\
&= y + (1 - tm_{k+1})(a_{k+1} - y) \\
&= a_{k+1} + m_{k+1} t(y - a_{k+1}).
\end{aligned}
$$

Let $p = t(y - a_{k+1})$. We can choose $h \in \mathbb{Z}$ such that $hm_{k+1} + q \ge 0$. Then,

$$y + (hm_{k+1} + q)m_1 \cdots m_k = a_{k+1} + m_{k+1}(p + hm_1 \cdots m_k),$$

which shows that

$$y + (hm_{k+1} + q)m_1 \cdots m_k \equiv a_{k+1} (\bmod\ (m_{k+1})).$$

From the induction hypothesis, we also have

$$y + (hm_{k+1} + q)m_1 \cdots m_k \equiv a_i (\bmod\ (m_i)),$$

which completes the proof.

Inductively Defined Sets

22. Show that if x, y, and z are variable names and 12, 6, and 3 are constants, then $((x+y) * 12 - (6+3))/z$ is in the set EXP defined in Example 4.3.1.

23. Give an inductive definition of the odd natural numbers similar to the definition of the even natural numbers given in Example 4.3.6.

24. Let S be the subset of $\mathbf{N} \times \mathbf{N}$ given by

   (1) $(0, n)$ is in S for all $n \in \mathbf{N}$;
   (2) if (n, m) is in S, then so is $(n, m + 1)$.

   (a) Show that $(4, 3) \in S$.
   (b) Prove that $S = \mathbf{N} \times \mathbf{N}$.

25. Let T be the subset of $\mathbf{N} \times \mathbf{N}$ given by

   (1) $(0, 0)$ is in T;
   (2) if (n, m) is in T, then so is $(n, m + 1)$;
   (3) if (n, m) is in T, then so is $(n + 1, m)$;
   (4) if (n, m) is in T, then so is $(n + 1, m + 1)$.

   (a) Give two essentially different proofs that $(4, 3) \in T$.
   (b) Prove that $T = \mathbf{N} \times \mathbf{N}$.
   (c) For each rule in the definition of T, what set would result if that rule were left out of the definition?

26. Let Σ be an alphabet and let w be a fixed member of Σ^*.

   (a) Give an inductive definition of the subset of Σ^* consisting of those strings that contain w as a substring.
   (b) Give an inductive definition of the subset of Σ^* consisting of those strings that contain w as a subsequence. (For example, if $w = 000$, then 010101 contains w as a subsequence, but not as a substring.)

27. Show that $aabbbbaa$ and $baba$ are both in the set S defined in Example 4.3.9.

28. (a) Show that the program E defined in Example 4.3.10 is in WP.

    (b) Using your intuitive understanding of what it means to run a while program, explain why program E will halt whenever it is run with x and y, both natural numbers.

    (c) Does program E always halt if x and y are arbitrary integers?

    (d) If program E is run with x and y, both natural numbers and at least one of them positive, what is the value of z (as a function of x and y) when the program halts?

29. (a) Write a while program that sets z equal to the integer square root of x if x is a natural number. (The integer square root of x is the unique natural number y such that $y^2 \le z < (y+1)^2$.) Assume that you can use multiplication and addition in your expressions, but not a sqrt function.

    (b) Repeat part (a), this time assuming that you can use addition but not multiplication in your expressions.

30. Show that
$$((x \cup (x \circ y))^* \cup (\phi^* \cup x))$$
is a regular expression over $\Sigma = \{x, y\}$.

Proof by Structural Induction

31. (a) Prove that the set you defined in Exercise 23 really is the set of odd natural numbers. (Take the definition of the odd natural numbers to be $\{2n + 1 \mid n \in N\}$.)

    (b) Give the principle of structural induction for the odd natural numbers that comes from your definition.

32. If E belongs to the set EXP defined in Example 4.3.1, let $L(E)$ denote the number of left parentheses in E and let $R(E)$ denote the number of right parentheses in E. Use structural induction to show that $L(E) = R(E)$ for every $E \in$ EXP.

33. In this exercise, we justify the claim made in the proof in Example 4.4.11.

    (a) Let n be a natural number, $n > 1$, and let $f : \{k \in N \mid 1 \le k \le n\} \to Z$. Suppose that for all j with $1 \le j < n$ we have $f(j + 1) \le f(j) + 1$. Prove that if $f(1) < 0$ and $f(n) > 0$, then there must be some k with $1 \le k \le n$ such that $f(k) = 0$.
    Hint: Let $k' = max\{j \mid 1 \le j \le n$ and $f(j) < 0\}$. Show that $k' < n$ and that if $k = k' + 1$ then $f(k) = 0$.

(b) Let $w \in \{a, b\}^*$ be a word that has the same number of a's as b's and whose first and last symbols are both a's. Show that there are strings u and v, neither one equal to λ , such that $N_a(u) = N_b(u)$, $N_a(v) = N_b(v)$, and $w = uv$.

Hint: Let $w = x_1 \cdots x_n$, where each x_i is in $\{a, b\}$. Define $h(a) = -1, h(b) = 1$ and define $f : \{k \in N \mid 1 \leq k \leq n\} \to Z$ by $f(k) = \sum_{j=1}^{k} h(x_j)$. Consider the values of $f(1), f(n), f(n-1)$ and apply part (a).

34. In this exercise, we introduce the idea of a properly balanced string of parentheses. Let $\Sigma = \{(,)\}$. We define BP, the set of *properly balanced strings of parentheses*, to be the subset of Σ^* given by the following inductive definition:

(1) λ is in BP;

(2) if w is in BP, then so is (w);

(3) if w_1 and w_2 are both in BP, then so is $w_1 w_2$.

(a) Prove that for every $w \in \Sigma^*$, $w \in$ BP if and only if w meets the following two conditions:

BP1. w has the same number of left and right parentheses;

BP2. every proper prefix of w has at least as many left parentheses as right parentheses.

Solution: First, we use structural induction to show that every element of BP satisfies both BP1 and BP2; λ certainly satisfies both of these conditions. Suppose that $w \in \Sigma^*$ satisfies both conditions. Then, (w) obviously satisfies BP1. The proper prefixes of (w) are $($, w' where w' is a proper prefix of w, and $(w$. Prefixes of the second form have more left parentheses than right since by induction hypothesis the proper prefix w' of w has at least as many left parentheses as right; and $(w$ has one more left parenthesis than right since w satisfies BP2. Finally, if w_1 and w_2 both satisfy BP1 and BP2, then $w_1 w_2$ also satisfies BP1 and has proper prefixes w_1' for all proper prefixes w_1' of w_1, w_1 itself, and $w_1 w_2'$ for all proper prefixes w_2' of w_2. Using the induction hypothesis, each of these proper prefixes has at least as many left parentheses as right parentheses.

To show the converse, we show by course-of-values induction on the natural number n that if $w \in \Sigma^*$ has length n and satisfies BP1 and BP2, then $w \in$ BP. Suppose that the result is true for all natural numbers less than n and that $w \in \Sigma^*$ has length n and satisfies BP1 and BP2. We consider two cases.

Case 1: w has a proper prefix u with the same number of left as right parentheses. Then, write $w = uv$. It is clear that v also has the same number of left as right parentheses. If u' is a proper prefix of u, then u' is a proper prefix of w, so it has at least as many left parentheses as right parentheses. If v' is a proper prefix of v, then uv' is a proper prefix of w, so it has at least as many left parentheses as right parentheses. Since u has the same number of left and right parentheses, v' must have at least as many left as right parentheses. Thus, u and v both satisfy BP1 and BP2 and have length less than n. By the induction hypothesis, u and v are both in BP, and hence, by the third rule in the definition of BP, $w \in$ BP.

Case 2: w does not have a proper prefix with the same number of left and right parentheses. If $w = \lambda$, then $w \in$ BP by the first rule in the definition of BP. Otherwise, w must begin with a left parenthesis (else w has ")" as a proper prefix) and must end with ")" (else w has a prefix with one more right parenthesis than left). Thus, we have $w = (u)$, and u satisfies BP1. If u' is a proper prefix of u, then $(u'$ is a proper prefix of w, and since we are in Case 2, $(u'$ must have more left parentheses than right, which means that u' has at least as many left parentheses as right. Thus, u satisfies both BP1 and BP2 and has length less than n. By the induction hypothesis, $u \in$ BP, and hence, by the second rule in the definition of BP, $w = (u) \in$ BP.

If $w \in \Sigma^*$, define $O(w) = \{i \mid 0 \le i < |w| \text{ and } w_i = (\}$ and $C(w) = \{i \mid 0 \le i < |w| \text{ and } w_i =)\}$.

(b) Prove that for every $w \in \Sigma^*$, $w \in$ BP if and only if there is a bijection $\alpha : O(w) \to C(w)$ such that for every $i \in O(w)$, $\alpha(i) > i$.

35. Let A be a set and let ρ be a relation on A. Let $\hat{\rho}$ be the new relation defined by

(1) Every pair in ρ is also in $\hat{\rho}$.

(2) If $(x, y) \in \rho$ and $(y, z) \in \hat{\rho}$, then $(x, z) \in \hat{\rho}$.

Show that $\hat{\rho}$ is equal to ρ^+, the transitive closure of ρ, and state the corresponding principle of induction for ρ^+.

36. Let ρ be a relation on a set A. Show that for each of the following definitions of a relation $\hat{\rho}$, $\hat{\rho}$ is equal to ρ^*, the reflexive, transitive closure of ρ.

(1)   i. For each $x \in A$, $(x, x) \in \hat{\rho}$.
   ii. Each pair in ρ is also in $\hat{\rho}$.
   iii. If (x, y) and (y, z) are in $\hat{\rho}$, then so is (x, z).

(2)   i. For each $x \in A$, $(x, x) \in \hat{\rho}$.

 ii. If $(y, w) \in \hat{\rho}$ and $(w, z) \in \rho$, then $(y, z) \in \hat{\rho}$.

(3)   i. For each $x \in A$, $(x, x) \in \hat{\rho}$.

 ii. If $(x, y) \in \rho$ and $(y, z) \in \hat{\rho}$, then $(x, z) \in \hat{\rho}$.

State the principle of induction that corresponds to each of the inductive definitions given above.

We present an algorithm for computing the transitive closure of a relation, known as *Warshall's algorithm*.

Consider a relation $\rho \subseteq M \times M$, where $M = \{x_1, \ldots, x_n\}$. For $k \in \mathbf{N}$, $k \geq 1$, let $W^{(k)} \in M_{n \times n}(B)$ be an $n \times n$ matrix whose entries belong to the set $B = \{0, 1\}$ such that $W_{ij}^{(k)} = 1$ if and only if in the digraph $G(\rho)$ there is a path from x_i to x_j whose vertices belong to the set $\{x_1, \ldots, x_k\}$, with the exception of the ends of the path (that is, with the exception of x_i and x_j). Define $W^{(0)}$ as being $M(\rho)$.

Notice that there is a path in the graph $G(\rho)$ from x_i to x_j if and only if $W_{ij}^{(n)} = 1$ since the interior vertices of any such path must come from the set $\{x_1, \ldots, x_n\}$. Therefore, the matrix of the transitive closure $M(\rho^+)$ coincides with $W^{(n)}$.

37. (a) Prove by induction on k that for $i, j, k \in \mathbf{N}$ and $0 \leq i, j, k \leq n$ we have

$$W_{ij}^{(k)} = W_{ij}^{(k-1)} \vee (W_{ik}^{(k-1)} \wedge W_{kj}^{(k-1)}).$$

(b) Let us remark that if $W_{ij}^{(k-1)} = 1$ then in any matrix $W^{(h)}$ the (i, j) entry remains equal to 1 for any $h \geq k$. Also, the entries of $W^{(k)}$ are determined by the the k-th columns and the k-th line of the matrix $W^{(k-1)}$. Justify that the following algorithm computes the matrix of the transitive closure of the relation $\rho \subseteq M \times M$, where $M = \{x_1, \ldots, x_n\}$.

Algorithm 4.14.1 Input: *A matrix* $M \in M_{n \times n}(\{0, 1\})$ *that represents the relation* ρ.

Output: *A matrix* $T \in M_{n \times n}(\{0, 1\})$ *that represents the relation* ρ^+.

Method: *Construct the sequence of* $n \times n$ *matrices*

$$W^{(1)}, \ldots, W^{(n)}$$

as follows:

 i. *Define* $W^{(1)} = M$.

 ii. *For k varying between 2 and n,*

  A. *if* $W_{ij}^{(k-1)} = 1$, *then* $W_{ij}^{(k)} = 1$;

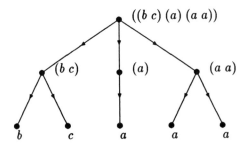

FIGURE 4.5. A tree for the list $((b\ c)\ (a)\ (a\ a))$.

    B. *determine the list $l_1, l_2, \ldots$ of all locations in the k-th column of $W^{(k-1)}$ where 1 occurs; determine the list $c_1, c_2, \ldots$ of all positions where 1 occurs in the k-th line of $W^{(k-1)}$;*

    C. *define $W_{l_p, c_q}^{(k)} = 1$ for all l_p and c_q in the previous lists.*

  iii. *Define $T = W^{(n)}$.*

  (c) Apply Warshall's algorithm to the relation ρ on the set $\{x_1, x_2, x_3, x_4\}$ given by

$$\rho = \{(x_1, x_2), (x_2, x_3), (x_3, x_4), (x_3, x_1)\}.$$

Recursive Definitions of Functions

Let A be a fixed set of symbols whose members are called *atoms*. We define inductively *the set of lists over A*, denoted by List(A), as a subset of the set of words over $A \cup \{(,)\}$, that is, as a set of words that can be written using the symbols of A and the left and right parentheses. To this end, we use the following rules:

1. if $a \in A$, then a is in List(A);
2. the word (), called *the null list*, is in List(A);
3. if $l_1, \ldots, l_n \in$ List(A), then $(l_1 \cdots l_n)$ belongs to List(A), for every $n \in \mathbf{P}$.

Suppose that $A = \{a, b, c\}$ is a set of atoms. We can easily see that (a) is a list by applying the first rule (which gives $a \in$ List(A)) and, then, by applying the third rule for $n = 1$ which implies $(a) \in$ List(A).

A more complicated expression such as $((b\ c)\ (a)\ (a\ a))$, appears to be a list over A because of the existence of the tree given in Figure 4.5.

A *proper list* is a list that is not null and is not an atom. We denote by List$_1(A)$ the set of all proper lists over A.

Lists are extensively used by such languages as LISP[1]; actually, we are using here a slightly simplified version of this notion.

38. (a) Prove that the following word $(((1))((1)(2))())$ is a list from List($\{1, 2\}$).

   (b) Prove that in any proper prefix of a list there are more left parentheses than right parentheses.

   (c) Prove that the definition of the lists over A satisfies the unique readability condition.

39. Let A be a set of atoms and consider the set of lists over A, List(A). Define the functions head : List(A) $\longrightarrow$ List(A) and tail : List(A) $\longrightarrow$ List(A) by the following recursive definitions[2]:

$$\text{head}(l) = \begin{cases} () & \text{if } l \notin \text{List}_1(A) \\ l_1 & \text{if } l = (l_1 \cdots l_n) \text{ for } n \geq 1, \end{cases}$$

and

$$\text{tail}(l) = \begin{cases} () & \text{if } l \notin \text{List}_1(A) \\ (l_2 \cdots l_n) & \text{if } l = (l_1 \cdots l_n) \text{ for } n \geq 1. \end{cases}$$

We need also the function join : List(A) $\times$ List(A) $\longrightarrow$ List(A) given by

$$\text{join}(l, l') = \begin{cases} a & \text{if } l' = a \in A \\ l & \text{if } l' = () \\ (l\, l_1 \cdots l_n) & \text{if } l' = (l_1 \cdots l_n). \end{cases}$$

40. Using the functions head and tail, write the definition of a function that returns the second element of a list, when such an element exists and () otherwise.

41. Let PS(l) be the set of proper sublists of the list $l \in$ List(A). Consider the following recursive definition of PS(l):

$$\text{PS}(l) = \begin{cases} \emptyset & \text{if } l \notin \text{List}_1(A) \\ \{\text{head}(l)\} \cup \text{PS}(\text{head}(l)) \cup \text{PS}(\text{tail}(l)) & \text{otherwise.} \end{cases}$$

Compute the proper sublists of the list $(((a)\ b\ c\ a)\ ((a\ b)\ b))$.

---

[1] LISP is a programming language developed in the late 1950s by John McCarthy at MIT. It was designed essentially for list processing (the name LISP is an acronym for LISt Processing), and it is a very important tool for artificial intelligence.

[2] The functions head and tail are similar, but not identical, to the LISP function CAR and CDR. The main difference is that CAR and CDR may be applied only to proper lists.

42. Consider the recursive definition of the function part : List(A) × List$(A) \longrightarrow$ **Bool** given below:

$$\text{part}(l, l') = \begin{cases} \textbf{F} & \text{if } l = () \\ \textbf{T} & \text{if } l \neq (), \text{head}(l) = l' \\ \text{part}(\text{tail}(l), l') \vee \text{part}(\text{head}(l), l'), & \text{otherwise.} \end{cases}$$

Prove that part(l, l') returns **T** if and only if l' occurs as a proper sublist of l.

43. Give a recursive definition of the function rev : List$(A) \longrightarrow$ List(A), where

$$\text{rev}(l) = \begin{cases} l & \text{if } l \notin \text{List}_1(A) \\ (l_n \cdots l_1) & \text{if } l = (l_1 \cdots l_n). \end{cases}$$

Constructors

44. (a) Prove the existence part of Theorem 4.6.31 from Theorem 4.6.37.

   Hint: Let U, f, B, C, A, G, H be as in Theorem 4.6.31. Define $H' : A \times C \to A$ by $H'(a, c) = H(a)$. Apply Theorem 4.6.37 with H' in place of H.

   (b) Convince yourself by a similar argument that the existence part of Theorem 4.6.48 can be derived from Theorem 4.6.50.

45. (a) Prove the existence part of Theorem 4.6.37 from Theorem 4.6.31.

   Hint: Let U, f, B, C, A, G, H be as in Theorem 4.6.37. Define $A' = A \times C$, $G' : B \to A'$ by $G'(x) = (G(x), x)$, and $H' : A' \to A'$ by $H'(a, c) = (H(a, c), f(c))$. Let $g' : C \to A \times C$ be the function given by applying Theorem 4.6.31 with A', G', H' in place of A, G, H. Let $p_1 : A \times C \to A$ and $p_2 : A \times C \to C$ be the usual projection functions. First, show by induction that for all $x \in C$, $p_2 g'(x) = x$. Then, define $g : C \to A$ by $g = p_1 \circ g'$. Show that g is the desired function.

   (b) Convince yourself that by a similar argument the existence part of Theorem 4.6.50 can be derived from Theorem 4.6.48.

46. (a) Prove the existence part of Theorem 4.6.37 from Theorem 4.6.41.

   Hint: Let U, f, B, C, G, H be as in Theorem 4.6.37. Let P be a one-element set $\{p\}$. Define $G' : P \times B \to A$ by $G'(p, b) = G(b)$ and $H' : P \times C \times A \to A$ by $H'(p, c, a) = H(c, a)$. Let $g' : P \times C \to A$ be the function obtained by applying Theorem 4.6.41 with G', H' in place of G, H. Define $g : C \to A$ by $g(c) = g'(p, c)$. Show that g is the desired function.

   (b) Convince yourself that by a similar argument the existence part of Theorem 4.6.50 can be derived from Theorem 4.6.52.

47. (a) Derive the existence part of Theorem 4.6.41 from Theorem 4.6.37
         by filling in this argument.
         Let U, f, B, C, P, A, G, H be as in Theorem 4.6.41. Let $U' = P \times U$, $B' = P \times C$, $C' = P \times C$ and define $f' : U' \to U'$ by $f'(p, x) = (p, f(x))$. Show that C' is the closure of B' under f' and that C' is freely generated by f' over B'. Define $H' : A \times C' \to A$, by $H'(a, (p, x)) = H(p, a, x)$. Let $g : P \times C = C' \to A$ be the function obtained from Theorem 4.6.37 with U', f', B', C', H' in place of U, f, B, C, H and show that g is the desired function.

     (b) Convince yourself that a similar argument can be used to derive
         Theorem 4.6.52 from Theorem 4.6.50.

48. (a) Derive Theorem 4.6.41 from Theorem 4.6.37 by filling in the
         following argument.
         Let U, f, B, C, P, A, G, H be as in Theorem 4.6.41. Let $A' = P \to A$ and define $G' : B \to A'$ by $G'(x)(p) = G(p, x)$. (To put this another way, $G'(x) = h$, where $h(p) = G(p, x)$.) Define $H' : A' \times C \to A'$ by $H'(h, x)(p) = H(p, h(p), x)$. Let $g' : C \to A'$ be obtained from Theorem 4.6.37 with A', G', H' in place of A, G, H. Define $g : P \times C \to A$ by $g(p, x) = g'(x)(p)$. Show that g is the desired function.

     (b) Convince yourself that by a similar argument Theorem 4.6.52
         can be derived from Theorem 4.6.50.

49. Consider the family of constructors $\mathcal{R} = \{R_p, R_q\}$ on the set of integers, where $R_p, R_q \subseteq \mathbf{Z} \times \mathbf{Z}$ are given by $R_p = \{(x, y) \mid x, y \in \mathbf{Z}, x - y \in \{p, -p\}\}$ and $R_q = \{(x, y) \mid x, y \in \mathbf{Z}, x - y \in \{q, -q\}\}$.

     (a) Determine the closure of the set $\{0\}$ under the set of constructors
         $\mathcal{R} = \{R_8, R_{12}\}$.

     (b) Prove that, in general, the closure of $\{0\}$ under $\{R_p, R_q\}$ equals
         the set of the multiples of d, where d is the greatest common
         divisor of p and q.

Let S, V be two finite sets. Let λ be the null word from V^*, the set of words over V, and consider a ternary relation $\rho \subseteq S \times (V \cup \{\lambda\}) \times S$ such that $(s, \lambda, s) \in \rho$ for any $s \in S$. We refer to the triple $T = (S, V, \rho)$ as a *transition system*. S, V, and ρ are called the *set of states*, the *set of inputs*, and the *transition relation* of T, respectively. The following construction plays an important role in automata theory.

Let $\mathrm{CONF}_T = S \times V^*$ be the *set of configurations* of T. Define the constructor R_T on CONF_T as

$$((s, xu), (s', u)) \in R_T, \text{ if and only if } (s, x, s') \in \rho.$$

The closure of configuration (s, w) under R_T will be denoted by $C_T(s, w)$.

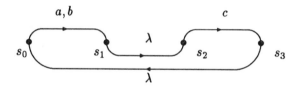

FIGURE 4.6. The digraph of the transition system T.

The language recognized by T through *the initial set of states* S_0 and the set of *final states* S_1 is the set of words

$$L(T, S_0, S_1) = \{w \mid w \in V^*, (s_1, \lambda) \in C_T(s_0, w) \text{ for } s_0 \in S_0, s_1 \in S_1\}.$$

Transition systems can be represented by directed graphs whose edges are labeled by subsets of the set $V \cup \{\lambda\}$. Namely, for a transition system $T = (S, V, \rho)$, we consider the graph $G_T = (S, U)$, where $(s, s') \in U$ if there is a triple $(s, x, s') \in \rho$. The arc (s, s') will be labeled by the set $\{x \mid x \in V \cup \{\lambda\}, (s, x, s') \in \rho\}$. We omit from this representation the loops labeled by λ.

If $L(u) \subseteq V \cup \{\lambda\}$ is the set of labels of the arc u, then a word $w \in V^*$ is a label of a path $(u_1, \ldots, u_n)$ in G_T $u \in L(u_1) \cdots L(u_n)$.

Consider the transition system $T = (\{s_0, s_1, s_2, s_3\}, \{a, b, c\}, \rho)$, where

$$\rho = \{(s_0, a, s_1), (s_0, b, s_1), (s_2, c, s_3), (s_1, \lambda, s_2), (s_3, \lambda, s_0)\}.$$

Its graph is represented in Figure 4.6.

The set of labels of the path $(s_0, s_1)(s_1, s_2)(s_2, s_3)$ is $\{a, b\}\{\lambda\}\{c\} = \{ac, bc\}$.

50. Let $S = \{s_0, s_1, s_2\}$ and $V = \{a, b\}$. Define the transition relation as $\rho = \{(s_0, a, s_1), (s_1, a, s_1), (s_1, b, s_2), (s_2, b, s_2)\}$.

   (a) Draw the graph of the transition system $T = (S, V, \rho)$.

   (b) Prove that in the transition system T, no symbol b may precede a in the word w in any configuration $(s, w) \in C_T(s_0, a^n b^m)$ $m, n \geq 1$.

   (c) Show that the language $L(T, \{s_0\}, \{s_2\})$ equals $\{a\}^+\{b\}^+$.

51. Let $T = (S, V, \rho)$ be a transition system. Prove that $w \in L(T, S_0, S_1)$ if and only if there is a path in the digraph G_T joining a state from S_0 to a state S_1 for which w is a label.

52. Let T, T' be two transition systems, $T = (S, V, \rho)$ and $T' = (S', V, \rho')$, where $S \cap S' = \emptyset$. Consider the languages $L = L(T, S_0, S_1)$ and

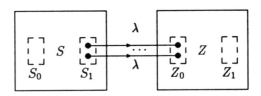

FIGURE 4.7. The transition system T''

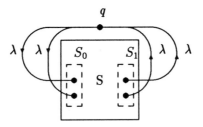

FIGURE 4.8. The transition system T_*

$L' = (T', Z_0, Z_1)$. Prove that the language LL' is represented in the transition system $T'' = (S \cup S', V, \rho \cup \rho' \cup \{(s_1, \lambda, z_0) \mid s_1 \in S_1, z_0 \in Z_0\})$ by the initial set of states S_0 and by the final set of states Z_1.

The construction defined by this exercise is given in Figure 4.7.

53. Consider a transition system $T = (S, V, \rho)$ and assume that $q \notin S$. If $L = L(T, S_0, S_1)$, prove that $L^* = L(T_*, \{q\}, \{q\})$, where T_* is the transition system $T_* = (S \cup \{q\}, V, \rho \cup \{(s_1, \lambda, q) \mid s_1 \in S_1\} \cup \{(q, \lambda, s_0) \mid s_0 \in S_0\} \cup \{(q, \lambda, q)\})$.

The construction of T_* is given in Figure 4.8.

54. Let $T_1 = (S_1, V, \rho_1)$ and $T_2 = (Z, V, \rho_2)$ be two transition systems such that $S \cap Z = \emptyset$. Suppose that $L_1 = L(T_1, S_0, S_1)$ and $L_2 = L(T_2, Z_0, Z_1)$. If $q_0, q_1 \notin S \cup Z$, consider the transition system $T_\cup$, represented in Figure 4.9.

Prove that $L_1 \cup L_2 = L(T_\cup, \{q_0\}, \{q_1\})$.

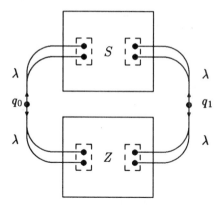

FIGURE 4.9. The transition system $T_\cup$.

Propositional Logic

55. Show that if p, q, and r are statement variables, then $(((p \wedge q) \to r) \to (p \to r))$ is in the set of formulas of propositional logic as introduced in Definition 4.8.2.

56. If φ is a formula of propositional logic (as given in Definition 4.8.2), let $s(\varphi)$ denote the number of occurrences of statement variables in φ and let $c(\varphi)$ denote the number of occurrences of binary connectives (i.e., $\vee, \wedge, \to, \leftrightarrow$) in φ. [Note: $s(\varphi)$ and $c(\varphi)$ count *occurrences*. Thus, if $\varphi = (\sim(((p \leftrightarrow q) \vee p) \vee (r \leftrightarrow s)))$, then $s(\varphi) = 5$ (not 4) and $c(\varphi) = 4$ (not 2).]

No matter what φ is, there is a fixed relationship that holds between $s(\varphi)$ and $c(\varphi)$. See if you can discover what this relationship is by examining some particular cases. Then, prove that this relationship holds for all φ by using structural induction.

57. If SV is the set of statement variables, then according to Definition 4.8.2, each formula of propositional logic is an element of $\mathrm{Seq}(SV \cup \{(,),\sim,\vee,\wedge,\leftarrow,\leftrightarrow\})$, and hence, it makes sense to talk about the length of such a formula. Some natural numbers are lengths of statement forms and some are not. Determine exactly which natural numbers can be obtained as lengths of statement forms and prove your answer.

58. A *substitution* is a function $s : SV \longrightarrow \mathcal{F}_{prop}$, i.e., a function from statement variables to formulas. Any substitution s determines a function $\bar{s} : \mathcal{F}_{prop} \longrightarrow \mathcal{F}_{prop}$, where for any formula φ, $\bar{s}(\varphi)$ is the

formula that results when each statement variable p in φ is replaced by $s(p)$. If φ is a formula and s is a substitution, then $\bar{s}(\varphi)$ is called a *substitution instance* of φ.

(a) Let s be a substitution. Give a recursive definition of the function $\bar{s}$.

(b) Let s be a substitution and let v be a truth assignment. Define a truth assignment v' by setting

$$v'(p) = \bar{v}(s(p))$$

for every statement variable p. Use structural induction to prove that, for every formula φ,

$$\bar{v}(\bar{s}(\varphi)) = \overline{v'}(\varphi).$$

(c) Use part (b) to prove that if φ is a tautology and s is a substitution, then $\bar{s}(\varphi)$ is also a tautology, (i.e., any substitution instance of a tautology is a tautology).

59. Use structural induction to show that if φ is a formula and v and w are two truth assignments that agree on all the statement variables that occur in φ (i.e., $v(p) = w(p)$ for all statement variables p that occur in φ), then $\bar{v}(\varphi) = \bar{w}(\varphi)$.

60. Prove that every proper initial segment of a formula of propositional logic contains more left parentheses than right parentheses.

Primitive Recursive and Partial Recursive Functions

61. Prove that the following functions are primitive recursive:

(a) pred : $\mathbf{N} \longrightarrow \mathbf{N}$, where $\mathrm{pred}(n) = 0$ if $n = 0$ and $\mathrm{pred}(n) = n-1$ if $n \geq 1$.

(b) monus : $\mathbf{N}^2 \longrightarrow \mathbf{N}$, where

$$\mathrm{monus}(m,n) = \begin{cases} m - n & \text{if } m \geq n \\ 0 & \text{otherwise.} \end{cases}$$

The value of $\mathrm{monus}(m,n)$ will be denoted by $m \mathbin{\dot{-}} n$.

62. A set L is primitive recursive if $L \subseteq \mathbf{N}^n$ and its charcteristic function $\chi_L : \mathbf{N}^n \longrightarrow \mathbf{N}$ is primitive recursive. If χ_L is partial recursive, then we refer to L as a recursive set.

Prove that if L, K are two recursive sets then so are $L \cup K, L \cap K$, and $L - K$.

63. Let $\rho \subseteq N \times N$ be a binary relation on N. Consider the function $f_\rho : N^2 \longrightarrow N$ defined by

$$f_\rho(x, y) = \begin{cases} 1 & \text{if } (x, y) \in \rho, \\ 0 & \text{otherwise.} \end{cases}$$

The function f_ρ is the characteristic function of the relation ρ considered as a subset of $\rho \subseteq N \times N$. The relation ρ is *primitive recursive* if f_ρ is primitive recursive.

(a) Prove that the relations $=, <, >, \leq, \geq, \neq$ are primitive recursive.

(b) If ρ, σ are primitive recursive, then so are $\rho \cup \sigma$, $\rho \cap \sigma$, and $\rho - \sigma$. What can be said about $\rho \circ \sigma$?

Solution: In order to prove the primitive recursiveness of "$<$," observe that we can write for the characteristic function $f_<$, $f_<(x, y) = \text{sgn}(y-x)$. Also, $f_\leq(x, y) = f_<(x, S(y))$, and this implies the primitive recursiveness of "$\leq$." Now, the primitive recursiveness of $f_=$ follows immediately from the fact that $f_=(x, y) = f_\leq(x, y)$ and $f_\leq(y, x)$.

64. Let $f : N^2 \longrightarrow N$ be a primitive recursive function. Define the function $g : N^2 \longrightarrow N$ by

$$g(x_1, x_2) = f(x_1, 0) + \cdots + f(x_1, x_2)$$

for every $x_1, x_2 \in N$. Prove that g is primitive recursive.

65. A collection of functions $\{f_i | 1 \leq i \leq p\}$, $f_i : N^n \longrightarrow N$, for $1 \leq i \leq p$, defines a partition on N^n if

$$\sum_{1 \leq i \leq p} f_i(x_1, \ldots, x_n) = 1$$

for every $(x_1, \ldots, x_n) \in N^n$. Note that for any $(x_1, \ldots, x_n) \in N^n$ exactly one of the numbers $f_i(x_1, \ldots, x_n)$ equals 1. Therefore, every $(x_1, \ldots, x_n) \in N^n$ belongs to exactly one of the sets

$$B_i = \{(x_1, \ldots, x_n) \in N^n | f_i(x_1, \ldots, x_n) = 1\}$$

for $1 \leq i \leq p$.

Consider a collection of functions $\{g_i | 1 \leq i \leq p\}$, $g_i : N^n \longrightarrow N$ for $1 \leq i \leq p$ and the function $h : N^n \longrightarrow N$ defined by

$$h(x_1, \ldots, x_n) = \sum_{1 \leq i \leq p} f_i(x_1, \ldots, x_n) g_i(x_1, \ldots, x_n)$$

for every $(x_1, \ldots, x_n) \in N^n$. This method of introducing new functions is known as *definition by cases*; each set B_i represents a case and every $(x_1, \ldots, x_n)$ belongs to exactly one such case.

Sometimes, the definition of the function h is given as

$$h(x_1, \ldots, x_n) = \begin{cases} g_1(x_1, \ldots, x_n) & \text{if } (x_1, \ldots, x_n) \in B_1 \\ \vdots \\ g_p(x_1, \ldots, x_n) & \text{if } (x_1, \ldots, x_n) \in B_p \end{cases}$$

for every $(x_1, \ldots, x_n) \in \mathbf{N}^n$.

Prove that if f_i, g_i are primitive recursive for $1 \le i \le p$ then h is also primitive recursive.

66. Suppose that $f : \mathbf{N}^{n+1} \longrightarrow \mathbf{N}$ and $g : \mathbf{N}^n \longrightarrow \mathbf{N}$. Define functions $h : \mathbf{N}^n \longrightarrow \mathbf{N}$ and $k : \mathbf{N}^n \longrightarrow \mathbf{N}$ by

$$h(x_1, \ldots, x_n) = \sum_{i=0}^{g(x_1, \ldots, x_n)} f(x_1, \ldots, x_n, i)$$

and

$$k(x_1, \ldots, x_n) = \prod_{i=0}^{g(x_1, \ldots, x_n)} f(x_1, \ldots, x_n, i)$$

for every $(x_1, \ldots, x_n) \in \mathbf{N}^n$. Prove that if f and g are primitive recursive then so are h and k.

67. Consider the functions $\max_n : \mathbf{N}^n \longrightarrow \mathbf{N}$ and $\min_n : \mathbf{N}^n \longrightarrow \mathbf{N}$ where $\max_n(x_1, \ldots, x_n)$ and $\min_n(x_1, \ldots, x_n)$ are the greatest and the least components of the n-tuple $(x_1, \ldots, x_n)$, respectively. Prove that for any $n \ge 1$ the functions $\max_n$ and $\min_n$ are primitive recursive.

Solution: We discuss only the case of $\max_n$. The proof is by induction on n. For $n = 1$, $\max_1$ is the identity function on $\mathbf{N}$, which is primitive recursive (since it coincides with the function p_1^1).

Suppose that the statement is true for $\max_n$ and consider the function $\max_{n+1}$.

We can define $\max_{n+1}$ by

$$\begin{aligned} \max_{n+1}(x_1, \ldots, x_n, 0) &= \max_n(x_1, \ldots, x_n), \\ \max_{n+1}(x_1, \ldots, x_n, y+1) &= h(x_1, \ldots, x_n, y, \\ &\qquad \max_{n+1}(x_1, \ldots, x_n, y)), \end{aligned}$$

where $h : \mathbf{N}^{n+2} \longrightarrow \mathbf{N}$ is given by

$$h(x_1, \ldots, x_{n+2}) = \begin{cases} x_{n+2} & \text{if } x_{n+1} < \max_n(x_1, \ldots, x_n) \\ x_{n+1} + 1 & \text{otherwise.} \end{cases}$$

Using the inductive hypothesis, it is not difficult to show that h is primitive recursive. This, in turn, implies the primitive recursiveness of $\max_{n+1}$.

68. Define the function **FORALL** : $\text{Seq}(\mathbf{N}) \longrightarrow \mathbf{N}$ by

$$\mathbf{FORALL}(s) = \begin{cases} 1 & \text{if } s = (s_0, \ldots, s_{n-1}) \text{ and} \\ & s_i > 0 \text{ for all } i, 0 \le i \le n-1 \\ 0 & \text{otherwise.} \end{cases}$$

Consider also the function **EXISTS** : $\text{Seq}(\mathbf{N}) \longrightarrow \mathbf{N}$ given by

$$\mathbf{EXISTS}(s) = \begin{cases} 1 & \text{if } s = (s_0, \ldots, s_{n-1}) \text{ and} \\ & s_i > 0 \text{ for some } i, 0 \le i \le n-1 \\ 0 & \text{otherwise.} \end{cases}$$

If $s = (s_0, \ldots, s_{n-1})$, then we use the notation $\mathbf{FORALL}_{i=0}^{n-1} s_i$ for **FORALL**(s) and $\mathbf{EXISTS}_{i=0}^{n-1} s_i$ for **EXISTS**(s).

(a) Define inductively the functions **FORALL** and **EXISTS**. Note that $\mathbf{FORALL}(\lambda) = 1$ and $\mathbf{EXISTS}(\lambda) = 0$.

(b) Suppose that $f : \mathbf{N}^{n+1} \longrightarrow \mathbf{N}$ and $g : \mathbf{N}^n \longrightarrow \mathbf{N}$. Define functions $h : \mathbf{N}^n \longrightarrow \mathbf{N}$ and $k : \mathbf{N}^n \longrightarrow \mathbf{N}$ by

$$h(x_1, \ldots, x_n) = \mathbf{EXISTS}_{i=0}^{g(x_1, \ldots, x_n)} f(x_1, \ldots, x_n, i),$$
$$k(x_1, \ldots, x_n) = \mathbf{FORALL}_{i=0}^{g(x_1, \ldots, x_n)} f(x_1, \ldots, x_n, i)$$

for every $(x_1, \ldots, x_n) \in \mathbf{N}^n$. Prove that if f and g are primitive recursive then so are h and k.

69. Let $f : \mathbf{N}^{n+1} \longrightarrow \mathbf{N}$ and $h : \mathbf{N}^n \longrightarrow \mathbf{N}$. Define the set $A_{(x_1, \ldots, x_n)}^h$ by

$$A_{(x_1, \ldots, x_n)}^{h,f} = \{ p \mid p \in \mathbf{N}, 0 \le p \le h(x_1, \ldots, x_n) \\ \text{and } f(x_1, \ldots, x_n, p) = 0 \},$$

where $x_1, \ldots, x_n \in \mathbf{N}$.
The function $(\mu_{\le h} f) : \mathbf{N}^n \longrightarrow \mathbf{N}$ is defined by

$$(\mu_{\le h} f)(x_1, \ldots, x_n) = \begin{cases} \min A_{(x_1, \ldots, x_n)}^{h,f} & \text{if } A_{(x_1, \ldots, x_n)}^{h,f} \ne \emptyset \\ 0 & \text{otherwise.} \end{cases}$$

In order to indicate that $g = (\mu_{\le h} f)$, we sometimes write

$$g(x_1, \ldots, x_n) = \mu y \le h(x_1, \ldots, x_n)[f(x_1, \ldots, x_n, y) = 0].$$

If $g = (\mu_{\le h} f)$, then we say that g is obtained from f by *bounded minimization*. Prove that if f and h are primitive recursive then so is $g = (\mu_{\le h} f)$.

Solution: Consider the function $t : \mathbf{N}^{n+1} \longrightarrow \mathbf{N}$ defined by

$$t(x_1, \ldots, x_n, y) = \mathbf{FORALL}_{i=0}^{y} f_{\neq}(f(x_1, \ldots, x_n), 0)$$

for every $(x_1, \ldots, x_n, y) \in \mathbf{N}^{n+1}$.

If $A_{(x_1,\ldots,x_n)}^{h,f} \neq \emptyset$ and $z = \min A_{(x_1,\ldots,x_n)}^{h,f}$, then, if $y < z$, we have $t(x_1, \ldots, x_n, y) = 1$. Also, if $y \geq z$, then $t(x_1, \ldots, x_n, y) = 0$. Therefore, we can write

$$z = \sum_{y=0}^{h(x_1,\ldots,x_n)} t(x_1, \ldots, x_n, y)$$

if $A_{(x_1,\ldots,x_n)}^{h,f} \neq \emptyset$. On the other hand, consider the function $r : \mathbf{N}^n \longrightarrow \mathbf{N}$ given by

$$r(x_1, \ldots, x_n) = \mathbf{EXISTS}_{y=0}^{h(x_1,\ldots,x_n)} f_{=}(f(x_1, \ldots, x_n, y), 0)$$

for every $(x_1, \ldots, x_n) \in \mathbf{N}^n$. Note that this function is primitive recursive and, also, that $A_{(x_1,\ldots,x_n)}^{h,f} \neq \emptyset$ if and only if $r(x_1, \ldots, x_n) = 1$. This allows us to define, by cases, the function $(\mu_{\leq h} f)$ as

$$
(\mu_{\leq h} f)(x_1, \ldots, x_n) \\
= \begin{cases} \sum_{y=0}^{h(x_1,\ldots,x_n)} t(x_1, \ldots, x_n, y), & \text{if } r(x_1, \ldots, x_n) = 1 \\ 0 & \text{otherwise,} \end{cases}
$$

which proves that $(\mu_{\leq h} f)$ is primitive recursive.

70. Define the function $\operatorname{div} : \mathbf{N}^2 \longrightarrow \mathbf{N}$ as

$$\operatorname{div}(x, y) = \begin{cases} \text{the largest } k \text{ such that } k < x/y & \text{if } y > 0 \\ 0 & \text{if } y = 0. \end{cases}$$

Prove that div is primitive recursive.

Hint: Use bounded minimization.

71. Consider the primitive recursive functions $g_0, \ldots, g_{k-1} : \mathbf{N}^n \longrightarrow \mathbf{N}$ and $h : \mathbf{N}^{n+2} \longrightarrow \mathbf{N}$. The function $f : \mathbf{N}^{n+1} \longrightarrow \mathbf{N}$ is defined starting from the functions $g_0, \ldots, g_{k-1}, h$ as

$$f(x_1, \ldots, x_n, 0) = g_0(x_1, \ldots, x_n),$$
$$\vdots$$
$$f(x_1, \ldots, x_n, k-1) = g_{k-1}(x_1, \ldots, x_n),$$
$$f(x_1, \ldots, x_n, y+1) = h(x_1, \ldots, x_n, y, f(x_1, \ldots, x_n, y)),$$

for every $x_1, \ldots, x_n, y \in \mathbf{N}$ and $y \geq k-1$. Prove that f is primitive recursive.

Solution: Consider the function $\bar{h} : \mathbf{N}^{n+2} \longrightarrow \mathbf{N}$ given by:

$$h(x_1, \ldots, x_n, x_{n+1}, x_{n+2})$$
$$= f_=(1, x_{n+1})g_1(x_1, \ldots, x_n) + \cdots$$
$$+ f_=(k-1, x_{n+1})g_{k-1}(x_1, \ldots, x_n)$$
$$+ f_\geq(k, x_{n+1})h(x_1, \ldots, x_n, x_{n+1}, x_{n+2}).$$

Note that f can be defined by primitive recursion from g_0 and $\bar{h}$.

Grammars

72. Prove that for each grammar $G = (N, T, S, P)$ there exists an equivalent grammar G' such that terminal symbols occur only in the right member of productions of the form (X, a) with $X \in N$ and $a \in T$.

    Moreover, if G is context-sensitive, length-increasing, or context-free, then so is G'.

73. Explain why in Theorem 4.10.21 the condition $\gamma \in T^*$ is essential.

74. Let $G = (\{S, X, Y, Y', Z, A, B\}, \{a, b\}, S, P)$ be a grammar, where P consists of the following set of rules:

$$
\begin{array}{ll}
S \rightarrow XYZ, & S \rightarrow \lambda, \\
XY \rightarrow aXA, & XY \rightarrow bXB, \\
Aa \rightarrow aA, & Ab \rightarrow bA, \\
Ba \rightarrow aB, & Bb \rightarrow bB, \\
AZ \rightarrow YaZ, & AZ \rightarrow Y'a, \\
BZ \rightarrow YbZ, & BZ \rightarrow Y'b, \\
aY \rightarrow Ya, & bY \rightarrow Yb, \\
aY' \rightarrow Y'a, & bY' \rightarrow Y'b, \\
XY' \rightarrow \lambda. &
\end{array}
$$

    Prove that $L(G) = \{uu \mid u \in \{a, b\}^*\}$.

    Hint: To show that $\{uu \mid u \in \{a, b\}^*\} \subseteq L(G)$ prove, by induction on $|u|$, that for every $u \in \{a, b\}^*$ we have $S \overset{*}{\underset{G}{\Rightarrow}} uXYuZ$ and $S \overset{*}{\underset{G}{\Rightarrow}} uu$.

    To show the converse inclusion, it is helpful to prove that if $S \overset{*}{\underset{G}{\Rightarrow}} \alpha$, then α has one of the following forms: $uXu'Yu''Z$, $uaXu'Au''Z$, $ubXu'Bu''Z$, $uXu'Y'u'''$, or uu, where $u, u', u'' \in \{a, b\}^*$ and $u'u'' = u$.

75. Prove that for every length-increasing grammar G there is an equivalent context-sensitive grammar G'.

    Solution: Let $G = (N, T, S, P)$ be a length-increasing grammar. Without restricting the generality, because of Exercise 72, we can assume that terminal symbols occur only in the right member of productions of the form (X, a) with $X \in N$ and $a \in T$.

Initially, we also assume that $(S, \lambda) \notin P$. Suppose that $P = P_1 \cup P_{li}$, where P_1 is the set of context-sensitive productions and P_{li} is the set of length-increasing rules of G.

If $P_{li} = \emptyset$, then we can take $G' = G$, and the argument is completed. Suppose, therefore, that $P_{li} \neq \emptyset$ and let $|P_{li}| = p$. We can apply an argument by induction on p.

Only the base step is discussed here. For $p = 1$, let

$$X_1 \cdots X_m \rightarrow Y_1 \cdots Y_n$$

be the unique length-increasing, non-context sensitive production of the grammar. Clearly, we have $1 < m \leq n$.

Consider m new nonterminals $Z_1, \ldots, Z_m$ and let P' be the following set of rules.

$$\begin{aligned}
X_1 X_2 \cdots X_m &\rightarrow Z_1 X_2 \cdots X_m \\
Z_1 X_2 \cdots X_m &\rightarrow Z_1 Z_2 \cdots X_m \\
&\vdots \\
Z_1 Z_2 \cdots Z_{m-1} X_m &\rightarrow Z_1 Z_2 \cdots Z_{m-1} Z_m Y_{m+1} \cdots Y_n \\
Z_1 Z_2 \cdots Z_{m-1} Z_m Y_{m+1} \cdots Y_n &\rightarrow Y_1 Z_2 \cdots Z_{m-1} Z_m Y_{m+1} \cdots Y_n \\
&\vdots \\
Y_1 Y_2 \cdots Y_{m-1} Z_m Y_{m+1} \cdots Y_n &\rightarrow Y_1 Y_2 \cdots Y_{m-1} Y_m Y_{m+1} \cdots Y_n .
\end{aligned}$$

Note that the rules from P' are all context-sensitive. The grammar $G' = (N \cup \{Z_1, \ldots, Z_m\}, T, S, P_1 \cup P')$ is context sensitive and is equivalent to the grammar G. Indeed, if the production

$$X_1 \cdots X_m \rightarrow Y_1 \cdots Y_n$$

is used in deriving $u \in L(G)$, then we can replace the application of this production by the application of the $2m$ productions of P', which shows that $L(G) \subseteq L(G')$.

Conversely, if $S \overset{*}{\underset{G'}{\Rightarrow}} u$ and we use the productions of P', then note that these productions can be applied only in the order listed above, and they must all be used. This application (in G') can be replaced by the application of the production $X_1 \ldots X_m \rightarrow Y_1 \ldots Y_n$ in G which proves that $L(G') \subseteq L(G)$.

We leave to the reader the rest of the argument.

76. Prove that the grammars G_1, G_2, G_3 are equivalent, where

(a) $G_1 = (\{S_0\}, \{a, b\}, S_0, P_1)$, where

$$P_1 = \{S_0 \rightarrow \lambda, S_0 \rightarrow aS_0b, S_0 \rightarrow bS_0a, S_0 \rightarrow S_0S_0\}.$$

(b) $G_2 = (\{S_0, S_1, S_2\}, \{a, b\}, S_0, P_2)$, where

$$\begin{aligned}
P_2 &= \{S_0 \rightarrow \lambda, S_0 \rightarrow S_1S_0S_2, \\
&\quad S_0 \rightarrow S_2S_0S_1, S_0 \rightarrow S_0S_0, \\
&\quad S_1 \rightarrow a, S_2 \rightarrow b\}.
\end{aligned}$$

(c) $G_3 = (\{S, S_0, S_1, S_2\}, \{a, b\}, S, P_3)$, where

$$\begin{aligned}
P_3 &= \{S \rightarrow \lambda, S \rightarrow S_0, S_0 \rightarrow S_0S_0, \\
&\quad S_0S_0 \rightarrow aS_2, S_0S_0 \rightarrow bS_1, S_1 \rightarrow a, \\
&\quad S_2 \rightarrow b, S_1S_0 \rightarrow S_0S_1, S_2S_0 \rightarrow S_0S_2\}.
\end{aligned}$$

Peano's Axioms

77. Let $\mathcal{S} = (N, 0, S)$ be a Peano structure and let addition, multiplication, and exponentiation be defined as in Definition 4.11.13. Prove that all of the following laws hold for $\mathcal{S}$. (Your proofs will use only the definitions of the operations and the third Peano axiom.)

(a) Associative law for $+$: For every $n, m, p \in N$, $(n + m) + p = n + (m + p)$.
    Hint: Use induction on p.

(b) For every $n \in N$, $S(0)n = n$.
    Hint: Use induction on n.

(c) Right distributive law: For every $n, m, p \in N$, $(n + m)p = np + mp$.
    Hint: Use induction on p and the commutativity and associativity of multiplication.

(d) Commutative law for $\times$: For every $n, m \in N$, $nm = mn$.
    Hint: Use induction on n. For the basis step, you will need a further induction on m. For the inductive step, use part (b) and the right distributive law.

(e) Left distributive law: For every $n, m, p \in N$, $n(m+p) = nm+np$.
    Hint: Use the right distributive law and commutativity of multiplication.

(f) Associativity of multiplication: For every $n, m, p \in N$, $(nm)p = n(mp)$.
    Hint: Use induction on p and the left distributive law.

(g) For all $n, m, p \in N$, $n^{m+p} = n^m n^p$.
    Hint: Use induction on p. For the basis step, use part (b) and commutativity of addition.

(h) For all $n, m, p \in N$, $(n^m)^p = n^{mp}$.

Hint: Use induction on p and the previous part.

78. Let $\mathcal{S} = (N, 0, S)$ be a Peano structure and let $\leq$ be defined as in Definition 4.11.13. Prove that the following properties hold. (Your proofs will use only the definition of $\leq$, properties from the previous exercise, and induction, so they will follow from the existence of the addition function and the induction axiom.)

(a) Reflexivity of $\leq$: For every $n \in N$, $n \leq n$.

(b) Transitivity of $\leq$: For every $n, m, p \in N$, if $n \leq m$ and $m \leq p$, then $n \leq p$.

(c) For every $n \in N$, $0 \leq n$.

(d) If $n \in N$ is $\neq 0$, then there is an $m \in N$ with $n = S(m)$.

(e) For every $n \in N$, $n \leq S(n)$.

(f) Totality of $\leq$: For every $n, m \in N$, either $n \leq m$ or $m \leq n$.

(g) For every $n, m \in N$, if $n \leq m$ and $n \neq m$, then $S(n) \leq m$.

(h) For every $n, m \in N$, $n \leq m$ if and only if $S(n) \leq S(m)$.

79. Let $\mathcal{S} = (N, 0, S)$ be a Peano structure and let $+$ and $\leq$ be defined as in Definition 4.11.13. Show that the following properties hold. (Unlike the previous two exercises, this exercise contains problems whose solutions require the use of the first two Peano axioms.)

(a) For all $n, m \in N$, $n + S(m) \neq n$.

Hint: Use induction on n.

(b) For all $n, m \in N$, if $n + m = n$, then $m = 0$.

(c) For all $n, m \in N$, if $n + m = 0$, then $m = 0$.

(d) Antisymmetry of $\leq$: For all $n, m \in N$, if $n \leq m$ and $m \leq n$, then $n = m$.

Hint: Use the last two parts.

(e) For all $n \in N$, $S(n) \neq n$.

Hint: Use induction on n.

80. In this exercise, we establish some facts about finite sets that we took for granted in Chapter 2. Let $\mathcal{S} = (N, 0, S)$ be a Peano structure.

(a) Show that if n and m are different elements of N, then there is no bijection between $\{i \in N \mid i < n\}$ and $\{i \in N \mid i < m\}$.

Hint: Use induction on n.

(b) Show that if $m \in N$, then there is a bijection between N and $N - \{m\}$.

(c) Show that for every $n \in N$, there is no bijection between N and $\{i \in N \mid i < n\}$.

   Hint: Use induction on n and part (b).

(d) Show that if $n < m$ then there is no injection from $\{i \in N \mid i < m\}$ to $\{i \in N \mid i < n\}$.

   Hint: Use part (a) and the Schröder-Bernstein theorem (Theorem 5.2.11).

(e) Show that if $n < m$, then there is no surjection from $\{i \in N \mid i < n\}$ to $\{i \in N \mid i < m\}$.

   Hint: Suppose that such a surjection exists and consider a left inverse.

81. Let $\mathcal{S} = (N, 0, S)$ be a Peano structure and let $\mathcal{S}' = (N, 0, S, \leq)$, where $\leq$ is defined as usual. Show that $\mathcal{S}'$ satisfies the conditions C1-C4 of Exercise 18 and therefore also satisfies all of the conditions I1-I3.

   Hint: Most of the work has been done in previous exercises.

82. Let $\mathcal{S} = (N, 0, S, \leq)$ be a structure that satisfies conditions C1-C4 of Exercise 18 and that also satisfies the equivalent conditions I1-I3. Show that $\mathcal{S}$ is a Peano structure.

83. Let $\mathcal{S}_1$ and $\mathcal{S}_2$ be the Peano structures of Examples 4.11.6 and 4.11.7. Describe explicitly the addition, multiplication, and exponentiation operations and the $\leq$ relation on these structures.

84. (a) For each of Peano's axioms, show that if $\mathcal{S}$ is a structure that satisfies that axiom and $\mathcal{S}_1$ is a structure that is isomorphic to $\mathcal{S}$ then $\mathcal{S}_1$ satisfies the same axiom. (Hence, if $\mathcal{S}$ is a Peano structure, then any structure isomorphic to $\mathcal{S}$ is also a Peano structure.)

   (b) Let $\mathcal{S}$ be a Peano structure and let $\mathcal{S}' = (N', 0', S')$ be any structure with $0' \in N'$ and S' a unary operation on N'. Prove that $\mathcal{S}'$ is a Peano structure if and only if the unique homomorphism from $\mathcal{S}$ to $\mathcal{S}'$ is an isomorphism.

85. Let $\mathcal{S} = (N, 0, S)$ be a structure with $0 \in N$ and S a unary operation on N. Suppose that for every structure $\mathcal{S}' = (N', 0', S')$ with $0' \in N'$ and S' a unary operation on N' there is a unique homomorphism from $\mathcal{S}$ to $\mathcal{S}'$. Prove that $\mathcal{S}$ is a Peano structure.

The next two exercises show how the natural numbers can be constructed in set theory.

86. We wish to construct the set N of natural numbers. Our intuition will be that the natural number n is to be an n element set. For $n = 0$,

this means that 0 must be $\emptyset$. For $n = 1$, there are many n element sets. We choose to make 1 be $\{0\} = \{\emptyset\}$. Similarly, we make 2 equal $\{0, 1\}$ so 2 is $\{\emptyset, \{\emptyset\}\}$. The general idea is to make $n + 1 = \{0, \ldots, n\}$.

(a) Continuing in this way, write out the definitions of the numbers 3 and 4.

Using this idea, we can write out any particular natural number (at least in theory), but we still cannot define N. If x is any set, define $s(x)$, the *successor* of x, to be $x \cup \{x\}$. Note that if $n = \{0, \ldots, n-1\}$, then $s(n) = n \cup \{n\} = \{0, \ldots, n-1\} \cup \{n\} = \{0, \ldots, n\}$, so our scheme for defining particular integers amounts to setting $0 = \emptyset$ and $n + 1 = s(n)$. (It is tempting to give a recursive "definition" of the natural numbers at this point, but that would be circular since the definition would involve recursion on the natural numbers.)

We define a set to be *inductive* if it contains $\emptyset$ and is closed under the successor operation.

(b) Show that the intersection of any nonempty family of inductive sets is inductive.

The *axiom of infinity* from set theory states that there is an inductive set. Let A be an inductive set and define the natural numbers N as $N = \bigcap\{B \subseteq A \mid B \text{ is inductive}\}$. Note that by the previous part, N is inductive.

(c) Show that if A' is an inductive set and we define $N' = \bigcap\{B \subseteq A' \mid B \text{ is inductive}\}$, then $N' = N$. (This shows that the definition of N does not depend on what set A is chosen.)

(d) Show that N is contained in every inductive set.

(e) Show that if A is an inductive subset of N then $A = N$.

87. In this exercise, we develop properties of the structure $(N, \emptyset, s)$ which was constructed in the last exercise. In particular, we will show that it is a Peano structure.

(a) Show that $(N, \emptyset, s)$ satisfies the first and third Peano axioms.

We call a set x *transitive* if every member of a member of x is itself a member of x, that is, for all sets y and z, if $y \in z$ and $z \in x$, then $y \in z$.

(b) Use the third Peano axiom to show that every element of N is a transitive set.

(c) Show that for every $n \in N$, $n \notin n$.

(d) Use the last two parts to show that $(N, \emptyset, s)$ satisfies the second Peano axiom (and hence, is a Peano structure).

Since $(N, \emptyset, s)$ is a Peano structure, Definition 4.11.13 gives us operations of addition, multiplication, and exponentiation and a total order $\leq$ on N.

(e) Prove that for any two natural numbers n and m, $n \leq m$ if and only if $n \subseteq n$ (hence, $n < m$ if and only if $n \subset m$).

(f) Prove that every element of a natural number is also a natural number.

(g) Show that for each $n \in N$, $n = \{m \in N \mid m < n\}$.

A structure $\mathcal{S} = (N, 0, S)$ with $0 \in N$ and S a unary operation on N is called an *induction structure* if it satisfies the third Peano axiom (Axiom 4.11.4), the induction axiom.

88. Let $\mathcal{S}$ be a Peano structure and let $\mathcal{S}' = (N', 0', S')$ be any structure with $0' \in N'$ and S' a unary operation on N'. Show that $\mathcal{S}'$ is an induction structure if and only if the unique homomorphism from $\mathcal{S}$ to $\mathcal{S}'$ is onto.

89. Let $\mathcal{S} = (N, 0, S)$ be a structure (with $0 \in N$ and S a unary operation on N). Show that $\mathcal{S}$ is an induction structure if and only if the following condition is met.

> For every structure $\mathcal{S}'$ (of the same type as $\mathcal{S}$) and homomorphism φ from $\mathcal{S}'$ to $\mathcal{S}$, φ is onto.

90. For each pair of natural numbers n, m with $m < n$, we define a structure $T_{n,m}$ by $T_{n,m} = (\{i \in N \mid i < n\}, 0, S_{n,m})$, where

$$S_{n,m}(i) = \begin{cases} S(i) & \text{if } S(i) < n \\ m & \text{if } S(i) = n \end{cases}$$

(here S is the successor operation for the natural numbers).

(a) Show that each $T_{n,m}$ is an induction structure.

(b) Show that if $(n, m), (n', m')$ are two distinct pairs of natural numbers with $n > m, n' > m'$, then $T_{n,m}$ and $T_{n',m'}$ are not isomorphic. Show also that no $T_{n,m}$ is isomorphic to the natural numbers.

(c) Show that if $\mathcal{S}$ is an induction structure, then either $\mathcal{S}$ is isomorphic to the natural numbers or else $\mathcal{S}$ is isomorphic to $T_{n,m}$ for some $n > m$.

91. (a) Show that for all $n > 0$, $T_{n,0}$ satisfies the second Peano axiom (Axiom 4.11.3).

(b) Show that if $n > m > 0$, then $T_{n,m}$ satisfies the first Peano axiom (Axiom 4.11.2).

(c) Conclude from the previous two parts and Exercises 84 and 90c that if S is an induction structure, then S satisfies at least one of the two other Peano axioms as well.

92. (a) Show that if $n < n'$, then there is no isomorphism from $T_{n,m}$ to $T_{n',m'}$.

    Hint: Use Exercises 80 and 89.

Thus, Theorem 4.11.11 does not hold for all induction structures. Since this theorem follows from Theorem 4.11.8, it follows that this latter theorem does not hold for every induction structure.

(b) Let $\varphi_{n,m}$ be the unique homomorphism from the natural numbers to $T_{n,m}$. Describe $\varphi_{n,m}$ explicitly.

(c) Classify exactly those pairs $(n, m), (n', m')$ such that there is a homomorphism from $T_{n,m}$ to $T_{n',m'}$.

93. (a) Show that if $S = (N, 0, S)$ is an induction structure, then there is a unique operation f_+ on N satisfying the equations of Definition 4.11.13.

    Hint: You cannot use Theorem 4.11.9 since Exercise 92 shows that this theorem is not true for all induction models. Instead, use induction on n to show that for each n there is a unique unary operation f_{+n} on N satisfying

    $$f_{+n}(0) = n,$$
    $$f_{+n}(S(m)) = S(f_{+n}(m)) \text{ for all } m \in N.$$

    Then, construct f_+ from the f_{+n}'s.

In any induction structure, we write $n + m$ for $f_+(n, m)$, and we define the $\leq$ relation as it was for a Peano structure. All of the properties of $+$ and $\leq$ given in Exercises 77 and 78 hold in an arbitrary induction structure since their proofs depended only on the definitions of $+$ and $\leq$ and the induction axiom.

(b) Show that if $S = (N, 0, S)$ is an induction structure, then there is a unique binary operation $f_\times$ on N satisfying the equations in Definition 4.11.13.

    Hint: Try an approach similar to that of the previous part.

All of the properties of multiplication given in Exercise 77 hold in an arbitrary induction structure as well.

94. For each of the structures $T_{n,m}$, describe explicitly what the addition and multiplication operations in the structure are.

95. Let $S = (N, 0, S)$ be a Peano structure and let $S' = (N', 0', S')$ be an induction structure. We write $+, \cdot$ for the operations in S and $+', \cdot'$ for the operations in S'. Let φ be the unique homomorphism form S to S'. Show that for any $n, m \in N$

$$\varphi(n + m) = \varphi(n) +' \varphi(m),$$
$$\varphi(n \cdot m) = \varphi(n) \cdot' \varphi(m).$$

96. (a) Show that if $n > 1$, then there is no binary operation $f_{\exp}$ on $T_{n,0}$ satisfying the equations of Definition 4.11.13.

   Hint: Suppose that such an operation $f_{\exp}$ exists. Using the fact that $0 = S_{n,0}(n-1)$, evaluate $f_{\exp}(0,0)$ in two different ways getting two different answers.

   Thus, an arbitrary induction structure need not have an exponential function.

   (b) For what n, m with $n > m$ does $T_{n,m}$ have an exponential function?

Fixed Points and Fixed Point Induction

97. Let (M, ρ) be a cpo. Prove that if p, q are admissible predicates then so are the predicates $p \wedge q$ and $p \vee q$, where

$$(p \wedge q)(t) = p(t) \wedge q(t),$$
$$(p \vee q)(t) = p(t) \vee q(t),$$

   for every $t \in M$.

   Solution: We discuss only the case of $p \vee q$. Let L be a chain in (M, ρ) such that for all $t \in L$, $(p \vee q)(t) = \mathbf{T}$. Let $L_p = \{x \mid x \in L, p(x) = \mathbf{T}\}$ and $L_q = \{x \mid x \in L, q(x) = \mathbf{T}\}$. It is easy to see that $L = L_p \cup L_q$ and, therefore, $\sup L = \sup\{\sup L_p, \sup L_q\}$, according to Exercise 37 of Chapter 3. Since $p(\sup L_p) = \mathbf{T}$, we have $p(\sup L) = \mathbf{T}$; similarly, $q(\sup L) = \mathbf{T}$, and this gives $(p \vee q)(\sup L) = \mathbf{T}$.

98. Consider the mapping $F : (\mathbf{N}_\perp \longrightarrow_c \mathbf{N}_\perp) \longrightarrow_c (\mathbf{N}_\perp \longrightarrow_c \mathbf{N}_\perp)$, defined by

$$F(f)(n) = \text{ if } x > 10 \text{ then } x - 10 \text{ else } f(f(x+11)).$$

   Let $f_0 : (\mathbf{N}_\perp \longrightarrow_c \mathbf{N}_\perp)$, defined by $f_0(n) = \perp$, for all $n \in N$. Define $f_i = F^i(f_0)$.

   (a) The functions f_i are given by

$$f_i(n) = \text{ if } n > 100 \text{ then } n - 10 \text{ else }$$
$$(\text{if } n > 101 - i \text{ then } 91 \text{ else } \perp)$$

for $1 \le i \le 11$ and

$$f_i(n) = \text{if } n > 100 \text{ then } n - 10 \text{ else}$$
$$(\text{if } n > 211 - 11i \text{ then } 91 \text{ else } \bot)$$

for $n \ge 11$.

(b) Prove that the least fixed point of F is the function $f : \mathbf{N}_\bot \longrightarrow_c \mathbf{N}_\bot$ defined by

$$f(n) = \begin{cases} n - 10 & \text{if } n > 100 \\ 91 & \text{otherwise,} \end{cases}$$

known as *McCarthy's 91 function*.

99. Let the mapping $G : (\mathbf{N}_\bot \longrightarrow_c \mathbf{N}_\bot) \longrightarrow_c (\mathbf{N}_\bot \longrightarrow_c \mathbf{N}_\bot)$ be defined by

$$G(f)(n) = \text{if } n = 0 \text{ then } 1 \text{ else } f(n + 1).$$

Prove that the following functions are all fixed points for G:

(a)

$$f_1(n) = \begin{cases} 1 & \text{if } n = 0 \\ \bot & \text{otherwise;} \end{cases}$$

(b)

$$f_2(n) = \begin{cases} 1 & \text{if } x = 0 \\ 2 & \text{otherwise;} \end{cases}$$

(c) $f_3(n) = 1$ for $n \in \mathbf{N}_\bot$.

100. Define the function DOM $: (\mathbf{N}_\bot \longrightarrow_c \mathbf{N}_\bot) \longrightarrow \mathcal{P}(\mathbf{N}_\bot)$ by DOM$(f) = \{n \mid n \in \mathbf{N}, f(n) \ne \bot\}$. Prove that the function DOM is continuous.

101. Prove that the function $F : (\mathbf{N}_\bot \longrightarrow \mathbf{N}_\bot) \longrightarrow (\mathbf{N}_\bot \longrightarrow \mathbf{N}_\bot)$, given by $F(f)(n) = \text{if } f(n) = a \text{ then } b \text{ else } a$, has no fixed point, where $a \ne b$, $a, b \in \mathbf{N}$.

102. Prove that continuity is sufficient, but not necessary, for the existence of a fixed point of a function.

Solution: Consider the function $F : (\mathbf{N}_\bot \longrightarrow \mathbf{N}_\bot) \longrightarrow (\mathbf{N}_\bot \longrightarrow \mathbf{N}_\bot)$, where

$$F(f)(n) = \text{if } f(n) = a \text{ then } a \text{ else } b,$$

where $a, b \in \mathbf{N}$ and $a \ne b$. Observe that F is not even monotonic, yet both strict extensions of the constant functions f_a and f_b (where $f_a(n) = a$ and $f_b(n) = b$ for all $n \in \mathbf{N}$) are fixed points.

103. Let $f : \mathbf{N}^2 \longrightarrow \mathbf{N}$ be Ackermann's function, introduced in Example 4.13.16.

For every $x, y \in \mathbf{N}$, prove the following.

(a) $f(x, y) > x$.
   Hint: Use induction on x.

(b) $f(x, y + 1) > f(x, y)$.
   Hint: Use induction on x and property 1.

(c) If $y > z$, then $f(x, y) > f(x, z)$.
   Hint: Define $u = y - z - 1$ and apply induction on u and property 2.

(d) $f(x + 1, y) > f(x, y + 1)$.
   Hint: Use induction on y.

(e) $f(x, y) > x$.

(f) If $x > u$, then $f(x, y) > f(u, y)$.

(g) $f(x + 2, y) > f(x, 2y)$.
   Hint: Apply induction on x.

4.15 Bibliographical Comments

Induction in its simpler forms is covered in almost every text on discrete mathematics. Our approach to structural induction and recursively defined functions is based on that in [End77], [HJ84], and [LS87]. A very good source for applications of induction in computer science is [Wan80]. Our treatment of fixpoint induction is based on [Man74] and [LS87]. The source of many of the exercises on Peano structures is [Hen60]. The reader interested in transition systems may consult [Har87]. Definitive references in formal languages are [Gin66] and [Sal73]quit .

5

Enumerability and Diagonalization

5.1   Introduction
5.2   Equinumerous Sets
5.3   Countable and Uncountable Sets
5.4   Enumerating Programs
5.5   Abstract Families of Functions
5.6   Exercises and Supplements
5.7   Bibliographical Comments

5.1   Introduction

One of the purposes of this chapter is to present a formal approach to the intuitive notion of the size of a set. The facts discussed here play a fundamental role in computer science; they serve to make precise the idea that there are more functions on the set of natural numbers than programs (in any programming language), and therefore, there exist functions that cannot be computed by any program in that programming language.

A focal point of this chapter is a proof method originating in set theory: the diagonalization method. Developments in theoretical computer science have made diagonalization an essential instrument for proving negative and limitative results in this field. We discuss diagonalization both in its original set-theoretical context and in the context of computability theory.

5.2   Equinumerous Sets

It follows from Corollary 4.2.7 that there is a bijection $f : \{a_0, \ldots, a_{m-1}\} \longrightarrow \{b_0, \ldots, b_{p-1}\}$ if and only if $m = p$, that is, if and only if the two sets involved have the same number of elements. In this section, our main objective is to extend this idea to arbitrary sets.

Definition 5.2.1 *Two sets M and P are called* equinumerous, *denoted by $M \sim P$, if there exists a bijection $f : M \longrightarrow P$.*

It is easy to verify the following properties for any sets M, P, and Q:

1. $M \sim M$;

2. if $M \sim P$, then $P \sim M$; and

3. if $M \sim P$ and $P \sim Q$, then $M \sim Q$.

The first property follows immediately from the fact that the mapping $1_M : M \longrightarrow M$ is a bijection. Further, if $f : M \longrightarrow P$ is a bijection, then $f^{-1} : P \longrightarrow M$ is also a bijection; therefore, if $M \sim P$ then, $P \sim M$. Finally, if $f : M \longrightarrow P$ and $g : P \longrightarrow Q$ are two bijections, then $gf : M \longrightarrow Q$ is also a bijection.

Example 5.2.2 Consider the set $\mathbf{E}$ of all even natural numbers. We have $\mathbf{E} \subset \mathbf{N}$. Nevertheless, we also have $\mathbf{N} \sim \mathbf{E}$ because of the fact that the mapping $f : \mathbf{N} \longrightarrow \mathbf{E}$ defined by $f(n) = 2n$ for every $n \in \mathbf{N}$ is a bijection.

Example 5.2.3 The set $\mathbf{Z}$ of integers is equinumerous with $\mathbf{N}$. Consider the mapping $g : \mathbf{N} \longrightarrow \mathbf{Z}$ defined by

$$g(n) = \begin{cases} n/2 & \text{if } n \text{ is even} \\ -(n+1)/2 & \text{if } n \text{ is odd.} \end{cases}$$

It is not difficult to see that g is a bijection, which proves that $\mathbf{Z} \sim \mathbf{N}$.

Example 5.2.4 Any two open intervals of the set of real numbers are equinumerous. Indeed, consider two intervals (a, b) and (c, d), where $a < b$ and $c < d$. Define $g : (a, b) \longrightarrow (c, d)$ by

$$g(x) = \frac{d-c}{b-a}x + \frac{bc-ad}{b-a}.$$

It is easy to see that g is a bijection between (a, b) and (c, d). Of course, this is true even if $(a, b) \subseteq (c, d)$ since we did not use the relative positions of the two intervals in our argument.

Furthermore, we can prove that $\mathbf{R} \sim (a, b)$ for any open interval (a, b). It suffices to consider the mapping $\tan : (-\frac{\pi}{2}, \frac{\pi}{2}) \longrightarrow \mathbf{R}$, which is a bijection, to obtain $(-\frac{\pi}{2}, \frac{\pi}{2}) \sim \mathbf{R}$. On the other hand, any open interval (a, b) is equinumerous with $(-\frac{\pi}{2}, \frac{\pi}{2})$, which gives $\mathbf{R} \sim (a, b)$.

The following theorem gives some evidence that two equinumerous sets do in fact have the same size.

Theorem 5.2.5 *Let M and P be sets such that $M \sim P$. Then, either M and P are both finite or they are both infinite.*

Proof: If one of M and P is finite and $M \sim P$, then it follows from Theorem 2.4.10 that the other of these two sets is also finite. ∎

In our next definition, we try to make precise the ideas of one set being at least as large as or strictly larger than another.

Definition 5.2.6 *Let M and P be two sets. M is* dominated *by P, written $M \preceq P$, if there exists an injection $f : M \longrightarrow P$.*

M is strictly dominated *by P, written $M \prec P$, if $M \preceq P$ and M is not equinumerous with P.*

Example 5.2.7 For any finite set M, we have $M \preceq \mathbf{N}$ since if M has n elements, $M = \{x_0, \ldots, x_{n-1}\}$, the mapping $f : M \longrightarrow \mathbf{N}$ defined by $f(x_i) = i$ for $0 \le i \le n - 1$ is injective.

Theorem 5.2.8 *For every $M, P, Q, R, M_1, \ldots, M_k, P_1, \ldots, P_k$, we have*

1. *$M \preceq M$;*

2. *if $M \preceq P$ and $P \preceq Q$, then $M \preceq Q$;*

3. *if $M \subseteq P$, then $M \preceq P$;*

4. *if $M \sim P$, then $M \preceq P$;*

5. *if $M_1 \preceq P_1, \ldots, M_k \preceq P_k$, then $M_1 \times \cdots \times M_k \preceq P_1 \times \cdots \times P_k$;*

6. *if $M \preceq P$, $Q \sim M$, and $R \sim P$, then $Q \preceq R$.*

Proof. The argument consists of direct applications of the definition of $\preceq$, and it is left to the reader. ∎

The connection between finiteness and domination is given in the next theorem.

Theorem 5.2.9 *Let M and P be sets such that $M \preceq P$. Then, if P is finite, so is M, and if M is infinite, so is P.*

Proof: Since $M \preceq P$, we have $M \sim K$ for some subset K of P. If P is finite, then by Theorem 4.2.11 so is K, but then by Theorem 5.2.5, M is finite.

The second part of the theorem follows immediately from the first. ∎

The next theorem shows the connection between cardinality and domination for finite sets.

Theorem 5.2.10 *Let M and P be finite sets. Then, $M \preceq P$ if and only if $|M| \le |P|$ and $M \prec P$ if and only if $|M| < |P|$.*

Proof: The theorem follows immediately from Corollary 4.2.9. ∎

Theorem 5.2.8 implies that if $M \sim P$ then $M \preceq P$ and $P \preceq M$. By the previous theorem, the converse of this statement is true if M and P are finite sets. We now show that the converse is true in general. This supports our intuition that $M \preceq P$ means that P is at least as big as M and $M \sim P$ means that M and P have the same size.

Theorem 5.2.11 (Schröder-Bernstein Theorem) *For all sets M and P, if $M \preceq P$ and $P \preceq M$, then $M \sim P$.*

Proof. According to the hypothesis of the theorem, we have two injective mappings $f : M \longrightarrow P$ and $g : P \longrightarrow M$.

Define $\Phi : \mathcal{P}(M) \longrightarrow \mathcal{P}(M)$ by $\Phi(K) = M - g(P - f(K))$. This mapping is monotonic since $K_1 \subseteq K_2$ implies $f(K_1) \subseteq f(K_2)$ (because of Theorem 2.5.6); hence, $P - f(K_2) \subseteq P - f(K_1)$, which gives $g(P - f(K_2)) \subseteq g(P - f(K_1))$. This, in turn, implies $M - g(P - f(K_1)) \subseteq M - g(P - f(K_2))$; hence, $\Phi(K_1) \subseteq \Phi(K_2)$.

Let K_0 be a subset of M such that $\Phi(K_0) = K_0$. Such a subset exists because of Theorem 3.9.10. For this set, we have

$$K_0 = \Phi(K_0) = M - g(P - f(K_0)). \tag{5.1}$$

Define now a mapping $h : M \longrightarrow P$ by

$$h(a) = \begin{cases} f(a) & \text{if } a \in K_0 \\ g^{-1}(a) & \text{if } a \in M - K_0. \end{cases}$$

Since g is an injection, g^{-1} is a function and if $a \in M - K_0$, then $a \in g(P - f(K_0))$, because of (5.1), so $g^{-1}(a)$ is defined. Thus, h is well-defined.

The mapping h is onto. In order to justify this claim, let $b \in P$. If $b \in f(K_0)$, then there is $a \in K_0$ such that $b = f(a)$. In this case, we have $h(a) = f(a) = b$. If $b \notin f(K_0)$, then $b \in P - f(K_0)$ implies $a = g(b) \notin K_0$ by (5.1), so $b = h(a)$ and this proves that every element of P is the image (by h) of an element of M.

The mapping h is injective. Suppose that $h(a) = h(a') = b$. If both $a, a' \in K_0$, then we have $f(a) = b = f(a')$ so $a = a'$, because of the fact that f is injective. If neither a nor a' are in K_0, then we have $a = g(b) = a'$. Finally, suppose that $a \in K_0$ and $a' \notin K_0$. Then, $b = h(a) = f(a)$, and $a' = g(h(a')) = g(b)$, so $a' = g(f(a)) \in g(f(K_0))$. Together with $a' \in M - K_0 = g(P - f(K_0))$, this contradicts the injectivity of g.

Thus, h is a bijection and $M \sim P$. ∎

Theorem 5.2.12 *If M, P, and Q are sets such that either $M \prec P \preceq Q$ or $M \preceq P \prec Q$, then $M \prec Q$.*

Proof: Suppose that $M \prec P \preceq Q$. Then, $M \preceq P \preceq Q$, so by Theorem 5.2.8, $M \preceq Q$. Since $P \preceq Q$, there must be an injection $g : P \longrightarrow Q$. If $M \sim Q$, say, $f : Q \longrightarrow M$ is a bijection, then fg is an injection from P to M, so $P \preceq M$. In combination with $M \preceq P$, we get from Theorem 5.2.11 that $M \sim P$, contradicting $M \prec P$. Thus, we do not have $M \sim Q$, and hence, $M \prec Q$.

If $M \preceq P \prec Q$, then a similar proof shows that $M \prec Q$. ∎

It is interesting to note that, while there is always an injective mapping $\Psi : M \longrightarrow \mathcal{P}(M)$ given by $\Psi(a) = \{a\}$, there is no such thing as a bijection between a set M and its power set $\mathcal{P}(M)$.

Theorem 5.2.13 *For every set M, $M \prec \mathcal{P}(M)$.*

Proof. Suppose that $f : M \longrightarrow \mathcal{P}(M)$ is a surjection. Define $K = \{a \in M \mid a \notin f(a)\}$. K is a subset of M, and therefore, there exists $a_0 \in M$ such that $f(a_0) = K$, since f is onto. We have either $a_0 \in K$ or $a_0 \notin K$.

In the first case, $a_0 \in K$ implies (by the definition of K) that $a_0 \notin f(a_0) = K$, which is a contradiction.

In the second case, $a_0 \notin K = f(a_0)$ amounts to $a_0 \in K$, which is, again, a contradiction, so f cannot exist.

Since there is no surjection from M to $\mathcal{P}(M)$, we cannot have $M \sim \mathcal{P}(M)$. By the remark just prior to the theorem, we do have $M \preceq \mathcal{P}(M)$. Thus, $M \prec \mathcal{P}(M)$. ∎

The set $\mathbf{N}$ is, in a certain sense, the least infinite set. We give a more precise formulation of this statement in the next theorem.

Theorem 5.2.14 *If M is an infinite set, then $\mathbf{N} \preceq M$.*

Proof. From the axiom of choice we infer the existence of a selective function f for M.

Define $h : \mathbf{N} \longrightarrow \mathcal{P}(M)$ recursively by

$$h(0) = \emptyset,$$
$$h(n+1) = h(n) \cup \{f(M - h(n))\}.$$

It is easy to show, by induction on n, that for any $n \in \mathbf{N}$, $h(n)$ is finite, and therefore, $M - h(n)$ is nonempty for every $n \in \mathbf{N}$. We also note that if $m < n$ then $h(m) \subset h(n)$.

Define $g : \mathbf{N} \longrightarrow M$ by $g(n) = f(M - h(n))$. Clearly, $h(n+1) = h(n) \cup \{g(n)\}$, $g(n) \notin h(n)$, and $g(n) \in h(p)$ for every $p > n$.

In order to prove the theorem, we need to show that g is injective. Consider two numbers $m, n \in \mathbf{N}$, with $m < n$. Then, $g(m) \in h(n)$, while $g(n) \notin h(n)$. This implies that $g(m) \neq g(n)$. ∎

Definition 5.2.15 *A set M is* countably infinite *if $M \sim \mathbf{N}$.*

Examples 5.2.2 and 5.2.3 show that $\mathbf{E}$ and $\mathbf{Z}$ are both countably infinite. We have seen that $\mathbf{N}$ is an infinite set. Thus, by Theorem 5.2.5 a set that is countably infinite is, in fact, infinite.

If M is countably infinite and $M \sim P$, then it is immediate that P is also countably infinite.

The following theorem could be obtained as a corollary of Theorem 5.2.14, but we give a slightly longer proof in order to avoid use of the axiom of choice.

Theorem 5.2.16 *If M is an infinite subset of $\mathbf{N}$, then M is countably infinite.*

Proof. Since $M \subseteq \mathbf{N}$, we can define a selective function f for M without using the axiom of choice by letting $f(K)$ be the least element of K for every nonempty subset K of M. Continuing now, as in the proof of Theorem 5.2.14, we obtain $\mathbf{N} \preceq M$.

Since $M \subseteq \mathbf{N}$, we also have $M \preceq \mathbf{N}$, and Theorem 5.2.11 implies that $M \sim \mathbf{N}$. $\blacksquare$

Theorem 5.2.17 *If M is a set such that $M \prec \mathbf{N}$, then M is finite.*

Proof. If $M \prec \mathbf{N}$, then $M \sim K$ for some subset K of $\mathbf{N}$. If K is infinite, then by the previous theorem, $K \sim \mathbf{N}$, so $M \sim \mathbf{N}$, contradicting $M \prec \mathbf{N}$. Thus, K is finite; hence, M is finite. $\blacksquare$

In Examples 5.2.2 and 5.2.3, we have shown that an infinite set can be equinumerous with one of its proper subsets. This turns to be a characteristic property of infinite sets.

Theorem 5.2.18 *A set is infinite if and only if it is equinumerous with one of its proper subsets.*

Proof. Let M be an infinite set. By Theorem 5.2.14, $\mathbf{N} \preceq M$; hence, we have an injection $f : \mathbf{N} \longrightarrow M$. Consider the function $g : M \longrightarrow M$ defined by

$$g(x) = \begin{cases} x & \text{if } x \notin f(\mathbf{N}) \\ f(n+1) & \text{if } x = f(n) \text{ for } n \in \mathbf{N}. \end{cases}$$

Note that g is well-defined because of the fact that f is injective. The range of g is $M - \{f(0)\}$, and g is clearly injective, which makes it a bijection between M and $M - \{f(0)\}$.

Conversely, if M is finite, then by Corollary 4.2.10, any injection from M to itself is a surjection, so there cannot be a bijection between M and one of its proper subsets. $\blacksquare$

5.3   Countable and Uncountable Sets

The property of sets we are about to study plays an essential role in computability.

Definition 5.3.1 *A set M is countable if it is finite or countably infinite. A set that is not countable is said to be uncountable.*

It follows immediately from this definition that the countably infinite sets are, indeed, the countable sets which are infinite.

We have already remarked that $\mathbf{Z}$ and $\mathbf{E}$, the set of even integers, are countably infinite and hence, these are examples of countable sets, as of course is $\mathbf{N}$ itself. On the other hand, by Theorem 5.2.13, $\mathbf{N} \prec \mathcal{P}(N)$

which implies that $\mathcal{P}(\mathbf{N})$ is neither finite nor countably infinite and hence, is uncountable. Further examples will be discussed later in this section, but first, we give some characterizations of countable sets.

Theorem 5.3.2 *For every set M, M is countable if and only if $M \preceq \mathbf{N}$.*

Proof. Let M be a countable set. If M is finite, then by Example 5.2.7 $M \preceq \mathbf{N}$. If M is countably infinite, we have $M \sim \mathbf{N}$, so $M \preceq \mathbf{N}$, as we observed before.

Conversely, assume that $M \preceq \mathbf{N}$. Then, either $M \prec \mathbf{N}$, or $M \sim \mathbf{N}$. In the former case, M is finite by Theorem 5.2.17, while in the latter case, M is countably infinite by definition, so in either case M is countable. ∎

Theorem 5.3.3 *If M is a nonempty set, we have $M \preceq \mathbf{N}$ if and only if there exists a surjection $f : \mathbf{N} \longrightarrow M$.*

Proof. Let M be a nonempty set such that $M \preceq \mathbf{N}$. There exists a injective mapping $g : M \longrightarrow \mathbf{N}$. Since $M \neq \emptyset$, Theorem 2.3.11 implies the existence of a left inverse $f : \mathbf{N} \longrightarrow M$ for g, which is necessarily a surjection (since g is a right inverse for f).

Conversely, let $f : \mathbf{N} \longrightarrow M$ be a surjection. Theorem 2.3.11 implies the existence of a right inverse $g : M \longrightarrow \mathbf{N}$ for f, but this part of that theorem uses the axiom of choice. We can define a right inverse g for f explicitly and avoid the use of the axiom of choice by defining, for every $a \in M$, $g(a)$ to be the least element of $f^{-1}(\{a\})$. Then, g has f as left inverse, so is an injection, and $M \preceq \mathbf{N}$. ∎

The previous theorem says that the countable sets are exactly those sets which can be enumerated as the range of a (not necessarily injective) infinite sequence, together with $\emptyset$. When we use a phrase such as "Let $M = \{s_0, s_1, \ldots\}$ be a nonempty countable set," we are implicitly using this fact to write an arbitrary nonempty countable set as the range of an infinite sequence. Of course, since M could be finite, we cannot assume that the sequence is injective. However, when we use a phrase such as "Let $M = \{s_0, s_1, \ldots\}$ be a countably infinite set," we will assume, unless the contrary is stated, that s is an injective sequence (just as when we write "Let $M = \{s_0, \ldots, s_{n-1}\}$ be a finite set," we implicitly assume that M has n elements and hence, that the sequence s lists the elements of M without repetition).

Corollary 5.3.4 *Let M be a nonempty set. The following three statements are equivalent:*

1. *M is countable;*

2. *$M \preceq \mathbf{N}$;*

3. *there is a surjection $f : \mathbf{N} \longrightarrow M$.*

Proof. The corollary follows immediately from Theorems 5.3.2 and 5.3.3.
∎

Corollary 5.3.5 *Let M be a countable set, and let P be any set. If any of the following conditions is met, then P is countable.*

1. *There is an injection $f : P \longrightarrow M$ (i.e., $P \preceq M$).*

2. *There is a surjection $g : M \longrightarrow P$.*

3. *$M \sim P$.*

Proof. Suppose that the first condition holds. Since M is countable, we have by Theorem 5.3.2 that $M \preceq \mathbf{N}$ and hence, by Part 2 of Theorem 5.2.8, $P \preceq \mathbf{N}$, so P is countable.

Now, suppose that $g : M \longrightarrow P$ is a surjection. If P is empty, then P is certainly countable. If $P \neq \emptyset$, then the existence of the surjection shows that M is not empty and hence, by Theorem 5.3.3, there is a surjection $h : \mathbf{N} \longrightarrow M$. The composition $gh : \mathbf{N} \longrightarrow P$ is also a surjection and hence, P is countable.

If the third condition holds, then so do both of the first two, so P is countable. ∎

Definition 5.3.6 *Let M be a set. A numbering for M is an injection $g : M \longrightarrow \mathbf{N}$. An enumeration of M is a surjection $f : \mathbf{N} \longrightarrow M$.*

Note that by Theorem 5.3.2, a set M is countable if and only if M has a numbering, and by definition M is countably infinite if and only if M has a bijective numbering. Also, by Theorem 5.3.3, a nonempty set is countable if and only if it has an enumeration.

The proof of Theorem 5.3.3 gives an explicit method for converting an enumeration $f : \mathbf{N} \longrightarrow M$ into a numbering $g : M \longrightarrow \mathbf{N}$ of M; namely, for each $a \in M$, $g(a)$ is the least natural number n such that $f(n) = a$. The resulting numbering will not be a bijection unless f was a bijection, and in this case g is f^{-1}.

Sometimes one wants to carry the process out in reverse, i.e., one has a numbering $g : M \longrightarrow \mathbf{N}$ of a nonempty set M and wants to produce an enumeration $f : \mathbf{N} \longrightarrow M$ of M. If g is bijective, then we can take $f = g^{-1}$. Otherwise, the method from the proof of the second part of Theorem 2.3.11 can be used. An element m_0 of M is chosen. (Usually m_0 is in some sense one of the simplest elements of M.) Then, f is defined by letting $f(n) = g^{-1}(n)$ for $n \in g(M)$ and $f(n) = m_0$ for those $n \in \mathbf{N}$ which are not in the range of g.

If M is a countably infinite set then, in many situations, it is advantagous to have a bijective numbering of M instead of just an ordinary one. If $g : M \longrightarrow \mathbf{N}$ is a numbering of an infinite set M, then $g(M)$ is an infinite subset of $\mathbf{N}$ and the proof of Theorem 5.2.16 provides a method to define an injection $f : \mathbf{N} \longrightarrow g(M)$; namely, $f(0)$ is the smallest element of $g(M)$, $f(1)$ the second smallest element of $g(M)$, and so on. (In other words, f enumerates the elements of $g(M)$ in increasing order.) It is easy to see that f is in fact a bijection, and hence, we obtain a new bijective numbering g' for M by defining $g' = f^{-1}g$. Explicitly, $g'(a) = n$ if $g(a)$ is the $n + 1$st largest element of $g(M)$. This provides an automatic process for converting a numbering of an infinite set into a bijective one, but even if the original numbering is a natural one, the resulting bijective numbering will in general be very unnatural.

We turn now to the question of what operations can be performed on countable sets so that the resulting set is countable.

Our first observation is quite easy to prove, yet yields many results.

Theorem 5.3.7 *Seq*($\mathbf{N}$) *is countable.*

Proof. Let $(p_0, p_1, \ldots)$ be the sequence of prime natural numbers listed in increasing order. Define a function $g : \text{Seq}(\mathbf{N}) \longrightarrow \mathbf{N}$ by

$$g\left((a_0, \ldots, a_{n-1})\right) = p_0^{a_0+1} \cdots p_{n-1}^{a_{n-1}+1}$$

for each sequence $(a_0, \ldots, a_{n-1})$ in Seq($\mathbf{N}$). (As usual, the empty product is 1.) If a and b are two different elements of Seq($\mathbf{N}$), then either there is an i with $a_i \neq b_i$ or else the two sequences have different lengths. In either case, $g(a)$ and $g(b)$ will be numbers with different prime factorizations and hence, by the uniqueness of the prime factorization for each positive natural number, $g(a) \neq g(b)$. Thus, g is an injection, and hence, by Theorem 5.3.2, Seq($\mathbf{N}$) is countable. ∎

Note that the numbering of Seq($\mathbf{N}$) defined in the previous theorem is not a bijection.

Corollary 5.3.8 *Let M be a set.*

1. *If M is countable, then Seq(M) is countable.*

2. *If M is countable, then $\mathcal{P}_{fin}(M)$, the collection of all finite subsets of M, is countable.*

3. *If $n \in \mathbf{N}$ and $(M_0, \ldots, M_{n-1})$ is a sequence of countable sets, then the Cartesian product $M_0 \times \cdots \times M_{n-1}$ is countable.*

4. *If M is countable and $n \in \mathbf{N}$, then M^n is countable.*

Proof. Let M be a countable set. Then, by Theorem 5.3.2 there is an injection $g : M \longrightarrow \mathbf{N}$.

To see (1), observe that in an obvious way, g induces an injection $f : \text{Seq}(M) \longrightarrow \text{Seq}(\mathbf{N})$ (namely, $f\left((a_0, \ldots, a_{n-1})\right) = (g(a_0), \ldots, g(a_{n-1}))$ for each sequence $(a_0, \ldots, a_{n-1})$ in Seq(M)). Thus, by Theorem 5.3.7 and Corollary 5.3.5, Seq(M) is countable.

To show (2), we define an injection $h : \mathcal{P}_{\text{fin}}(M) \longrightarrow \text{Seq}(M)$ as follows. For each finite subset K of M, let $h(K)$ be the sequence of length $|K|$ obtained by listing the elements of K in increasing order according to their numbering under g. Since we have just seen that $\text{Seq}(M)$ is countable, it follows that $\mathcal{P}_{\text{fin}}(M)$ is countable.

For part (3), let $n \in \mathbf{N}$ and let $(M_0, \ldots, M_{n-1})$ be a sequence of countable sets. Then, for each i, $0 \leq i \leq n - 1$, $M_i \preceq \mathbf{N}$, so by Theorem 5.2.8, part (5), $M_0 \times \ldots \times M_{n-1} \preceq \mathbf{N}^n$. Since $\mathbf{N}^n \subseteq \text{Seq}(\mathbf{N})$ and $\text{Seq}(\mathbf{N})$ is countable, $\mathbf{N}^n$ is countable and hence, $M_0 \times \ldots \times M_{n-1}$ is countable.

Part (4) follows immediately from part (3). ∎

In the previous corollary, we have shown that certain sets built up from countable ones are countable, but have not specified when the set constructed is finite and when it is countably infinite. A finer analysis, which makes this distinction, can be obtained using the results in Exercise 5.

Since the numbering of $\text{Seq}(\mathbf{N})$ given in Theorem 5.3.7 is not bijective, none of the numberings implicit in the proof of the previous Corollary are bijective either.

Example 5.3.9 We can prove now that the set $\mathbf{Q}$ of rational numbers is countable. If $r \in \mathbf{Q}$, we can determine uniquely p and q such that $r = \frac{p}{q}$, where p, q are relatively prime integers and $q > 0$. The mapping $f : \mathbf{Q} \longrightarrow \mathbf{Z} \times \mathbf{Z}$ given by $f(r) = (p, q)$ is clearly an injection; hence, $\mathbf{Q} \preceq \mathbf{Z} \times \mathbf{Z}$.

Since $\mathbf{Z}$ is countable, so is $\mathbf{Z} \times \mathbf{Z}$, because of Corollary 5.3.8. Consequently, $\mathbf{Q}$ is countable.

Our next observations are quite trivial. If $\mathcal{C}$ is any nonempty collection of sets, then $\bigcap \mathcal{C}$ is a subset of every element of $\mathcal{C}$. Thus, if even one of the elements of $\mathcal{C}$ is countable, $\bigcap \mathcal{C}$ will be countable.

Similarly, $M - P \subseteq M$, so if M is countable and P is any set, $M - P$ will be countable.

A more interesting question is when unions of countable sets are countable. We first show that the union of finitely many countable sets is countable.

Theorem 5.3.10 Let $\mathcal{C} = \{M_0, \ldots, M_{n-1}\}$ be a finite collection of countable sets. Then, the union $\bigcup \mathcal{C} = \bigcup \{M_k \mid 0 \leq k \leq n - 1\}$ is countable.

Proof. For $n = 0$ or $n = 1$, the result is immediate. We proceed by induction on $n \geq 2$.

For the basis step, $n = 2$, consider countable sets M_0 and M_1. If either M_0 or M_1 is empty, then $M_0 \cup M_1$ is clearly countable. Therefore, we assume that both M_0 and M_1 are nonempty.

There are two surjections $f_0 : \mathbf{N} \longrightarrow M_0$ and $f_1 : \mathbf{N} \longrightarrow M_1$. This allows us to consider the mapping $f : \mathbf{N} \longrightarrow M_0 \cup M_1$, defined by

$$f(n) = \begin{cases} f_0(\frac{n-1}{2}) & \text{if } n \text{ is odd} \\ f_1(\frac{n}{2}) & \text{if } n \text{ is even.} \end{cases}$$

Note that $M_0 = f(\mathbf{N} - \mathbf{E})$, while $M_1 = f(\mathbf{E})$. Therefore, $M_0 \cup M_1 = f(\mathbf{N})$; hence, f is onto, which means that $M_0 \cup M_1$ is countable.

For the inductive step, assume that $n \geq 2$ and the union of any collection of n countable sets is countable. Since

$$\bigcup\{M_k \mid 0 \leq k \leq n\} = \bigcup\{M_k \mid 0 \leq k \leq n - 1\} \cup M_n,$$

in view of the inductive hypothesis and of the basis step, we may conclude that $\bigcup\{M_k \mid 0 \leq k \leq n\}$ is countable. ∎

The next theorem shows that the union of a countable collection of countable sets is countable, which immediately implies the previous theorem. We proved the previous theorem separately because that proof did not involve the axiom of choice, while the proof of the following theorem necessarily does.

Theorem 5.3.11 *Let C be a countable collection of countable sets. The set $M = \bigcup C$, the union of all the sets in C, is countable.*

Proof. If $C = \emptyset$ or $C = \{\emptyset\}$, the result is immediate. In any other case, we may assume that C is a nonempty countable collection of nonempty countable sets since, otherwise, we can replace C by $C - \{\emptyset\}$ without modifying the set M.

Let $C = \{M_0, M_1, \ldots, M_n, \ldots\}$, and for each $n \in \mathbf{N}$, let S_n be the set of all surjective mappings from $\mathbf{N}$ to M_n. By Theorem 5.3.3, we have $S_n \neq \emptyset$ for every $n \in \mathbf{N}$ and, by the axiom of choice, there is a choice function f for the collection $\{S_n \mid n \in \mathbf{N}\}$. For each $n \geq 0$, let $h_n = f(S_n)$.

Define a mapping $h : \mathbf{N} \times \mathbf{N} \longrightarrow M$ by $h(n, m) = h_n(m)$. Clearly, h is surjective. By Corollary 5.3.5, M is countable. ∎

We will now show how to give natural bijective numberings for some of the infinite sets which we have already shown to be countable, using numberings which are not bijections.

First we look at $\mathrm{Seq}(M)$ where M is finite. If $M = \emptyset$, then $\mathrm{Seq}(M) = \{\lambda\}$ is finite. If M is nonempty, then it is an alphabet, and the following example gives a natural bijective numbering.

Example 5.3.12 Let $V = \{a_1, \ldots, a_k\}$ be an alphabet. We define a bijection $h : V^* \longrightarrow \mathbf{N}$ from $V^* = \mathrm{Seq}(V)$, the set of all words over V, to $\mathbf{N}$. If $w = a_{i_0} a_{i_1} \cdots a_{i_{m-1}}$, where $1 \leq i_0, \cdots, i_{m-1} \leq k$, is a word of length m then

$$h(w) = i_0 k^{m-1} + i_1 k^{m-2} + \cdots + i_{m-1}.$$

For instance, if $V = \{a_1, a_2, a_3\}$, then

$$h(a_2 a_1 a_2 a_3) = 2 \cdot 3^3 + 1 \cdot 3^2 + 2 \cdot 3^1 + 3 = 72.$$

We can think of $h(w)$ as being the number which $i_0 \ldots i_{m-1}$ represents in base k. Note however that the "digits" $i_0, \ldots, i_{m-1}$ vary between 1 and k, not 0 and $k - 1$ as they would in the usual base k representation.

We leave to the reader the task of proving that h is a bijection. (See Exercise 28.)

Now, we turn to bijective numberings for $\text{Seq}(M)$, $\mathcal{P}_{\text{fin}}(M)$, and M^n when M is countably infinite and $n > 0$. There is no loss in generality in taking $M = \mathbf{N}$.

First we consider $\text{Seq}(\mathbf{N})$. We get a bijective numbering for this set by making only a small change to the numbering defined in Theorem 5.3.7.

Example 5.3.13 We define a bijection $g : \text{Seq}(\mathbf{N}) \longrightarrow \mathbf{N}$. As before, let $(p_0, p_1, \dots)$ be the sequence of prime positive integers listed in increasing order. Let $g(\lambda) = 0$, and define

$$g((a_0, \dots, a_{n-1})) = p_0^{a_0} p_1^{a_1} \cdots p_{n-2}^{a_{n-2}} p_{n-1}^{a_{n-1}+1} - 1$$

for each nonnull sequence $(a_0, \dots, a_{n-1})$ in $\text{Seq}(\mathbf{N})$. It is clear that if a is a nonnull sequence, then $g(a) \neq 0$, so λ is the only sequence mapped to 0 by g. If $a = (a_0, \dots, a_{n-1})$ and $b = (b_0, \dots, b_{m-1})$ are two nonnull sequences both mapped to the same number x by g, then p_{n-1} is the largest prime which divides $x+1$, as is p_{m-1}, (because of the plus one in the final exponent in the definition of g) and so $n = m$. Furthermore, a_{n-1} and b_{n-1} both are one less than the highest power of p_{n-1} which divides $x + 1$, so are equal, while for $0 \leq i < n - 1$, a_i and b_i are both equal to the largest power of p_i which divides $x + 1$. Thus, $a = b$ and g is an injection.

Clearly 0 is in the range of g. If $x > 0$ then $x + 1 \geq 2$, so we may write $x + 1 = p_0^{b_0} \cdots p_{n-1}^{b_{n-1}}$ for some $n > 0$ and $b_0, \dots, b_{n-1}$ with $b_{n-1} > 0$. Then, $g((b_0, \dots, b_{n-2}, b_{n-1} - 1)) = x$, so g is onto.

Next we consider $\mathcal{P}_{\text{fin}}(\mathbf{N})$.

Example 5.3.14 We define a bijection $g : \mathcal{P}_{\text{fin}}(\mathbf{N}) \longrightarrow \mathbf{N}$ by defining

$$g(\{a_0, \dots, a_{n-1}\}) = 2^{a_0} + 2^{a_1} + \cdots + 2^{a_{n-1}}$$

for each finite subset $\{a_0, \dots, a_{n-1}\}$ of $\mathbf{N}$, where $(a_0, \dots, a_{n-1})$ is a listing of the elements of the set in increasing order.

Supplement 30 shows that g is a bijection.

Now, we give bijective numberings for $\mathbf{N}^k$ for $k \geq 1$. For $k = 1$, if we take our usual identification of $\mathbf{N}^1$ with $\mathbf{N}$, then $1_{\mathbf{N}}$ is the desired bijection. In the next two examples, we give the two standard bijective numberings for $\mathbf{N} \times \mathbf{N}$ and later we give a recursive process to obtain the remaining bijections.

Example 5.3.15 Define $f : \mathbf{N} \times \mathbf{N} \longrightarrow \mathbf{N}$ by $f(m, n) = 2^m(2n + 1) - 1$ for every $m, n \in \mathbf{N}$. This mapping is injective because, if $f(m, n) = f(p, q)$, we have $2^m(2n + 1) = 2^p(2q + 1)$. This implies immediately $2^m = 2^p$ and $2n + 1 = 2q + 1$; hence, $m = p$ and $n = q$. It is easy to see that f is surjective. If $r \in \mathbf{N}$, we can decompose $r + 1$ as a product of powers of prime numbers:

$$r + 1 = 2^{m_1} 3^{m_2} \ldots.$$

This allows us to write $r = 2^{m_1} k - 1$, where k is an odd number. By writing $k = 2h + 1$, we obtain $f(m_1, h) = r$, which shows that f is surjective. Therefore, f is a bijection.

There are many possible bijections between $\mathbf{N} \times \mathbf{N}$ and $\mathbf{N}$. A different bijection is introduced in the next example.

Example 5.3.16 Consider the function $f : \mathbf{N} \times \mathbf{N} \longrightarrow \mathbf{N}$ defined by

$$f(m, n) = m + \frac{(m + n)(m + n + 1)}{2}.$$

The motivation for this definition can be seen in Figure 5.1. Every diagonal line corresponds to a set of points

$$D_k = \{(p, q) \in \mathbf{N} \times \mathbf{N} \mid p + q = k\}.$$

The set D_k contains $k + 1$ points. In order to find the place of a point (m, n) in the numbering suggested by Figure 5.1, observe that the diagonal D_{m+n} is preceded by the diagonals $D_0, \ldots, D_{m+n-1}$, which contain a total of

$$\frac{(m + n)(m + n + 1)}{2}$$

points. In addition, (m, n) is the $(m + 1)$-st point on the diagonal D_{m+n}, and since the numbering starts from 0, the place of (m, n) in the numbering is indeed $f(m, n)$.

The previous examples suggest the following.

Definition 5.3.17 A pairing function *is a bijection* $f : \mathbf{N} \times \mathbf{N} \longrightarrow \mathbf{N}$.

If we have a pairing function $f : \mathbf{N} \times \mathbf{N} \longrightarrow \mathbf{N}$ and $f(p, q) = n$, then, given n, we should be able to determine both components of the pair (p, q) that is mapped by f to n. The following definition introduces an appropriate device.

Definition 5.3.18 A pairing triple *is a triple of functions* (f, p_1, p_2) *such that* $f : \mathbf{N} \times \mathbf{N} \longrightarrow \mathbf{N}$, $p_1 : \mathbf{N} \longrightarrow \mathbf{N}$, $p_2 : \mathbf{N} \longrightarrow \mathbf{N}$ *and* $f(p, q) = n$ *if and only if* $p_1(n) = p$ *and* $p_2(n) = q$ *for every* $p, q, n \in \mathbf{N}$. *We refer to* p_1, p_2 *as the* projections *of the pairing triple.*

In any pairing triple (f, p_1, p_2), f is a bijection, that is, a pairing function. Indeed, since $n = f(p_1(n), p_2(n))$ for every $n \in \mathbf{N}$, it follows that f is surjective. To show that f is injective, assume that $f(p, q) = f(u, v) = n$. We have $p = p_1(n) = u$ and $q = p_2(n) = v$; hence, $(p, q) = (u, v)$, which shows that f is also one-to-one.

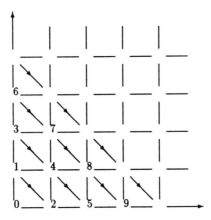

FIGURE 5.1. Definition of the function f.

Conversely, for every pairing function f, we can define $p_1, p_2 : N \longrightarrow N$ by $p_1 = p_1^2 f^{-1}$ and $p_2 = p_2^2 f^{-1}$ so that (f, p_1, p_2) is a pairing triple. Here, the projections $p_1^2, p_2^2 : N \times N \longrightarrow N$ are given by $p_1^2(m, n) = m$ and $p_2^2(m, n) = n$ for every $(m, n) \in N \times N$.

Example 5.3.19 Let us define the pairing triple involving the function f introduced in Example 5.3.16. Let $n \in N$. We need to determine initially the set D_k to which the n-th point belongs. In other words, we need to find the number $k \in N$ for which

$$\frac{k(k+1)}{2} \leq n < \frac{(k+1)(k+2)}{2}.$$

The first inequality implies

$$k \leq \frac{\sqrt{1+8n}-1}{2},$$

while the second implies

$$k > \frac{\sqrt{1+8n}-3}{2}.$$

Since

$$\frac{\sqrt{1+8n}-1}{2} - \frac{\sqrt{1+8n}-3}{2} = 1,$$

there is exactly one integer in the interval

$$\left(\frac{\sqrt{1+8n}-3}{2}, \frac{\sqrt{1+8n}-1}{2} \right],$$

namely,

$$k = \left\lfloor \frac{\sqrt{1+8n} - 1}{2} \right\rfloor . \tag{5.2}$$

It is easy to see then that

$$p_1(n) = n - \frac{k(k+1)}{2},$$
$$and$$
$$p_2(n) = k - p_1(n) = \frac{k(k+3)}{2} - n,$$

where k has been determined from equation (5.2).

If the pairing function $f : \mathbf{N} \times \mathbf{N} \longrightarrow \mathbf{N}$ is monotonic in both its arguments (with respect to the usual order), we shall refer to the pairing triple (f, p_1, p_2) as a *monotonic pairing triple*. Note that, because of the fact that f is a bijection, f is actually a strictly monotonic function in each variable.

Example 5.3.20 The function f from Example 5.3.15 is monotonic in each variable. Thus, the corresponding pairing triple (f, p_1, p_2) is monotonic.

Monotonic pairing triples have certain useful properties mentioned in the next theorem.

Theorem 5.3.21 *Let* (f, p_1, p_2) *be a monotonic pairing triple. We have*

1. $f(0,0) = 0$;

2. $p_1(n) \le n$ *and* $p_2(n) \le n$ *for every* $n \in \mathbf{N}$.

Proof. Since f is a bijection, there exists $(p, q) \in \mathbf{N} \times \mathbf{N}$ such that $f(p, q) = 0$; on the other hand, the monotonicity of f implies $f(0, 0) \le f(p, q) = 0$, and therefore, we obtain $f(0, 0) = 0$.

To prove the second part of the theorem, we shall prove that

$$p \le f(p, q)$$

for every $p, q \in \mathbf{N}$.

For $p = 0$ this fact is clearly true. Suppose now that the statement is true for p. We have

$$
\begin{array}{rcll}
p & \le & f(p, q) & \text{(the inductive hypothesis),} \\
f(p, q) & < & f(p+1, q) & \text{(the strict monotonicity of } f\text{),} \\
p+1 & \le & f(p+1, q) & \text{(from the previous two steps).}
\end{array}
$$

A similar inductive argument shows that $q \le f(p, q)$ for every $p, q \in \mathbf{N}$. ∎

In our next example, we show how to start from an arbitrary pairing function f and construct recursively a family of functions $\{f_k \mid k \ge 1\}$ such that each f_k is a bijection from $\mathbf{N}^k$ to $\mathbf{N}$. We shall refer to the functions $\{f_k \mid k \ge 1\}$ as the *functions associated* with f.

Example 5.3.22 Let $f : \mathbf{N} \times \mathbf{N} \longrightarrow \mathbf{N}$ be a pairing function and define $f_1 = 1_{\mathbf{N}}$.

Suppose now that $k \geq 1$ and we have a bijection $f_k : \mathbf{N}^k \longrightarrow \mathbf{N}$. Define $f_{k+1} : \mathbf{N}^{k+1} \longrightarrow \mathbf{N}$ by

$$f_{k+1}(x_1, \ldots, x_k, x_{k+1}) = f_k(x_1, \ldots, x_{k-1}, f(x_k, x_{k+1})), \qquad (5.3)$$

for every $x_1, \ldots, x_k, x_{k+1} \in \mathbf{N}$.

(Note that $f_2 = f$ according to this definition.) We leave it to the reader to prove that f_{k+1} is also a bijection.

A useful function which can be derived from the functions f_k is given in the following

Definition 5.3.23 *Let* f *be a pairing function. The* uniform projection function for f is the function $\pi_f : \mathbf{N}^3 \longrightarrow \mathbf{N}$, *where*

$$\pi_f(i, k, n) = \begin{cases} 0 & \text{if } k = 0, \\ x_1 & \text{if } f_k(x_1, \ldots, x_i, \ldots, x_k) = n \text{ and } i < 1 \leq k, \\ x_i & \text{if } f_k(x_1, \ldots, x_i, \ldots, x_k) = n \text{ and } 1 \leq i \leq k, \\ x_k & \text{if } f_k(x_1, \ldots, x_i, \ldots, x_k) = n \text{ and } 1 \leq k < i, \end{cases}$$

for every $i, k, n \in \mathbf{N}$.

If f is understood from the context, we denote its uniform projection function π_f simply by π. Properties of uniform projection functions are further discussed in Supplements 38 and 39.

We have already seen that $\mathrm{Seq}(\mathbf{N})$ is countable and have in fact given two numberings for this set, one of which is a bijection. In the next example, we give yet another numbering for $\mathrm{Seq}(\mathbf{N})$ which we will take as our official numbering for finite sequences of integers. We use this numbering in the sequel because it is closely connected with the family of bijections $\{f_k \mid k \geq 1\}$ associated with a particular pairing function.

Example 5.3.24 Let $f : \mathbf{N} \times \mathbf{N} \longrightarrow \mathbf{N}$ be the mapping introduced in Example 5.3.16 and let $\{f_k \mid k \geq 1\}$ be the family of functions associated with f. Consider the mapping gnseq $: \mathrm{Seq}(\mathbf{N}) \longrightarrow \mathbf{N}$ defined by

$$\mathrm{gnseq}((n_0, \ldots, n_{k-1})) = f(k, f_k(n_0, \ldots, n_{k-1})),$$

for every $k \in \mathbf{P}$ and $(n_0, \ldots, n_{k-1}) \in \mathbf{N}^k$; for the null sequence, we define gnseq by $\mathrm{gnseq}(\lambda) = 0$. It is easy to see that gnseq is an injection. However, gnseq is not a surjection since no number $f(0, m)$ with $m > 0$ is in its range.

We can define a slight variation g of gnseq which is a bijection by letting $g((n_0, \ldots, n_{k-1}))$ be $f(k - 1, f_k(n_0, \ldots, n_{k-1})) + 1$ if $k \geq 1$, and be 0 if $k = 0$. We find it more convenient to use the slightly simpler function gnseq in the sequel, even though it is not a bijection.

The number gnseq(s) will be referred to as the *Gödel number*[1] of s.

We now turn to a brief study of uncountable sets. We first give two easily derived methods for showing that sets are uncountable.

Corollary 5.3.25 *If P is an uncountable set and $P \preceq M$, then M is uncountable.*

Proof. This follows immediately from the first part of Corollary 5.3.5. ∎

Corollary 5.3.26 *Let M and P be two sets such that M is uncountable and P is countable. Then, $M - P$ is uncountable.*

Proof. Since $M = P \cup (M - P)$, $M - P$ must be uncountable; otherwise, M would be countable by Theorem 5.3.10. ∎

Uncountable sets can be naturally constructed starting from countable ones. One such example, already mentioned, is the set $\mathcal{P}(\mathbf{N})$. A more general statement is given below.

Theorem 5.3.27 *If M is an infinite set, then $\mathcal{P}(M)$ is uncountable.*

Proof. Suppose that $\mathcal{P}(M)$ is countable. We have

$$M \prec \mathcal{P}(M) \preceq \mathbf{N}$$

because of Theorems 5.2.13 and 5.3.2. By Theorem 5.2.12, this implies $M \prec \mathbf{N}$; hence, M is finite because of Theorem 5.2.17, which is a contradiction. ∎

Example 5.3.28 The set of languages over an alphabet V, $\mathcal{P}(V^*)$, is uncountable.

Theorem 5.3.29 *Let P be a set that has more than one element. If M is an infinite set, then the set of functions $M \longrightarrow P$ is uncountable.*

Proof. Let p_0, p_1 be be two fixed and otherwise arbitrary elements of P. Then, $M \longrightarrow \{p_0, p_1\} \subseteq M \longrightarrow P$, so if we can show that the former set is uncountable, then it will follow from the third part of Theorem 5.2.8 and Corollary 5.3.25 that $M \longrightarrow P$ is uncountable. Now, it is clear that $M \longrightarrow \{p_0, p_1\} \sim M \longrightarrow \{0, 1\}$ and by Theorem 2.3.16 this latter set is

---

[1]Gödel numberings are named after the logician Kurt Gödel in whose work they played an important part. Gödel was born on April 28, 1906 in Brno, Czechoslovakia, and died on January 14, 1978 in Princeton, New Jersey. He was associated with the University of Vienna, first as a student and then as a lecturer, from 1924 to 1939. From 1940 until his death he was affiliated with the Institute for Advanced Study in Princeton. Gödel was the most outstanding logician of the twentieth century. His work, including his results on the undecidability of formal systems, settled fundamental open questions, influenced greatly subsequent work, and has had a profound effect on our understanding of the foundations of mathematics and the limitations of computation.

equinumerous with $\mathcal{P}(M)$. By Theorem 5.3.27, $\mathcal{P}(M)$ is uncountable, so $M \longrightarrow \{p_0, p_1\}$ is also uncountable. ∎

An immediate consequence of this fact is that $\text{ISeq}(\{0,1\}) = \mathbf{N} \longrightarrow \{0,1\}$, the set of infinite sequences over $\{0,1\}$, is uncountable.

We can also prove that $\text{ISeq}(\{0,1\})$ is uncountable using an instructive direct approach. Suppose that $\text{ISeq}(\{0,1\})$ is countable. In this case, we can enumerate its elements, and we have

$$\text{ISeq}(\{0,1\}) = \{s_0, s_1, \ldots, s_n, \ldots\}.$$

Consider the following table listing the components of the infinite sequences $s_0, s_1, \ldots, s_n, \ldots$.

Sequence	0	1	...	n	...
s_0	$s_0(0)$	$s_0(1)$	...	$s_0(n)$	...
s_1	$s_1(0)$	$s_1(1)$	...	$s_1(n)$	...
⋮	⋮	⋮	...	⋮	...
s_n	$s_n(0)$	$s_n(1)$	...	$s_n(n)$	...
⋮	⋮	⋮	...	⋮	...

We now define a new sequence starting from the elements located on the diagonal of this table. Let s be given by

$$s(k) = \begin{cases} 0 & \text{if } s_k(k) = 1 \\ 1 & \text{if } s_k(k) = 0. \end{cases}$$

Note that s is an infinite $\{0,1\}$-sequence, so $s \in \text{ISeq}(\{0,1\})$. On the other hand, s is different from any element of the list $\{s_0, s_1, \ldots, s_n, \ldots\}$ because of the fact that $s(k) \neq s_k(k)$ for every $k \in \mathbf{N}$. This contradiction proves that $\text{ISeq}(\{0,1\})$ is not countable.

The proof technique used here is referred to as *diagonalization* and it appears frequently in mathematics and theoretical computer science. We use this method of proof even when there is no direct "graphical" connection to a table. If we formulate the same argument using subsets of $\mathbf{N}$, we assume that we have a listing of all subsets of $\mathbf{N}$: $\{K_0, K_1, \ldots, K_n, \ldots\}$, and we construct a set K of natural numbers different from any set K_n on this list by defining K as

$$K = \{n \in \mathbf{N} \mid n \notin K_n\}.$$

This is exactly the method we used in proving Theorem 5.2.13 (if we take $M = \mathbf{N}$ and let $f : \mathbf{N} \longrightarrow \mathcal{P}(\mathbf{N})$ be given by $f(n) = K_n$), which justifies calling the argument used in that proof diagonalization.

The countability considerations developed in this section have an important consequence for computability theory. From Theorem 5.3.29, we can infer that the set $\mathbf{N} \longrightarrow \mathbf{N}$ is uncountable. On the other hand, the set P

of all programs (written in some programming language L) that compute one-argument functions from $\mathbf{N} \longrightarrow \mathbf{N}$ is countable since any program of L is essentially a word over the basic alphabet of L. Therefore, $P \prec \mathbf{N} \longrightarrow \mathbf{N}$ and this implies the existence in $\mathbf{N} \longrightarrow \mathbf{N}$ of functions that cannot be computed by *any* program of L.

5.4 Enumerating Programs

In this section, we present the syntax and semantics of a very simple programming language (VSPL) that will allow us to illustrate some of the mathematical techniques used in computability theory. As the reader will soon realize, VSPL is not a real programming language. It lacks constructions that implement data structures and input and output facilities, and its control structures are rather primitive. Yet, what makes this language interesting is precisely its simplicity, which makes it convenient to use as a mathematical tool.

VSPL is built starting from the LOOP language introduced by Meyer and Ritchie [MR67b, MR67a]. Further references are [CB72, BL74]. The basic ingredients of this language are *variables* and *instructions*. We assume that there are infinitely many variables each of which is a nonnull string of symbols over some fixed alphabet V. The set of variables is VAR. The variables of VAR are denoted by $w_0, \ldots, w_n, \ldots$, where $h(w_0) < \cdots < h(w_n) < \ldots$. Here, $h : V^* \longrightarrow \mathbf{N}$ is the bijection introduced in Example 5.3.12. In addition, we use special symbols that do not belong to V:

$$\leftarrow \quad + \quad 1 \quad 0 \quad \textbf{goto} \quad \textbf{loop} \quad \textbf{endloop}.$$

The *basic alphabet* V_b is obtained by adding the special symbols to V.

Variables of VSPL programs assume their values in $\mathbf{N}$.

The instructions of the language are defined as follows.

1. For every $v \in$ VAR, $v \leftarrow v + 1$ is an instruction.

2. For every $v \in$ VAR, $v \leftarrow 0$ is an instruction.

3. For every $v \in$ VAR, **loop** v is an instruction.

4. For every $u \in \{1\}^*$, **goto** u is an instruction.

5. The symbol **endloop** is an instruction.

An instruction **goto** u will be denoted by **goto** l, where $l = |u|$; for instance, we denote **goto** 111 by **goto** 3.

If an instruction contains the symbols $\leftarrow$ and $+$, it must be $v \leftarrow v + 1$ with v uniquely determined (as consisting of all symbols to the left of $\leftarrow$). On the other hand, if an instruction contains $\leftarrow$ but no $+$, then it must be $v \leftarrow 0$ with v uniquely determined.

Note that if an instruction contains **loop** or **goto** then it must be **loop** v and **goto** l, respectively, with v and l uniquely determined in each case. In any other case, the instruction is **endloop**. The above remarks give the unique readability property for instructions.

As usual, we will denote the length of a finite sequence P of instructions by $|P|$.

When we display a program, we usually put one instruction on each line. Sometimes we will want to save space by putting several instructions on a single line of text. When we do this we will separate instructions with semicolons. This is just a notational device. The semicolon is not in the basic alphabet.

Let P be a sequence of instructions satisfying the following conditions.

Pi. P contains the same number of **loop** instructions as **endloop** instructions.

Pii. Every suffix of P contains at least as many **endloop** instructions as **loop** instructions.

Note that every prefix of P contains at least as many **loop** instructions as **endloop** instructions. Consider the sets

$$L_P = \{i | 0 \le i \le |P| - 1 \text{ and } P(i) \text{ is a } \textbf{loop} \text{ instruction}\}$$
$$\text{and}$$
$$K_P = \{i | 0 \le i \le |P| - 1 \text{ and } P(i) \text{ is an } \textbf{endloop} \text{ instruction}\}.$$

Because of (Pi), $|L_P| = |K_P|$. If $P(i)$ is a **loop** instruction then, since the suffix of P beginning with that **loop** has at least as many **endloops** as **loops**, there must be a shortest segment of P beginning with $P(i)$ that has at least as many **endloops** as **loops**. This segment will in fact have the same number of **loops** as **endloops** and will end with an **endloop** statement $P(l)$. This **endloop** is said to be *associated* with the corresponding **loop** and we define $\alpha_P : L_P \longrightarrow K_P$ by $\alpha_P(i) = l$ where $P(l)$ is associated with $P(i)$.

Theorem 5.4.1 *The mapping* $\alpha_P : L_P \longrightarrow K_P$ *defined above is a bijection such that* $i < \alpha_P(i)$ *and* $i < j < \alpha_P(i)$ *implies* $\alpha_P(j) < \alpha_P(i)$ *for every* $i, j \in L_P$.

Proof. The mapping α_P is injective, that is, an **endloop** cannot be associated with two distinct **loop** instructions. Indeed, if the same **endloop** corresponds to both **loop** v_1 and **loop** v_2, then we have the same number of **loop** instructions and **endloop** instructions in both segments S_1 and S_2.

$$\overbrace{\text{loop } v_1; \ldots; \underbrace{\text{loop } v_2; \ldots; \text{endloop}.}_{S_2}}^{S_1}$$

This would imply that the segment S, defined by

$$\underbrace{\textbf{loop } v_1; \ldots; \textbf{loop } v_2,}_{S}$$

contains the same number of **loop** instructions and **endloop** instructions, contradicting the fact that **loop** v_1 corresponds to the given **endloop**. This proves that α_P is injective. Since $|L_P| = |K_P|$, we also have that α_P is a bijection.

It is immediate from the definition that $i < \alpha_P(i)$ for all $i \in L_P$.

Suppose that $P(i)$ and $P(j)$ are two loop instructions, $i < j < \alpha_P(i) = l$, and $\alpha_P(j) = m$. Assume that $l < m$. If X, Y, Z are the sequences defined by

$$P(i); \underbrace{\ldots}_{X}; P(j); \underbrace{\ldots}_{Y}; P(l); \underbrace{\ldots}_{Z}; P(m),$$

denote by $x_{loop}, y_{loop}, z_{loop}$ and $x_{end}, y_{end}, z_{end}$ the number of **loop** and **endloop** instructions that occur in X, Y, Z, respectively.

Since $\alpha_P(i) = l$, $x_{loop} + 1 > x_{end}$ and $x_{loop} + y_{loop} + 2 = x_{end} + y_{end} + 1$, from which it follows that $y_{end} > y_{loop}$. On the other hand, since $\alpha_P(j) = m$, $y_{loop} + 1 > y_{end}$, so $y_{loop} \geq y_{end}$. This contradiction shows that $\alpha_P(j) < l$. ∎

Note that if α_P satisfies the conditions of Theorem 5.4.1 then

$$i < j < \alpha_P(i) \text{ implies } i < j < \alpha_P(j) < \alpha_P(i) \qquad (5.4)$$

for every $i, j \in L_P$.

Theorem 5.4.1 suggests the following.

Definition 5.4.2 *Let P be a sequence of instructions satisfying conditions (Pi) and (Pii).*

The body *of a* **loop** *instruction that occurs in P is the sequence Z of instructions situated between the* **loop** *and its associated* **endloop**:

$$\textbf{loop} \underbrace{\ldots}_{Z} \textbf{endloop}.$$

Definition 5.4.3 *A VSPL program is a sequence of instructions P satisfying conditions (Pi)-(Piii), where (Pi), (Pii) are as defined before Theorem 5.4.1 and (Piii) is the following condition:*

If **goto** *l is an instruction of P, then P contains at least $l+1$ instructions.*

In other words, a VSPL program is a sequence of instructions containing properly nested **loops** and **endloops**, where every **goto** instruction has a line of the program to which it can jump. The set of all programs is denoted by $PROG$.

Now that we have given the syntax of our programming language, we give semantics for it. We will give an operational semantics for the language; that is, we give a mathematical description of how the execution of a program is carried out in a step-by-step manner.

We begin with a set of counter variables

$$COUNT = \{c_i | i \in \mathbf{N}\},$$

with $COUNT \cap VAR = \emptyset$. Each **loop** instruction of a program has an associated counter variable. If **loop** v is the $i+1$-st instruction of a program, then the associated counter variable is c_i.

We introduced the notion of environment of a program in Exercise 30 of Chapter 2. Here, we redefine this notion as needed for defining the semantics of VSPL.

Definition 5.4.4 *The* set of states *STATES is defined by*

$$STATES = VAR \longrightarrow \mathbf{N}.$$

The set of environments is the set

$$ENV = (VAR \cup COUNT) \longrightarrow \mathbf{N}.$$

If e is an environment and $v \in \text{VAR} \cup \text{COUNT}$, then $e(v)$ is the value of v in e.

Definition 5.4.5 *The* set of configurations *of a program* P *is the set* $CONFIG_P$ *given by*

$$CONFIG_P = \{0, \dots, |P|\} \times ENV.$$

The set of terminal configurations *of P is the set $TCONFIG_P$ given by*

$$TCONFIG_P = \{|P|\} \times ENV.$$

The set of initial configurations *of P is the set $ICONFIG_P$, where*

$$ICONFIG_P = \{(0, e) | e(u) = 0 \text{ for every } u \in COUNT\}.$$

Our intention is to use configurations to represent the instruction of the program that is executed at a certain moment and the values of the variables at that moment.

The function introduced below allows us to compute the configuration that succeeds a given configuration. We remind the reader of the notation $[a \rightarrow b]f$ introduced in Section 2.3, which refers to the function obtained from a function f by making its value on a be b.

Let $\mathbf{next}_P : \text{CONFIG}_P \rightsquigarrow \text{CONFIG}_P$ be the partial function (where $\text{Dom}(\mathbf{next}_P) = \text{CONFIG}_P - \text{TCONFIG}_P$) defined as follows for $(i, e) \in \text{CONFIG}_P - \text{TCONFIG}_P$:

1. If $P(i)$ is $v \leftarrow v + 1$, then

$$\textbf{next}_P(i, e) = (i + 1, [v \rightarrow e(v) + 1]s). \tag{5.5}$$

2. If $P(i)$ is $v \leftarrow 0$, then

$$\textbf{next}_P(i, e) = (i + 1, [v \rightarrow 0]e). \tag{5.6}$$

3. If $P(i)$ is **goto** l, then

$$\textbf{next}_P(i, e) = (l, e'), \tag{5.7}$$

where e' is obtained from e by setting to 0 all counter variables c_j such that $P(i)$ is in the body of a **loop** instruction $P(j)$ but the $l + 1$-st instruction is not in this body and is not the associated **endloop**.

4. If $P(i)$ is **loop** v, then

$$\textbf{next}_P(i, e) = \begin{cases} (i + 1, [c_i \rightarrow e(v)]e) & \text{if } e(v) > 0 \\ (j + 1, e) & \text{otherwise,} \end{cases} \tag{5.8}$$

where j is the line number of the **endloop** associated with $P(i)$.

5. If $P(i)$ is **endloop**, then

$$\textbf{next}_P(i, e) = \begin{cases} (k + 1, [c_k \rightarrow e(c_k) - 1]e) \text{ if } e(c_k) > 1 \\ (i + 1, [c_k \rightarrow 0]e) \text{ otherwise.} \end{cases} \tag{5.9}$$

Here, $\alpha_P(k) = i$, that is, k is the line number of the **loop** associated with the **endloop**. Note that an **endloop** instruction resets to 0 the counter variable of the associated **loop** whenever it equals 1.

The function $\textbf{next}_P$ is not defined for terminal configurations. This means that P halts when entering such a configuration.

Let s be a state of a VSPL program P. We define an environment e_s determined by s as

$$e_s(v) = \begin{cases} s(v) & \text{if } v \in VAR \\ 0 & \text{if } v \in COUNT. \end{cases}$$

Definition 5.4.6 *The meaning of a program P is the partial mapping $M_P : STATES \rightsquigarrow STATES$, where $M_P(s) = s'$ if there is $k \in \mathbf{N}$ such that $\textbf{next}_P^k(0, e_s) = (|P|, e')$ and $s' = e' \upharpoonright VAR$.*

Note that M_P is well-defined since there is at most one $k \in \mathbf{N}$ such that $\textbf{next}_P^k(0, e_s)$ is a terminal configuration. This is true because no terminal configuration is in the domain of $\textbf{next}_P$. If $\textbf{next}_P^k(0, e_s)$ is a terminal configuration, then we call k the number of steps in the execution of P on input s.

If s is a state that is not in $\text{Dom}(M_P)$, then there is no k such that $\text{next}_P^k(0, e_s)$ is a terminal configuration. In other words, the program will cycle indefinitely.

Having given a precise semantics, we can now prove precisely results about program behavior. Such proofs inevitably involve induction. In the proof of the next lemma, we give all the details of such an argument. Later, we leave such details for the reader to supply.

Lemma 5.4.7 *Let P be a VSPL program and let s be a state such that $\text{next}_P^k(0, e_s)$ is defined and equals (i, e).*

If $P(j)$ is a loop statement such that $P(i)$ is not in the body of the loop beginning at $P(j)$, and $P(i)$ is not the endloop corresponding to $P(j)$, then $e(c_j) = 0$.

Proof. The argument is by induction on k. For $k = 0$, the statement holds trivially because of the definition of initial configuration.

Suppose that the result is true for $k \geq 0$ and that $\text{next}_P^{k+1}(0, e_s) = (i, e)$, $P(j)$ is a loop statement, $\alpha_P(j) = l$, and either $i \leq j$ or $i > l$. Let $\text{next}_P^k(0, e_s) = (h, e')$. We need to consider several cases, depending on the position and the nature of $P(h)$.

1. If $P(h)$ is not a goto, a loop, or an endloop, then the value of the counter variable c_j remains the same in the configuration (i, e) as it was in the previous configuration (h, e'), that is, $e(c_j) = e'(c_j) = 0$.

2. Suppose now that $P(h)$ is a goto i statement. If $j < h < l$, then (5.7) implies that $e(c_j) = 0$. Otherwise, we have $e(c_j) = e'(c_j) = 0$, by the inductive hypothesis.

3. If $P(h)$ is a loop statement and $h \neq j$, note that we cannot have $j < h < l$, since this would imply $j < \alpha_P(h) < l$. In view of the fact that the line executed after the loop statement $P(h)$ is either $h + 1$ or $\alpha_P(h) + 1$, this would imply that line i is in the body of the loop $P(j)$ or coincides with the endloop $P(\alpha_P(j))$. Therefore, the loop $P(h)$ is located outside the body of the loop $P(j)$ and, by the induction hypothesis, $e'(c_j) = 0$, which gives $e(c_j) = 0$.

4. If $h = j$, then we have either $i = j + 1$ or $i = l + 1$. The first case contradicts the hypothesis. In the second case $e(c_j) = e'(c_j) = 0$.

5. Assume now that $P(h)$ is an endloop. Note that if $h = \alpha_P(j)$, then we must have $i = h + 1$ and, by (5.9), we have $e(c_j) = 0$.

    If $P(h)$ were located inside the body of the loop $P(j)$, $P(i)$ would also be included in this body or it would be the endloop $P(\alpha_P(j))$, which contradicts the hypothesis made on $P(i)$. The only case left is when $P(h)$ is outside the body of $P(j)$, $h \neq l$, and then we have $e(c_j) = e'(c_j) = 0$. ∎

If we concatenate two VSPL programs P and Q, the **gotos** in the resulting sequence that originate in Q will no longer redirect the execution of the new program to the proper instructions. Therefore, we need a special concatenation operation for programs called program concatenation.

If S is a sequence of VSPL instructions, we denote by $S^{[k]}$ the sequence obtained by replacing every instruction **goto** l by **goto** $l + k$.

Let P, Q be two VSPL programs. Consider the sequence of VSPL instructions R obtained by concatenating the sequences P and $Q^{[|P|]}$. We leave to the reader the task of verifying that R satisfies conditions (Pi)-(Piii), and hence, is a program.

Definition 5.4.8 *The VSPL program obtained by* program concatenation *from VSPL programs P and Q is the VSPL program R introduced above.*

The program concatenation of P and Q will be denoted by

$$
\begin{array}{c}
P \\
\bullet \\
Q
\end{array}
$$

It is easily seen that program concatenation is associative and hence, we may write concatenations of arbitrary finite sequences of programs unambiguously.

Theorem 5.4.9 *If R is the program concatenation of two VSPL progams P and Q, then $M_R = M_Q M_P$.*

Proof. Consider a state s. If $\mathbf{next}_P^k(0, e_s)$ is defined, then

$$\mathbf{next}_R^k(0, e_s) = \mathbf{next}_P^k(0, e_s).$$

Thus, if $s \notin \mathrm{Dom}(M_P)$, then $s \notin \mathrm{Dom}(M_R)$, and if $s \in \mathrm{Dom}(M_P)$ say, $M_P(s) = s'$ and $\mathbf{next}_P^k(0, e_s) = (|P|, e')$, where $e'\!\upharpoonright\!\mathrm{VAR} = s'$, then $\mathbf{next}_R^k(0, e_s) = (|P|, e')$. By Lemma 5.4.7 we have $e' = e_{s'}$.

If (n, e) is a configuration for Q, let $(n, e)^+ = (|P| + n, e^+)$, where

$$
e^+(x) = \begin{cases}
e(x) & \text{if } x \in \mathrm{VAR} \\
e(c_{i-|P|}) & \text{if } x = c_i, i \geq |P| \\
0 & \text{if } x = c_i, i < |P|.
\end{cases}
$$

If $\mathbf{next}_Q^l(0, e_{s'})$ is defined, then $\mathbf{next}_R^{k+l}(0, e_s)$ is defined and

$$\mathbf{next}_R^{k+l}(0, e_s) = (\mathbf{next}_Q^l(0, e_{s'}))^+.$$

Thus, if $s' \notin \mathrm{Dom}(M_Q)$, then $s \notin \mathrm{Dom}(M_R)$ while if $s' = M_P(s) \in \mathrm{Dom}(M_Q)$ and $\mathbf{next}_Q^l(0, e_{s'}) = (|Q|, e'')$ with $M_Q(s') = e''\!\upharpoonright\!\mathrm{VAR} = s''$, then

$$\begin{aligned}
\mathbf{next}_R^{k+l}(0, e_s) &= (\mathbf{next}_Q^l(0, e_{s'}))^+ \\
&= (|Q|, e'')^+ \\
&= (|P| + |Q|, e''),
\end{aligned}$$

so $M_R(s) = s'' = M_Q(M_P(s))$. ∎

Theorem 5.4.10 *Let P be a VSPL program. Consider the sequence of instructions S defined by*

$$S \left\{ \begin{array}{l} \textbf{loop } v \\ P^{[1]} \\ \textbf{endloop} \end{array} \right.$$

The sequence S is a program and its meaning M_S is given by

$$M_S(s) = M_P^{s(v)}(s). \tag{5.10}$$

Proof. It is easy to see that S is a VSPL program. Let e be an environment. For every $r \in \mathbf{N}$, define an environment $\Psi_r(e)$ by

$$\Psi_r(e)(x) = \left\{ \begin{array}{ll} e(x) & \text{if } x \in \text{VAR} \\ e(c_{i-1}) & \text{if } x = c_i, i > 0 \\ r & \text{if } x = c_0. \end{array} \right.$$

We have two claims.

Claim 1: If $k \in \mathbf{N}$, $e \in \text{ENV}$ and $\mathbf{next}_P^k(0, e) = (m, e')$, then for all $r \in \mathbf{N}$,

$$\mathbf{next}_S^k(1, \Psi_r(e)) = (m+1, \Psi_r(e')).$$

This can be proven by induction on k, and the argument is left to the reader.

Claim 2: If $s \in \text{STATES}$, $0 \le l < s(v)$ and $M_P^l(s) = s_l$, then, for some t_l, we have

$$\mathbf{next}_S^{t_l}(0, e_s) = (1, \Psi_{s(v)-l}(e_{s_l})).$$

The argument is by induction on l, using Claim 1, and it is also left to the reader.

Now, suppose that $M_P^{s(v)}(s)$ is undefined. Then, there is some l, $0 \le l < s(v)$ such that $M_P^l(s) = s_l$ and $M_P(s_l)$ is undefined. By Claim 2, for some $t_l \in \mathbf{N}$, we have

$$\mathbf{next}_S^{t_l}(0, e_s) = (1, \Psi_{s(v)-l}(e_{s_l})).$$

Since $M_P(s_l)$ is undefined, for all k, $\mathbf{next}_P^k(0, e_{s_l})$ is defined. Hence, by Claim 1, for all k,

$$\mathbf{next}_S^{t_l+k}(0, e_s) = \mathbf{next}_S^k(1, \Psi_{s(v)-l}(e_{s_l}))$$

$\mathbf{next}_S^t(0, e_s) = (1, \Psi_1(e_{s_{s(v)-1}}))$. Since $M_P(s_{s(v)-1}) = s_{s(v)}$ is defined, for some k, $\mathbf{next}_P^k(0, e_{s_{s(v)-1}}) = (|P|, e_{s_{s(v)}})$. By Claim 1, we have

$$\begin{aligned}
\mathbf{next}_S^{t+k}(0, e_s) &= \mathbf{next}_S^k(1, \Psi_1(e_{s_{s(v)-1}})) \\
&= (|P| + 1, \Psi_1(e_{s_{s(v)}})),
\end{aligned}$$

so $\mathbf{next}_S^{t+k+1}(0, e_s) = (|P| + 2, \Psi_0(e_{s_{s(v)}}))$ and $M_S(s) = s_{s(v)} = M_P^{s(v)}(s)$.
∎

Example 5.4.11 For the one-instruction program P,

$$v \leftarrow v + 1,$$

we have $M_P(s) = [v \rightarrow s(v) + 1]s$, for every state s.

Consider the following program:

> **loop** u
> $\quad v \leftarrow v + 1$
> **endloop**

We assume that $u \neq v$.

Suppose that initially we have $s(v) = 0$ and $s(u) = b$. By Theorem 5.4.10, the program ends in state $z = M_P^b(s)$. We have $z(v) = b$. This shows that the effect of the program is to "copy" the value of u into v, provided that the program starts from a state s in which the value of v is 0. Note that we can always arrange this to happen if we consider a slightly larger program:

> $v \leftarrow 0$
> **loop** u
> $\quad v \leftarrow v + 1$
> **endloop**

We shall denote this program by $v \leftarrow u$. This is nothing but shorthand for the real thing, and it is a useful notational device.

Example 5.4.12 A VSPL program such as

> $w \leftarrow 0$
> $w \leftarrow w + 1$
> $w \leftarrow w + 1$
> $w \leftarrow w + 1$
> $w \leftarrow w + 1$
> $w \leftarrow w + 1$

will simply be denoted by $w \leftarrow 5$.

We use the programs introduced in this section to define a certain class of functions.

Let $\bar{x} = (x_1, \ldots, x_r) \in \mathbf{N}^r$, where $r \in \mathbf{N}$. Define a state $s_{\bar{x}}$ by

$$s_{\bar{x}}(w_k) = \begin{cases} x_{k+1} & \text{if } 0 \leq k \leq r-1 \\ 0 & \text{if } k \geq r. \end{cases}$$

Definition 5.4.13 *The r-argument function $f_P^{(r)}$ computed by the VSPL program P (for $r \in \mathbf{N}$) is the partial function $f : \mathbf{N}^r \rightsquigarrow \mathbf{N}$, whose domain is $\{\bar{x} \in \mathbf{N}^r \mid s_{\bar{x}} \in Dom(M_P)\}$, and for each $\bar{x} \in Dom(f_P^{(r)})$,*

$$f_P^{(r)}(\bar{x}) = M_P(s_{\bar{x}})(w_0).$$

Note that the first r variables $w_0, \ldots, w_{r-1}$ serve as input variables and that w_0 also serves as output variable. When $\bar{x} \notin Dom(f_P^{(r)})$, if P starts with the state $s_{\bar{x}}$, then P will never halt. Observe that there is no requirement that the variables $w_0, \ldots, w_{r-1}$ occur in the program P. This allows us to use the same program to compute several functions (for different values for r).

Let $\mathcal{F}_{VSPL}$ be the class of partial functions computable by VSPL programs.

Definition 5.4.14 *Let $r \in \mathbf{N}$ and let $f : \mathbf{N}^r \rightsquigarrow \mathbf{N}$. We say that a VSPL program P strongly computes f if for all $(x_1, \ldots, x_r) \in \mathbf{N}^r$ and all states s such that $s(w_i) = x_{i+1}$ for all i, $0 \leq i \leq r-1$, $M_P(s)$ is defined if and only if $(x_1, \ldots, x_r) \in Dom(f)$ and if $(x_1, \ldots, x_r) \in Dom(f)$, then $M_P(s)(w_0) = f(x_1, \ldots, x_r)$.*

Note that if P strongly computes the r-argument function f, then $f = f_P^{(r)}$, but more is true. Given r inputs in $w_0, \ldots, w_{r-1}$, P computes the correct answer no matter what values are assigned initially to the other variables.

Theorem 5.4.15 *If $f : \mathbf{N}^r \rightsquigarrow \mathbf{N}$ is in $\mathcal{F}_{VSPL}$, then there is a VSPL program which strongly computes f.*

Proof. Let $f = f_P^{(r)}$ for P a VSPL program. Let $v_0, \ldots, v_{m-1}$ consist of all the variables in P which are not among $w_0, \ldots, w_{r-1}$, listed in order according to their code number. In addition, if $r = 0$ and w_0 does not appear in P, include w_0 in this list.

Let Q be the VSPL program

$$v_0 \leftarrow 0; \ldots; v_{m-1} \leftarrow 0,$$

and let P' be the program concatenation of Q and P. Then by Theorem 5.4.9, $M_{P'} = M_P M_Q$. If $\bar{x} = (x_1, \ldots, x_r) \in \mathbf{N}^r$ and s is a state such that $s(w_i) = x_{i+1}$ for all i, $0 \leq i \leq r-1$, then $M_Q(s)$ and $s_{\bar{x}}$ agree on all variables which occur in P. Thus, by Exercise 55, $M_{P'}(s) = M_P(M_Q(s))$ is defined if and only if $M_P(s_{\bar{x}})$ is defined, and, by assumption, this happens if and only if $f(\bar{x})$ is defined. Suppose that $f(\bar{x})$ is defined. If w_0 appears in P then, by the same exercise, $M_{P'}(s)(w_0) = M_P(M_Q(s))(w_0) = M_P(s_{\bar{x}})(w_0) = f(\bar{x})$. If w_0 does not appear in P, then $M_{P'}(s)(w_0) = M_P(M_Q(s))(w_0) = M_Q(s)(w_0) = s_{\bar{x}}(w_0) = M_P(s_{\bar{x}})(w_0) = f(\bar{x})$. ∎

Example 5.4.16 The function pred : $\mathbf{N} \longrightarrow \mathbf{N}$ defined by

$$\text{pred}(n) = \begin{cases} 0 & \text{if } n = 0 \\ n - 1 & \text{if } n > 0, \end{cases}$$

belongs to $\mathcal{F}_{VSPL}$.

Consider the following VSPL program:

$$R \begin{cases} \textbf{loop } w_0 \\ \quad w_2 \leftarrow w_1 \\ \quad w_1 \leftarrow w_1 + 1 \\ \textbf{endloop} \\ w_0 \leftarrow w_2 \end{cases}$$

Suppose that we have state s before executing Q:

$$Q \begin{cases} w_2 \leftarrow w_1 \\ w_1 \leftarrow w_1 + 1 \end{cases}$$

If $s(w_1) = a$ and $s(w_2) = b$, then, after executing the first instruction, we obtain state s', where $s'(w_1) = a$ and $s'(w_2) = a$; after the execution of the second instruction, we have state s'', where $s''(w_1) = a + 1$ and $s''(w_2) = a$. The meaning of this fragment, M_Q, transforms state s into state $\bar{s} = M_Q(s)$, where

$$\bar{s}(w) = \begin{cases} s(w) & \text{if } w \neq w_1, w_2 \\ s(w_1) + 1 & \text{if } w = w_1 \\ s(w_1) & \text{if } w = w_2. \end{cases}$$

Note that the value assigned to w_2 is one less than the value assigned to w_1.

Using Theorem 5.4.10, the meaning of the fragment P,

$$P \begin{cases} \textbf{loop } w_0 \\ \quad w_2 \leftarrow w_1 \\ \quad w_1 \leftarrow w_1 + 1 \\ \textbf{endloop} \end{cases}$$

is M_P, where $M_P(s) = M_Q^{s(w_0)}(s)$. In other words, if $M_P(s) = s'$, then we have

$$s'(w) = \begin{cases} s(w) & \text{if } w \neq w_1, w_2 \\ s(w_1) + s(w_0) & \text{if } w = w_1 \\ s(w_1) + s(w_0) - 1 & \text{if } w = w_2, \end{cases}$$

when $s(w_0) \geq 1$. If $s(w_0) = 0$, then $s' = s$.

Let P_1 be the program

$$w_0 \leftarrow w_2$$

The meaning of P_1 is given by

$$M_{P_1}(s) = [w_0 \rightarrow s(w_2)]s.$$

The meaning of the entire program is given by

$$M_R(s) = M_{P_1}(M_P(s)).$$

Suppose that a computation of R begins in state s_0, where $s_0(w) = n$ if $w = w_0$ and $s_0(w) = 0$, otherwise. Then, $s_1 = M_P(s_0)$ is given by

$$s_1(w) = \begin{cases} n & \text{if } w = w_0 \\ n & \text{if } w = w_1 \\ \text{pred}(n) & \text{if } w = w_2 \\ 0 & \text{otherwise.} \end{cases}$$

Therefore, the program will halt in state $s_2 = M_{P_1}(s_1)$, where $s_2(w_0) = \text{pred}(n)$, which shows that P computes the function pred.

Note that the program

$$R' \begin{cases} w_1 \leftarrow 0 \\ w_2 \leftarrow 0 \\ \textbf{loop } w_0 \\ \quad w_2 \leftarrow w_1 \\ \quad w_1 \leftarrow w_1 + 1 \\ \textbf{endloop} \\ w_0 \leftarrow w_2 \end{cases}$$

is the program which is obtained from R using the method of the proof of Theorem 5.4.15 and hence, R' strongly computes pred. (In fact, the instruction $w_2 \leftarrow 0$ is not needed for R' to strongly compute pred, but $w_1 \leftarrow 0$ is needed.)

Example 5.4.17 The function $f : \mathbf{N}^2 \longrightarrow \mathbf{N}$ given by $f(x,y) = x + y$ for $x, y \in \mathbf{N}$ belongs to $\mathcal{F}_{VSPL}$.

Indeed, consider the following program.

$$\textbf{loop } w_1$$
$$w_0 \leftarrow w_0 + 1$$
$$\textbf{endloop}$$

If s is the initial state of the program, then the variable w_0 is incremented by 1 $s(w_1)$ times, which, of course, amounts to storing in w_0 the sum of the initial values of w_0 and w_1.

Example 5.4.18 Consider the following VSPL program:

$$\textbf{loop } w_0$$
$$w_1 \leftarrow 1$$
$$\textbf{endloop}$$
$$w_0 \leftarrow w_1$$

This program computes the function sgn defined in Example 4.9.6.

Theorem 5.4.19 $\mathcal{F}_{VSPL}$ *is closed under composition. In other words, if* $f : \mathbf{N}^n \rightsquigarrow \mathbf{N}$ *and* $g_i : \mathbf{N}^m \rightsquigarrow \mathbf{N}$ *for* $1 \leq i \leq n$ *belong to* $\mathcal{F}_{VSPL}$, *then the function* $f(g_1, \ldots, g_n) : \mathbf{N}^m \rightsquigarrow \mathbf{N}$ *defined by*

$$f(g_1, \ldots, g_n)(x_1, \ldots, x_m) = f(g_1(x_1, \ldots, x_m), \ldots g_n(x_1, \ldots, x_m))$$

belongs to $\mathcal{F}_{VSPL}$.

Proof. Suppose that $P, P_1, \ldots, P_n$ strongly compute $f, g_1, \ldots, g_n$, respectively, and let $v_0, v_1, \ldots, v_{m-1}, z_1, \ldots, z_n$ be distinct variables that do not appear in $P_1, \ldots, P_n$ and are different from $w_0, \ldots, w_{\max(m,n)-1}$. Let Q be the following program:

$$v_0 \leftarrow w_0; \ldots; v_{m-1} \leftarrow w_{m-1}$$
$$\bullet$$
$$P_1$$
$$\bullet$$
$$z_1 \leftarrow w_0$$
$$w_0 \leftarrow v_0; \ldots; w_{m-1} \leftarrow v_{m-1}$$
$$\bullet$$
$$P_2$$
$$\bullet$$
$$z_2 \leftarrow w_0$$
$$w_0 \leftarrow v_0; \ldots; w_{m-1} \leftarrow v_{m-1}$$
$$\bullet$$
$$P_3$$
$$\bullet$$
$$\vdots$$
$$\bullet$$
$$z_{n-1} \leftarrow w_0$$
$$w_0 \leftarrow v_0; \ldots; w_{m-1} \leftarrow v_{m-1}$$
$$\bullet$$
$$P_n$$
$$\bullet$$
$$z_n \leftarrow w_0$$
$$w_0 \leftarrow z_1; \ldots; w_{n-1} \leftarrow z_n$$
$$\bullet$$
$$P$$

It is not hard to show, using Theorem 5.4.9 and Exercise 55, that Q computes $f(g_1, \ldots, g_n)$. ∎

Example 5.4.20 We can show now that the test function $for : \mathbf{N}^2 \longrightarrow \mathbf{N}$ defined by

$$for(x, y) = \begin{cases} 1 & \text{if } x > 0 \text{ or } y > 0 \\ 0 & \text{otherwise,} \end{cases}$$

belongs to $\mathcal{F}_{VSPL}$. To this end, observe that we can write $for(x, y) = \text{sgn}(x + y)$; since $\mathcal{F}_{VSPL}$ is closed with respect to composition, it follows that $for \in \mathcal{F}_{VSPL}$.

Theorem 5.4.21 $\mathcal{F}_{VSPL}$ *is closed under primitive recursion. In other words, if* $g : \mathbf{N}^m \rightsquigarrow \mathbf{N}$, *and* $h : \mathbf{N}^{m+2} \rightsquigarrow \mathbf{N}$ *belong to* $\mathcal{F}_{VSPL}$, *then the function* $f : \mathbf{N}^{m+1} \rightsquigarrow \mathbf{N}$ *defined by*

$$\begin{aligned} f(x_1, \ldots, x_m, 0) &= g(x_1, \ldots, x_m), \\ f(x_1, \ldots, x_m, n+1) &= h(x_1, \ldots, x_m, n, f(x_1, \ldots, x_m, n)), \end{aligned}$$

belongs to $\mathcal{F}_{VSPL}$.

Proof. Let P and Q be VSPL programs which strongly compute g and h respectively. Let $v_0, \ldots, v_{m-1}, v$ and w be distinct variables which do not appear in either P or Q and are different from $w_0, \ldots, w_m$.

Let R be the VSPL program:

$$R \begin{cases} v_0 \leftarrow w_0; \ldots; v_{m-1} \leftarrow w_{m-1} \\ w \leftarrow w_m \\ \bullet \\ P \\ \bullet \\ v \leftarrow 0 \\ \bullet \\ \textbf{loop } w \\ T^{[1]} \\ \textbf{endloop} \end{cases}$$

Here, T is the program:

$$T \begin{cases} w_{m+1} \leftarrow w_0; w_m \leftarrow v \\ w_0 \leftarrow v_0; \ldots; w_{m-1} \leftarrow v_{m-1} \\ \bullet \\ Q \\ \bullet \\ v \leftarrow v + 1 \end{cases}$$

Using Theorem 5.4.9 and Exercise 55, it is not hard to show that if s is any state such that $s(w_0) = f(s(v_0), \ldots, s(v_{m-1}), s(v))$, then $M_T(s)$ is defined if and only if $f(s(v_0), \ldots, s(v_{m-1}), s(v) + 1)$ is defined and if $s' = M_T(s)$ is defined, then $s'(v_i) = s(v_i)$ for all i, $0 \le i \le m-1$, $s'(v) = s(v)+1$ and $s'(w_0) = f(s'(v_0), \ldots, s'(v_{m-1}), s'(v))$.

Using this remark, Theorems 5.4.9 and 5.4.10, and Exercise 55, it is now fairly easy to show that R computes f. ∎

Theorem 5.4.22 *If $f : \mathbf{N}^{n+1} \rightsquigarrow \mathbf{N}$ belongs to $\mathcal{F}_{VSPL}$, then so does the function $\mu f : \mathbf{N}^n \rightsquigarrow \mathbf{N}$ obtained from f by minimization.*

Proof. We remind the reader that $\mu f(r_1, \ldots, r_n) = q$ if q is the least natural number such that $(r_1, \ldots, r_n, q') \in \mathrm{Dom}(f)$ for all q' with $0 \le q' \le q$ and $f(r_1, \ldots, r_n, q) = 0$, if such a q exists; otherwise, $\mu f(r_1, \ldots, r_n)$ is undefined.

Suppose that P is a VSPL program which strongly computes f and that $v_0, \ldots, v_{n-1}, v$ are distinct variables which do not appear in P and are different from $w_0, \ldots, w_n$. Let Q be obtained by appending to P the lines

> **loop** w_0
> $w_0 \leftarrow v_0; \ldots; w_{n-1} \leftarrow v_{n-1}$
> $v \leftarrow v + 1$
> $w_n \leftarrow v$
> **goto** 0
> **endloop**
> $w_0 \leftarrow v$

It is easy to show that Q is a VSPL program. Let R be the program concatenation of the initialization program

$$v_0 \leftarrow w_0; \ldots; v_{n-1} \leftarrow w_{n-1}$$

with Q, so that we have

$$R \begin{cases} v_0 \leftarrow w_0; \ldots; v_{n-1} \leftarrow w_{n-1} \\ \bullet \\ P \\ \textbf{loop } w_0 \\ w_0 \leftarrow v_0; \ldots; w_{n-1} \leftarrow v_{n-1} \\ v \leftarrow v + 1 \\ w_n \leftarrow v \\ \textbf{goto } 0 \\ \textbf{endloop} \\ w_0 \leftarrow v \end{cases}$$

The **goto** 0 in Q transfers control to the first line of P. The **loop** w_0 never terminates normally; it serves as an **if** statement checking the value of w_0 produced by P. Note that Q is not the program concatenation of P with the lines following it, nor is the loop just after P in the form of

Theorem 5.4.10. Thus, neither of Theorems 5.4.9 and 5.4.10 is of much use in analyzing Q. It is not to hard to prove the following statement, however. Suppose that s is a state such that $s(v_i) = s(w_i)$ for all i, $0 \le i \le n-1$, and $s(v) = s(w_n)$. If $f(s(w_0), \ldots, s(w_{n-1}), s(v))$ is undefined, then $M_Q(s)$ is undefined. If $f(s(w_0), \ldots, s(w_{n-1}), s(v)) = 0$, then $M_Q(s)$ is defined, and $M_Q(s)(w_0) = s(v)$. If $f(s(w_0), \ldots, s(w_{n-1}), s(v)) > 0$, then there is a state s' such that $s'(v_i) = s'(w_i) = s(w_i)$ for all i, $0 \le i \le n-1$, $s'(v) = s'(w_n) = s(v)+1$, and $M_Q(s) = M_Q(s')$.

We leave it to the reader to use this observation to show that R computes μf. ∎

Theorem 5.4.23 *Every partial recursive function is VSPL-computable.*

Proof. The initial functions Z, p_i^n and S are all VSPL-computable. Indeed, the VSPL programs that compute these functions are

$$w_0 \leftarrow 0$$

for the zero function Z,

$$w_0 \leftarrow w_{i-1}$$

for the projection p_i^n with $i > 1$, the empty program (that is, the empty sequence of VSPL statements) for p_1^n, and

$$w_0 \leftarrow w_0 + 1$$

for the successor function S.

Since the set of partial recursive functions is defined as the closure of the set of initial functions under composition, primitive recursion and minimization, the theorem follows immediately from Theorems 5.4.19, 5.4.21 and 5.4.22. ∎

We define a one-to-one mapping $gn : PROG \longrightarrow \mathbf{N}$ in such a manner that it will be easy to reconstitute the program P starting from $gn(P)$.

For the remainder of the chapter, we denote by (f, p_1, p_2) a fixed, monotonic pairing triple such that f and its projections p_1, p_2 are primitive recursive functions (see Exercise 35).

Definition 5.4.24 *The code $d(S)$ of an instruction S is*

$$d(S) = \begin{cases} f(1,p) & \text{if } S \text{ is } w_p \leftarrow w_p + 1 \\ f(2,p) & \text{if } S \text{ is } w_p \leftarrow 0 \\ f(3,p) & \text{if } S \text{ is } \mathbf{loop}\ w_p \\ f(4,l) & \text{if } S \text{ is } \mathbf{goto}\ l \\ f(5,0) & \text{if } S \text{ is } \mathbf{endloop}. \end{cases}$$

Note that any two distinct instructions have distinct codes. The mapping gn, defined on $PROG$, is given by

$$gn(P) = gnseq(d(S_0), \ldots, d(S_{n-1})),$$

where $P = S_0; \ldots; S_{n-1}$.

In view of its definition, it is clear that gn is one-to-one. Moreover, starting from the code gn(P), we can easily reconstitute P. If $m = $ gn(P), then $\mathrm{p}_1(m)$ gives the length of P, and $\pi(i, \mathrm{p}_1(m), \mathrm{p}_2(m))$ will give us the code of the i-th instruction, for $1 \leq i \leq \mathrm{p}_1(m)$.

We call gn(P) the *Gödel number* of P and we denote by $GPROG$ the set gn($PROG$) of all Gödel numbers of programs from $PROG$.

VSPL programs can be indexed by their Gödel numbers. Namely, if $x \in$ gn($PROG$), then P_x is the VSPL program whose Gödel number is x.

Not all natural numbers are Gödel numbers of programs, and this creates certain technical problems. Therefore, we shall adopt the following convention: if $x \notin$ gn($PROG$), then P_x will be the ever cycling program

goto 0

This convention makes every natural number the Gödel number of a VSPL program.

The partial function $f_{P_x}^{(n)} : \mathbf{N}^n \rightsquigarrow \mathbf{N}$ will be denoted by $f_x^{(n)}$.

An example of a function that does not belong to $\mathcal{F}_{VSPL}$ originates in the *halting problem* for VSPL programs.

Definition 5.4.25 *Let P be a VSPL program. P halts on input m if $m \in Dom(f_P^{(1)})$.*

There is no VSPL program that is able to decide, starting from the Gödel number x_1 of a program P and a number $x_2 \in \mathbf{N}$, whether or not P_{x_1} halts on x_2. In order to make a formal statement of this property of VSPL programs, consider the function $h : \mathbf{N} \times \mathbf{N} \longrightarrow \mathbf{N}$ defined by

$$h(x_1, x_2) = \begin{cases} 1 & \text{if } x_2 \in \mathrm{Dom}(f_{x_1}^{(1)}) \\ 0 & \text{otherwise.} \end{cases}$$

Theorem 5.4.26 *The function h does not belong to $\mathcal{F}_{VSPL}$.*

Proof. We present an argument by diagonalization. Suppose that there is a VSPL program Q that computes h. This implies the existence of a VSPL program R that will compute the function $g : \mathbf{N} \longrightarrow \mathbf{N}$, defined by

$$g(z) = \begin{cases} 1 & \text{if } P_z \text{ halts on input } z \\ 0 & \text{otherwise.} \end{cases}$$

Indeed, R can be obtained from Q by adding one extra instruction:

$$R \begin{cases} w_1 \leftarrow w_0 \\ \bullet \\ Q \end{cases}$$

Starting from R, consider the program P:

$$P \begin{cases} R \\ \textbf{loop } w_0 \\ \textbf{goto } |R| + 1 \\ \textbf{endloop} \end{cases}$$

Let $z = \text{gn}(P)$ and let us apply P to its own Gödel number. If P halts on input z, then upon exiting R, the value of the variable w_0 would be 1. This, however, implies that P will cycle forever on the $(|R| + 1)$-st instruction **goto** $|R| + 1$ and P never halts on z.

Assume now that P does not halt on z. This happens only if R assigns the value 0 in w_0. The definition of R tells us that this means that the program P halts on its own Gödel number. Since both situations yield contradictions, we conclude that R does not exist and, therefore, Q does not exist. ∎

5.5   Abstract Families of Functions

The purpose of this section is to present an abstract version of the families of functions that are computable by certain classes of programs. We use this section to provide the reader with additional exposure to proofs by diagonalization.

Definition 5.5.1 *An* abstract family of functions *(AFF) is a family of partial functions*

$$\mathbf{F} = \{f_i^{(n)} | n \in \mathbf{N}, i \in \mathbf{N}\},$$

where $f_i^{(n)} : \mathbf{N}^n \rightsquigarrow \mathbf{N}$ *for every* $n, i \in \mathbf{N}$.

The definition of AFF implies that every AFF is countable.

Example 5.5.2 Let $\mathbf{G}$ be the abstract family consisting of the functions $g_i^{(n)} : \mathbf{N}^n \rightsquigarrow \mathbf{N}$, where $g_i^{(n)}(x_1, \ldots, x_n) = x_1 + \cdots + x_n + i$ for every $n, i \in \mathbf{N}$. Note that this family contains such functions as $g_0^{(2)}(x_1, x_2) = x_1 + x_2$ and $g_0^{(1)}(x_1) = x_1$ for $x_1, x_2 \in \mathbf{N}$. All members of this family are total functions.

Example 5.5.3 Consider the family of functions $f_i^{(n)}$, where $f_i^{(n)}$ is the n-argument function computed by the VSPL program P with $\text{gn}(P) = i$, as defined in Section 5.4. This abstract family contains both total and nontotal functions.

In the sequel, we are going to use the constant functions $K_k^n : \mathbf{N}^n \longrightarrow \mathbf{N}$, with $n, k \in \mathbf{N}$, defined by

$$K_k^n(x_1, \ldots, x_n) = k$$

for every $x_1, \ldots, x_n \in \mathbf{N}$.

Example 5.5.4 Consider the AFF **POL** that consists of all multivariable polynomials with coefficients in **N**. Any polynomial $g : \mathbf{N}^n \longrightarrow \mathbf{N}$ can be written as

$$g(x_1, \ldots, x_n) = \sum_{(a_1, \ldots, a_n) \in A} c_{a_1, \ldots, a_n} x_1^{a_1} \cdots x_n^{a_n}$$

for every $x_1, \ldots, x_n \in \mathbf{N}$, where the sum extends over a finite subset A of $\mathbf{N}^n$, and $c_{a_1, \ldots, a_n} \neq 0$ for every $(a_1, \ldots, a_n) \in A$. Each such polynomial uniquely determines the sets A and $\{c_{a_1, \ldots, a_n} | (a_1, \ldots, a_n) \in A\}$. We index g using the codes for the $(n+1)$-tuples $(c_{a_1, \ldots, a_n}, a_1, \ldots, a_n)$. Consider the function $f_{n+1} : \mathbf{N}^{n+1} \longrightarrow \mathbf{N}$ introduced in Example 5.3.22. Suppose that g has r monomials and let

$$m_l = c_{a_{1l}, \ldots, a_{nl}} x_1^{a_{1l}} \cdots x_n^{a_{nl}}$$

be the l-th monomial of g (where l is determined by the place of the n-tuple $(a_1, \ldots, a_n)$ in the lexicographic order of the poset $(\mathbf{N}^n, \preceq)$) for $1 \leq l \leq r$. Let

$$b_l = f_{n+1}(c_{a_{1l}, \ldots, a_{nl}}, a_{1l}, \ldots, a_{nl})$$

be "the code" of m_l. The index of g will be defined as the code of the sequence $(b_1, \ldots, b_r)$ (as introduced in Example 5.3.24). If $A = \emptyset$, then $g = K_0^n$, and the index of g is 0.

If i is not the code of a polynomial, we define $g_i^n = K_0^n$.

Now, we introduce or recall some special notations for functions that we need in this section.

We use the notation $p_m^n : \mathbf{N}^n \longrightarrow \mathbf{N}$ to designate the projection

$$p_m^n(x_1, \ldots, x_n) = x_m$$

for all $x_1, \ldots, x_n \in \mathbf{N}$ and $1 \leq m \leq n$.

For each $n \in \mathbf{N}$, we have $\emptyset : \mathbf{N}^n \rightsquigarrow \mathbf{N}$ and $\text{Dom}(\emptyset) = \emptyset$. When we want to emphasize that we are considering $\emptyset$ as a partial function on $\mathbf{N}^n$, we will write it as $\emptyset^n$.

The successor function $S : \mathbf{N} \longrightarrow \mathbf{N}$ is given by $S(n) = n + 1$ for $n \in \mathbf{N}$ and the *discriminator function* $D : \mathbf{N}^4 \longrightarrow \mathbf{N}$ is defined by

$$D(m, n, p, q) = \begin{cases} m & \text{if } p = q \\ n & \text{otherwise.} \end{cases}$$

The *collection of basic functions* consists of the projections, the constant functions, and the discriminator function.

For an AFF **F**, the set of functions from **F** that have m arguments will be denoted by $\mathbf{F}_m$.

Definition 5.5.5 *Let* **F** *be an AFF. The* universal function *for* $\mathbf{F}_m$ *is the partial function* $u_m : \mathbf{N}^{m+1} \rightsquigarrow \mathbf{N}$ *such that* $(i, x_1, \ldots, x_m) \in Dom(u_m)$ *if and only if* $(x_1, \ldots, x_m) \in Dom(f_i^{(m)})$ *and*

$$u_m(i, x_1, \ldots, x_m) = f_i^{(m)}(x_1, \ldots, x_m)$$

for every $(x_1, \ldots, x_m) \in Dom(f_i^{(m)})$.

F *is a* universal family *if the universal function* u_m *for* $\mathbf{F}_m$ *is in* $\mathbf{F}_{m+1}$ *for every* $m \in \mathbf{N}$.

Example 5.5.6 The AFF introduced in Example 5.5.2 is universal. Indeed, note that $g_0^{(n+1)}$ is the universal function for $\mathbf{G}_n$ since

$$g_0^{(n+1)}(i, x_1, \ldots, x_n) = g_i^{(n)}(x_1, \ldots, x_n)$$

for every $i, x_1, \ldots, x_n \in \mathbf{N}$.

Furthermore, Exercise 67 shows that the collection of partial recursive functions (which is $\mathcal{F}_{VSPL}$) is a universal AFF.

Definition 5.5.7 *An AFF* **F** *is* closed under composition *if from* $h \in \mathbf{F}_m$, *with* $m \geq 1$, *and* $g_1, \ldots, g_m \in \mathbf{F}_n$, *it follows that* $h(g_1, \ldots, g_m) \in \mathbf{F}_n$.

Using a diagonalization argument we can prove the following.

Theorem 5.5.8 *A universal AFF that is closed under composition and contains the projections and the successor function cannot consist entirely of total functions.*

Proof. Let **F** be a universal AFF and let $S \in \mathbf{F}_1$ be the successor function. In view of the above hypothesis, we have $f = S(u_2(p_1^1, p_1^1)) \in \mathbf{F}_1$. We have $f(n) = u_2(n, n) + 1 = f_n^{(1)}(n) + 1$ for $n \in Dom(f_n^{(1)})$.

Suppose that every member of **F** is a total function. Since $f \in \mathbf{F}_1$, there is $k \in \mathbf{N}$ such that $f = f_k^{(1)}$, and since f is total, $f(k)$ is defined. However, this generates a contradiction since we also have $f_k^{(1)}(k) = f(k) = f_k^{(1)}(k) + 1$, and therefore, such a family cannot consist entirely of total functions. ∎

A consequence of Theorem 5.5.8 is given in the following.

Corollary 5.5.9 *Let* **F** *be a universal AFF that is closed under composition and contains the constant functions, the projections, and the successor function. Then* **F** *contains the empty function.*

Proof. From the proof of Theorem 5.5.8, it follows that **F** contains a nontotal, one-argument function, $f : \mathbf{N} \rightsquigarrow \mathbf{N}$. Let k be a number, $k \in \mathbf{N} - Dom(f)$. Since **F** is closed under composition, we have $h = f \circ K_k^0 \in \mathbf{F}$. Note that, $h() = f(K_k^0()) = f(k)$ and therefore, $h = \emptyset^0$ is the empty function. ∎

The halting problem discussed in Section 5.4 can be rephrased for AFFs. The proof is by diagonalization.

Theorem 5.5.10 *Let* **F** *be a universal AFF that is closed under composition and contains the constant functions, the projections, the successor function and the discriminator function. There is no function* $h : \mathbf{N}^2 \longrightarrow \mathbf{N}$ *in* $\mathbf{F}_2$ *such that*

$$h(x, y) = \begin{cases} 1 & \text{if } y \in Dom(f_x^{(1)}) \\ 0 & \text{otherwise.} \end{cases}$$

Proof. From Corollary 5.5.9, we have $\emptyset^1 \in \mathbf{F}_1$. Therefore, we have $\emptyset^1 = f_i^{(1)}$ for some $i \in \mathbf{N}$. Also, since $K_0^1 \in \mathbf{F}_1$, we can write $K_0^1 = f_j^{(1)}$.

Suppose that $h \in \mathbf{F}_2$ and consider the function $f : \mathbf{N} \rightsquigarrow \mathbf{N}$ defined by

$$f(x) = u_2(D(j, i, h(x, x), 0), x)$$

for every $x \in \mathbf{N}$. Note that we have $f \in \mathbf{F}_1$ since we can write

$$f = u_2(D(K_j^1, K_i^1, h(p_1^1, p_1^1), K_0^1), p_1^1).$$

From the definition of f we obtain

$$\begin{aligned} f(x) &= u_2(D(j, i, h(x, x), 0), x) \\ &= \begin{cases} u_2(j, x) & \text{if } h(x, x) = 0 \\ u_2(i, x) & \text{if } h(x, x) = 1 \end{cases} \\ &= \begin{cases} 0 & \text{if } h(x, x) = 0 \\ \text{undefined} & \text{if } h(x, x) = 1. \end{cases} \end{aligned}$$

As customary in proofs by diagonalization, suppose that c is an index for f, that is, $f = f_c^{(1)}$. The previous equalities give

$$\begin{aligned} f(c) &= \begin{cases} 0 & \text{if } h(c, c) = 0 \\ \text{undefined} & \text{if } h(c, c) = 1 \end{cases} \\ &= \begin{cases} 0 & \text{if } c \notin Dom(f_c^{(1)}) \\ \text{undefined} & \text{if } c \in Dom(f_c^{(1)}) \end{cases} \\ &= \begin{cases} 0 & \text{if } c \notin Dom(f) \\ \text{undefined} & \text{if } c \in Dom(f), \end{cases} \end{aligned}$$

which is a contradiction. We conclude that $h \notin \mathbf{F}_1$. ∎

Corollary 5.5.11 *Let* **F** *be a universal AFF that is closed under composition and contains the constant functions, the projections, the successor function and the discriminator function. The function* $k : \mathbf{N} \longrightarrow \mathbf{N}$, *defined by*

$$k(x) = \begin{cases} 1 & \text{if } x \in Dom(f_x^{(1)}) \\ 0 & \text{otherwise,} \end{cases}$$

does not belong to $\mathbf{F}_1$.

Proof. Let us remark that $k(x) = h(x, x)$ for $x \in \mathbf{N}$ and that the previous proof shows that a contradiction can be derived from the assumption that $k \in \mathbf{F}_1$. ∎

Let P be a program written in some programming language. Assume that we use P for computing the values of some function $f : \mathbf{N}^{m+n} \rightsquigarrow \mathbf{N}$. Furthermore, suppose that f belongs to some AFF $\mathbf{F} = \{f_j^{(l)} | l, j \in \mathbf{N}\}$ and that i is a subscript for f, that is, $f = f_i^{(m+n)}$. If we fix the first m arguments of f to be $a_1, \ldots, a_m$, we obtain a new function $g : \mathbf{N}^n \rightsquigarrow \mathbf{N}$, where

$$g(x_1, \ldots, x_n) = f(a_1, \ldots, a_m, x_1, \ldots, x_n),$$

for $x_1, \ldots, x_n \in \mathbf{N}$. Note that if $\mathbf{F}$ contains the projections and the constant functions and it is closed with respect to composition, we have $g \in \mathbf{F}_n$. In general, when an AFF family is defined in relation to a programming language, we use some code of the program as the subscript for the function computed by that program (see Example 5.5.3). If this is the case for the AFF $\mathbf{F}$, then, since $g \in \mathbf{F}_m$, we can write $g = f_j^{(n)}$, where j is the code of a program Q that computes g. We are interested in those AFFs that contain functions $s_{m,n} : \mathbf{N}^{m+1} \longrightarrow \mathbf{N}$ which allow us to write $j = s_{m,n}(i, a_1, \ldots, a_m)$.

Definition 5.5.12 *An AFF has the smn property if, for every $m, n \in \mathbf{N}$, there exists a total function $s_{m,n} : \mathbf{N}^{m+1} \longrightarrow \mathbf{N}$ such that for every $f_i^{(m+n)} \in \mathbf{F}_{m+n}$ we have*

$$f_i^{(m+n)}(x_1, \ldots, x_m, x_{m+1}, \ldots, x_{m+n}) = f_{s_{m,n}(i, x_1, \ldots, x_m)}^{(n)}(x_{m+1}, \ldots, x_{m+n})$$

for every $x_1, \ldots, x_{m+n} \in \mathbf{N}$.

Definition 5.5.13 *A standard AFF is a universal AFF that has the smn property, is closed with respect to composition, and contains the basic functions.*

Note that, according to Corollary 5.5.9, any standard AFF contains the empty function.

Example 5.5.14 Exercise 70 implies that $\mathcal{F}_{VSPL}$ has the smn property. This shows that the AFF of partial recursive functions is a standard AFF.

An application of the smn property is given below.

Theorem 5.5.15 *Let $\mathbf{F}$ be a standard AFF. There exists a function $c \in \mathbf{F}_2$ such that*

$$f_{c(i,j)}^{(1)} = f_i^{(1)} \circ f_j^{(1)}$$

for every $i, j \in \mathbf{N}$.

Proof. Consider the function $g : \mathbf{N}^3 \rightsquigarrow \mathbf{N}$, defined by

$$g = u_2(p_1^3, u_2(p_2^3, p_3^3)),$$

which belongs to $\mathbf{F}_3$. We have $g = f_i^{(3)}$ for some $i \in \mathbf{N}$ and

$$g(x_1, x_2, x_3) = u_2(x_1, u_2(x_2, x_3))$$

for $x_1, x_2, x_3 \in \mathbf{N}$. Applying the definition of universal function, we obtain

$$g(x_1, x_2, x_3) = f_{x_1}^{(1)}(f_{x_2}^{(1)}(x_3)).$$

On the other hand, an application of the *smn* property allows us to write

$$f_{s_{2,1}(i,x_1,x_2)}^{(1)}(x_3) = f_{x_1}^{(1)}(f_{x_2}^{(1)}(x_3)),$$

which shows that we can define c as

$$c(x_1, x_2) = s_{2,1}(i, x_1, x_2)$$

for $x_1, x_2 \in \mathbf{N}$. ∎

Definition 5.5.16 *Let $\mathbf{F}$ be an AFF, and let $n \in \mathbf{N}$. A set $A \subseteq \mathbf{N}^n$ is an n-ary $\mathbf{F}$-set if its characteristic function χ_A belongs to $\mathbf{F}_n$. A set A is an $\mathbf{F}$-set if it is an n-ary $\mathbf{F}$-set for some n.*

Theorem 5.5.17 *Let $\mathbf{F}$ be a standard AFF, and let $A, B \subseteq \mathbf{N}^n$ be two n-ary $\mathbf{F}$-sets. Then, $A \cup B$, $A \cap B$, and $\mathbf{N}^n - A$ are also n-ary $\mathbf{F}$-sets.*

Proof. The characteristic function of the complement $C = \mathbf{N}^n - A$ can be expressed as

$$\chi_C(x) = D(1, 0, \chi_A(x), 0),$$

which shows that C is an $\mathbf{F}$-set.

Consider the function f defined by

$$
\begin{aligned}
f(x) &= D(\chi_A(x), 0, \chi_A(x), \chi_B(x)) \\
&= \begin{cases} \chi_A(x) & \text{if } \chi_A(x) = \chi_B(x) \\ 0 & \text{otherwise,} \end{cases} \\
&= \begin{cases} 1 & \text{if } \chi_A(x) = \chi_B(x) = 1 \\ 0 & \text{otherwise} \end{cases}
\end{aligned}
$$

for every $x \in \mathbf{N}^n$. This is precisely the characteristic function of the set $A \cap B$, and this proves that $A \cap B$ is an $\mathbf{F}$-set.

$A \cup B$ is an $\mathbf{F}$-set because the intersection and the complement of two $\mathbf{F}$-sets is an $\mathbf{F}$-set. ∎

Note that, for any $n \in \mathbf{N}$, $\emptyset$ and $\mathbf{N}^n$ are n-ary $\mathbf{F}$-sets for any standard AFF, since their characteristic functions are K_0^n and K_1^n, respectively.

Theorem 5.5.18 *Consider an AFF family* $\mathbf{F} = \{f_i^{(n)} | n, i \in \mathbf{N}\}$ *and the set* $T(\mathbf{F})$, *defined by*

$$T(\mathbf{F}) = \{n \in \mathbf{N} | f_n^{(1)} \text{ is total}\}.$$

For any standard AFF $\mathbf{F}$, *the set* $T(\mathbf{F})$ *is not an* $\mathbf{F}$-*set.*

Proof. We have to prove that the characteristic function $t : \mathbf{N} \longrightarrow \mathbf{N}$ of $T\mathbf{F})$ is not a member of $\mathbf{F}_1$ for any standard AFF $\mathbf{F}$.

Consider the function $g : \mathbf{N}^2 \rightsquigarrow \mathbf{N}$ given by

$$g = K_0^1(u_2(p_1^2, p_1^2)).$$

For $p, q \in \mathbf{N}$, we have

$$g(p, q) = K_0^1(u_2(p, p)) = \begin{cases} 0 & \text{if } p \in \text{Dom}(f_p^{(1)}) \\ \text{undefined} & \text{otherwise.} \end{cases}$$

Since $\mathbf{F}$ is a standard family, we have $g \in \mathbf{F}_2$; hence, $g = f_i^{(2)}$ for some $i \in \mathbf{N}$. Using the *smn* property we have

$$g(p, q) = f_i^{(2)}(p, q) = f_{s_{1,1}(i,p)}^{(1)}(q),$$

for every $p, q \in \mathbf{N}$. Consider the total function $\ell : \mathbf{N} \longrightarrow \mathbf{N}$ given by $\ell(p) = s_{1,1}(i, p)$ for every $p \in \mathbf{N}$. We have

$$f_{\ell(p)}^{(1)} = \begin{cases} K_0^1 & \text{if } p \in \text{Dom}(f_p^{(1)}) \\ \emptyset^1 & \text{otherwise.} \end{cases}$$

Assume that the function t is a member of the family, $t \in \mathbf{F}_1$. Since $\mathbf{F}$ is closed under composition, the total function $t\ell$ is also a member of $\mathbf{F}_1$.

We claim that $t\ell = k$, where k is the function introduced in Corollary 5.5.11. Indeed, if $p \in \text{Dom}(f_p^{(1)})$, the function $f_{\ell(p)}^{(1)}$ is total (being equal to K_0^1), and therefore, $t(\ell(p)) = 1 = k(p)$. On the other hand, if $p \notin \text{Dom}(f_p^{(1)})$, then $t(\ell(p)) = 0 = k(p)$, since $f_{\ell(p)}^{(1)} = \emptyset^1$. This is a contradiction since $t\ell \in \mathbf{F}_1$ and $k \notin \mathbf{F}_1$, according to Corollary 5.5.11. ∎

For any standard AFF $\mathbf{F}$, the set $K(\mathbf{F})$, defined by

$$K(\mathbf{F}) = \{x \in \mathbf{N} \mid x \in \text{Dom}(f_x^{(1)})\}$$

is not an $\mathbf{F}$-set. This follows immediately from the fact that $\chi_{K(\mathbf{F})}$ coincides with the function k considered in Corollary 5.5.11.

In general, a function f that belongs to an AFF $\mathbf{F}$ may have many subscripts; this corresponds to the fact that a function may be computed by several programs. From the point of view of a program P whose Gödel number is i, a function $f_i^{(n)}$ describes the input-output behavior of P.

Therefore, it is interesting to examine those collections of programs whose input-output behaviors are described by functions that belong to certain families of functions.

For any AFF **F** and collection of functions $\mathcal{C} \subseteq \mathbf{F}_1$, define the *index set of* $\mathcal{C}$ as $P(\mathcal{C}) = \{i \in \mathbf{N} | f_i^{(1)} \in \mathcal{C}\}$.

Theorem 5.5.19 (Rice's Theorem) *If* **F** *is a standard AFF, and* $P(\mathcal{C})$ *is an* **F**-*set, then we have either* $P(\mathcal{C}) = \mathbf{N}$ *or* $P(\mathcal{C}) = \emptyset$.

Proof. Suppose that $\emptyset \subset P(\mathcal{C}) \subset \mathbf{N}$ and that $P(\mathcal{C})$ is an **F**-set. In this case, $\mathcal{C}$ is neither empty nor the entire set $\mathbf{F}_1$. We may assume that the empty function $\emptyset^1$ is not a member of $\mathcal{C}$; if this is not the case, we can replace $\mathcal{C}$ in the subsequent argument by $\mathbf{F}_1 - \mathcal{C}$.

Let $f_i^{(1)}$ be a function from $\mathcal{C}$. Consider the function $g : \mathbf{N}^2 \rightsquigarrow \mathbf{N}$ defined by

$$g(x_1, x_2) = D(u_2(i, x_2), 0, u_2(x_1, x_1), u_2(x_1, x_1))$$

for $x_1, x_2 \in \mathbf{N}$. Since g is a member of $\mathbf{F}_2$, we have $g = f_j^{(2)}$ for some $j \in \mathbf{N}$, and we can write

$$f_j^{(2)}(x_1, x_2) = \begin{cases} f_i^{(1)}(x_2) & \text{if } x_1 \in \text{Dom}(f_{x_1}^{(1)}) \\ \text{undefined} & \text{otherwise.} \end{cases}$$

Applying the *smn* property, we obtain

$$f_{s_{1,1}(j, x_1)}^{(1)}(x_2) = \begin{cases} f_i^{(1)}(x_2) & \text{if } x_1 \in \text{Dom}(f_{x_1}^{(1)}) \\ \text{undefined} & \text{otherwise,} \end{cases}$$

which, in turn, implies

$$f_{s_{1,1}(j, x_1)}^{(1)} = \begin{cases} f_i^{(1)} & \text{if } x_1 \in \text{Dom}(f_{x_1}^{(1)}) \\ \emptyset^1 & \text{otherwise.} \end{cases}$$

Consider the function $h : \mathbf{N} \rightsquigarrow \mathbf{N}$, defined by $h(x) = s_{1,1}(j, x)$. In view of the choice made for the function $f_i^{(1)}$, we have $h(x) \in P(\mathcal{C})$ if and only if $x \in K(\mathbf{F})$. In terms of characteristic functions this amounts to

$$\chi_{K(\mathbf{F})} = \chi_{P(\mathcal{C})} \circ h,$$

which implies that $K(\mathbf{F})$ is an **F**-set. Since we know that this is not the case, it follows that $P(\mathcal{C})$ is not an **F**-set either. ∎

Rice's theorem is a powerful tool for proving that certain sets are not **F**-sets. For instance, Theorem 5.5.18 follows as an immediate consequence (see part (1) of the next Corollary).

Corollary 5.5.20 *For every standard AFF* **F** *the following sets are not* **F**-*sets.*

1. $P_1 = \{i \in \mathbf{N} | f_i^{(1)} \text{ is total }\}$.

2. $P_2 = \{i \in \mathbf{N} | Dom(f_i^{(1)}) = \emptyset\}$.

3. $P_3 = \{i \in \mathbf{N} | Dom(f_i^{(1)}) \text{ is infinite}\}$.

4. $P_4 = \{i \in \mathbf{N} | f_i^{(1)}(x_0) = y_0\}$ for some fixed x_0, y_0 in $\mathbf{N}$.

5. $P_5 = \{i \in \mathbf{N} | f_i^{(1)} = f\}$ for some fixed function $f \in \mathbf{F}_1$.

6. $P_6 = \{i \in \mathbf{N} | y_0 \in Ran(f_i^{(1)})\}$ for some fixed $y_0 \in \mathbf{N}$.

Let $f \in \mathbf{F}_1$ be a total function. We shall prove that if $\mathbf{F}$ is a standard AFF then there exists a subscript i such that $f_i^{(1)} = f_{f(i)}^{(1)}$. Such a subscript is called a *fixed point* of f. If i is a fixed point of the function f, this means that both i and $f(i)$ denote the same function.

This result has some interesting consequences. If $\mathbf{F}$ is an AFF then we can list all functions that belong to $\mathbf{F}_1$ as

$$f_0^{(1)}, f_1^{(1)}, \ldots, f_n^{(1)}, \ldots. \tag{5.11}$$

Since $\mathbf{F}$ contains the successor function S and it is closed with respect to composition, it also contains the function $f : \mathbf{N} \longrightarrow \mathbf{N}$, defined by $f(x) = x + k$. Since f has a fixed point, it follows that there is $i \in \mathbf{N}$ such that $f_i^{(1)} = f_{f(i)}^{(1)}$ or, equivalently, that $f_i^{(1)} = f_{i+k}^{(1)}$. This shows that for each $k \geq 1$ there are identical functions located k positions apart in list (5.11).

Theorem 5.5.21 (Recursion Theorem) *Let* $\mathbf{F}$ *be a standard AFF. If* $f \in \mathbf{F}_1$ *is a total function then there is* $i \in \mathbf{N}$ *such that* $f_i^{(1)} = f_{f(i)}^{(1)}$.

Proof. Consider the function $g \in \mathbf{F}_2$ defined by

$$g(x, y) = u_2(u_2(x, x), y)$$

for $x, y \in \mathbf{N}$. We have

$$g(x, y) = \begin{cases} f_{f_x^{(1)}(x)}^{(1)}(y) & \text{if } f_x^{(1)}(x) \text{ is defined} \\ \text{undefined} & \text{otherwise.} \end{cases}$$

If $g(x, y) = f_r^{(2)}(x, y)$, then, applying the *smn* property, we have $g(x, y) = f_{s_{1,1}(r,x)}^{(1)}(y)$. Let $l \in \mathbf{F}_1$ be the total function determined by

$$l(x) = s_{1,1}(r, x)$$

for every $x \in \mathbf{N}$. We have

$$f_{l(x)}^{(1)} = \begin{cases} f_{f_x^{(1)}(x)}^{(1)} & \text{if } f_x^{(1)}(x) \text{ is defin ed} \\ \emptyset^1 & \text{otherwise.} \end{cases}$$

Note that the function $h = f \circ l$ is total since both f and l are total. We write $h = f_m^{(1)}$, since $h \in \mathbf{F}_1$. Let $n = l(m)$. We claim that n is a fixed point of f.

Indeed, we have

$$f_n^{(1)} = f_{l(m)}^{(1)} = f_{f_m^{(1)}(m)}^{(1)}$$

since $f_m^{(1)}(m) = h(m)$ is defined. Furthermore, in view of the definition of h, we have

$$f_n^{(1)} = f_{h(m)}^{(1)} = f_{f(l(m))}^{(1)} = f_{f(n)}^{(1)},$$

which proves that n is a fixed point for f. ∎

5.6 Exercises and Supplements

Equinumerous Sets

1. Let M and P be sets such that $M \preceq P$. Prove that

   (a) $\mathcal{P}(M) \preceq \mathcal{P}(P)$;

   (b) $\mathcal{M}(M) \preceq \mathcal{M}(P)$ (here $\mathcal{M}(M)$ is the collection of all multisets over M as introduced in Chapter 1);

   (c) $M \times Q \preceq P \times Q$ for all sets Q;

   (d) $M \cup Q \preceq P \cup Q$ for all sets Q such that $M \cap Q = P \cap Q = \emptyset$.

   Conclude that if $M \sim P$, then $\mathcal{P}(M) \sim \mathcal{P}(P)$, $\mathcal{M}(M) \sim \mathcal{M}(P)$, $M \times Q \sim P \times Q$ for all sets Q, and $M \cup Q \sim P \cup Q$ for all sets Q such that $M \cap Q = P \cap Q = \emptyset$.

2. Using the axiom of choice, show that if ρ is an equivalence on a set M, then $\{[x]_\rho \mid x \in M\} \preceq M$.

3. Without using the axiom of choice, show that if M and P are sets such that $\mathbf{N} \preceq M \cup P$, then either $\mathbf{N} \preceq M$ or $\mathbf{N} \preceq P$.

4. Using the axiom of choice, show that if M is a nonempty set and P is any set, then $M \preceq P$ if and only if there is a surjection $f : P \longrightarrow M$.

   (Theorem 5.3.3 is a special case of this exercise, which can be proven without the axiom of choice.)

   Hint: Use Theorem 2.3.11.

5. Let M be a set, n a natural number and $(M_0, \ldots, M_{n-1})$ be a sequence of n sets. Show that

   (a) $\mathrm{Seq}(M)$ is finite if and only if $M = \emptyset$.

   (b) $\mathcal{P}_{\mathrm{fin}}(M)$ is finite if and only if M is finite.

(c) $M_0 \times \cdots \times M_{n-1}$ is finite if and only if either

    i. there is an i, $0 \leq i \leq n - 1$, such that $M_i = \emptyset$, or

    ii. for every i, $0 \leq i \leq n - 1$, M_i is finite.

(d) M^n is finite if and only if either $n = 0$ or M is finite.

Hint: You may find Theorem 5.2.9 useful for showing that certain sets are infinite. For example, to show that $\mathrm{Seq}(M)$ is infinite when M is nonempty, let a be some element of M, and define an injection $f : \mathbf{N} \longrightarrow \mathrm{Seq}(M)$ by defining $f(n)$ to be the sequence of length n each of whose entries equals a. Since $\mathbf{N}$ is infinite, Theorem 5.2.9 shows that $\mathrm{Seq}(M)$ is infinite.

6. Using the axiom of choice, generalize the third part of the previous exercise by showing that for any family of sets $\{M_i \mid i \in I\}$ indexed by an arbitrary set I, $\prod_{i \in I} M_i$ is finite if and only if either

    (a) there is an $i \in I$ such that $M_i = \emptyset$, or

    (b) there is a finite subset J of I such that for every $j \in J$, M_j is finite, and for every $i \in I - J$, M_i is a singleton.

7. Show that if $\mathcal{C}$ is a collection of sets, then $\bigcup \mathcal{C}$ is finite if and only if $\mathcal{C}$ is finite and every element of $\mathcal{C}$ is finite.

8. Consider a sequence of sets $M_0, \ldots, M_{n-1}$ such that $M_k \preceq M_{k+1}$ for $0 \leq k \leq n - 2$ and $M_{n-1} \preceq M_0$. Prove that $M_p \sim M_q$ for $0 \leq p, q \leq n - 1$.

9. Let M, P, f, g and Φ be as in Theorem 5.2.11 and its proof.

    (a) Prove that we have $\Phi(M) = M$ if and only if f is a surjection. Also, $\Phi(\emptyset) = \emptyset$ if and only if g is a surjection.

    (b) Prove that if K and H are fixed points of Φ, then so are $K \cup H$ and $K \cap H$.

10. Let M and P be two sets, and let $f : M \longrightarrow P$ and $g : P \longrightarrow M$ be two injections. Show that for every subset K of M, $M - K \subseteq \mathrm{Ran}(g)$. Also, the mapping $h : M \longrightarrow P$ defined by

$$h(a) = \begin{cases} f(a) & \text{if } a \in K \\ g^{-1}(a) & \text{if } a \in M - K, \end{cases}$$

is a bijection if and only if K is a fixed point of the mapping Φ defined in the proof of Theorem 5.2.11.

Hint: One half of the argument is contained in the proof of Theorem 5.2.11.

11. Let M and P be two sets and let $f : M \longrightarrow P$ and $g : P \longrightarrow M$ be two injective mappings. For each $a \in M$ we define a sequence s^a as follows.

$$s_0^a = a,$$

$$s_{n+1}^a = \begin{cases} g^{-1}(s_n^a) & \text{if } s_n^a \text{ is defined, } n \text{ is even} \\ & \text{and } s_n^a \in \text{Ran}(g) \\ f^{-1}(s_n^a) & \text{if } s_n^a \text{ is defined, } n \text{ is odd,} \\ & \text{and } s_n^a \in \text{Ran}(f) \\ \text{undefined} & \text{otherwise.} \end{cases}$$

Similarly, for each $b \in P$ we define a sequence t^b as follows.

$$t_0^b = b,$$

$$t_{n+1}^b = \begin{cases} f^{-1}(t_n^b) & \text{if } t_n^b \text{ is defined, } n \text{ is even} \\ & \text{and } t_n^b \in \text{Ran}(f) \\ g^{-1}(t_n^b) & \text{if } t_n^b \text{ is defined, } n \text{ is odd,} \\ & \text{and } t_n^b \in \text{Ran}(g) \\ \text{undefined} & \text{otherwise.} \end{cases}$$

(a) Prove that for all $a \in M$, $t^{f(a)} = (a) \cdot s^a$; similarly, for all $b \in P$, $s^{g(b)} = (b) \cdot t^b$.

Let M_M, M_P and M_∞ be defined as

$$\begin{aligned} M_M &= \{a \in M \,|\, s^a \text{ is of even finite length }\}, \\ M_P &= \{a \in M \,|\, s^a \text{ is of odd finite length }\}, \\ M_\infty &= \{a \in M \,|\, s^a \text{ is infinite }\}. \end{aligned}$$

Similarly, P_P, P_M, P_∞ are introduced as

$$\begin{aligned} P_P &= \{b \in P \,|\, t^b \text{ is of even finite length }\}, \\ P_M &= \{b \in P \,|\, t^b \text{ is of odd finite length }\}, \\ P_\infty &= \{b \in P \,|\, t^b \text{ is infinite }\}. \end{aligned}$$

(b) Prove that $f {\upharpoonright} M_M$ is a bijection between M_M and P_M, $f {\upharpoonright} M_\infty$ is a bijection between M_∞ and P_∞, $g {\upharpoonright} P_P$ is a bijection between P_P and M_P, and $g {\upharpoonright} P_\infty$ is a bijection between P_∞ and M_∞.

(Thus, if $K = M_M$ or $K = M_M \cup M_\infty$ and $h : M \longrightarrow P$ is defined as in Exercise 10, then h is a bijection. This gives another proof of the Schröder-Bernstein Theorem where we explicitly construct two fixed points of the mapping Φ as defined in our original proof.)

12. (a) Let $f : M \longrightarrow P$ and $g : P \longrightarrow M$ be injections, and let $(Q_0, \ldots, Q_n, \ldots)$ be the infinite sequence of subsets of M defined recursively by $Q_0 = M - \mathrm{Ran}(g)$ and $Q_{n+1} = g(f(Q_n))$, for $n \in \mathbf{N}$. Prove that, in the notation of the previous exercise, $M_M = \bigcup_{n \in \mathbf{N}} Q_n$.

(b) Show directly (i.e., without referring to the previous exercise) that if K is defined to be the union of the Q_n's and h is defined as in Exercise 10, then h is a bijection.

(Motivation for this proof of Theorem 5.2.11 can be given as follows. As a first attempt to construct h, we might let it equal g^{-1} wherever we can (i.e., on $g(P)$) and let it equal f elsewhere (i.e., on $Q_0 = M - g(P)$). This suggested h will not be an injection because every element of $f(Q_0)$ will be the image under h of an element of Q_0 and also of an element of $Q_1 = g(f(Q_0))$. We can remedy this problem by redefining h on Q_1 to be the same as f, but now elements of $f(Q_1)$ are the image under h of an element of Q_1 and also of $Q_2 = g(f(Q_1))$. We fix this by redefining h on Q_2 and so on. We are led to let K be the union of the Q_n's and define $h \upharpoonright K = f \upharpoonright K$ and $h \upharpoonright (M - K) = g^{-1} \upharpoonright (M - K)$.)

13. Let $f : M \longrightarrow P$ and $g : P \longrightarrow M$ be injections and let $\Phi : \mathcal{P}(M) \longrightarrow \mathcal{P}(M)$ be, as in the proof of Theorem 5.2.11.

(a) Prove that Φ is continuous (and not just monotonic as claimed in the proof of Theorem 5.2.11).

(b) According to the previous part, the least fixed point of Φ can be obtained as $\bigcup_{n \in \mathbf{N}} \Phi^n(\emptyset)$. Prove, by induction on n, that for every $n \in \mathbf{N}$

$$\Phi^n(\emptyset) = \bigcup_{i < n} Q_i$$

where the sets Q_i are as defined in the previous exercise.

Conclude that if we take the least fixed point of Φ in our proof of the Schröder-Bernstein Theorem, then we get the same bijection from M to P that was constructed in the last two exercises.

14. Consider the following statement:

$$\text{If } Q \subseteq P \subseteq M \text{ and } M \sim Q \text{ then } M \sim P. \tag{5.12}$$

(a) Show how to derive (5.12) easily from Theorem 5.2.11.

(b) Conversely, show how Theorem 5.2.11 can be derived easily from (5.12).

Hint: If $f : M \longrightarrow P$ and $g : P \longrightarrow M$ are injections, then $g(f(M)) \subseteq g(P) \subseteq M$ and $M \sim g(f(M))$ via gf.

(c) We can prove (5.12) as follows. If $Q \subseteq P \subseteq M$ and $k : M \longrightarrow Q$ is a bijection, define a sequence of sets recursively by $D_0 = M - P$ and $D_{n+1} = k(D_n)$ for $n \in \mathbf{N}$. Let $K = \bigcup_{n \in \mathbf{N}} D_n$, and define l by

$$l(a) = \begin{cases} k(a) & \text{if } a \in K \\ a & \text{if } a \in M - K. \end{cases}$$

Show that l is a bijection from M to P.

(d) Let $f : M \longrightarrow P$ and $g : P \longrightarrow M$ be injections. Then, $g(f(M)) \subseteq g(P) \subseteq M$ and $k : M \longrightarrow g(f(M))$ is a bijection when $k = gf$. Let $l : M \longrightarrow g(P)$ be the bijection which results from the above proof of (5.12). Then, $h = g^{-1}l$ is a bijection from M to P. Show that h is the same bijection that was constructed in the last three exercises.

15. (a) Without using the axiom of choice, show by induction that if M is an infinite set, then for all $n \in \mathbf{N}$, $\{0, \ldots, n-1\} \preceq M$ (and hence, for all finite sets P, $P \preceq M$).

   (b) Explain why the argument you gave in (a) cannot be used to prove Theorem 5.2.14 without the axiom of choice.

16. Show that in the proof of Theorem 5.2.16, when the explicitly defined selective function f for M is used, the resulting injection $g : \mathbf{N} \longrightarrow M$ is in fact a bijection (and hence, the final paragraph of the proof is actually unneccessary).

Countable and Uncountable Sets

17. Let M be a countable set.

   (a) Let $g : M \longrightarrow \mathbf{N}$ be a numbering for M. Define a relation ρ_g on M by $\rho_g = \{(a, b) \in M^2 \mid g(a) \le g(b)\}$. Prove that ρ_g is a linear ordering on M such that every element of M has only finitely many predecessors under ρ_g.

   (b) Conversely, suppose that ρ is a linear ordering on M such that each element of M has only finitely many predecessors under ρ. Define $g_\rho : M \longrightarrow \mathbf{N}$ by

$$g_\rho(a) = |\{b \in M \mid b \rho a \text{ and } b \ne a\}|$$

for all $a \in M$. Show that g_ρ is a numbering of M.

Let NUMB_M be the set of all numberings of M and let L_M be the set of all linear orderings on M such that each element of M has only finitely many predecessors under the ordering. Define $\Phi : \text{NUMB}_M \longrightarrow L_M$ by $\Phi(g) = \rho_g$ for all $g \in \text{NUMB}_M$, and define $\Psi : L_M \longrightarrow \text{NUMB}_M$ by $\Psi(\rho) = g_\rho$ for all $\rho \in L_M$.

(c) Show that Φ is a left inverse for Ψ (i.e., $\Phi(\Psi(\rho)) = \rho$ for all $\rho \in L_M$), but that in general Φ is not a right inverse for Ψ. (Thus, Φ is onto but not one-to-one and Ψ is one-to-one but not onto.)

(d) Show that the range of Ψ is the set of numberings g of M with the following property: If $0 \leq n \leq m$ and $m \in g(M)$, then $n \in g(M)$ (i.e., there are no gaps in the range of g). In particular, if M is countably infinite, then the range of Ψ is the set of bijective numberings of M.

It follows that if we restrict our attention to numberings with the property just given, then the two processes of the first two parts of the exercise are inverse to each other.

(e) Suppose that M is countably infinite. Let g be a numbering of M, and let g' be the bijective numbering of M obtained from g by using the method of page 347. Show that $g' = \Psi(\Phi(g))$.

Thus, g' is the unique bijective numbering whose induced ordering $\rho_{g'}$ is the same as the ordering ρ_g induced by the original numbering g.

18. Let $g : \mathrm{Seq}(\mathbf{N}) \longrightarrow \mathbf{N}$ be the function defined in Theorem 5.3.7. Give a simple characterization in terms of the prime factorization of a number which determines whether or not the number is in the range of g.

19. Let g be as in the previous exercise, and let g' be the bijective numbering which results from applying the process described on page 347. What is $g'((2, 1, 0))$?

20. Define a function $f : \mathrm{Seq}(\mathbf{N}) \longrightarrow \mathbf{N}$ by

$$ f((a_0, \ldots, a_{n-1})) = p_0^{a_0} \cdots p_{n-1}^{a_{n-1}} - 1 $$

for each sequence $(a_0, \ldots, a_{n-1})$ in $\mathrm{Seq}(\mathbf{N})$ where $(p_0, p_1, \ldots)$ is the sequence of prime natural numbers listed in increasing order.

(a) Show that f is a surjection, but not an injection.

(b) Show that for each $n \in \mathbf{N}$, the restriction of f to $\mathbf{N}^n$ is one-to-one.

(c) Let T be the subset of $\mathrm{Seq}(\mathbf{N})$ consisting of those sequences which do not end with a 0. Show that the restriction of f to T is a bijection from T to $\mathbf{N}$.

21. Prove that the following sets are countable:

(a) the set of one-variable polynomials with rational coefficients;

(b) the set of algebraic numbers;[2]

(c) the set of all formulas of propositional logic;

(d) the set of all partial recursive functions.

22. Let K be a subset of $\mathbf{R}$ that is well-ordered under the usual order relation. Prove that K is countable.

    Hint: We define a function $f : K \longrightarrow \mathbf{Q}$. Let $x \in K$, and consider the set $G_x = \{y \in K \mid x < y\}$. If $G_x \neq \emptyset$, then G_x has a least element y_x, and we can let $f(x)$ be a rational number in (x, y_x). (For definiteness, choose $f(x)$ with smallest possible denominator, and if there are several candidates with the same denominator, choose the one with the smallest numerator, when written in lowest terms.) If $G_x = \emptyset$, then x is the greatest element of K, and we can define $f(x)$ as an arbitrary rational number greater than x. You need to show that f is one-to-one.

23. Let M be a set such that there exists a linear ordering on M. Without using the axiom of choice, show that the union of any countable collection of finite subsets of M is countable.

    Conclude without using the axiom of choice that the union of any countable collection of finite sets of real numbers is countable.

24. Consider a set of positive real numbers K. Suppose that there a number $a \in \mathbf{R}$ such that for any finite subset J of K, the sum of the numbers in J is less than or equal to a. Prove that K is countable.

    Solution: Consider the countable collection of sets $\{Q_n \mid n \in \mathbf{N}\}$, where $Q_n = \{x \in \mathbf{R} \mid \frac{1}{2^{n+1}} < x \leq \frac{1}{2^n}\}$ for $n \geq 0$ and the set $P = \{x \in \mathbf{R} \mid x > 1\}$. Note that none of these sets can contain more than a finite number of elements of K. Indeed, if $U_n = Q_n \cap K$ would be an infinite set it would suffice to consider a set containing $2^{n+2}\lceil a \rceil$ elements of U_n. The sum of these elements would exceed

    $$\frac{1}{2^{n+1}} 2^{n+2} \lceil a \rceil \geq 2 \lceil a \rceil \geq a,$$

    which would contradict our assumption about K. Consequently, K is a countable union of finite sets of reals and, therefore, it is countable.

25. Show that if M and P are countable sets, then so is $M \oplus P$.

26. Let M and P be two sets such that M is countable and P is countably infinite. Prove that $M \cup P \sim P$.

---

[2] An algebraic number is a real number that is the root of some polynomial with integer coefficients.

27. Prove that every countably infinite set can be obtained as the union of a countably infinite disjoint family of countably infinite sets.

28. Let $V = \{a_1, \ldots, a_k\}$ be an alphabet, and let $h : V^* \longrightarrow \mathbf{N}$ be the function defined in Example 5.3.12. For each $m \in \mathbf{N}$, define $I_m = \{n \in \mathbf{N} \mid \frac{k^m - 1}{k - 1} \leq n < \frac{k^{m+1} - 1}{k - 1}\}$.

    (a) Show that $\{I_m \mid m \in \mathbf{N}\}$ is a partition of $\mathbf{N}$.

    (b) Show that for each $m \in \mathbf{N}$, the restriction of h to V^m is a bijection between V^m and I_m.

    Hint: First argue that the restriction maps into I_m and that it is one-to-one, then use a counting argument to show that it is onto.

    (c) Conclude from the first two parts that h is a bijection.

29. Let $V = \{a_0, \ldots, a_{k-1}\}$ be an alphabet. Define $f : V^* \longrightarrow \mathbf{N}$ by

    $$f(a_{i_0} \ldots a_{i_{m-1}}) = i_0 k^{m-1} + i_1 k_{m-2} + \cdots + i_{m-1}.$$

    (Note that here the indices i_j vary from 0 to $k - 1$, not 1 to k.) Prove that f is a surjection but is not an injection. Show however that for all $m \in \mathbf{N}$, the restriction of f to V^m is an injection.

30. Let $\mathrm{ASeq}(\mathbf{N})$ and $\mathrm{NDSeq}(\mathbf{N})$ be the set of ascending sequences and the set of nondescending sequences over $\mathbf{N}$, respectively. A sequence $(a_0, \ldots, a_{n-1})$ is in $\mathrm{ASeq}(\mathbf{N})$ if $a_0 < a_2 < \cdots < a_{n-1}$ and it is in $\mathrm{NDSeq}(\mathbf{N})$ if $a_0 \leq a_1 \leq \cdots \leq a_{n-1}$.

    Prove by giving explicit bijections that

    (a) $\mathrm{Seq}(\mathbf{N}) \sim \mathrm{NDSeq}(\mathbf{N})$,

    (b) $\mathrm{NDSeq}(\mathbf{N}) \sim \mathrm{ASeq}(\mathbf{N})$,

    (c) $\mathrm{ASeq}(\mathbf{N}) \sim \mathcal{P}_{\mathrm{fin}}(\mathbf{N})$,

    (d) $\mathcal{P}_{\mathrm{fin}}(\mathbf{N}) \sim \mathbf{N}$.

    Solution: (a) Consider the mapping $f : \mathrm{Seq}(\mathbf{N}) \longrightarrow \mathrm{NDSeq}(\mathbf{N})$ given by $f(a_0, \ldots, a_{n-1}) = (a_0, a_0 + a_1, \ldots, a_0 + a_1 + \cdots + a_{n-1})$ for every sequence $(a_0, \ldots, a_{n-1})$. Note that $f(\lambda) = \lambda$. It is clear that $(a_0, a_0 + a_1, \ldots, a_0 + a_1 + \cdots + a_{n-1})$ is a nondescending sequence and that f is an injection. Morever, if $(b_0, \ldots, b_{n-1}) \in \mathrm{NDSeq}(\mathbf{N})$, we have $f(a_0, \ldots, a_{n-1}) = (b_0, \ldots, b_{n-1})$, where $a_0 = b_0$ and $a_i = b_i - b_{i-1}$ for $1 \leq i \leq n - 1$, which shows that f is a bijection.

    (b) Let $(a_0, \ldots, a_{n-1})$ be a nondescending sequence. Define $h(a_0, a_1, \ldots, a_{n-1})$ to be $(b_0, b_1, \ldots, b_{n-1})$, where $b_j = a_j + j$ for $0 \leq j \leq n - 1$. Since $b_j - b_{j-1} = a_j - a_{j-1} + 1 \geq 1$, we have $b_j > b_{j-1}$ for $1 \leq j \leq n - 1$. This shows that $(b_0, b_1, \ldots, b_{n-1})$ is indeed an ascending sequence. Since $a_j = b_j - j$, it follows that h is a one-to-one mapping. Furthermore, observe that if $(b_0, b_1, \ldots, b_{n-1})$ is an ascending sequence, then $b_j \geq j$ for $0 \leq j \leq n - 1$. This allows us to define a sequence of natural numbers $(a_0, \ldots, a_{n-1})$ by $a_j = b_j - j$ for $0 \leq j \leq n - 1$. Note that $a_j - a_{j-1} = b_j - b_{j-1} - 1 \geq 0$ (because of the fact that $b_j > b_{j-1}$); hence, $a_j \geq a_{j-1}$ and $(a_0, \ldots, a_{n-1})$ is a nondescending sequence with $h(a_0, \ldots, a_{n-1}) = (b_0, \ldots, b_{n-1})$. This proves that h is onto, and, therefore, it is the desired bijection.

(c) Define $k : \text{ASeq}(\mathbf{N}) \longrightarrow \mathcal{P}_{\text{fin}}(\mathbf{N})$ by letting $k(a_0, \ldots, a_{n-1}) = \{a_0, \ldots, a_{n-1}\}$ for every ascending sequence $(a_0, \ldots, a_{n-1})$. Then, $n = |k(a_0, \ldots, a_{n-1})|$, and for each i, $0 \leq i \leq n - 1$, a_i is the $i + 1$st largest element of $k(a_0, \ldots, a_{n-1})$ so the sequence $(a_0, \ldots, a_{n-1})$ can be recovered from the value of k on that sequence, which shows that k is injective. If $K \in \mathcal{P}_{\text{fin}}(\mathbf{N})$ then $K = k(a_0, \ldots, a_{n-1})$ where $(a_0, \ldots, a_{n-1})$ is the sequence listing the elements of K in increasing order, so k is onto.

(d) Define $l : \mathcal{P}_{\text{fin}}(\mathbf{N}) \longrightarrow \mathbf{N}$ by $l(S) = 2^{a_0} + \cdots + 2^{a_{n-1}}$ where $(a_0, \ldots, a_{n-1})$ is the sequence listing the elements of S in increasing order. We have $l(\emptyset) = 0$, and if we have $l(S) = x$ for some $S \in \mathcal{P}_{\text{fin}}(\mathbf{N})$ and $x \in \mathbf{N}$, then $2x = l(\{a + 1 \mid a \in S\})$ and $2x + 1 = l(\{a + 1 \mid a \in S\} \cup \{0\})$. It follows from this that l is onto. If $l(S) = 0$, then $S = \emptyset$. If $l(S) = x > 0$, then the least element a_0 of S is the greatest number z such that 2^z divides x evenly. If $2^{a_0} = x$, then $S = \{a_0\}$. Otherwise, we recover a_1, the second largest element of S, as being the largest number z such that 2^z divides $x - 2^{a_0}$ evenly and so forth. These considerations (made precise by an induction proof) show that l is one-to-one.

31. Determine the pairing functions $f : \mathbf{N}^2 \longrightarrow \mathbf{N}$ for the numbering mechanisms suggested by Figure 5.2.

32. Let k be a natural number which is greater than 1. We generalize the pairing function of Example 5.3.15 as follows. Let $p_0, \ldots, p_{k-2}$ be the first $k - 1$ primes listed in increasing order, and let $(s_0, s_1, \ldots)$ be the sequence of positive integers which are relatively prime to $p_0 \cdots p_{k-2}$ listed in increasing order. Define $g_k : \mathbf{N}^k \longrightarrow \mathbf{N}$ by

$$g_k((a_0, \ldots, a_{k-1})) = p_0^{a_0} p_1^{a_1} \cdots p_{k-2}^{a_{k-2}} s_{a_{k-1}} - 1$$

for all $(a_0, \ldots, a_{k-1}) \in \mathbf{N}^k$.

(a) Show that g_2 is the pairing function f defined in Example 5.3.15.

(b) Show that each g_k is a bijection.

33. In this exercise, we generalize the pairing function of Example 5.3.16. Let $\leq_L$ denote lexicographic order on $\text{Seq}(\mathbf{N})$. For each $k \geq 2$, we

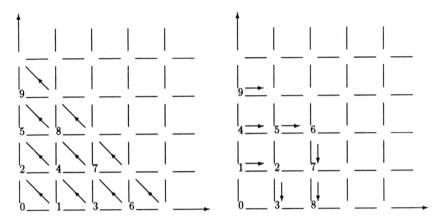

FIGURE 5.2. Other definitions of pairing functions.

define a relation ρ_k on $\mathbf{N}^k$. If $a = (a_0, \ldots, a_{k-1})$ and $b = (b_0, \ldots, b_{k-1})$ are two elements of $\mathbf{N}^k$, then we define $a\rho_k b$ to hold if and only if either

$$\sum_{i=0}^{k-1} a_i < \sum_{i=0}^{k-1} b_i$$

or

$$\sum_{i=0}^{k-1} a_i = \sum_{i=0}^{k-1} b_i \text{ and } a \leq_L b.$$

(a) Show that for each $k \geq 2$, ρ_k is a linear order on $\mathbf{N}^k$ such that each element of $\mathbf{N}^k$ has only finitely many predecessors.

For each $k \geq 2$, let g_k be the bijective numbering of $\mathbf{N}^k$ obtained from ρ_k as in Exercise 17.

(b) Show that g_2 is the pairing function defined in Example 5.3.16.

34. Let $S = \bigcup_{k \geq 1}(\mathbf{N}^k \longrightarrow \mathbf{N})$ and let f be a pairing function. Define a function $H_f : S \longrightarrow S$ such that if $\{f_k \mid k \geq 1\}$ are the functions associated with f then we have

$$f_1 = 1_{\mathbf{N}},$$
$$f_{k+1} = H_f(f_k),$$

for all $k \geq 1$.

35. Prove the following properties of pairing functions and pairing triples:

(a) The pairing function introduced in Example 5.3.16 is primitive recursive.

(b) If $f : \mathbf{N}^2 \longrightarrow \mathbf{N}$ is primitive recursive then so are each of the functions f_k introduced in Example 5.3.22.

(c) If $(f, \mathbf{p}_1, \mathbf{p}_2)$ is a monotonic pairing triple and f is primitive recursive then both $\mathbf{p}_1$ and $\mathbf{p}_2$ are primitive recursive. We refer to such a triple as a monotonic primitive recursive pairing triple or, for short, as an *mprp-triple*.

Hint: Observe that for a monotonic primitive recursive triple we can write

$$\mathbf{p}_1(n) = (\mu m \le n)[\mathbf{EXISTS}_{p=0}^{n} f(m, p) = n]$$

for $n \in \mathbf{N}$, which shows that $\mathbf{p}_1$ is primitive recursive.

36. Let $f : \mathbf{N}^2 \longrightarrow \mathbf{N}$ be a pairing function and let $\{f_n \mid n \ge 1\}$ be the family of functions generated by f. Prove that

$$f_{k+i}(x_1, \ldots, x_{k+i}) = f_k(x_1, \ldots, x_{k-1}, f_{i+1}(x_k, \ldots, x_{k+i})),$$

for every $k, i \in \mathbf{P}$.

Solution: The argument is by induction on i. For $i = 1$ we retrieve equality (5.3) from Example 5.3.22. Let us assume now that the equality holds for i. We have

$$
\begin{aligned}
&f_{k+i+1}(x_1, \ldots, x_{k+i+1}) \\
&= f_{k+i}(x_1, \ldots, x_{k+i-1}, f(x_{k+i}, x_{k+i+1})) \\
&= f_k(x_1, \ldots, x_{k-1}, f_{i+1}(x_k, \ldots, f(x_{k+i}, x_{k+i+1}))) \\
&= f_k(x_1, \ldots, x_{k-1}, f_{i+2}(x_k, \ldots, x_{k+i}, x_{k+i+1})),
\end{aligned}
$$

which concludes the argument.

37. Let $\{f_k \mid k \in \mathbf{P}\}$ be the collection of functions associated with a pairing function f.

(a) Show that if $f = f_2$ is monotonic then so is f_k for any $k \ge 3$.

(b) Show that if f_2 is monotonic and $1 \le m < k$ then

$$f_k(x_1, \ldots, x_m, 0, \ldots, 0) = f_{m+1}(x_1, \ldots, x_m, 0),$$

for every $x_1, \ldots, x_m \in \mathbf{N}$.

Solution: We give a solution for the second part of the problem. The argument is by induction on k.

For $k = 2$, we have $m = 1$ and the equality is trivially true. Assume that the equality holds for $k \ge 2$ and consider $m \in \mathbf{P}$ such that

$m < k + 1$. We distinguish two cases: $m < k$ and $m = k$. In the first case, we have

$$
\begin{aligned}
f_{k+1}(x_1, \ldots, x_m, 0, \ldots, 0) &= f_k(x_1, \ldots, x_m, 0, \ldots, f_2(0,0)) \\
&= f_k(x_1, \ldots, x_m, 0, \ldots, 0) \\
&= f_{m+1}(x_1, \ldots, x_m, 0),
\end{aligned}
$$

where the second equality follows from the monotonicity of f_2. In the second case, we have a trivial equality.

38. Give a recursive definition of the uniform projection function introduced in Definition 5.3.23. Prove that if (f, p_1, p_2) is an mprp triple then π is a primitive recursive function.

Hint: Prove that for $k \geq 1$:

$$
\pi(i, k+1, n) = \begin{cases}
\pi(i, k, n) & \text{if } i < k, \\
p_1(\pi(k, k, n)) & \text{if } i = k \\
p_2(\pi(k, k, n)) & \text{if } i > k.
\end{cases}
$$

The primitive recursiveness of π follows from Supplement 71 of Chapter 4, Definition 5.3.23, and the above recursive definition.

39. Prove the following properties of the uniform projection function π corresponding to an mprp triple (f, p_1, p_2):

   (a) For any $i, k, n \in \mathbf{N}$, $\pi(i, k, n) \leq n$.

   (b) Let $\ell : \mathbf{N}^2 \longrightarrow \mathbf{N}$ be the function defined by

$$
\ell(k, n) = f_{k+1}(n, \ldots, n),
$$

   for every $k, n \in \mathbf{N}$. Prove that ℓ is primitive recursive and that $\pi(i, k+1, \ell(k, n)) = n$ for every $i, k, n \in \mathbf{N}$.

   Hint: The primitive recursiveness of ℓ follows from the existence of the following recursive definition:

$$
\begin{aligned}
\ell(0, n) &= n, \\
\ell(k+1, n) &= f(n, \ell(k, n)).
\end{aligned}
$$

40. Let $L : \mathbf{N} \longrightarrow \mathbf{N}$ be the function defined by $L(n) = \ell(n, n)$ for every $n \in \mathbf{N}$, where ℓ is the function defined in Exercise 39. If $s, t, r \in \mathrm{Seq}(\mathbf{N})$ and r is the concatenation of s and t, prove that

$$
\mathrm{gnseq}(r) \leq L(\mathrm{gnseq}(s) + \mathrm{gnseq}(t)).
$$

Solution: If both s and t are the null sequence then so is r and the inequality is trivially true. Therefore, we assume that at least one of

s and t is not null. Let $u = (u_0, \ldots, u_{n-1}) \in \text{Seq}(\mathbf{N})$. If $p = \text{gnseq}(u)$, the properties of mprp triples allow us to infer that $p \geq n$ and also that $p \geq u_j$ for every j, $0 \leq j \leq n - 1$. This allows us to write for $s = (s_0, \ldots, s_{n-1})$ and $t = (t_0, \ldots, t_{m-1})$, where $p = \text{gnseq}(s)$ and $q = \text{gnseq}(t)$:

$$\begin{aligned}
\text{gnseq}(r) &= f(n + m, f_{n+m}(s_0, \ldots, s_{n-1}, t_0, \ldots, t_{m-1})) \\
&= f_{n+m+1}(n + m, s_0, \ldots, s_{n-1}, t_0, \ldots, t_{m-1}) \\
&\leq f_{n+m+1}(p + q, \ldots, p + q) \\
&\leq f_{p+q+1}(p + q, \ldots, p + q) = L(p + q).
\end{aligned}$$

41. Let GSEQ be the set of all Gödel numbers for sequences of natural numbers:

$$\text{GSEQ} = \{n \in \mathbf{N} \mid n = \text{gnseq}(s) \text{ for some } s \in \text{Seq}(\mathbf{N})\}.$$

Prove that GSEQ is a primitive recursive set.

Solution: We have $n \in \text{GSEQ}$ if $p_1(n) = 0$ implies $n = 0$. Consequently, we can write $\chi_{GSEQ}(n) = f_{\to}(f_=(p_1(n), 0), f_=(n, 0))$ for every $n \in \mathbf{N}$, which shows that χ_{GSEQ} is primitive recursive.

42. Prove that there is a primitive recursive function catseq : $\mathbf{N}^2 \longrightarrow \mathbf{N}$ such that if $s, t \in \text{Seq}(\mathbf{N})$ and $\text{gnseq}(s) = p$, $\text{gnseq}(t) = q$ then $\text{catseq}(p, q)$ is the Gödel number of the concatenation of s and t.

Hint: We can define catseq using bounded minimization:

$$\begin{aligned}
\text{catseq}(p, q) = \ &(\mu r \leq L(p + q))[\chi_{GSEQ}(r) = 1 \text{ and} \\
&p_1(r) = p_1(p) + p_1(q) \text{ and} \\
&\text{FORALL}_{j=0}^{p_1(p)-1}(\pi(j + 1, p_1(p), p_2(p)) \\
&= \pi(j + 1, p_1(r), p_2(r))) \text{ and} \\
&\text{FORALL}_{j=0}^{p_1(q)-1}(\pi(j + 1, p_1(q), p_2(q)) \\
&= \pi(j + 1 + p_1(p), p_1(r), p_2(r)))].
\end{aligned}$$

43. Using Theorem 5.2.14 (and hence, the axiom of choice), prove that for any set M, M is uncountable if and only if $\mathbf{N} \prec M$.

44. Let M be a set. Prove that if $\mathcal{P}(M)$ is infinite then it is uncountable.

45. Prove that $(0, 1)$ is uncountable. Conclude that every open interval (c, d) of the set $\mathbf{R}$ of real numbers is uncountable and $\mathbf{R}$ is also uncountable.

Solution: We know that the set $\text{ISeq}(\{0, 1\})$ of infinite sequences over $\{0, 1\}$ is uncountable. Let us consider the set S that consists of the following infinite sequences:

(a) the sequences s_0, s_1 where $s_0(n) = 0$ and $s_1(n) = 1$ for every $n \in \mathbf{N}$;

(b) any infinite sequence s for which there is a number n_s such that $s(n_s) = 1$ and $s(m) = 0$ for every $m > n_s$.

S is clearly countable. Therefore, $\text{ISeq}(\{0,1\}) - S$ is uncountable (because of Corollary 5.3.26).

We prove that $(0,1) \sim \text{ISeq}(\{0,1\}) - S$ by giving a bijection $f : (0,1) \longrightarrow \text{ISeq}(\{0,1\}) - S$. If $x \in (0,1)$ then x has a unique infinite binary expansion $x = 0.x_0 x_1 \ldots$ such that for every $k \in \mathbf{N}$ there is $h > k$ with $x_h = 1$. For example, 0.5 can be written in binary as $0.100 \ldots$. This, of course, does not qualify as an expansion in the sense introduced above. Rather, we adopt the equivalent representation $0.011 \ldots$. We define $f(x) = (x_0, x_1, \ldots, x_n, \ldots)$ for every $x \in (0,1)$. The reader will easily verify that f is a bijection. Since $\text{ISeq}(\{0,1\}) - S$ is uncountable, it follows that so is $(0,1)$.

The existence of the bijection $g : (0,1) \longrightarrow (c,d)$ given by $g(x) = (d-c)x + c$ (see Example 5.2.4) proves that (c,d) is uncountable for $c < d$. The uncountability of $\mathbf{R}$ also follows from Example 5.2.4.

46. Using the axiom of choice, show that for any family of sets $\{M_i \mid i \in I\}$ indexed by an arbitrary set I, $\prod_{i \in I} M_i$ is countable if and only if either

    (a) there is an $i \in I$ such that $M_i = \emptyset$, or

    (b) there is a finite subset J of I such that for every $j \in J$, M_j is countable and for every $i \in I - J$, M_i is a singleton.

    (Compare this condition with the one given in Exercise 6.)

47. (a) Using the axiom of choice, show that if $\mathcal{C}$ is a disjoint collection of sets then $\bigcup \mathcal{C}$ is countable if and only if $\mathcal{C}$ is countable and every element of $\mathcal{C}$ is countable.

    (Compare this condition with the one in Exercise 7 but note that here we assume that $\mathcal{C}$ is a disjoint collection of sets.)

    (b) Give an example of an uncountable collection $\mathcal{C}$ such that $\bigcup \mathcal{C}$ is countable.

48. Let $\{c_0, \ldots, c_{n-1}\}$ be the set of all n male inhabitants of a small town. Among them, there is one barber, and the town council orders that the barber will shave all male inhabitants who do not shave themselves and will not shave any male inhabitant who shaves himself. Prove that the town council ordinance is unenforceable since it puts the barber in an impossible situation.

Solution: If the barber shaves himself, then the ordinance says that he should not shave himself, but if he does not shave himself then the ordinance says that he should.

We can use diagonalization to help understand the situation. Consider an $n \times n$ table with $0, 1$-entries; each row and each column correspond to one of the male inhabitants of the town. The element s_{ij} situated in row i and column j is 1 if and only if c_i shaves c_j. Note that the element located in the diagonal of the table and in row i, s_{ii}, equals 1 if and only if c_i shaves himself.

	c_0	c_1	$\ldots$	c_{n-1}
c_0	1	0	$\ldots$	1
c_1	0	0	$\ldots$	0
$\vdots$	$\vdots$	$\vdots$	$\ldots$	$\vdots$
c_{n-1}	0	0	$\ldots$	0

Consider now the sequence $b = (b_0, \ldots, b_{n-1})$ where, for $0 \leq i \leq n-1$, $b_i = 1$ if $s_{ii} = 0$ and $b_i = 0$ if $s_{ii} = 1$. According to the town council ordinance, the row of the table corresponding to the barber must be b, but, by diagonalization, b is different from every row of the table, so there can be no barber in the town obeying the ordinance. (If the barber did not himself live in the town, the ordinance could be enforced.)

49. Consider the poset $(\mathbf{N}, \leq)$ and the set of monotonic functions $\mathbf{N} \longrightarrow_m \mathbf{N}$. Prove that $\mathbf{N} \longrightarrow_m \mathbf{N}$. is uncountable.

    Hint: Use a diagonalization argument.

50. Let r and t be natural numbers such that $t \geq 1$. The notions of partial recursive and primitive recursive function can be extended to $\mathbf{N}^r \rightsquigarrow \mathbf{N}^t$ and $\mathbf{N}^r \longrightarrow \mathbf{N}^t$, respectively, as follows. A function $f : \mathbf{N}^r \rightsquigarrow \mathbf{N}^t$ is partial recursive (primitive recursive) if $p_j^t f$ is partial recursive (primitive recursive) for every j, $1 \leq j \leq t$.

    (a) Suppose that $f : \mathbf{N}^p \rightsquigarrow \mathbf{N}^q$ and $g : \mathbf{N}^q \rightsquigarrow \mathbf{N}^r$ with $p, q, r \in \mathbf{N}$ and $q, r \geq 1$ are partial recursive (primitive recursive). Show that $gf : \mathbf{N}^p \rightsquigarrow \mathbf{N}^r$ is partial recursive (primitive recursive).

    Definition by primitive recursion can also be extended. Namely, let $r, t \in \mathbf{N}$ and suppose that $g : \mathbf{N}^r \rightsquigarrow \mathbf{N}^t$ and $h : \mathbf{N}^{r+1+t} \rightsquigarrow \mathbf{N}^t$. The function $k : \mathbf{N}^{r+1} \rightsquigarrow \mathbf{N}^t$ is defined by primitive recursion from g and h if

    $$k(x_1, \ldots, x_r, 0) = g(x_1, \ldots, x_r),$$
    $$k(x_1, \ldots, x_r, x+1) = h(x_1, \ldots, x_r, x, p_1^t k(x_1, \ldots, x_r, x),$$
    $$\ldots, p_t^t k(x_1, \ldots, x_r, x)),$$

for all $x_1, \ldots, x_m, x$ in $\mathbf{N}$.

(b) Prove that if $g : \mathbf{N}^r \rightsquigarrow \mathbf{N}^t$ and $h : \mathbf{N}^{r+1+t} \rightsquigarrow \mathbf{N}^t$ with $t \geq 1$ are partial recursive (primitive recursive) functions and k is obtained from g, h by primitive recursion, then k is also partial recursive (primitive recursive).

If $p, q, r \in \mathbf{N}$ and $f : \mathbf{N}^p \rightsquigarrow \mathbf{N}^q$, $g : \mathbf{N}^p \rightsquigarrow \mathbf{N}^r$, we define $\langle f, g \rangle :$ $\mathbf{N}^p \rightsquigarrow \mathbf{N}^{q+r}$ by $\mathrm{Dom}(\langle f, g \rangle) = \mathrm{Dom}(f) \cap \mathrm{Dom}(g)$ and

$$\langle f, g \rangle(x_1, \ldots, x_p) = f(x_1, \ldots, x_p) \cdot g(x_1, \ldots, x_p),$$

for $(x_1, \ldots, x_p) \in \mathrm{Dom}(f \times g)$.

(c) Show that if $f : \mathbf{N}^p \rightsquigarrow \mathbf{N}^q$, $g : \mathbf{N}^p \rightsquigarrow \mathbf{N}^r$, and $h : \mathbf{N}^p \rightsquigarrow \mathbf{N}^t$ then $\langle \langle f, g \rangle, h \rangle = \langle f, \langle g, h \rangle \rangle$.

(Thus, the operation taking f and g to $\langle f, g \rangle$ is associative and we may use notations such as $\langle f_0, \ldots, f_{n-1} \rangle$ unambiguously as long as the functions all take the same number of arguments.)

(d) Let $f : \mathbf{N}^p \rightsquigarrow \mathbf{N}^q$ and $g : \mathbf{N}^p \rightsquigarrow \mathbf{N}^r$ be two partial recursive (primitive recursive) functions, where $q, r \geq 1$. Prove that $\langle f, g \rangle :$ $\mathbf{N}^p \rightsquigarrow \mathbf{N}^{q+r}$ is partial recursive (primitive recursive).

Let $f : \mathbf{N}^p \rightsquigarrow \mathbf{N}^q$ and $g : \mathbf{N}^r \rightsquigarrow \mathbf{N}^t$ be two partial functions. Define the function $f \times g : \mathbf{N}^{p+r} \rightsquigarrow \mathbf{N}^{q+t}$ by $\mathrm{Dom}(f \times g) = \{(x_1, \ldots, x_{p+r}) \mid (x_1, \ldots, x_p) \in \mathrm{Dom}(f) \text{ and } (x_{p+1}, \ldots, x_{p+r}) \in \mathrm{Dom}(g)\}$ and

$$f \times g(x_1, \ldots, x_{p+r}) = f(x_1, \ldots, x_p) \cdot g(x_{p+1}, \ldots, x_{p+r}),$$

for every $(x_1, \ldots, x_{p+r}) \in \mathrm{Dom}(f \times g)$.

(e) Prove that if $f : \mathbf{N}^p \rightsquigarrow \mathbf{N}^q$, $g : \mathbf{N}^r \rightsquigarrow \mathbf{N}^t$, and $h : \mathbf{N}^u \rightsquigarrow \mathbf{N}^v$ then

$$f \times (g \times h) = (f \times g) \times h.$$

(Thus, $\times$ is also associative.)

(f) Prove that if $f : \mathbf{N}^p \rightsquigarrow \mathbf{N}^q$ and $g : \mathbf{N}^r \rightsquigarrow \mathbf{N}^t$ with $q, t \geq 1$ are partial (primitive) recursive, then so is $f \times g$.

Suppose that $k : \mathbf{N}^r \rightsquigarrow \mathbf{N}^r$. Define a function $k^\# : \mathbf{N}^{r+1} \rightsquigarrow \mathbf{N}^r$ by

$$k^\#(x_1, \ldots, x_r, x) = k^x(x_1, \ldots, x_r),$$

for $x_1, \ldots, x_r, x \in \mathbf{N}$. (We use here the notation of Example 4.5.24.)

(g) Prove that if $k : \mathbf{N}^r \rightsquigarrow \mathbf{N}^r$ with $r \geq 1$ is partial (primitive) recursive then so is $k^\#$.

Solution: To see (a), note that for all i with $1 \le i \le r$ we have

$$p_i^r(gf) = (p_i^r g)(p_1^q f, \ldots, p_q^q f).$$

Since each projection of f and g is partial recursive (primitive recursive), it follows that each projection of gf is also partial recursive (primitive recursive).

In order to prove (b), consider the function $K : \mathbf{N}^{r+1} \rightsquigarrow \mathbf{N}$ defined by $K = f_t k$, where $f_t : \mathbf{N}^t \longrightarrow \mathbf{N}$ is a primitive recursive bijection (as shown to exist by Exercise 35). Define the functions $G : \mathbf{N}^r \rightsquigarrow \mathbf{N}$ and $H : \mathbf{N}^{r+2} \rightsquigarrow \mathbf{N}$ by

$$
\begin{aligned}
G(x_1, \ldots, x_r) &= f_t(g(x_1, \ldots, x_r)), \\
H(x_1, \ldots, x_r, y, z) &= f_t(h(x_1, \ldots, x_r, y, \pi(1, t, z), \ldots, \pi(t, t, z))),
\end{aligned}
$$

for $x_1, \ldots, x_r, y, z \in \mathbf{N}$. It is easy to see that both G and H are partial recursive (primitive recursive) if g and h are. For instance, in the case of G we can write

$$G = f_t(p_1^t g, \ldots, p_t^t g),$$

and, since all functions $p_j^t g$ are partial (primitive) recursive, then so is G. The function $K : \mathbf{N}^{r+1} \rightsquigarrow \mathbf{N}$ can be defined from G and H by primitive recursion and, therefore, it is partial (primitive) recursive. Note that we have

$$p_j^t k(x_1, \ldots, x_r, x) = \pi(j, t, K(x_1, \ldots, x_r, x)),$$

which shows (using Exercise 38) that $p_j^t k$ is partial (primitive) recursive for $1 \le j \le t$. This implies that k is partial (primitive) recursive.

Parts (c) through (f) are easy and are left to the reader.

In order to prove (g), let $i_r : \mathbf{N}^r \longrightarrow \mathbf{N}^r$ and $h : \mathbf{N}^{r+1+r} \rightsquigarrow \mathbf{N}^r$ be given by $i_r(x_1, \ldots, x_r) = (x_1, \ldots, x_r)$ and

$$h(x_1, \ldots, x_r, x, z_1, \ldots, z_r) = k(z_1, \ldots, z_r),$$

for every $x_1, \ldots, x_r, x, z_1, \ldots, z_r \in \mathbf{N}$. Then, for each i, $1 \le i \le r$, we have $p_i^r i_r = p_i^r$ and $p_i^r h = p_i^r k(p_{r+2}^{2r+1}, \ldots, p_{2r+1}^{2r+1})$, so i_r is primitive recursive and h is partial (primitive) recursive. Note that

$$
\begin{aligned}
k^\#(x_1, \ldots, x_r, 0) &= i_r(x_1, \ldots, x_r) \text{ and} \\
k^\#(x_1, \ldots, x_r, x+1) &= h(x_1, \ldots, x_r, x, p_1^r k^\#(x_1, \ldots, x_r, x), \\
&\qquad \ldots, p_r^r k^\#(x_1, \ldots, x_r, x)),
\end{aligned}
$$

which is a definition by primitive recursion of $k^\#$ starting from i_r and h. This shows that $k^\#$ is partial (primitive) recursive.

51. Given a function $\ell : \{1,\ldots,n\} \longrightarrow \{1,\ldots,m\}$ define the function $\ell^\diamond : \mathbf{N}^m \longrightarrow \mathbf{N}^n$ by

$$\ell^\diamond(x_1,\ldots,x_m) = (x_{\ell(1)},\ldots,x_{\ell(n)}),$$

for every $(x_1,\ldots,x_m) \in \mathbf{N}^m$. Prove that for every ℓ, $\ell^\diamond$ is primitive recursive.

Solution: Observe that for every j, $1 \le j \le n$, the function $p_j^n \ell^\diamond$ equals the projection $p_{\ell(j)}^m$.

52. Another way to extend the notions of primitive recursive and partial recursive function to functions with range in $\mathbf{N}^m$ with $m > 1$ would be to give an inductive definition similar to the ones given in Definitions 4.9.3 and 4.9.10 for the primitive and partial recursive functions with range in $\mathbf{N}$. We show here how this can be done.

Let $S = \bigcup\{\mathbf{N}^m \longrightarrow \mathbf{N}^q \mid m \in \mathbf{N} \text{ and } q \in \mathbf{P}\}$. We define a subset $\mathcal{I}$ of S by the following inductive definition.

(1) Every initial function (as in Definition 4.9.2) is in $\mathcal{I}$.

(2) If $f : \mathbf{N}^p \longrightarrow \mathbf{N}^q$ and $g : \mathbf{N}^p \longrightarrow \mathbf{N}^r$ are both in $\mathcal{I}$ then so is $\langle f,g \rangle$ (as defined in Exercise 50).

(3) If $f : \mathbf{N}^p \longrightarrow \mathbf{N}^q$ and $g : \mathbf{N}^q \longrightarrow \mathbf{N}^r$ are both in $\mathcal{I}$ then so is gf.

(4) If f and g are both in $\mathcal{I}$ and h is obtained from f and g by primitive recursion (in the sense of Exercise 50) then h is in $\mathcal{I}$.

Similarly, let $T = \bigcup\{\mathbf{N}^m \rightsquigarrow \mathbf{N}^q \mid m \in \mathbf{N} \text{ and } q \in \mathbf{P}\}$ and define a subset $\mathcal{J}$ of T inductively using the same four rules as in the definition of $\mathcal{I}$ (but with "$\longrightarrow$" replaced by "$\rightsquigarrow$") plus this additional rule:

(5) If $f : \mathbf{N}^{m+1} \rightsquigarrow \mathbf{N}$ is in $\mathcal{J}$ and $g : \mathbf{N}^m \rightsquigarrow \mathbf{N}$ is obtained from f by minimization (as in Definition 4.9.9) then g is in $\mathcal{J}$.

Prove that $\mathcal{I}$ is the set of primitive recursive functions and $\mathcal{J}$ is the set of partial recursive functions as defined in Exercise 50.

Hint: First show by structural induction that every element of $\mathcal{I}$ is primitive recursive and every element of $\mathcal{J}$ is partial recursive. (Most of the work for this part has been done in Exercise 50.) For the converse, show by structural induction on Definitions 4.9.3 and 4.9.10 that if $f : \mathbf{N}^m \longrightarrow \mathbf{N}$ is primitive recursive ($g : \mathbf{N}^m \rightsquigarrow \mathbf{N}$ is partial recursive) then $f \in \mathcal{I}$ ($g \in \mathcal{J}$). Then, note that for any $f : \mathbf{N}^m \rightsquigarrow \mathbf{N}^q$ with $q > 0$, we have $f = \langle p_1^q f,\ldots,p_q^q f \rangle$.

53. Using an inductive definition, it is possible to extend the notions of primitive recursive and partial recursive function one step further to include functions in $\mathbf{N}^m \rightsquigarrow \mathbf{N}^0$.

Let $S' = \bigcup\{\mathbf{N}^m \longrightarrow \mathbf{N}^q \mid m, q \in \mathbf{N}\}$ and let $T' = \bigcup\{\mathbf{N}^m \rightsquigarrow \mathbf{N}^q \mid m, q \in \mathbf{N}\}$. Define subsets $\mathcal{I}'$ and $\mathcal{J}'$ of S' and T' respectively by taking the definitions of $\mathcal{I}$ and $\mathcal{J}$ from the last exercise and adding to each definition the following rule.

For each $n \in \mathbf{N}$, p_0^n, the (unique) function in $\mathbf{N}^n \longrightarrow \mathbf{N}^0$, is in $\mathcal{I}'$ ($\mathcal{J}'$).

Prove that $\mathcal{I}' \cap S = \mathcal{I}$ and $\mathcal{J}' \cap T = \mathcal{J}$ (where S and T are as in the previous exercise). (Thus, $\mathcal{I}'$ and $\mathcal{J}'$ provide generalizations of the definitions of primitive recursive and partial recursive function given earlier.)

Hint: It is immediate that $\mathcal{I} \subseteq \mathcal{I}' \cap S$ and $\mathcal{J} \subseteq \mathcal{J}' \cap T$. It is also not hard to show by structural induction on the definition of $\mathcal{J}'$ that every member of $\mathcal{J}' \cap S$ is primitive recursive. Showing the remaining inclusion is trickier. To this end, if $m \in \mathbf{N}$ and $f : \mathbf{N}^m \rightsquigarrow \mathbf{N}^0$, define $\psi_f : \mathbf{N}^m \rightsquigarrow \mathbf{N}$ by

$$\psi_f(x_1, \ldots, x_m) = \begin{cases} 1 & \text{if } (x_1, \ldots, x_m) \in \mathrm{Dom}(f) \\ \text{undefined} & \text{otherwise.} \end{cases}$$

Show by structural induction on the definition of $\mathcal{J}'$ that if $f : \mathbf{N}^m \rightsquigarrow \mathbf{N}^q$ is in $\mathcal{J}'$ then f is partial recursive if $q > 0$ and ψ_f is partial recursive if $q = 0$.

Enumerating Programs

54. Show that the function computed by the program

$$R \begin{cases} \textbf{loop } w_0 \\ w_0 \leftarrow w_1 \\ w_1 \leftarrow w_1 + 1 \\ \textbf{endloop} \end{cases}$$

is the same as the function computed by the program R given in Example 5.4.16.

55. Prove the following assertions.

(a) Let x be a variable that does not occur in the VSPL program P. For all states s, if $M_P(s)$ is defined then $M_P(s)(x) = s(x)$.

(b) If s, s' are two states of a VSPL program P such that $s(w) = s'(w)$ for every w that occurs in P then $M_P(s)$ is defined if and only if $M_P(s')$ is defined, and if both are defined, $M_P(s)(w) = M_P(s')(w)$ for every variable w that occurs in P.

56. Prove that the set

$$\text{INSTR} = \{n \in \mathbf{N} \mid n = d(S) \text{ for some VSPL instruction}\}$$

is primitive recursive.

Hint: Note that

$$\text{INSTR} = \{n \in \mathbf{N} \mid 1 \le \text{p}_1(n) \le 5 \text{ and} \\ (\text{p}_1(n) = 5 \to \text{p}_2(n) = 0)\}.$$

57. Let $\gamma : \mathbf{N}^2 \longrightarrow \mathbf{N}$ be defined by $\gamma(i, k) = \text{gn}(w_i \leftarrow k)$, where $w_i \leftarrow k$ was introduced in Example 5.4.12. Prove that γ is primitive recursive.

    Solution: The code of the statement $w_i \leftarrow 0$ is $f(2, i)$. Therefore, for $k = 0$ we have $\gamma(i, 0) = f(1, f(2, i))$.

    Note that

$$\begin{aligned} \gamma(i, k) &= f(k+1, f_{k+1}(f(2, i), f(1, i), \ldots, f(1, i))), \\ &= f(k+1, f(f(2, i), f_k(f(1, i), \ldots, f(1, i)))). \end{aligned}$$

    Hence

$$\text{p}_2(\text{p}_2(\gamma(i, k))) = f_k(f(1, i), \ldots, f(1, i))).$$

    This allows us to write

$$\gamma(i, k+1) = f(k+2, f(f(2, i), f(f(1, i), \text{p}_2(\text{p}_2(\gamma(i, k)))))),$$

    which shows that γ is, indeed, primitive recursive.

58. Consider the functions

$$\begin{aligned} \text{length} &: \mathbf{N} \longrightarrow \mathbf{N}, \\ \text{instr} &: \mathbf{N}^2 \longrightarrow \mathbf{N}, \\ \text{var} &: \mathbf{N} \longrightarrow \mathbf{N}, \\ \text{type} &: \mathbf{N} \longrightarrow \mathbf{N}, \\ \text{varprog} &: \mathbf{N} \longrightarrow \mathbf{N}, \end{aligned}$$

    defined by

$$\begin{aligned} \text{length}(p) &= \text{p}_1(p), \\ \text{instr}(p, i) &= \pi(i, \text{length}(p), \text{p}_2(p)), \\ \text{var}(m) &= \begin{cases} \text{p}_2(m) & \text{if } 1 \le \text{p}_1(m) \le 3 \\ 0 & \text{otherwise,} \end{cases} \\ \text{type}(m) &= \text{p}_1(m) \\ \text{varprog}(p) &= \max_{1 \le i \le \text{length}(p)} \text{var}(\text{instr}(p, i)) \end{aligned}$$

    for every $p, i, m \in \mathbf{N}$. Prove that:

(a) If p is the Gödel number of a VSPL program P, length(p) gives the number of statements in P.

(b) If p is the code of a VSPL program P and $1 \le i \le$ length(p), then instr(p, i) gives the code $d(S_{i-1})$ of the i-th instruction S_{i-1} of P.

(c) var(m) gives the subscript of the variable v used by the instruction whose code is m, where $1 \le p_1(m) \le 3$.

(d) type(m) can be used to determine the type of the instruction having code m, as introduced in Definition 5.4.24.

(e) If gn$(P) = p$ for a nonempty VSPL program P then varprog(p) gives the largest subscript of a variable used by P.

Prove that the functions length, instr, var, type, varprog are all primitive recursive.

59. States, environments, and configurations of VSPL programs can be encoded as numbers. The interest of such encodings resides in the possibility of expressing the effect of VSPL instructions and, eventually, of VSPL programs as partial functions on $\mathbf{N}$.

We remind the reader that the pairing function f is assumed to be primitive recursive and, therefore, so are its associated projections p_1, p_2 and all functions f_n, for $n \ge 1$ (see Exercise 35).

To encode states and environments of VSPL programs, we need to encode both the values of the variables and the values of the counter variables. Let P be a VSPL program, where gn$(P) = p$. The *code of a state s* of P is given by

$$\mathrm{gns}_P(s) = f_{m+1}(s(w_0), \ldots, s(w_m)),$$

where $m = \mathrm{varprog}(p)$.

The set of counter variables available to P is given by

$$\{c_i \mid 0 \le i \le \mathrm{length}(p) - 1\}.$$

In order to define the code of an environment e of P, consider the sequence of values of counter variables:

$$c_e = (e(c_0), \ldots, e(c_{|P|-1})).$$

The code of e is

$$\mathrm{gne}_P(e) = f_3(p, \mathrm{gns}_P(s), f_{\mathrm{length}(p)}(c_e)),$$

where s is the state that corresponds to e.

The code of a configuration $C = (k, e)$ of P is

$$\text{gnc}_P(C) = f_3(\text{gn}(P), k, \text{gne}_P(e)).$$

Prove the existence of the primitive recursive functions

$$\text{valvar} : \mathbf{N}^3 \longrightarrow \mathbf{N} \text{ and valcount} : \mathbf{N}^3 \longrightarrow \mathbf{N}$$

such that if $p = \text{gn}(P)$ and $n = \text{gn}(e)$ (where P is a VSPL program and e is an environment of P), then $\text{valvar}(p, n, j)$ gives the value of the variable w_j in e for $0 \leq j \leq \text{varprog}(p)$, while $\text{valcount}(p, n, i)$ will return the value of the counter variable c_i corresponding to line i, where $0 \leq i \leq \text{length}(p) - 1$.

Solution: We have

$$\text{valvar}(p, n, j) = \pi(j + 1, \text{varprog}(p) + 1, \pi(2, 3, n))$$

and

$$\text{valcount}(p, n, i) = \pi(i + 1, \text{length}(p), \pi(3, 3, n))$$

for every p, n, i, k that can be interpreted as above.

60. Show that for every $m \in \mathbf{N}$ there is a primitive recursive function $\text{ugns}^{(m)} : \mathbf{N}^{m+1} \longrightarrow \mathbf{N}$ such that if P is a VSPL program with $\text{gn}(P) = p$ then for all $x_0, \ldots, x_{m-1} \in \mathbf{N}$ we have

$$\text{ugns}^{(m)}(p, x_0, \ldots, x_{m-1}) = \text{gns}_P(s),$$

where $s(w_i) = x_i$ for $0 \leq i \leq m - 1$ and $s(w_i) = 0$ when $i \geq m$.

Solution: Let $q(p) = \text{varprog}(p) + 1$. We have

$$\text{ugns}^{(m)}(p, x_0, \ldots, x_{m-1}) =$$
$$\begin{cases} f_{q(p)}(x_0, \ldots, x_{q(p)-1}) & \text{if } q(p) - 1 < m \\ f_{q(p)}(x_0, \ldots, x_{m-1}, 0, \ldots, 0) & \text{if } q(p) - 1 \geq m. \end{cases}$$

Let $M^{(m)} : \mathbf{N}^{m+1} \longrightarrow \mathbf{N}$ be the primitive recursive function given by

$$M^{(m)}(p, x_0, \ldots, x_{m-1}) = \ell(\text{varprog}(p), x_0 + \ldots + x_{m-1}).$$

Using this function we then have

$$\text{ugns}^{(m)}(p, x_0, \ldots, x_{m-1}) =$$
$$(\mu y \leq M^{(m)}(p, x_0, \ldots, x_{m-1}))[\textbf{FORALL}_{j=0}^{\text{varprog}(p)}$$
$$(j < m \rightarrow \pi(j + 1, q(p), y) = x_j) \textbf{ and}$$
$$(j \geq m \rightarrow \pi(j + 1, q(p), y) = 0)]$$

which shows that $\text{ugns}^{(m)}$ is primitive recursive.

61. Prove the existence of the following four primitive recursive functions:

$$\text{varincr}, \text{varzero}, \text{countdecr} : \mathbf{N}^3 \longrightarrow \mathbf{N}$$

and countset : $\mathbf{N}^4 \longrightarrow \mathbf{N}$, where $\text{varincr}(p, n, i)$ and $\text{varzero}(p, n, i)$ give the Gödel number of the environment e' of the VSPL program P (where $\text{gn}(P) = p$) obtained from e (where $\text{gne}_P(e) = n$) by applying the monus to the value of the variable w_i or after setting the value of w_i to 0, respectively when $0 \leq i \leq \text{varprog}(p)$.

Similarly,

$$\text{countdecr}(p, n, k) \text{ and } \text{countset}(p, n, k, l)$$

give the Gödel number of the environment e' obtained from e after applying monus to the value of the counter c_k or resetting c_k to l, respectively.

62. Let P be a VSPL program with $\text{gn}(P) = p$ and e is an environment with $\text{gne}_P(e) = n$. Prove that there is a primitive recursive function

$$\text{countinstr} : \mathbf{N}^5 \longrightarrow \mathbf{N},$$

such that if $p = f_n(r_1, \ldots, r_n)$ and $r_k \in \text{INSTR}$ for $1 \leq k \leq n$ then $\text{countinstr}(p, n, t, i, j)$ gives the number of instruction codes of type t (as introduced in Definition 5.4.24) that occur in the interval $r_i, \ldots, r_j$ when $1 \leq i \leq j \leq n$ and $1 \leq t \leq 5$.

Solution: A recursive definition of this function is given by

$$\text{countinstr}(p, n, t, i, 0) = 0,$$

$$\text{countinstr}(p, n, t, i, j+1) = \begin{cases} \text{countinstr}(p, n, t, i, j) + 1 \\ \quad \text{if type}(\pi(j + 1, n, p)) = t \\ \quad \text{and } i \leq j + 1 \leq n \\ \text{countinstr}(p, n, t, i, j) \\ \quad \text{otherwise} \end{cases}$$

which shows that countinstr is primitive recursive.

63. Prove that there are two primitive recursive functions assoc : $\mathbf{N}^2 \longrightarrow \mathbf{N}$ and invassoc : $\mathbf{N}^2 \longrightarrow \mathbf{N}$ such that, if p is the code of a VSPL program P, where $\hat{P} = S_0; \ldots; S_m$, S_i is a **loop** instruction and S_j is its associated **endloop** instruction, then $\text{assoc}(p, i) = j$ and $\text{invassoc}(p, j) = i$.

Solution: We have:

$$\text{assoc}(p, i) =$$
$$(\mu y < \text{length}(p) + 1)[\text{type}(\text{instr}(p, i)) = 3, y > i \text{ and}$$
$$\text{countinstr}(\mathbf{p}_2(p), \text{length}(p), 3, i + 1, y) =$$
$$\text{countinstr}(\mathbf{p}_2(p), \text{length}(p), 5, i + 1, y)] \dot{-} 1].$$

and

$$\text{invassoc}(p, j) = (\mu y < j)[\text{assoc}(p, y) = j],$$

for every $p, i, j \in \mathbf{N}$. Here, " $\dot{-}$ " is the notation introduced in Exercise 61 of Chapter 4 for the monus function.

64. Prove that the set GPROG of those numbers that are Gödel numbers for VSPL programs is primitive recursive.

Solution: We have $r \in$ GPROG if the following conditions are satisfied:

P0. $r \in GSEQ$ and

$$\mathbf{FORALL}_{j=0}^{\mathbf{p}_1(r)-1}[\chi_{INSTR}(\pi(j+1, \mathbf{p}_1(r), \mathbf{p}_2(r))) = 1].$$

P1. r encodes a sequence of instruction codes that contains the same number of **loop** instructions as **endloop** instructions:

$$\text{countinstr}(\mathbf{p}_2(r), \mathbf{p}_1(r), 3, 1, \mathbf{p}_1(r)) = \\ \text{countinstr}(\mathbf{p}_2(r), \mathbf{p}_1(r), 5, 1, \mathbf{p}_1(r)).$$

P2. Every suffix of the sequence encoded by r contains at least as many **endloop** instructions as **loop** instructions:

$$\mathbf{FORALL}_{j=0}^{\mathbf{p}_1(r)-1}\text{countinstr}(\mathbf{p}_2(r), \mathbf{p}_1(r), 3, j+1, \mathbf{p}_1(r)) \\ \leq \text{countinstr}(\mathbf{p}_2(r), \mathbf{p}_1(r), 5, j+1, \mathbf{p}_1(r)).$$

P3. Every **goto** instruction that occurs in the sequence points towards an instruction that occurs in the sequence:

$$\mathbf{FORALL}_{j=0}^{\mathbf{p}_1(r)-1} \mathbf{p}_1(\pi(j+1, \mathbf{p}_1(r), \mathbf{p}_2(r))) = 4 \\ \rightarrow (1 \leq \mathbf{p}_2(\pi(j+1, \mathbf{p}_1(r), \mathbf{p}_2(r))) \leq \mathbf{p}_1(r)).$$

GPROG is a primitive recursive set as an intersection of the four primitive recursive sets whose characteristic functions are defined by (P0)-(P3).

65. Prove that there are three primitive recursive functions

$$\begin{align} \text{nextline} &: \mathbf{N}^3 \longrightarrow \mathbf{N} \\ \text{nextenv} &: \mathbf{N}^3 \longrightarrow \mathbf{N} \\ \text{follconf} &: \mathbf{N}^3 \longrightarrow \mathbf{N} \end{align}$$

such that if p is the Gödel number of a VSPL program P, n is the Gödel number of an environment e and i is the current statement of P being executed then we have:

(a) nextline$(p, i, n) = j$ if P will continue its execution with the j-th statement.

(b) nextenv$(p, i, n) = m$ if P, after being in configuration (i, e), enters environment e' where gne$_P(e') = m$.

(c) If P is in configuration C with gnc$_P(C) = k$, then, after m steps, the program P will be in configuration C', where gnc$_P(C') = $ follconf(p, k, m).

Solution: The function nextline can be defined as follows:

i) nextline$(p, i, n) = i + 1$ if type$(instr(p, i)) \in \{1, 2\}$.

ii) If type$(instr(p, i)) = 4$, then

$$\text{nextline}(p, i, n) = p_2(instr(p, i)).$$

iii) If type$(instr(p, i)) = 3$, then

$$\text{nextline}(p, i, n) =$$
$$\begin{cases} i + 1 & \text{if valvar}(p, n, \text{var}(instr(p, i))) > 0 \\ \text{assoc}(p, i) + 1 & \text{otherwise.} \end{cases}$$

iv) If type$(instr(p, i)) = 5$, then if invassoc$(p, i) = j$ we have

$$\text{nextline}(p, i, n) = \begin{cases} j + 1 & \text{if valcount}(p, n, j) > 1 \\ i + 1 & \text{otherwise.} \end{cases}$$

v) In any other case, nextline$(p, i, n) = p$.

The existence of this definition proves that nextline is primitive recursive.

A primitive recursive definition of follconf can be given as follows: follconf$(p, k, 0) = k$, and

$$\text{follconf}(p, k, m + 1) = f_3(p, \text{nextline}(p, \pi(2, 3, k'), \pi(3, 3, k')),$$
$$\text{nextenv}(p, \pi(2, 3, k'), \pi(3, 3, k'))),$$

where $k' = \text{follconf}(p, k, m)$.

66. Prove that there exists a partial recursive function time : $\mathbf{N}^2 \leadsto \mathbf{N}$ such that time(p, m) is defined if and only if there is a VSPL program P with gn$(P) = p$ and a state s with gns$_P(s) = m$ such that $s \in$ Dom(M_P), and if time(p, m) is defined, then time$(p, m) = k$, where k is the number of steps in the execution of P on s.

Solution: Note that we can write

$$\text{time}(p, m) =$$
$$\mu n[\chi_{\text{GPROG}}(p) = 1 \text{ and } \pi(2, 3, \text{follconf}(p, l, n)) = \text{length}(p)],$$

where $l = f_3(p, 0, \text{gne}_P(e_s))$ is the code of the initial configuration that corresponds to the state s,

$$\text{gne}_P(e_s) = f_3(p, m, 0).$$

67. For every $m \in \mathbf{N}$ define $u_m : \mathbf{N}^{m+1} \rightsquigarrow \mathbf{N}$ such that

$$u_m(p, x_1, \ldots, x_m) = f_p^{(m)}(x_1, \ldots, x_m),$$

if $(x_1, \ldots, x_m) \in \text{Dom}(f_p^{(m)})$ and is undefined, otherwise. The function u_m is called the VSPL m-universal function.

Show that

(a) For each $m \in \mathbf{N}$, u_m is partial recursive.

(b) For each $m \in \mathbf{N}$, u_m can be obtained from primitive recursive functions by applying minimization only once.

Solution: (a) Let $(x_1, \ldots, x_m) \in \mathbf{N}^m$. Define the state s_0 by

$$s_0(w_i) = \begin{cases} x_{i+1} & \text{if } 0 \le i \le m - 1 \\ 0 & \text{if } m \le i. \end{cases}$$

If $\text{gn}(P) = p$ then $\text{ugns}^{(m)}(p, x_1, \ldots, x_m) = \text{gns}_P(s_0)$ and, if we define $l : \mathbf{N}^{m+1} \longrightarrow \mathbf{N}$ by

$$l(p, x_1, \ldots, x_m) = f_3(p, 0, f_3(p, \text{ugns}^{(m)}(p, x_1, \ldots, x_m), 0))$$

then $l(p, x_1, \ldots, x_m)$ gives the Gödel number of of the initial configuration that corresponds to s_0. If $f_p^{(m)}(x_1, \ldots, x_m)$ is defined then the Gödel number of the final configuration entered by P is given by

$$\text{fc}(p, x_1, \ldots, x_m) =$$
$$\text{follconf}(p, l(p, x_1, \ldots, x_m), \text{time}(p, \text{ugns}^{(m)}(p, x_1, \ldots, x_m))).$$

If $f_p^{(m)}(x_1, \ldots, x_m)$ is undefined then, $\text{fc}(p, x_1, \ldots, x_m)$ is undefined.

If p is not the Gödel number of a program, then $\text{fc}(p, x_1, \ldots, x_m)$ is undefined.

This allows us to write

$$u_m(p, x_1, \ldots, x_m)$$
$$= \text{valvar}(p, \pi(3, 3, \text{fc}(p, x_1, \ldots, x_m)), 0),$$

for $p, x_1, \ldots, x_m \in \mathbf{N}$ which proves that u_m is partial recursive.

(b) Note that minimization is used only once, in the definition of time.

68. Prove that the class of VSPL-computable functions equals the class of partial recursive functions.

    Solution: Theorem 5.4.23 states that the class of partial recursive functions is included in the class of VSPL-computable functions. Exercise 67 implies that that every VSPL-computable function is partial recursive.

69. Prove the existence of a primitive recursive function

    $$\text{catprogs} : \mathbf{N}^2 \longrightarrow \mathbf{N}$$

    such that if P, Q are two VSPL programs with $\text{gn}(P) = p$ and $\text{gn}(Q) = q$ then $\text{catprogs}(p, q)$ gives the Gödel number of the program concatenation of P and Q.

    Solution: Note that we cannot directly use the function catseq since, when two VSPL programs P and Q are concatenated, we need to modify the **goto** instructions of Q, according to Definition 5.4.8. Therefore, we need to define a function $f : \mathbf{N}^2 \longrightarrow \mathbf{N}$, such that if $q = \text{gn}(Q)$, and $k \in \mathbf{N}$ then $r = f(q, k)$ gives the Gödel number of $Q^{[k]}$.

    It is easy to see that if $q \in \text{GPROG}$ and if $r = f(q, k)$ then $r < \ell(q, f(5, q + k))$. This allows us to define the function f by bounded minimization:

    $$f(q, k) =$$
    $$(\mu r \le \ell(q, f(5, q + k)))[\text{p}_1(r) = \text{length}(q)$$
    and
    $$\textbf{FORALL}_{j=0}^{\text{length}(q)-1}(\text{type (instr } (q, j + 1)) = 4 \rightarrow$$
    $$\text{p}_2(\pi(j + 1, \text{p}_1(r), \text{p}_2(r))) = \text{p}_2(\text{instr}(q, j + 1)) + k$$
    and
    $$\text{p}_1(\pi(j + 1, \text{p}_1(r), \text{p}_2(r))) = 4))$$
    and
    $$(\text{type (instr } (q, j + 1)) \ne 4 \rightarrow$$
    $$\pi(j + 1, \text{p}_1(r), \text{p}_2(r)) = \text{instr}(q, j + 1))],$$

    for $q, k \in \mathbf{N}$ which proves that f is primitive recursive.

    We define now the function catprogs as

    $$\text{catprogs}(p, q) = \text{catseq}(p, f(q, \text{length}(p))),$$

    for every $p, q \in \mathbf{N}$. It is clear that this function is primitive recursive.

70. Prove that for every $m \in \mathbf{P}$ and $n \in \mathbf{N}$ there exists a primitive recursive function

    $$s_{m,n} : \mathbf{N}^{m+1} \longrightarrow \mathbf{N}$$

such that for every $f_p^{(m+n)} \in \mathbf{F}_{VSPL}$ we have

$$f_p^{(m+n)}(x_1, \ldots, x_m, x_{m+1}, \ldots, x_{m+n})$$
$$= f_{s_{m,n}(p, x_1, \ldots, x_m)}^{(n)}(x_{m+1}, \ldots, x_{m+n})$$

for every $x_1, \ldots, x_{m+n} \in \mathbf{N}$.

Solution: If $p \notin \text{GPROG}$ we define $s_{m,n}(p, x_1, \ldots, x_m) = p$ for every $m \in \mathbf{P}$ amd $n, x_1, \ldots, x_m \in \mathbf{N}$.

The argument is by induction on $m \geq 1$. For $m = 1$ we fix the first argument x_1 of the function $f_p^{(1+m)}$. Let $r = \text{varprog}(p)$, where $p = \text{gn}(P)$. Consider the program R defined by

$$R \begin{cases} w_n \leftarrow w_{n-1} \\ w_{n-1} \leftarrow w_{n-2} \\ \vdots \\ w_2 \leftarrow w_1 \\ w_1 \leftarrow w_0 \\ w_0 \leftarrow x_1 \\ P[4n + x_1 + 1] \end{cases}$$

Prior to executing R, we place the values $x_2, \ldots, x_{n+1}$ in the variables $w_0, \ldots, w_{n-1}$, respectively. The first $n - 1$ instructions shift these values into $w_1, \ldots, w_n$, and eventually, $w_0 \leftarrow x_1$ fixes the value of w_0 to x_1. Thus, we have

$$f_R^{(n)}(x_2, \ldots, x_{1+n}) = f_P^{(1+n)}(x_1, \ldots, x_{1+n}),$$

for every $x_2, \ldots, x_{1+n} \in \mathbf{N}$. If $r = \text{gn}(R)$ we can easily express r as a function of $p = \text{gn}(P)$ using a primitive recursive function. Let

$$r_k = \text{gnseq}(f(2, k+1), f(3, k), f(1, k+1), f(5, 0))$$

be the Gödel number of the program $w_{k+1} \leftarrow w_k$ for $0 \leq k \leq n - 1$ and let $\gamma(0, x_1)$ be the Gödel number of $w_0 \leftarrow x_1$ (see Exercise 57). The Gödel number of R is

$$r = \text{gn}(R) = \text{catprogs}(r_{n-1}, (\text{catprogs}(r_{n-2}, \ldots \\ (\text{catprogs}(r_0, \text{catprogs}(\gamma(0, x_1), p))) \ldots))).$$

This proves the existence of the primitive recursive function $s_{1,n}$, given by

$$s_{1,n}(p, x_1) = \\ \begin{cases} \text{catprogs}(r_{n-1}, (\text{catprogs}(r_{n-2}, \ldots \\ \quad (\text{catprogs}(r_0, \text{catprogs}(\gamma(0, x_1), p))) \ldots)) \\ \quad \text{if } p \in \text{GPROG}, \\ p, \text{otherwise} \end{cases}$$

for every $p, x_1 \in \mathbf{N}$.

Suppose now that for $m \geq 1$ there exist primitive recursive functions $s_{m,n}$ for every $n \in \mathbf{N}$. Since

$$f_p^{(m+1+n)}(x_1, \ldots, x_m, x_{m+1}, \ldots, x_{m+1+n})$$
$$= f_{s_{m,1+n}(p,x_1,\ldots,x_m)}^{(1+n)}(x_{m+1}, \ldots, x_{m+1+n}),$$

for every $x_1, \ldots, x_{m+1+n} \in \mathbf{N}$, using again the existence of $s_{1,n}$ we can write

$$f_p^{(m+1+n)}(x_1, \ldots, x_m, x_{m+1}, \ldots, x_{m+1+n})$$
$$= f_{s_{m,1+n}(p,x_1,\ldots,x_m)}^{(1+n)}(x_{m+1}, \ldots, x_{m+1+n})$$
$$= f_{s_{1,n}(s_{m,n}(p,x_1,\ldots,x_m),x_{m+1})}^{(n)}(x_{m+2}, \ldots, x_{m+1+n}),$$

which allows us to define $s_{m+1,n}$ as

$$s_{m+1,n}(p, x_1, \ldots, x_{m+1}) = s_{1,n}(s_{m,n}(p, x_1, \ldots, x_m), x_{m+1}),$$

for every $n, p, x_1, \ldots, x_{m+1} \in \mathbf{N}$.

71. The notion of function computable by a VSPL program can be extended as follows. Let $r, t \in \mathbf{N}$, where $t \geq 1$. The r-argument, t-component partial function computed by a VSPL program P is the partial function $f_P^{(r,t)} : \mathbf{N}^r \rightsquigarrow \mathbf{N}^t$, whose domain is $\{\bar{x} \in \mathbf{N}^r | s_{\bar{x}} \in \text{Dom}(M_P)\}$, and for each $\bar{x} \in \text{Dom}(f_P^{(r,t)})$,

$$f_P^{(r,t)}(\bar{x}) = (M_P(s_{\bar{x}})(w_0), \ldots, M_P(s_{\bar{x}})(w_{t-1})).$$

If $\text{gn}(P) = p$ we denote $f_P^{(r,t)}$ by $f_p^{(r,t)}$. If p is not the Gödel number of a program, then $f_p^{(r,t)}$ is the empty function.

(a) Let $r, u, v \in \mathbf{N}$ be such that $1 \leq u \leq v$. If $p_i^v : \mathbf{N}^v \longrightarrow \mathbf{N}$ are the projection functions, $1 \leq i \leq v$, prove that

$$f_P^{(r,u)}(\bar{x}) = \langle p_1^v, \ldots, p_u^v \rangle f_P^{(r,v)}(\bar{x}),$$

for every $\bar{x} \in \mathbf{N}^r$.

(b) Let P be a VSPL program such that $\text{gn}(P) = p$ and let $j, m \in \mathbf{N}$ be such that $m \geq \max\{j, \text{varprog}(p) + 1\}$. If R is the program

$$\begin{array}{l} \textbf{loop } w_j \\ p^{[1]} \\ \textbf{endloop} \end{array}$$

prove that

$$f_R^{(m,m)} = (f_P^{(m,m)})\#\ell^\diamond,$$

where $\ell : \{1, \ldots, m+1\} \longrightarrow \{1, \ldots, m\}$ is given by $\ell(i) = i$ for $1 \leq i \leq m$ and $\ell(m+1) = j$.

72. Let VSPL $-$ {goto} be the set of VSPL programs that do not use goto. Similarly, VSPL $-$ {goto, loop} is the set of VSPL programs that do not use goto, **loop** and **endloop**.

    (a) Let P be a sequence of VSPL instructions that does not contain any **gotos** and let

    $$L_P = \{i|0 \leq i < |P| \text{ and } P(i) \text{ is a } \textbf{loop} \text{ instruction}\},$$
    $$K_P = \{i|0 \leq i < |P| \text{ and } P(i) \text{ is an } \textbf{endloop} \text{ instruction}\}.$$

    P is a VSPL $-$ {goto} program if and only if there exists a bijection $\alpha : L_P \longrightarrow K_P$ such that $i \leq \alpha(i)$ for every $i \in L_P$.

    (b) Let P be a VSPL $-$ {goto} program. If $P(i)$ is a **loop** instruction and $\alpha_P(i) = j$ then the body Q of the **loop** $P(i)$: $P(i+1); \ldots; P(j-1)$ is a VSPL $-$ {goto} program.

    (c) Let P be a VSPL $-$ {goto} program. If P contains a **loop** instruction then P can be written as

    $$P = \begin{cases} R \\ \textbf{loop } w_k \\ Q \\ \textbf{endloop} \\ T, \end{cases}$$

    where R is a VSPL$-${goto, loop} program and Q, T are VSPL$-${goto} programs.

Solution: (a) Let α be a bijection between L_P and K_P. It is clear that condition (Pi) is satisfied. Let $U = P(l); \ldots; P(n-1)$ be a suffix of $\hat{P}$. If $L_U = \{i \in L_P | l \leq i \leq n-1\}$ and $K_U = \{i \in K_P | l \leq i \leq n-1\}$ note that $\beta = \alpha \lceil L_P$ is an injective mapping between L_U and K_U because of the fact that $i < \alpha(i)$ for every $i \in L_U$. Therefore, $|L_U| \leq |K_U|$, which shows that P satisfies also condition (Pii).

Conversely, Theorem 5.4.1 gives the existence of the desired bijection, if P is a VSPL $-$ {goto} program.

(b) Theorem 5.4.1 implies that the mapping $\alpha : L_Q \longrightarrow K_Q$ and given by $\alpha(l) = \alpha_P(l + i + 1)$ is a bijection between L_Q and K_Q. By part a), Q is a VSPL $-$ {goto} program.

(c) Let $P(i)$ be the first **loop** that occurs in P. Define R, Q, T as

$$\hat{R} = P(0); \ldots; P(i-1)$$
$$\hat{Q} = P(i+1); \ldots; P(j-1)$$
$$\hat{T} = P(j+1); \ldots; P(n-1),$$

where $n = |P|$ and $j = \alpha_P(i)$. Note that no **loop** or **endloop** occurs in R. Therefore, R satisfies vacuously conditions (Pi) and (Pii) and, thus is a VSPL $-$ {**goto, loop**} program. From b) we obtain that Q is a VSPL $-$ {**goto**} program. Finally, the mapping $\alpha' : L_T \longrightarrow K_T$, given by $\alpha'(l) = \alpha_P(l+j+1)$ for every $l \in L_T$ is a bijection between L_T and K_T such that $l < \alpha'(l)$ for every $l \in L_T$. So, by a), T is a VSPL $-$ **goto** program.

73. Let P, Q be two VSPL $-$ {**goto**} programs, where $gn(P) = p$ and $gn(Q) = q$. Prove that $f_R^{(m,m)} = f_Q^{(m,m)} f_P^{(m,m)}$, where

$$m \geq \max\{\text{varprog}(p), \text{varprog}(q)\} + 1$$

and R is the concatenation of P and Q.

74. Let P be a VSPL program, where $gn(P) = p$ and $\text{varprog}(p)+1 = m$. Prove that if $f_P^{(m,m)}$ is partial recursive or primitive recursive then so is $f_P^{(u,v)}$ for every $u, v \in \mathbf{N}$, with $v \geq 1$.

    Solution: There are six cases to consider depending on the relative positions of u, v and m. If $u \leq v \leq m$ we have

$$p_l^v f_P^{(u,v)}(x_1, \ldots, x_u) = p_l^m f_P^{(m,m)}(x_1, \ldots, x_u, 0, \ldots, 0),$$

    for every $(x_1, \ldots, x_u) \in \mathbf{N}^u$ and for every l, $1 \leq l \leq u$, which shows that $f_P^{(u,v)}$ is indeed primitive recursive. We leave the remaining cases to the reader.

75. Prove that the set of primitive recursive functions equals the set of functions computable by VSPL $-$ {**goto**} programs.

    Solution: We prove initially that any function $f_P^{(r,m)}$ computed by a VSPL $-$ {**goto**} program P is primitive recursive. The argument is by induction on $|P|$.

    For $|P| = 0$, P is the empty sequence and we have $f_P^{(r,m)}(x_1, \ldots, x_r) = (x_1, \ldots, x_m)$, if $m \leq r$ or $f_P^{(r,m)}(x_1, \ldots, x_r) = (x_1, \ldots, x_r, 0, \ldots, 0)$, if $m > r$, which shows that $f_P^{(r,m)}$ is primitive recursive.

    When $|P| = 1$, P can be either $w_j \leftarrow 0$ or $w_j \leftarrow w_j + 1$. There are six possible cases to consider depending on the relative positions of the numbers r, m and j. For example, if P is $w_j \leftarrow w_j + 1$ and $j \leq r \leq m$ we have

$$f_P^{(r,m)}(x_1, \ldots, x_r) = (x_1, \ldots, x_{j-1}, x_j + 1, x_{j+1}, \ldots, x_r, 0, \ldots, 0),$$

    for every $(x_1, \ldots, x_r) \in \mathbf{N}^r$. Again, the verification of the primitive recursiveness is immediate. We leave the other five cases as exercises for the reader.

Suppose now that $n > 1$ and the functions computed by any VSPL $-$ {**goto**} program of length less than n is primitive recursive. Let P be a VSPL $-$ {**goto**} program with $|P| = n$, $\mathrm{gn}(P) = p$ and $m = \mathrm{varprog}(p) + 1$. If P does not contain any **loop** instruction, P is obtained by concatenating a one-line program P_1 (which consists of its first instruction) to a program Q with $|Q| = n - 1$. From Exercise 73 we have $f_P^{(m,m)} = f_Q^{(m,m)} f_{P_1}^{(m,m)}$; hence, $f_P^{(q,q)}$ is primitive recursive and so is $f_P^{(r,t)}$ for any $r, t \in \mathbf{N}$, where $t \geq 1$.

If P contains **loop** instructions then, by Exercise 72 P can be written as

$$
P = \begin{cases}
R \\
\textbf{loop } w_j \\
Q \\
\textbf{endloop} \\
T,
\end{cases}
$$

where R, Q, T are VSPL $-$ {**goto**} programs of length less than n. By Exercises 73 and 71, if $\ell : \{1, \ldots, m + 1\} \longrightarrow \{1, \ldots, m\}$ is the function given by $\ell(i) = i$ for $i \leq m$ and $\ell(m + 1) = j$ we have

$$
f_P^{(m,m)} = f_T^{(m,m)} (f_Q^{(m,m)})^{\#} \ell^{\diamond} f_R^{(m,m)},
$$

which shows, by inductive hypothesis, that $f_P^{(m,m)}$ is indeed, primitive recursive. Therefore, for all $r, t \in \mathbf{N}$ (with $t \geq 1$) the functions $f_P^{(r,t)}$ are primitive recursive.

In order to prove that every primitive recursive function is computed by a VSPL $-$ {**goto**} program observe that the initial functions have this property, as shown in the proof of Theorem 5.4.23. Further, if we start from VSPL $-$ {**goto**} programs that compute primitive recursive functions $f : \mathbf{N}^n \longrightarrow \mathbf{N}$ and $g_i : \mathbf{N}^m \longrightarrow \mathbf{N}$ for $1 \leq i \leq n$ then the program that computes their composition $f(g_1, \ldots, g_m)$ is, again, a VSPL $-$ {**goto**} program as shown by the proof of Theorem 5.4.19.

If the construction of Theorem 5.4.21 is applied to VSPL $-$ {**goto**} programs that compute the functions $g : \mathbf{N}^m \longrightarrow \mathbf{N}$ and $h : \mathbf{N}^{m+2} \longrightarrow \mathbf{N}$ the resulting program which computes the function obtained from g and h by primitive recursion is also a VSPL $-$ {**goto**} program. This concludes our argument.

Abstract Families of Functions

76. Let $\mathbf{F}$ be a universal AFF that contains the base functions and is closed under composition. Prove that $\mathbf{F}$ contains a partial function $f : \mathbf{N} \rightsquigarrow \mathbf{N}$ such that

$$
\emptyset \subset \mathrm{Dom}(f) \subset \mathbf{N}.
$$

77. Give a direct proof of the fact that for no $n \in \mathbf{P}$ is the universal function for $\mathbf{POL}_n$ a member of $\mathbf{POL}_{n+1}$, where $\mathbf{POL}$ is the AFF introduced in Example 5.5.4.

78. Derive Rice's theorem from the recursion theorem.

    Solution: Let $\mathbf{F}$ be a standard AFF and let $\mathcal{C}$ be a collection of functions. Assume that $\emptyset \subset P(\mathcal{C}) \subset \mathbf{N}$ and that $P(\mathcal{C})$ is an $\mathbf{F}$-set. There are $i \in P(\mathcal{C})$ and $j \notin P(\mathcal{C})$ and this allows us to consider the function $h : \mathbf{N} \rightsquigarrow \mathbf{N}$ defined by

    $$h(x) = D(j, i, \chi_{P(\mathcal{C})}(x), 1)$$

    for every $x \in \mathbf{N}$. Note that h is a total function that belongs to $\mathbf{F}_1$ since $P(\mathcal{C})$ was assumed to be an $\mathbf{F}$-set. The recursion Theorem implies that h has a fixed point n, that is, $f_n^{(1)} = f_{h(n)}^{(1)}$.

    Suppose that $n \in P(\mathcal{C})$, that is, $f_n^{(1)} \in \mathcal{C}$. This gives

    $$h(n) = D(j, i, \chi_{P(\mathcal{C})}(n), 1) = D(j, i, 1, 1) = j,$$

    so $f_{h(n)}^{(1)} \notin \mathcal{C}$, which is contradictory in view of the fact that n is a fixed point for h. Similarly, the assumption $n \notin P(\mathcal{C})$ also yields a contradiction, and this shows that $P(\mathcal{C})$ is not a $\mathbf{F}$-set.

79. Prove that for any standard AFF there is a total function h such that $f_{h(m)}^{(1)}(n) = f_n^{(1)}(m)$ for every $m, n \in \mathbf{N}$. Show that this implies the existence of $p, q \in \mathbf{N}$ such that $f_p^1(q) = f_q^{(1)}(p)$.

80. Let $\mathbf{F}$ be an AFF and let A, B be two sets, where $A, B \subseteq \mathbf{N}$. A is $\mathbf{F}$-reducible to B if there is a function f in $\mathbf{F}_1$ such that $x \in A$ if and only if $f(x) \in B$ for all $x \in \mathbf{N}$.

    Prove the following.

    (a) If $\mathbf{F}$ is closed under composition, A is $\mathbf{F}$-reducible to B, and B is an $\mathbf{F}$-set, then A is also an $\mathbf{F}$-set.

    (b) The set $K(\mathbf{F})$ is $\mathbf{F}$-reducible to the set $T(\mathbf{F})$, for any standard AFF $\mathbf{F}$.

    (c) If $\mathcal{C}$ is a collection of functions from $\mathbf{F}_1$, where $\mathbf{F}$ is a standard AFF, such that $P(\mathcal{C}) \neq \emptyset$ and $P(\mathcal{C}) \neq \mathbf{N}$, then $K(\mathbf{F})$ is $\mathbf{F}$-reducible to $P(\mathcal{C})$.

81. Define $g_i^{(n)} : \mathbf{N}^n \longrightarrow \mathbf{N}$ as the function $f_i^{(n)}$ if P is a VSPL $-$ {goto} program, with $\mathrm{gn}(P) = i$, and as the constant function K_0^n, otherwise. This allows us to consider the AFF $\mathbf{PRF} = \{g_i^{(n)} | n \in \mathbf{N}, i \in \mathbf{N}\}$. It is clear, because of Exercise 75 that $\mathbf{PRF}$ coincides with the class of primitive recursive functions.

    Prove that $\mathbf{PRF}$ is not a universal AFF.

5.7 Bibliographical Comments

The main topics of this chapter, enumerability and diagonalization, are treated in references on set theory ([End77, Ham82, HJ84, Sup60]). Proofs by diagonalization in computability theory are frequently used in standard references such as [Hen77, BL74, MY79]; Section 5.5 is based on [Hen77] and [MY79].

References

[BCLF81] M. Behzad, G. Chartrand, and L. Lesniak-Foster. *Graphs and Digraphs*. Wadsworth International Group, Belmont, California, 1981.

[Ber73] C. Berge. *Graphs and Hypergraphs*. North-Holland, Amsterdam, 1973.

[Bir73] G. Birkoff. *Lattice Theory*. American Mathematical Society, Providence, Rhode Island, third edition, 1973.

[BL74] W. S. Brainerd and L. H. Landweber. *Theory of Computation*. John Wiley, New York, 1974.

[Bol85] B. Bollobas. *Graph Theory: An Introductory Course*. Springer-Verlag, New York, 1985.

[Can55] G. Cantor. *Contributions to the Founding of the Theory of Transfinite Numbers*. Dover Publications, New York, 1955. translated by P. Jourdain.

[CB72] R. L. Constable and A. B. Borodin. Subrecursive programming languages, part i: Efficiency and program structure. *Journal of the Association for Computing Machinery*, 19:526–569, 1972.

[Cha77] G. Chartrand. *Graphs as Mathematical Models*. Wadsworth International Group, Belmont, California, 1977.

[End77] H. B. Enderton. *Elements of Set Theory*. Academic Press, New York, 1977.

[Eve79] S. Even. *Graph Algorithms*. Computer Science Press, Rockville, Maryland, 1979.

[Gin66] S. Ginsburg. *The Mathematical Theory of Context-Free Languages*. McGraw-Hill, New York, 1966.

[Hal87] P. R. Halmos. *Naive Set Theory*. Springer-Verlag, New York, 1987.

[Ham82] A. G. Hamilton. *Numbers, Sets, and Axioms — The Apparatus of Mathematics*. Cambridge University Press, Cambridge, 1982.

417

[Har78]   M. A. Harrison. *Introduction to Formal Language Theory.* Addison-Wesley, Reading, Massachsetts, 1978.

[Hen60]   L. Henkin. On mathematical induction. *American Mathematical Monthly,* 67:323–338, 1960.

[Hen77]   F. Hennie. *Introduction to Computability.* Addison-Wesley, Reading, 1977.

[HJ84]   K. Hrbacek and T. Jech. *Introduction to Set Theory.* Marcel Decker, New York, second edition, 1984.

[LS87]   J. Loeckx and K. Sieber. *The Foundations of Program Verification.* Wiley-Teubner Series in Computer Science. John Wiley and B. G. Teubner, Chichester, West Anglia, and Stuttgart, second edition, 1987.

[Mai83]   D. Maier. *The Theory of Relational Databases.* Computer Science Press, Rockville, Maryland, 1983.

[Man74]   Z. Manna. *Mathematical Theory of Computation.* McGraw-Hill, New York, 1974.

[MB67]   S. MacLane and G. Birkoff. *Algebra.* Macmillan, New York, 1967.

[MR67a]   A. R. Meyer and D. M. Ritchie. The complexity of loop programs. In *Proceedings of the ACM National Meeting,* pages 465–469, 1967.

[MR67b]   A. R. Meyer and D. M. Ritchie. Computational complexity and program structure. IBM Research Paper RC-1817, IBM Watson Research Center, Yorktown Heights, New York, 1967.

[MY79]   M. Machtey and P. Young. *An Introduction to the General Theory of Algorithms.* Elsevier North-Holland, New York, 1979.

[Roi90]   J. Roitman. *Introduction to Modern Set Theory.* John Wiley & Sons, New York, 1990.

[RR85]   H. Rubin and J. Rubin. *Equivalents of the Axiom of Choice.* North-Holland, Amsterdam, second edition, 1985.

[Sal73]   A. Salomaa. *Formal Languages.* Academic Press, New York, 1973.

[Sup60]   P. C. Suppes. *Axiomatic Set Theory.* Van Nostrand, New York, 1960.

[Ull88]   J. D. Ullman. *Database and Knowledge-Base Systems,* volume 1. Computer Science Press, Rockville, Maryland, 1988.

[vH67]   Jean van Heijenoort, editor. *From Frege to Gödel, A Source Book in Mathematical Logic, 1879-1931.* Harvard University Press, Cambridge, Massachusetts, 1967.

[Wan80]   M. Wand. *Induction, Recursion and Programming.* North-Holland, Amsterdam, 1980.

Index

A

absorption laws 75
abstract family of functions 374
    smn property for 378
    standard 378
    universal 376
Ackermann's function 300
addition in a Peano structure 288
addressing mapping 109
alphabet 45
    word on an 45
ancestor of a vertex 61
axiom of choice 157
axiom of infinity 332

B

bag 12
basic functions 375
basis rules 200
biconditional 259
bijection 33
body of a **loop** 359
bounded minimization 325
bound
    greatest lower 133
    greatest upper 133

C

canonical injection 72
cardinality of a finite set 44
Cartesian product 8
    of a family of sets 67
Cayley table 76
chain 135
    maximal 163
choice function 157
Chomsky hierarchy 272
Church-Rosser property 119
circuit 54

closure of a family of sets under
    a family of constructors
    255
code of an instruction 372
code 113
collection of sets 2
compatible functions 107
    family of 107
compatible relation 86
complement 10
complete subset of a cpo 150
composition of functions 70
computation in relational algebra
    91
concatenation of VSPL programs
    363
conclusion of a rule 200
conditional 258
    conclusion of a 258
conjugate words 117
conjunction 258
connective 258
    binary 258
    n-ary 258
constructor, n-ary 232
    subset closed under an 233
continuous function 151
contradiction 263
countable set 344
countably infinite 343
cpo 148

D

database schema 83
De Morgan's laws 75
definition by cases 323
descendant of a vertex 61
diagonalization 356

digraph 52
    arcs of a 52
    connected components of a 119
    converse 53
    directed acyclic 56
    functional 115
    partial 54
    path in a 54
        elementary 54
discriminator function 375
disjoint sum 72
disjunction 258
division in relational algebra 91

E
entities, class of 166
enumeration of a set 346
environment of a program 109
equinumerous sets 339
equivalence class 65
equivalence relation 64
erasure production 270
exponentiation in a Peano struc-
        ture 288

F
family of sets closed under a con-
        structor 255
Fibonacci sequence 195
formula of propositional logic 260
freely generated family of sets 256
freely generated set 236 program
        366
function 27
    between two sets 31
    characteristic 39
    computed by a VSPL program
        366
    empty 28
    extension of a 38
    kernel of a 64
    natural extension of a 147
    partial 31
    partial recursive 269
    preimage of a set under a 49

    restriction of a 38
    signum 266
    total 31
    zero-ary 70

G
Gödel number of a program 373
Gödel number of a sequence 355
grammar 270
    context-free 273
        ambiguous 282
        unambiguous 282
    context-sensitive 273
    derivation in a 271
    λ-free 270
    length-increasing 273

H
halting problem 373
Hasse diagram 130
hypotheses of a rule 201
hypothesis of a conditional 258

I
identifiers 109
identity 76
image of a set 49
image of a set 31
immediate ancestor 61
immediate descendant 60
immediate predecessor of a vertex
        53
immediate successor of a vertex
        53
index set of a collection of func-
        tions 381
induction structure 333
inductive definition of a set 200
inductive set 332
inductive step 180
infimum 133
infinite ascending sequence 137
infinite descending sequence 137
infinite sequence in a relation 290
injection 33
involution 115

J

joinable tuples 89

K

Kleene closure 47

L

lambda notation 32
language 46
    catenatively independent 112
    context-free 273
    context-sensitive 273
    empty 46
    regular 273
    transition system 319
leaf of a tree 61
left identity 76
left inverse 36, 78
leftmost derivation 280
length of a derivation 271
level of a tree 59
linear order 135
loop invariant 190
lower bound 131

M

mapping 31
mathematical induction 177
matrix product 79
matrix 48
maximal element 134
McCarthy's 91 function 336
minimal element 134
monotonic mapping 144
monotonic primitive recursive triple
    393
mprp-triple 393
multiplication in a Peano struc-
    ture 288
multiset 12
    multiplicity of an element in
    a 12

N

negation 259
numbering of a set 346

O

one-to-one correspondance 33
operation 75
    associative 75
    commutative 75
    idempotent 75
    partial 78

P

pair 8
    ordered 8
    unordered 8
pairing function 351
    associated with a 353
pairing triple 351
    monotonic 353
    monotonic primitive recursive
        393
    projections of a 351
partial correctness of a program
    190
partial order (*see posets*) 128
partially ordered set 128
partition 65
Peano structure 286
Peano's axioms 285
poset 128
    artinian 137
    dual of a 138
    flat 134
    greatest element of a 132
    least element of a 132
    noetherian 137
posets 128
    coalesced sum of 150
    isomorphic 145
    morhpism of 144
    product of 142
    strict morphism of 144
    sum of a family of 141
    well-founded 137
    well-ordered 136
positive closure 47
primitive recursion 264
primitive recursive functions 265

primitive recursive relation 323
primitive recursive set 322
principle of foundation 17
product of languages 46
    monotonicity of 47
projection mapping 71
proposition 257

Q
quotient of relations 91
quotient set 66

R
recursive definition of a function
    218
recursive function 270
relation 24
    acyclic 56
    antisymmetric 62
    asymmetric 62
    circular 119
    domain of a 24
    inverse of a 25
    irreflexive 62
    n-ary 69
    one-to-one 27
    product 26
    range of a 24
    reflexive 62
    restriction of a 30
    symmetric 62
    total 28
    transitive
relational database 83
    relation difference in a 86
    relation intersection in a 86
    relation join in a 89
    relation product in a 86
    relation projection in a 87
relation on a set 24
relations
    product of 26, 86
    union of 86
renaming of a table 85
right identity 76
right inverse 36

right inverse 78
rooted tree 58

S
selection 87
selective function 157
selective set 157
semijoin operation 124
sequence
    empty 41
    finite 41
    infinite 42
    length of a finite 41
    null 41
    reversal of a 110
separate sum 150
sequences
    concatenation of sequences 42
    product of 42
set 1
    dominated by another 341
    elements of a 1
    finite 44
    infinite 44
    members of a 1
    numbering of a 346
    property of elements of a 7
    recursive 322
    strictly dominated by another
        341
set included in another set 5
set of configurations of a program
    360
set of environments 360
set of initial configurations of a
    program 360
set of terminal configurations of a
    program 360
simple path 54
simultaneous inductive definition
    248
star closure 47
statement 257
statement variables 259
store of a program 109

strict function 146
strict inclusion 6
strict monotonic function 144
strict order 128
strict partial order 128
strong components of a graph 120
subgraph 53
subset closed under a function 230
subset of a set 5
subtree of a tree 61
sum of two multisets 20
supremum 133
surjection 33
symbols 45
symmetric difference 19

T
table 83
    heading of a 83
    schema of a 83
tautology 263
taxonomy 166
    inheritance property of 167
test functions 267
topological sorting 56
total correctness of a program 191
total order 135
transition system 318
    transition relation of a 318
transitive reduction of a partial
    order 130
transitive set 332
transpose of a matrix 114
truth function 259
truth values 39
tuple join 89
tuple projection 87

type of a relation from A to B 51
type over a set 254
typed constructor over a family of
    sets 255

U
uncountable set 344
uniform projection function 354
uniform rooted tree 116
unique readability condition 218,
    250
upper bound 131

V
vertex height in a tree 58
vertices of a directed graph 52

W
walk 119
well-founded set 290
well-ordering principle for natu-
    ral numbers 179
word 46
    infix of a 46
    interpretation of a 111
    null 45
    prefix of a 46
    proper infix of a 46
    proper prefix of a 46
    proper suffix of a 46
    square-free 111
    substring of a 46
    subword of a 46
    suffix of a 46

Z
Zorn's lemma 159

Texts and Monographs in Computer Science

continued

R.T. Gregory and E.V. Krishnamurthy
Methods and Applications of Error-Free Computation
1984. XII, 194 pages, 1 illus.

David Gries, Ed.
Programming Methodology: A Collection of Articles by Members of IFIP WG2.3
1978. XIV, 437 pages, 68 illus.

David Gries
The Science of Programming
1981. XV, 366 pages

Micha Hofri
Probabilistic Analysis of Algorithms
1987. XV, 240 pages, 14 illus.

A.J. Kfoury, Robert N. Moll, and Michael A. Arbib
A Programming Approach to Computability
1982. VIII, 251 pages, 36 illus.

E.V. Krishnamurthy
Error-Free Polynomial Matrix Computations
1985. XV, 154 pages

David Luckham
**Programming with Specifications: An Introduction to ANNA, A Language
for Specifying Ada Programs**
1990. XVI, 418 pages, 20 illus.

Ernest G. Manes and Michael A. Arbib
Algebraic Approaches to Program Semantics
1986. XIII, 351 pages

Robert N. Moll, Michael A. Arbib, and A.J. Kfoury
An Introduction to Formal Language Theory
1988. X, 203 pages, 61 illus.

Helmut A. Partsch
Specification and Transformation of Programs
1990. XIII, 493 pages, 44 illus.

Franco P. Preparata and Michael Ian Shamos
Computational Geometry: An Introduction
1988. XII, 390 pages, 231 illus.

Brian Randell, Ed.
The Origins of Digital Computers: Selected Papers, 3rd Edition
1982. XVI, 580 pages, 126 illus.

Thomas W. Reps and Tim Teitelbaum
The Synthesizer Generator: A System for Constructing Language-Based Editors
1989. XIII, 317 pages, 75 illus.

Texts and Monographs in Computer Science

continued

Thomas W. Reps and Tim Teitelbaum
The Synthesizer Generator Reference Manual, 3rd Edition
1989. XI, 171 pages, 79 illus.

Arto Salomaa and Matti Soittola
Automata-Theoretic Aspects of Formal Power Series
1978. X, 171 pages

J.T. Schwartz, R.B.K. Dewar, E. Dubinsky, and E. Schonberg
Programming with Sets: An Introduction to SETL
1986. XV, 493 pages, 31 illus.

Alan T. Sherman
VLSI Placement and Routing: The PI Project
1989. XII, 189 pages, 47 illus.

Santosh K. Shrivastava, Ed.
Reliable Computer Systems
1985. XII, 580 pages, 215 illus.

William M. Waite and Gerhard Goos
Compiler Construction
1984. XIV, 446 pages, 196 illus.

Niklaus Wirth
Programming in Modula-2, 4th Edition
1988. II, 182 pages

Study Edition

Edward Cohen
Programming in the 1990s: An Introduction to the Calculation of Programs
1990. XV, 265 pages